AF832030

Advances in Biochemistry in Health and Disease

Series Editor: Naranjan S. Dhalla
Winnipeg, Manitoba, Canada

Volume 1: S. K. Cheema (ed), *Biochemistry of Atherosclerosis*

Volume 2: S. W. Schaffer and M-Saadeh Suleiman (eds), *Mitochondria: The Dynamic Organelle*

Volume 3: A. K. Srivastava and M. B. Anand-Srivastava (eds), *Signal Transduction in the Cardiovascular System in Health and Disease*

Volume 4: B. Ostadal and N. S. Dhalla (eds), *Cardiac Adaptations-Molecular Mechanisms*

Volume 5: B. I. Jugdutt and N. S. Dhalla (eds), *Cardiac Remodeling-Molecular Mechanisms*

Volume 6: Jawahar L. Mehta and N. S. Dhalla (eds), *Biochemical Basis and Therapeutic Implications of Angiogenesis*

Volume 7: S. Chakraborti and N. S. Dhalla (eds), *Proteases in Health and Disease*

Volume 8: S. Chakraborti and N. S. Dhalla (eds), *Role of Proteases in Cellular Dysfunction*

Volume 9: B. Turan and N. S. Dhalla (eds), *Diabetic Cardiomyopathy: Biochemical and Molecular Mechanisms*

Volume 10: P. S. Tappia and N. S. Dhalla (eds), *Phospholipases in Health and Disease*

Volume 11: G. D. Lopaschuk and N. S. Dhalla (eds), *Cardiac Energy Metabolism in Health and Disease*

Volume 12: K. Ahmed, O.-G. Issinger, and R. Szyszka (eds), *Protein Kinase CK2 Cellular Function in Normal and Disease States*

Volume 13: I. M. C. Dixon and J. Wigle (eds), *Cardiac Fibrosis and Heart Failure: Cause or Effect?*

Volume 14: S. Chakraborti and N. S. Dhalla (eds), *Regulation of Ca^{2+}-ATPases, V-ATPases and F-ATPases*

Volume 15: S. Chakraborti and N. S. Dhalla (eds), *Regulation of Membrane Na^{+} -K + ATPase*

More information about this series at http://www.springer.com/series/7064

Ricardo Jorge Gelpi • Alberto Boveris
Juan José Poderoso
Editors

Biochemistry of Oxidative Stress

Physiopathology and Clinical Aspects

Springer

Editors
Ricardo Jorge Gelpi
School of Medicine
University of Buenos Aires
Buenos Aires, Argentina

Alberto Boveris
School of Pharmacy and Biochemistry
University of Buenos Aires
Buenos Aires, Argentina

Juan José Poderoso
Instituto de Inmunología, Genética y Metabolismo (INIGEM, UBA–CONICET), Laboratorio de Metabolismo del Oxígeno, Hospital Universitario
Universidad de Buenos Aires
Buenos Aires, Argentina

Advances in Biochemistry in Health and Disease
ISBN 978-3-319-45864-9 ISBN 978-3-319-45865-6 (eBook)
DOI 10.1007/978-3-319-45865-6

Library of Congress Control Number: 2016957889

© Springer International Publishing Switzerland 2016
This work is subject to copyright. All rights are reserved by the Publisher, whether the whole or part of the material is concerned, specifically the rights of translation, reprinting, reuse of illustrations, recitation, broadcasting, reproduction on microfilms or in any other physical way, and transmission or information storage and retrieval, electronic adaptation, computer software, or by similar or dissimilar methodology now known or hereafter developed.
The use of general descriptive names, registered names, trademarks, service marks, etc. in this publication does not imply, even in the absence of a specific statement, that such names are exempt from the relevant protective laws and regulations and therefore free for general use.
The publisher, the authors and the editors are safe to assume that the advice and information in this book are believed to be true and accurate at the date of publication. Neither the publisher nor the authors or the editors give a warranty, express or implied, with respect to the material contained herein or for any errors or omissions that may have been made.

Printed on acid-free paper

This Springer imprint is published by Springer Nature
The registered company is Springer International Publishing AG
The registered company address is: Gewerbestrasse 11, 6330 Cham, Switzerland

Preface

This book was thought as a reference text for postgraduate courses and for its use in research laboratories that perform studies on the biochemistry and physiology of oxidative stress with its ongoing extension to physiopathology and clinical aspects, in this latter case considering a series of human diseases. Although the idea of oxidative stress has been linked to more than a hundred diseases, the book deals with its application to cardiovascular diseases, neurodegenerative and neurological diseases, metabolic syndrome, and cancer. Originally, in 1985, Helmut Sies in the book *Oxidative Stress* (H. Sies, ed., Academic Press, London, p. 1–8) coined the concept of oxidative stress as a situation of "unbalance with either an excessive production of oxidants or a decrease in antioxidants that lead to cellular damage." The concept was a striking success and was immediately accepted and widely used both to design experimental studies and to explain obtained results. At the middle of 2016, after more than 30 years, the concept keeps its original strength and applicability. The entry "oxidative stress" in PubMed produces more than 160000 references. After the recognition of the deep changes that occur in the redox state of intracellular thiols in the oxidative stress situations, mainly due to the contributions of D.P. Jones (*Am J Physiol Cell Physiol*, 2008, **295**, 849–868), the definition of oxidative stress incorporated the concept of the redox state of intracellular thiols (-SH) and disulfides (-SS-). The intracellular concentration of both reduced glutathione (GSH) and oxidized glutathione (GSSG) is about 2–5 mM, with 90–95 % in the reduced state; the ratio of reduced glutathione/glutathione disulfide (GSH/GSSG) is taken as a valid measurement of the whole intracellular redox sate. The current views about the general regulation of cellular metabolism are that a series of regulatory proteins interact to keep cellular homeostasis and cell division. Most of these regulatory proteins have thiols and disulfides which are absolutely required for keeping the molecular conformation that is essential for their physiological regulatory function and effects. Many of these proteins exert their regulatory function by redox reactions involving their -SH/-SS- groups. The current definition by Sies and Jones (*Encyclopedia of Stress*, 2007, Fink G., ed., vol. 3, Elsevier, Amsterdam, p. 45–48) is that "oxidative stress is a situation in which there is an imbalance between oxidants and antioxidants in favor of the oxidants, leading to a disruption of redox signaling and control." This volume opens with the perspective chapter "The concept of oxidative stress after 30 years" by Helmut Sies.

Many of the authors of the chapters in the present volume develop their scientific activities in South America. One of the aims of this book is its use as a reference text in post-graduate courses and research laboratories in Latin America countries. It is worth recalling that distinguished scientists from this geographical area, from Brazil, Peru, and Argentina, participated in the development of ideas and concepts that later were the thresholds and the tributaries of the concept of oxidative stress. Alvaro Ozório de Almeida started in Rio de Janeiro in 1934–1941 the research on the clinical uses of hyperbaric oxygen, at first in the treatment of gangrene and cancer. Peruvian Eleazar Guzman Barron, working at the University of Chicago in the 1950s, established that glutathione and other similar thiols are highly sensitive to oxidation by ionizing radiation. Rebeca Gerschman, born in Argentina and disciple of Bernardo Houssay, working at the University of Rochester developed the "Gerschman theory" in 1954, in which she postulated that (a) the intermediates of the partial reduction of oxygen are the common molecular mechanism of oxygen and X-radiation toxicities, (b) an increase in oxygen partial pressure or a decrease in antioxidant defense leads equally to cell damage, and (c) aging is produced by the chronic effect of oxygen toxicity. The first two points are today fully accepted in the scientific community, with the second one reformulated in the form of Sies oxidative stress concept that is fully accepted and in evolution. The third point, oxygen toxicity and aging, is still in a sort of debate. However, an integration of the Gerschman theory and of the Harman theory (aging by free radical damage) is the preferred theory by the researchers in the field.

This book on the "biochemistry of oxidative stress" was born in one of the visits of Prof. N.S. Dhalla to Buenos Aires at the beginning of 2014, in which he was invited to a coffee and associated chat that the three editors have every Saturday at La Biela, a coffee shop and a Buenos Aires landmark. Prof. Dhalla got curious about why these three people meet every week. The answer was that we have common research interests and that frequently we publish scientific papers together. The existence of common research interests among these three full professors is somehow unexpected, because Ricardo Gelpi is professor of pathology and Juan J. (Teddy) Poderoso is professor of internal medicine, both of them at the School of Medicine of the University of Buenos Aires, and Alberto Boveris is professor of physical chemistry at the School of Pharmacy and Biochemistry of the same university. The present volume is one of the products of those coffees at La Biela in addition to the mentioned joint publications. The coffee and chat meetings (and the similar drinks and chat meetings) are indeed a Spanish tradition called *tertulia* (no English word for that) that is a social gathering in which friends joint to discuss varied matters, usually in the salons of hotels and clubs.

The book has four parts and 27 chapters keeping in mind the concept of oxidative stress and its application to human disease. Special comments will be made here for the chapters that deal with data from human patients. Part I is a general section devoted to oxidative stress and to the process of respiration, the basic cornerstone of mammalian biochemistry and cell physiology. This part opens with the chapter on oxidative stress by H. Sies. Other chapters consider the evolution of aerobic life, the role of mitochondria as source of the species of the partial reduction of oxygen, the mitochondrial production of nitric oxide, and the general metabolism of nitric oxide and peroxynitrite in mammalian cell biochemistry and physiology. There is a chapter on the recently recognized intercellular mitochondrial transfer. The part closes with a chapter by Gonzales et al. on the adaptation of human life to high altitude, dealing with the Peruvian population living at the altitude in the Andes and comparing it with the other two stable high-altitude populations in Ethiopia and Tibet. Part II is devoted to cardiovascular diseases with emphasis in ischemia-reperfusion and in the corresponding cardiac protection provided by thioredoxin and by preconditioning. The thioredoxin protection in ischemia-reperfusion is analyzed in transgenic mice in the chapter by D'Annunzio, Gelpi and colleagues. This part also has a chapter by Milei et al. with biochemical data on antioxidant supplementation in elderly cardiovascular hypertensive patients and on heart biopsy histological analysis in cardioplegic arrest/reperfusion in human cardiac surgery. Part III is devoted to neuronal function, neurodegeneration and neurodegenerative diseases, and their relationship to oxidative stress. A couple of most interesting chapters provide data involving patients, such as the one by Giulivi and colleagues, where autism in children is considered as a metabolic neurological disorder. Other chapter by Repetto and Boveris deals with the plasma markers of neurodegenerative diseases in human patients. Part IV contains chapters on metabolic syndrome and cancer. In this latter case, there is a chapter by J.J. Poderoso that considers cancer in the light of the Warburg theory of cancer as a mitochondrial phenomenon and that emphasizes in oxidative stress as the cause that leads to mutations and to cancer development. In another chapter, Merstelmann analyzes the recent success of immunotherapy in human cancer treatment. The editors of the scientific journal Science have chosen "cancer immunotherapy" as the Breakthrough of the Year in 2013 (Science 342:1432–1433, 2013).

Buenos Aires, Argentina
July 2016

Ricardo Jorge Gelpi
Alberto Boveris
Juan José Poderoso

Contents

Part II Cardiovascular Diseases

Part I
General Aspects

Chapter 1
The Concept of Oxidative Stress After 30 Years

Helmut Sies

Abstract *Oxidative stress is an imbalance between oxidants and antioxidants in favor of the oxidants, leading to a disruption of redox signaling and control and/or molecular damage* (Sies H, Jones DP, Oxidative stress. In: Fink G (eds) Encyclopedia of stress. 2nd edn, vol 3, Elsevier, Amsterdam, pp 45–48, 2007). The concept of oxidative stress, first formulated in 1985, is presented and discussed in the context of current developments. The role of hydrogen peroxide in oxidative stress and redox signaling has come into focus, with attempts to explore spatio-temporal control. Special aquaporins serve as peroxiporins. Research on molecular redox switches governing oxidative stress responses is in full bloom. On the more practical side, cautious use of terminology and methods regarding the so called ROS, reactive oxygen species, is recommended. The major role in antioxidant defense is fulfilled by antioxidant enzymes, not by small-molecule antioxidant compounds. The field of oxidative stress research embraces chemistry, biochemistry, cell biology, physiology and pathophysiology, all the way to health and disease research, ultimately providing a scientific basis for a modern redox medicine.

Keywords Oxidative stress • Redox homeostasis • Stress responses • Redox code • Hydrogen peroxide • Redox signalling • Molecular redox switches • Redox medicine

H. Sies (✉)
Institute of Biochemistry and Molecular Biology I, University of Düsseldorf, Düsseldorf, Germany

Leibniz Research Institute of Environmental Medicine, Heinrich-Heine-University Düsseldorf, Universitätsstrasse 1, D-40225 Düsseldorf, Germany

College of Science, King Saud University, Riyadh, Saudi Arabia
e-mail: sies@uni-duesseldorf.de

© Springer International Publishing Switzerland 2016
R.J. Gelpi et al. (eds.), *Biochemistry of Oxidative Stress*, Advances in Biochemistry in Health and Disease 16, DOI 10.1007/978-3-319-45865-6_1

1 Introduction

The concept of oxidative stress has been introduced for research in redox biology and medicine in 1985, now over 30 years ago, in an introductory chapter [1] in the book entitled *Oxidative Stress*. A concurrent comprehensive review entitled "Biochemistry of Oxidative Stress" presented the knowledge on prooxidants and antioxidants and their endogenous and exogenous sources and metabolic sinks [2]. Since then, redox biology as a research area has found fulminant development in a wide range of disciplines, starting from chemistry and radiation biology through biochemistry and cell physiology all the way into general biology and medicine. A recent commentary dealt with the merits and pitfalls of this concept: "Among the merits is the notion which is elicited by the two terms, namely (i) that aerobic metabolism is a steady-state characterized by a redox balance, as denoted by the term oxidative (and implicitly, reductive), and (ii) that strains in the balance can occur, as denoted by the term "stress". This latter concept evokes the occurrence of biological stress responses" [3]. Research on oxidative stress responses is in full bloom, concerning the functioning of central master switches, for example NF-kappaB or Nrf2/Keap1 in eukaryotes, or OxyR in prokaryotes. Much of the redox signaling occurs through molecular thiol redox switches, with significance of particularly reactive cysteines in specialized proteins [4, 5]. Regarding pitfalls of the concept, the term oxidative stress has been over-stressed, both in the public perception and in research circles. Indeed, the very first sentence in Ref. [1] is: "As a biochemist, one may wonder whether Selye's term should be stressed as it is in the present context". At the beginning of 2016, the number of hits in PubMed for "oxidative stress" is more than 1,000 entries per month, currently totalling over 153,000.

2 Concept of "Oxidative Stress"

The definition of oxidative stress as a global concept, in 1985, was [1]: *A disturbance in the prooxidant-antioxidant balance in favor of the former*. It was an important point, from the beginning, that there is a diversity of prooxidants and a diversity of antioxidants, operating with vastly different chemical and biological reactivities. Likewise, it may be mentioned that there is also a huge diversity of molecular targets: DNA, RNA, proteins, lipids, carbohydrates and other biomolecules. Prooxidants include free-radical species and non-radical species generated by enzymes or non-enzymatically, and antioxidants include powerful enzymes and also low-molecular mass compounds. A noteworthy insight was the perception that oxidation-reduction (redox) reactions in living cells are utilized in fundamental processes of redox regulation, collectively termed "redox signaling" and "redox control". The concept of oxidative stress was updated in 2007 to include the role of redox signaling [6]: *Oxidative stressis an imbalance between oxidants and antioxidants in favor of the oxidants, leading to a disruption of redox signaling and control and/or molecular damage*. The main idea underlying this global concept is that in living biological systems there is a maintenance of a redox balance, that is also called redox homeostasis. This is not an equilibrium as defined in thermodynamics, but it is a steady-state, away from thermodynamic equilibrium, *i.e.* a non-equilibrium [7]. As for a global definition, specific redox systems are not directly and specifically addressed, but they need to in given cases and conditions. Numerous redox systems of different nature coexist in cells and tissues, and not all of them are directly related or connected, contingent on the presence of redox catalysts, i.e. redox-active enzymes or redox-active compounds.

3 The Redox Code

These relationships have recently been conceptualized under four principles which, together, make up the "*Redox Code*" [8]: The *first* principle is the use of the reversible electron accepting and donating properties of nicotinamide in NAD and NADP to provide organization of metabolism, operating at near-equilibrium. Substrate oxidations are linked to reduction of NAD^+ and $NADP^+$, which in turn are linked to ATP production, catabolism and anabolism, respectively. The *second* principle is that metabolism is linked to structure through kinetically controlled redox switches in the proteome, which determine tertiary structure, macromolecular interactions and trafficking, activity and function. The abundance of proteins and reactivity of sulfur switches with oxidants vary over several orders of magnitude to support specificity in biological processes. The *third* principle is that of redox sensing, in that activation/deactivation cycles of redox metabolism, especially involving H_2O_2, support spatio-temporal sequencing in the differentiation and life cycles of cells and organisms.

The *fourth* principle is that redox networks form an adaptive system to respond to environment from microcompartments through subcellular systems to the levels of cell and tissue organization. This adaptive redox network structure is required to maintain health in a changing environment and, if functionally impaired, contributes to disease and organism failure. The term "Redox Code" applies to the redox organisation of cells, tissues and organisms. It is not confined to mammalian cells, and it extends ultimately to all living matter.

4 Specific Forms of Oxidative Stress

Given the enormous variety and range of prooxidant and antioxidant enzymes and compounds and of targets, as mentioned above, subforms of oxidative stress were identified [6], as shown in Table 1.1. Attempts were made to introduce intensity scales ranging from physiological oxidative stress to excessive and toxic oxidative burden [9]. A useful introduction to oxidative stress in biomedical and biological research is available, collecting basic concepts, definitions and currently employed methods used in this field [10].

Table 1.1 Oxidative Stress: definition, specific forms, and classification according to intensity

Category	Term	References
Definition	Oxidative stress	[1, 3, 6]
Specific form	Nutritional oxidative stress	
	Dietary oxidative stress	
	Postprandial oxidative stress	
	Physiological oxidative stress	
	Photooxidative stress	
	Ultraviolet (UV-A, UV-B)	
	Infrared-A	
	Radiation-induced oxidative stress	
	Nitrosative stress	
	Reductive stress	
Related terms	Oxidant stress, Pro-oxidant stress	
	Oxidative stress status (OSS)	
Classification	Basal oxidative stress	[9]
	Low intensity oxidative stress	
	Intermediate intensity oxidative stress	
	High intensity oxidative stress	

Modified from Refs. [3] and [6]

5 Role of Hydrogen Peroxide in Oxidative Stress and Redox Signaling

Hydrogen peroxide, the two-electron reduction product of oxygen, was identified in 1970 as a normal metabolite under aerobic conditions in living cells [11], occurring at about 10 nM intracellular concentration [12]. A major contributor is the mitochondrial respiratory chain [13, 14], notably Complexes I and III, but also Complex II [15]. Important further mitochondrial sources of hydrogen peroxide are given by several dehydrogenases, notably 2-oxoacid dehydrogenases [16, 17], as reviewed in [18]. In liver, H_2O_2 is produced at a rate of 50 nmol/min × g of tissue, which is about 2 % of total oxygen uptake in the steady state of physiological conditions [19]. Metabolically generated H_2O_2 emerged from recent research as a central hub in redox signaling and oxidative stress (see [20]). Hydrogen peroxide is well suited for redox signaling [21], and the role of peroxiredoxins is of paramount importance [22]. A novel aspect of hydrogen peroxide signaling relates to its diffusion properties. While it has long been assumed that hydrogen peroxide diffusion through lipid-containing membranes occurs at a sufficient rate, the discovery of facilitated diffusion through specific aquaporins provides for spatio-temporal control [23]. This discovery opened an exciting field on membrane transport of hydrogen peroxide by peroxiporins [24–26] and its significance for redox signaling [27–29]. See also Ref. [20] for discussion.

6 Molecular Redox Switches

The term "oxidative stress" encompasses the implicit notion of *adaptation*, coming from the general association of stress with stress response. This goes back to Selye's concept of stress as the "general adaptation syndrome" [30, 31]. The role of molecular *redox switches* came into focus in recent years, foremost the dynamic role of cysteines in proteins, opening the field of the redox proteome, currently flourishing because of advances in mass spectrometry and imaging methodology [32]. A bridge between phosphorylation/dephosphorylation and protein cysteine reduction/oxidation is given by the redox sensitivity of critical cysteinyl residues in protein phosphatases, opening the molecular pathway for signaling cascades as fundamental processes throughout biology. Much is yet to be learned in this area of research, as pointed out in a critical review on the impact of thiol peroxidases in redox regulation [33].

What was particularly exciting to many researchers was the discovery of master switch systems [34], prominent examples being OxyR in bacteria [35] and NFkappaB [36] and Nrf2/Keap1 [37] in higher organisms. That batteries of enzyme activities are mustered by activation of gene transcription through a "simple" redox signal is still an exciting strategy [38]. Much of current effort in redox biology is addressed towards these response systems. Obviously, medical and pharmacological intervention attempts are a consequence.

7 Redox Medicine

There are implications of the disruption of redox homeostasis for health and disease processes [39]. As outlined recently [8], the very functioning of cell metabolism is governed by redox processes. Fine-tuning of reaction cascades involves redox-active metabolites such as hydrogen peroxide (see above), nitric oxide, hydrogen sulfide and carbon monoxide. Biomedical research has addressed these aspects in overwhelming detail, beyond the scope of the present article. Just to name a few major topics: aging [40, 41], neurodegenerative diseases [42, 43], cardiovascular diseases/atherosclerosis [44], wound healing [45], cancer [46], immunology [47], and metabolic diseases such as diabetes [48].

There has been considerable effort in identifying potential therapeutic targets, biomarkers and pharmacological and clinical drugs applicable in redox medicine, as recently compiled [49–51].

Nutrition is another huge area impinging on health and disease, with important redox implication. Carbohydrate metabolism, in particular insulin signaling and postprandial oxidative stress, have a major impact [52–54]. Likewise, micronutrients and bioactives play a role. One example is the micronutrient element, selenium, which is present in selenoproteins in the form of selenocysteine. The selenoproteome in humans comprises 25 members with various functions [55, 56]. Viral and bacterial infections are often associated with deficiencies in micronutrients, including selenium, and the selenium status may affect the function of cells in both innate and adaptive immunity [57]. Our group has addressed other aspects of selenium homeostasis and health [58–60], a large field of current interest.

Physical exercise is another lifestyle factor with implications to redox processes. Principles for integrating reactive species into exercise physiology have been examined [61].

8 Cautionary Words

In contemporary research, to talk simply of "exposing cells or organs to oxidative stress" should be discouraged. Instead, the exact molecular conditions and specific compounds need to be identified. Even more important, in transposing redox considerations into medicine, concrete molecular oxidation-reduction descriptions are to be preferred.

A related pitfall is the use of ROS, which stands for reactive oxygen species, or RNS, which stands for reactive nitrogen species, where again the chemical species involved should be focused and defined whenever it is possible. This issue has been dealt with in detail recently [62]; this guide to free radical research terminology and methodology is highly recommended for perusal by researchers.

This "one-size-fits-all" mentality pervades also into the analytic methodology: measuring the so called "total antioxidant capacity (TAC)" in a blood plasma sample will not give useful information on the state of the organism, and should be discouraged [63]. Rather, individual relevant enzyme activities and patterns of antioxidant molecules need to be assessed.

The use of genetically encoded fluorescent protein indicators has permitted noninvasive studies in cells and organs, an early landmark being the hydrogen peroxide probe HyPer [64]. This field has developed impressively [65, 66], and specificity and sensitivity are being improved [67]. However, these probes need to be carefully applied with appropriate controls and calibration.

9 Conclusions

Useful as the term "oxidative stress" may be in research, there has been an inflationary development in research circles and more so in the medical field and, even more than that, in the public usage outside scientific endeavors. Obviously, a general term describing a global condition cannot be meant to depict specific spatiotemporal chemical relationships in detail and in specific cells or organ conditions. Recent progress in methodology has permitted insight in these topics.

Thus, the subcellular compartmentation of redox processes and redox components and their regulation is being studied at a new level in mammalian cells [68].

Oxidative stress and stress responses can now be studied in a more refined fashion at molecular detail. This gives us hope for exciting new insight and discoveries for the next 30 years of oxidative stress research.

Acknowledgements I gratefully acknowledge the input and friendship of many colleagues in shaping ideas in this multidisciplinary field, gathered under the umbrella of the Society for Free Radical Research International (SFRRI) and related organisations such as the Oxygen Club of California (OCC).

I also am thankful for the research support by the National Foundation of Cancer Research (NFCR), Bethesda, MD, USA, and to the Deutsche Forschungsgemeinschaft and the Alexander-von-Humboldt Foundation, Bonn, Germany.

References

1. Sies H (1985) Oxidative stress: introductory remarks. In: Sies H (ed) Oxidative stress. Academic, London, pp 1–8
2. Sies H (1986) Biochemistry of oxidative stress. Angew Chem Int Ed 25:1058–1071
3. Sies H (2015) Oxidative stress: a concept in redox biology and medicine. Redox Biol 4:180–183
4. Herrmann JM, Becker K, Dick TP (2015) Highlight: dynamics of thiol-based redox switches. Biol Chem 396:385–387
5. Riemer J, Schwarzländer M, Conrad M, Herrmann JM (2015) Thiol switches in mitochondria: operation and physiological relevance. Biol Chem 396:465–482
6. Sies H, Jones DP (2007) Oxidative stress. In: Fink G (ed) Encyclopedia of stress, vol 3, 2nd edn. Elsevier, Amsterdam, pp 45–48
7. Prigogine I (1978) Time, structure, and fluctuations. Science 201:777–785
8. Jones DP, Sies H (2015) The redox code. Antioxid Redox Signal 23:734–746
9. Lushchak VI (2014) Free radicals, reactive oxygen species, oxidative stress and its classification. Chem Biol Interact 224C:164–175
10. Breitenbach M, Eckl P (2015) Introduction to oxidative stress in biomedical and biological research. Biomolecules 5:1169–1177
11. Sies H, Chance B (1970) The steady state level of catalase compound I in isolated hemoglobin-free perfused rat liver. FEBS Lett 11:172–176
12. Chance B, Sies H, Boveris A (1979) Hydroperoxide metabolism in mammalian organs. Physiol Rev 59:527–605
13. Boveris A, Oshino N, Chance B (1972) The cellular production of hydrogen peroxide. Biochem J 128:617–630
14. Yin F, Boveris A, Cadenas E (2014) Mitochondrial energy metabolism and redox signaling in brain aging and neurodegeneration. Antioxid Redox Signal 20:353–371
15. Bleier L, Wittig I, Heide H et al (2015) Generator-specific targets of mitochondrial reactive oxygen species. Free Radic Biol Med 78:1–10
16. Goncalves RL, Quinlan CL, Perevoshchikova IV et al (2015) Sites of superoxide and hydrogen peroxide production by muscle mitochondria assessed ex vivo under conditions mimicking rest and exercise. J Biol Chem 290:209–227
17. Quinlan CL, Goncalves RL, Hey-Mogensen M et al (2014) The 2-oxoacid dehydrogenase complexes in mitochondria can produce superoxide/hydrogen peroxide at much higher rates than complex I. J Biol Chem 289:8312–8325
18. Mailloux RJ (2015) Teaching the fundamentals of electron transfer reactions in mitochondria and the production and detection of reactive oxygen species. Redox Biol 4:381–398
19. Oshino N, Chance B, Sies H, Bücher T (1973) The role of H_2O_2 generation in perfused rat liver and the reaction of catalase compound I and hydrogen donors. Arch Biochem Biophys 154:117–131

20. Sies H (2014) Role of metabolic H_2O_2 generation: redox signaling and oxidative stress. J Biol Chem 289:8735–8741
21. Forman HJ, Maiorino M, Ursini F (2010) Signaling functions of reactive oxygen species. Biochemistry 49:835–842
22. Rhee SG, Woo HA (2011) Multiple functions of peroxiredoxins: peroxidases, sensors and regulators of the intracellular messenger H_2O_2 and protein chaperones. Antioxid Redox Signal 15:781–794
23. Henzler T, Steudle E (2000) Transport and metabolic degradation of hydrogen peroxide in Chara corallina: model calculations and measurements with the pressure probe suggest transport of H_2O_2 across water channels. J Exp Bot 51:2053–2066
24. Bienert GP, Schjoerring JK, Jahn TP (2006) Membrane transport of hydrogen peroxide. Biochim Biophys Acta 1758:994–1003
25. Bienert GP, Moller AL, Kristiansen KA et al (2007) Specific aquaporins facilitate the diffusion of hydrogen peroxide across membranes. J Biol Chem 282:1183–1192
26. Bienert GP, Chaumont F (2014) Aquaporin-facilitated transmembrane diffusion of hydrogen peroxide. Biochim Biophys Acta 1840:1596–1604
27. Hara-Chikuma M, Satooka H, Watanabe S et al (2015) Aquaporin-3-mediated hydrogen peroxide transport is required for NF-kappaB signalling in keratinocytes and development of psoriasis. Nat Commun 6:7454
28. Hara-Chikuma M, Watanabe S, Satooka H (2016) Involvement of aquaporin-3 in epidermal growth factor receptor signaling via hydrogen peroxide transport in cancer cells. Biochem Biophys Res Commun 471(4):603–609. doi:10.1016/j.bbrc.2016.02.010
29. Watanabe S, Moniaga CS, Nielsen S, Hara-Chikuma M (2016) Aquaporin-9 facilitates membrane transport of hydrogen peroxide in mammalian cells. Biochem Biophys Res Commun 471(1):191–197, pii: MCB.00971-15
30. Selye H (1936) A syndrome produced by diverse nocious agents. Nature 138:32
31. Selye H (1976) Forty years of stress research: principal remaining problems and misconceptions. Can Med Assoc J 115:53–56
32. Go YM, Jones DP (2013) The redox proteome. J Biol Chem 288:26512–26520
33. Flohé L (2016) The impact of thiol peroxidases on redox regulation. Free Radic Res 50:126–142
34. D'Autreaux B, Toledano MB (2007) ROS as signalling molecules: mechanisms that generate specificity in ROS homeostasis. Nat Rev Mol Cell Biol 8:813–824
35. Christman MF, Morgan RW, Jacobson FS, Ames BN (1985) Positive control of a regulon for defenses against oxidative stress and some heat-shock proteins in Salmonella typhimurium. Cell 41:753–762
36. Schreck R, Rieber P, Baeuerle PA (1991) Reactive oxygen intermediates as apparently widely used messengers in the activation of the NF-kappa B transcription factor and HIV-1. EMBO J 10:2247–2258
37. Itoh K, Chiba T, Takahashi S et al (1997) A Nrf2/small Maf heterodimer mediates the induction of phase II detoxifying enzyme genes through antioxidant response elements. Biochem Biophys Res Commun 236:313–322
38. Espinosa-Diez C, Miguel V et al (2015) Antioxidant responses and cellular adjustments to oxidative stress. Redox Biol 6:183–197
39. Kehrer JP, Klotz LO (2015) Free radicals and related reactive species as mediators of tissue injury and disease: implications for health. Crit Rev Toxicol 45:765–798
40. Hohn A, König J, Grune T (2013) Protein oxidation in aging and the removal of oxidized proteins. J Proteomics 92:132–159
41. Mouchiroud L, Houtkooper RH, Auwerx J (2013) NAD(+) metabolism: A therapeutic target for age-related metabolic disease. Crit Rev Biochem Mol Biol 48:397–408
42. Chang CJ, Cravatt BF, Johnson DS et al (2014) Molecular medicine and neurodegenerative diseases. Chem Soc Rev 43:6668–6671

43. Lin MT, Beal MF (2006) Mitochondrial dysfunction and oxidative stress in neurodegenerative diseases. Nature 443:787–795
44. Stocker R, Keaney JF Jr (2004) Role of oxidative modifications in atherosclerosis. Physiol Rev 84:1381–1478
45. Cordeiro JV, Jacinto A (2013) The role of transcription-independent damage signals in the initiation of epithelial wound healing. Nat Rev Mol Cell Biol 14:249–262
46. Gorrini C, Harris IS, Mak TW (2013) Modulation of oxidative stress as an anticancer strategy. Nat Rev Drug Discov 12:931–947
47. Nathan C, Cunningham-Bussel A (2013) Beyond oxidative stress: an immunologist's guide to reactive oxygen species. Nat Rev Immunol 13:349–361
48. Watson JD (2014) Type 2 diabetes as a redox disease. Lancet 383:841–843
49. Casas AI, Dao VT, Daiber A et al (2015) Reactive oxygen-related diseases: therapeutic targets and emerging clinical indications. Antioxid Redox Signal 23:1171–1185
50. Frijhoff J, Winyard PG, Zarkovic N et al (2015) Clinical relevance of biomarkers of oxidative stress. Antioxid Redox Signal 23:1144–1170
51. Dao VT, Casas AI, Maghzal GJ et al (2015) Pharmacology and clinical drug candidates in redox medicine. Antioxid Redox Signal 23:1113–1129
52. Rains JL, Jain SK (2011) Oxidative stress, insulin signaling, and diabetes. Free Radic Biol Med 50:567–575
53. Ceriello A, Ihnat M (2010) Oxidative stress is, convincingly, the mediator of the dangerous effects of glucose variability. Diabet Med 27:968
54. Sies H, Stahl W, Sevanian A (2005) Nutritional, dietary and postprandial oxidative stress. J Nutr 135:969–972
55. Labunskyy VM, Hatfield DL, Gladyshev VN (2014) Selenoproteins: molecular pathways and physiological roles. Physiol Rev 94:739–777
56. Böck A, Flohé L, Köhrle J (2007) Selenoproteins - biochemistry and clinical relevance. Biol Chem 388:985–986
57. Steinbrenner H, Al-Quraishy S, Dkhil MA et al (2015) Dietary selenium in adjuvant therapy of viral and bacterial infections. Adv Nutr 6:73–82
58. Steinbrenner H (2013) Interference of selenium and selenoproteins with the insulin-regulated carbohydrate and lipid metabolism. Free Radic Biol Med 65:1538–1547
59. Steinbrenner H, Sies H (2013) Selenium homeostasis and antioxidant selenoproteins in brain: implications for disorders in the central nervous system. Arch Biochem Biophys 536:152–157
60. Steinbrenner H, Speckmann B, Sies H (2013) Toward understanding success and failures in the use of selenium for cancer prevention. Antioxid Redox Signal 19:181–191
61. Margaritelis NV, Cobley JN, Paschalis V et al (2016) Principles for integrating reactive species into in vivo biological processes examples from exercise physiology. Cell Signal 28:256–271
62. Forman HJ, Augusto O, Brigelius-Flohe R et al (2015) Even free radicals should follow some rules: a guide to free radical research terminology and methodology. Free Radic Biol Med 78:233–235
63. Pompella A, Sies H, Wacker R et al (2014) The use of total antioxidant capacity as surrogate marker for food quality and its effect on health is to be discouraged. Nutrition 30:791–793
64. Belousov VV, Fradkov AF, Lukyanov KA et al (2006) Genetically encoded fluorescent indicator for intracellular hydrogen peroxide. Nat Methods 3:281–286
65. Meyer AJ, Dick TP (2010) Fluorescent protein-based redox probes. Antioxid Redox Signal 13:621–650
66. Zhang X, Gao F (2015) Imaging mitochondrial reactive oxygen species with fluorescent probes: current applications and challenges. Free Radic Res 49:374–382
67. Ezerina D, Morgan B, Dick TP (2014) Imaging dynamic redox processes with genetically encoded probes. J Mol Cell Cardiol 73:43–49
68. Kaludercic N, Deshwal S, Di Lisa F (2014) Reactive oxygen species and redox compartmentalization. Front Physiol 5:285

Chapter 2
The Evolution of the Earth and Its Atmosphere

Juan José Poderoso

Abstract Earth is believed to have formed about 4.5 billion years ago. In the first 500 million years ago (Ma), a dense atmosphere emerged from vapor and gases that were expelled during degassing of the interior of the planet. These gases may have consisted of hydrogen, water vapor, methane, and carbon oxides. Prior to 3.5 billion years ago the atmosphere probably consisted of carbon dioxide, carbon monoxide, water, nitrogen, and hydrogen. The hydrosphere was formed 4 billion years ago from the condensation of water vapor, resulting in oceans of water in which sedimentation occurred. The most important feature of the ancient environment was the absence of free oxygen, which only began to persist in the atmosphere in small quantities about 50 Ma before the start of the great oxygenation event. This mass oxygenation of the atmosphere resulted in a rapid buildup of free oxygen. The rate of oxygen production by photosynthesis was slower in the Precambrian, and the concentrations were less than 10 % of the oxygen of today and probably fluctuated greatly. These fluctuations in oxygen concentration had little effect on life. The presence of oxygen allowed life with new opportunities. The origin of the mitochondria is related to the evolution of the earth. Mitochondria contain their own DNA, which is circular as in the bacteria. Mitochondrial ribosomes and transfer RNA molecules are also similar to those of bacteria, as are the components of their membrane. These observations led different researchers to propose an extracellular origin for mitochondria. The endosymbiotic hypothesis for the origin of mitochondria suggests that mitochondria are descended from specialized bacteria that somehow survived the endocytosis in another species of prokaryote or some other cell type, and became incorporated into the cytoplasm, providing a considerable evolutionary advantage.

Keywords Earth evolution • Oxygen • Atmosphere evolution • Bacteria • Origin of mitochondria

J.J. Poderoso (✉)
Instituto de Inmunología, Genética y Metabolismo (INIGEM, UBA–CONICET),
Laboratorio de Metabolismo del Oxígeno, Hospital Universitario,
Universidad de Buenos Aires, Buenos Aires, Argentina
e-mail: jpoderos@fmed.uba.ar

© Springer International Publishing Switzerland 2016

R.J. Gelpi et al. (eds.), *Biochemistry of Oxidative Stress*, Advances in Biochemistry in Health and Disease 16, DOI 10.1007/978-3-319-45865-6_2

1 Evolution of the Earth: Geologic Timeline

The geologic time scale is a system of measurement commonly used by earth scientists that relates rock strata to time, providing a rough history of geology and life on Earth, through the fossil record. The geologic timeline is vast, stretching back to the formation of the Earth approximately 4.5 billion years ago to the present days and into the future until the destruction of the Earth. It is divided into various eons, eras, periods and other denominations to make its study more practical and distinguish unique and defined time periods.

1.1 Precambrian

Precambrian is an informal term in geology and paleontology for the time from the formation of the Earth, about 4500 Ma ago, to the Cambrian explosion of life that occurred 542 Ma ago, which marked the beginning of the Paleozoic Era and the Cambrian Period. Little is known about the Precambrian, despite the fact that it constitutes the vast majority (about 88 %) of our planet history. The Precambrian saw the origin of life, the evolution of photosynthesis, the transformation of Earth's atmosphere into its modern form, and the creation of iron ore. It includes several distinct geologic eons: the Hadean, the Arcean and the Proterozoic.

The Hadean eon saw the formation of the Earth from about 4500 to 4000 Ma ago, the formation of the moon and the initial, hostile state of the Earth with heavy volcanism and poisoned waters when they existed. Very few traces of this time period remain due to geologic processes, and the few that persist are in scattered locations such as Australia and Greenland.

The Archean is the second eon of the early Earth (4000 to 2500 Ma ago), and consisted of higher volcanism and a near total lack of oxygen in the atmosphere. By the end of the period, the planet would have cooled significantly, allowing more familiar processes to act for the formation of actual continents and continent bases to begin. The most significant part of this time period would be that life began in this era.

The Proterozoic is the final eon of the Precambrian, lasting from 2500 to 542 Ma ago. This eon marks the beginning of life in a way we would understand it, however, not in simple terms, and the creatures that would ultimately make up the Cambrian life explosion that first evolved in this time period. Some of the most significant events of this time period would have been the first glacial periods, including a hypothesized time when all except the equator was locked under ice (the snowball earth hypothesis) and the change in the atmosphere as oxygen rose from almost nothing to the current percentage, as well as triggering a mass extinction of life forms to whom oxygen was a poison in the same way that carbon dioxide is to modern creatures.

1.2 Paleozoic

The Paleozoic Era is a geological era that started 542 Ma ago and finished 252 Ma ago, encompassing the Cambrian, Ordovician, Silurian, Devonian, Carboniferous, and Permian geological periods. It is marked in the beginning by the emergence of soft shelled life, and near the end of the Era, marked by complex plants (the first modern ones), insects, fish and small reptiles. Sea levels reached approximately 200 m higher than today, due to the warmer climate at the end of the period. However due to continental drift it became much warmer in some places and much colder in areas where life was located. This era ended with the greatest known mass extinction in history. The Cambrian was a geologic period of time that lasted from 542 to 488 Ma ago. It is distinguished from the preceding Pre-Cambrian period by a spectacular increase in the number of living organisms; specifically, the emergence of multicellular life that possessed hard exoskeletons. The Ordovician lasted from 488 to 443 Ma ago, and was the second period of the Paleozoic era. Some interesting facts include: the sea levels were ~180 m higher than today; southern continents were grouped into one named Gondwana; first land plant spores appeared approximately 470 Ma ago; ocean temperatures may have reached 45 °C; and a life extinction is thought to have been caused by an ice age, that was started by another extinction event as well. The Silurian is a period of geologic time that lasted from approximately 443 to 416 Ma ago. This period saw the first appearance of coral reefs and jawed fish in the fossil record, in addition to widespread terrestrial plant life. The Devonian period lasted from approximately 416 to 359 Ma ago and was marked by the formation of large continents, which would later conglomerate into Pangaea. It also witnessed extensive colonization of the land by plants, and the emergence of the first terrestrial animals, such as insects and other arthropods. By the end of the period, the first amphibians had appeared in the fossil record. The end of the Devonian period was marked by a mass extinction event. The Carboniferous lasted from approximately 359 to 299 Ma ago. Its name is derived from the numerous coal beds that were laid down during this time. The supercontinent Pangaea formed during the Carboniferous period. Animal colonization of the land increased, with the diversification of arthropods and amphibians. It was during the Carboniferous that most of today's so-called "fossil fuel" deposits were laid down. The massive amounts of oxygen in the atmosphere enabled arthropods to grow to enormous sizes. The Permian extended from 290 Ma ago to 250 Ma ago. The start of the Permian was cool and dry; however, by its end the temperature on Earth reached levels higher than at any time since the Cambrian explosion. The continents were in one giant landmass called Pangaea. Creationists claim that the split of Pangaea was a result of the flood caused by high volcanic activity and meteorites. The Permian mass extinction was the worst extinction within the most recent 600 Ma of Earth's history. It happened somewhere between 251 and 248 Ma ago, marking the transition from the Paleozoic to the Mesozoic era. This event is estimated to have killed off 90–95 % of sea life and about 70 % of land organisms. Various theories have been put forth on what caused this event including an asteroid or comet impact, volcanic activity, radiation from a nearby supernova, global warming, or a combination of any or all of the above factors.

1.3 Mesozoic

The Triassic is the name given to a geological time period which extended from 251 to 199 Ma ago. The Triassic is the start of the Mesozoic Era. It followed the largest extinction event known, with 95 % of life dying. It is the first geological period in which dinosaurs appear. The Jurassic lasted from 199.6 Ma ago to 145.5 Ma ago. This period is named for the Jura mountains in Europe, where rocks from this era can be found. Dinosaurs were the dominant animal group in this period, particularly the large sauropods. The Cretaceous lasted from approximately 145 to 65 Ma ago. It was preceded by the Jurassic period and followed by the Paleogene period. It was the last period of the "Age of Reptiles": dinosaurs, pterosaurs and sea reptiles were the dominant global fauna. The end of the Cretaceous period was marked by a mass extinction. It is believed to be caused by a combination of a meteorite impact and volcanic activity, and resulted in the extinction of the dinosaurs, in addition to the pterosaurs, many marine reptiles and other organisms. This is the most famous and widely known mass extinction, though it isn't the largest or most recent. The largest event is considered to be the Permian–Triassic extinction event, and the most recent may be well occurring now.

1.4 Cenozoic

The Cenozoic Era is a geologic period that includes the last 65 Ma. It begins with the mass extinction of the dinosaurs 65 Ma ago. The Cenozoic Era is divided into Paleogene and Neogene periods. Given the relatively recent time frame, the Cenozoic period is the best understood of all geological time periods including everything from weather features, temperature, geological changes, changes in life forms including detailed evolutions and extinctions. The Paleogene lasted from approximately 65 to 23 Ma ago. The Neogene was a geologic period that lasted from approximately 23 to 2.6 Ma ago. Alternatively, it has been argued that the current Quaternary period should be included within the Neogene period. This latter period followed the Paleogene period and was the second period of the Cenozoic era. During the Neogene period, the continents roughly assumed their present positions and the world's flora and fauna evolved into forms of which the majority would be comparable with modern species. The genus *Homo*, which would eventually give rise to humans, also appeared at the end of the Neogene.

1.5 Quaternary

The Quaternary is the current geologic time period, which began approximately 2.6 Ma ago. It is divided into two epochs: the older Pleistocene (2.6 million to 11700 years ago) and the current Holocene (11700 years ago to the present). The Quaternary is of particular importance to humanity as this is the time when *Homo sapiens* evolved.

2 The Origin of Mitochondria

Based on the characteristics of the genome it contains, the mitochondria results from unquestioned bacterial ancestry, specifically originated from the phylum alpha-Proteobacteria. According to the endosymbiont hypothesis, the idea is that mitochondria evolved from bacterial progenitors by endosymbiosis within an eukaryotic host cell. Yet mitochondrial genome evolution has taken radically different pathways in diverse eukaryotic lineages, and the organelle itself is increasingly viewed as a genetic and functional mosaic, with the bulk of the mitochondrial proteome having an evolutionary origin outside alpha-proteobacteria. New data reanalyzed our views regarding mitochondrial evolution. The question was raised of whether the mitochondria originated after the eukaryotic cell arose, as assumed in the classical endosymbiont hypothesis, or whether this organelle and the proteobacteria began at the same time. In 1970, Lynn Margulis published "The Origin of Eukaryotic Cells", an influential book that effectively revived the idea that mitochondria and chloroplasts evolved from free-living bacteria via symbiosis within a eukaryotic host cell [1]. The discovery in the 1960s of DNA within these organelles together with the recognition that they contain a translation system distinct from that of the cytosol were two of the observations that Margulis updated in support of the endosymbiont hypothesis of organelle origins. During her career, Margulis argued that symbiosis is a potent but largely unrecognized and unappreciated force in evolution. The development in DNA cloning and sequencing in the 1970s and 1980s opened the way to the detailed characterization of mitochondrial genome and genes, and the generation of key molecular data that were definitive instruments in affirming the bacterial origin of the mitochondrial and plastid genomes. Over the past decades, numerous author reviews have documented in detail the biochemical, molecular and cell biological data conducting to the endosymbiont hypothesis of organelle origins [2–4]. Various endosymbiotic models proposed over the years have been comprehensively criticized [5].

2.1 How Did the Mitochondrial Endo Symbiogenesis Occurred?

An alternative view, that the host cell for the mitochondrial endosymbiosis was a prokaryote, and specifically an archaeon, not a eukaryote, has recently gained in prominence [6]. The "hydrogen hypothesis"" [7] is perhaps the best known example of a symbiogenesis scenario. This scheme proposes that eukaryotes arose through symbiotic association of an anaerobic, strictly hydrogen-dependent, archaebacterium (the host) with a eubacterium (the symbiont) that was able to respire, but generated molecular hydrogen as a waste product of anaerobic heterotrophic metabolism. The dependence of the host upon molecular hydrogen produced by the symbiont is put forward as the selective principle that forged the common ancestor of eukaryotic cells. Thus, the hydrogen hypothesis "proposes that the origins of the heterotrophic organelle (the symbiont) and the origins of the eukaryotic lineage are identical" [7]. A corollary of the hydrogen hypothesis and other symbiogenesis scenarios is that the complexity of the eukaryotic cell and its defining features emerged after the mitochondrial symbiosis, rather than before. Several arguments can be advanced against a symbiogenesis scenario for the origin of mitochondria [6]. Endocytosis (an eukaryotic hallmark) has long been considered an essential function for incorporating a bacterial symbiont, although it is the case that bacterial endosymbioses (e.g., gamma-proteobacteria inside beta-proteobacteria) that have been documented [8]. In addition, assuming an a-proteobacterial symbiont as the mitochondrial progenitor in a partnership that simultaneously gave rise to this organelle and the rest of the eukaryotic cell, one might expect that any eubacterial genetic signal in the nuclear genome would be overwhelmingly alpha-proteobacterial. However, although an alpha-proteobacterial signal does, in fact, predominate, in any given eukaryotic lineage collectively more bacterial-type genes appear to derive from diverse non-alpha-proteobacterial lineages or fail to affiliate robustly with any specific bacterial phylum. Nevertheless, it is possible that ancestral lineages contributing to a bacterial–archaeal symbiogenesis might have possessed genomes already "scrambled" to a certain extent by horizontal gene transfer.

A prominent feature of the hydrogen hypothesis is its claim to account simultaneously for the origins of both aerobic and anaerobic energy metabolism in eukaryotes, the assumption being that both pathways were contained in and contributed to the hybrid cell by the alpha-proteobacterial partner partner. It is supposed that the two pathways would have been differentially expressed in the free-living bacterial symbiont when it encountered the appropriate environmental conditions. The hypothesis further posits that genes for aerobic respiration were lost in those eukaryotic lineages in which the mitochondrion was converted to a mitochondrion-related organelle, some types of which (e.g., hydrogenosome) function in anaerobic energy metabolism. The hydrogen hypothesis predicts that genes of anaerobic energy metabolism in mitochondrion-related organelles should have been inherited vertically throughout eukaryotes from a common ancestor, clustering as a monophyletic lineage together with alpha-proteobacteria in phylogenetic reconstructions.

However, a rigorous study of the phylogenetic distributions and histories of proteins involved in anaerobic pyruvate metabolism in eukaryotes has not provided support for this prediction [9]. Rather, mitochondrion-related organelles and the enzymatic machinery they contain for anaerobic energy metabolism appear to reflect a high degree of independent and convergent evolution.

Lane and Martin [10] have argued, from a consideration of the energetic of genome complexity, that because eukaryotes encode and express a substantially larger number of proteins than do prokaryotes, this increased expression demands a level of cellular energy that only the mitochondrion is able to satisfy. Accordingly, these investigators view the mitochondrion as the sine qua non of eukaryotic genomic and cellular complexity, concluding, rather definitively, that "the host for mitochondria was a prokaryote." On balance, it would seem that a symbiogenesis scenario (bacterial endosymbiont in an archaeal host) better accommodates the accumulated data relevant to the question of mitochondrion origin than does an archezoan scenario (bacterial endosymbiont in an amitochondriate but essentially eukaryotic host). However, as emphasized above, the latter scenario cannot be entirely discounted at this point. Each scenario raises its own set of issues that are difficult to rationalize without resorting to ad hoc explanations. In the end, each faces the conundrum that there is no straightforward and compelling way to discern how similar the genomes of the proposed prokaryotic ancestors of the eukaryotic cell were to their modern day descendants.

3 Geological History of Oxygen

Before photosynthesis evolved, Earth's atmosphere had no free oxygen (O_2). Photosynthetic prokaryotic organisms that produced O_2 as a waste product lived long before the first build-up of free oxygen in the atmosphere, perhaps as early as 3.5 billion years ago [11]. The oxygen they produced would have been rapidly removed from the atmosphere by weathering of reducing minerals, most notably iron. This "mass rusting" led to the deposition of iron oxide on the ocean floor, forming banded iron formations. Oxygen only began to persist in the atmosphere in small quantities about 50 Ma before the start of the Great Oxygenation Event. This mass oxygenation of the atmosphere resulted in rapid buildup of free oxygen. At current rates of primary production, today concentration of oxygen could be produced by photosynthetic organisms in 2000 years [12]. In the absence of plants, the rate of oxygen production by photosynthesis was slower in the Precambrian, and the concentrations of O_2 attained were less than 10 % of today levels and probably fluctuated greatly; oxygen may even have disappeared from the atmosphere again around 1900 Ma ago [13]. These fluctuations in oxygen concentration had little direct effect on life, with mass extinctions not observed until the appearance of complex life around the start of the Cambrian period, 541 Ma ago [14]. The presence of O_2 provided life with new opportunities. Aerobic metabolism is more

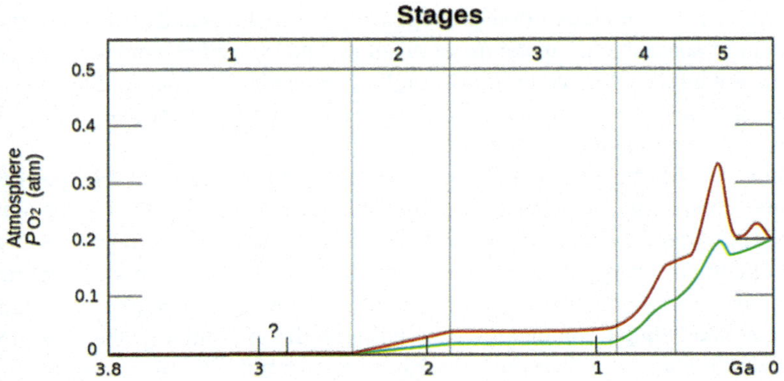

Fig. 2.1 O_2 build-up in the earth's atmosphere. *Red* and *green lines* represent the range of the estimates while time is measured in billions of years ago (Ga) [16]. Stage 1 (3.85–2.45 Ga): Practically no O_2 in the atmosphere. Stage 2 (2.45–1.85 Ga): O_2 produced, but absorbed in oceans & seabed-rocks. Stage 3 (1.85–0.85 Ga): O_2 starts to gas out of the oceans, but is absorbed by land surfaces and by formation of the ozone layer. Stages 4 & 5 (0.85 Ga–present): O_2 sinks filled, the gas accumulated Ref. [17]

efficient than anaerobic pathways, and the presence of oxygen undoubtedly created new possibilities for life to explore [15] (Fig. 2.1).

Since the start of the Cambrian period, atmospheric oxygen concentrations have fluctuated between 15 % and 35 % of atmospheric volume [18]. The maximum of 35 % was reached towards the end of the Carboniferous period (about 300 Ma ago), a peak which may have contributed to the large size of insects and amphibians at that time. Whilst human activities, such as the burning of fossil fuels, have an impact on relative carbon dioxide concentrations, their impact on the much larger concentration of oxygen is less significant [19].

3.1 Effects on Life

The concentration of atmospheric oxygen is often cited as a possible contributor to large-scale evolutionary phenomena, such as the origin of the multicellular Ediacara biota, the Cambrian explosion, the trends in animal body size, and other extinction and diversification events [15]. The large size of insects and amphibians in the Carboniferous period, where oxygen reached 35 % of the atmosphere, has been attributed to the limiting role of diffusion in these organisms metabolism. It has been proposed that this mechanism would only apply to insects. However, the biological basis for this correlation is not firm, and many lines of evidence show that oxygen concentration is not size-limiting in modern insects. Interestingly, there is no significant correlation between atmospheric oxygen and maximum body size elsewhere in the geological record. Ecological constraints can better explain the diminutive size of post-Carboniferous dragonflies; for instance, the appearance of flying competitors such as pterosaurs, birds and bats [15].

Rising oxygen concentrations have been cited as a driver for evolutionary diversification, although the physiological arguments behind such arguments are questionable, and a consistent pattern between oxygen concentrations and the rate of evolution is not clearly evident. The most celebrated link between oxygen and evolution occurs at the end of the last of the snowball glaciations, where complex multicellular life is first found in the fossil record. Under low oxygen concentrations, regular "nitrogen crises" could render the ocean inhospitable to life. Significant concentrations of oxygen were just one of the prerequisites for the evolution of complex life. Models based on uniformitarian principles (i.e. extrapolating present-day ocean dynamics into deep time) suggest that such a concentration was only reached immediately before metazoa first appeared in the fossil record. Further, anoxic or otherwise chemically "nasty" oceanic conditions that resemble those supposed to inhibit macroscopic life re-occur at intervals through the early Cambrian, and also in the late Cretaceous, with no apparent impact on lifeforms at these times. This might suggest that the geochemical signatures found in ocean sediments reflect the atmosphere in a different way before the Cambrian, perhaps as a result of the fundamentally different mode of nutrient cycling in the absence of planktivory [14].

4 Conclusion

The appearance of oxygen in the atmosphere is closely related to Earth's evolution. Only knowing, albeit briefly, the different stages of Earth's evolution is that we can know and understand the circumstances that made oxygen appear in the atmosphere in sufficient quantities and compatible with life a billion years ago. Prior to that, oxygen had been detected but it had been absorbed either by the oceans or the land. In direct relation to Earth's evolution and the appearance of oxygen is the origin of mitochondria. These organelles, whose main function is to regulate cell energy production were originally bacteria, that is, prokaryotes. They symbiotically incorporated to eukaryotic cells and from there, the process of functional adaptation begins and continues to our days. Thus we can conclude that the appearance of oxygen in the atmosphere and the origin of mitochondria are directly related to Earth's evolution and the origin of life.

References

1. Margulis L (1970) Origin of eukaryotic cells: evidence and research implications for a theory of the origin and evolution of microbial, plant, and animal cells on the precambrian earth. Yale University Press, New Haven, 349 pp

2. Gray MW, Doolittle WF (1982) Has the endosymbiont hypothesis been proven? Microbiol Rev 46:1–42
3. Andersson SG, Kurland CG (1999) Origins of mitochondria and hydrogenosomes. Curr Opin Microbiol 2:535–541
4. Bullerwell CE, Gray MW (2004) Evolution of the mitochondrial genome: protist connections to animals, fungi and plants. Curr Opin Microbiol 7:528–534
5. Martin W, Hoffmeister M, Rotte C, Henze K (2001) An overview of endosymbiotic models for the origins of eukaryotes, their ATP-producing organelles (mitochondria and hydrogenosomes) and their heterotrophic lifestyle. Biol Chem 382:1521–1539
6. Koonin EV (2010) The origin and early evolution of eukaryotes in the light of phylogenomics. Genome Biology 11:209, genomebiology.com/2010/11/5/209
7. Martin W, Müller M (1998) The hydrogen hypothesis for the first eukaryote. Nature 392:37–41
8. von Dohlen CD, Kohler S, Alsop ST, McManus WR (2004) Mealybug beta-proteobacterial endosymbionts contain gamma-proteobacterial symbionts. Nature 412:433–436
9. Hug LA, Stechmann A, Roger AJ (2010) Phylogenetic distributions and histories of proteins involved in anaerobic pyruvate metabolism in eukaryotes. Mol Biol Evol 27:311–324
10. Lane N, Martin W (2010) The energetics of genome complexity. Nature 467:929–934
11. Dutkiewicz A, Volk H, George SC et al (2006) Biomarkers from Huronian oil-bearing fluid inclusions: an uncontaminated record of life before the Great Oxidation Event. Geology 34:437–440
12. Dole M (1965) The natural history of oxygen. J Gen Physiol 49S:5–27
13. Frei R, Gaucher C, Poulton SW, Canfield DE (2009) Fluctuations in Precambrian atmospheric oxygenation recorded by chromium isotopes. Nature 461:250–253
14. Butterfield NJ (2007) Macroevolution and macroecology through deep time. Palaeontology 50:41–55
15. Butterfield NJ (2009) Oxygen, animals and oceanic ventilation: an alternative view. Geobiology 7:1–7
16. Wikipedia, The Free Encyclopedia. Geological history of oxygen. https://en.wikipedia.org/w/index.php?title=Geological_history_of_oxygen&oldid=730363681
17. Holland HD (2006) The oxygenation of the atmosphere and oceans. Phil Trans R Soc B 361:903–915
18. Berner RA (1999) Atmospheric oxygen over Phanerozoic time. Proc Natl Acad Sci U S A 96:10955–10957
19. Emsley J (2001) Oxygen. Nature's building blocks: an A-Z guide to the elements. Oxford University Press, Oxford, pp 297–304

Chapter 3
Mitochondria Are the Main Cellular Source of O_2^-, H_2O_2 and Oxidative Stress

Alberto Boveris and Marisa G. Repetto

Abstract Mitochondria reduce about 1–2 % of the O_2 consumed in the tissues to O_2^- that is dismutated in the mitochondrial matrix by the Mn-SOD reaction to O_2 and H_2O_2. O_2^- as a charged and non permeable species is confined into the mitochondrial matrix where is kept at a steady state level of 10^{-10} M. After dismutation, non charged H_2O_2 freely diffuses to the cytosol, where it is kept at about 10^{-7} M by catalase and glutathione peroxidase. In the cytosol H_2O_2 encounters Fe^{2+} (and Cu^+), suffers homolysis by the Fenton/Haber-Weiss reaction, and produces the highly reactive $HO^•$. This radical immediately abstracts one hydrogen atom from unsaturated fatty acids and starts the process of lipoperoxidation, in an open and non-equilibrium situation as long there are unsaturated fatty acids and O_2. The free-radical mediated oxidations of phospholipids, proteins and nucleic acids are a consequence of aerobic life. Increased oxidations define the oxidative stress situation. Then, mitochondria are the main cellular source of O_2^-, of H_2O_2 and of oxidative stress in the cell. The cellular metabolisms of O_2^-, H_2O_2, NO and $ONOO^-$ are integrated and faster rates of free-radical mediated reactions are considered the molecular mechanisms of pathological processes and of aging.

Keywords Superoxide radical • Hydrogen peroxide • Hydroxyl radical • Oxidative stress • Haber-Weiss reaction • Steady states • Pathology molecular mechanisms • Aging

A. Boveris (✉) • M.G. Repetto
Department of General Chemistry, Faculty of Pharmacy and Biochemistry, University of Buenos Aires, Buenos Aires, Argentina

Institute of Biochemistry and Molecular Medicine (UBA-CONICET), School of Pharmacy and Biochemistry, University of Buenos Aires, Junin 956, Buenos Aires, Argentina
e-mail: aboveris@ffyb.uba.ar

© Springer International Publishing Switzerland 2016

R.J. Gelpi et al. (eds.), *Biochemistry of Oxidative Stress*, Advances in Biochemistry in Health and Disease 16, DOI 10.1007/978-3-319-45865-6_3

1 Introduction. The Partial Reduction of Oxygen and the Reactive Oxygen Species

In 1954, the Argentine scientist Rebeca Gerschman published the famous article "Oxygen poisoning and x-irradiation: a mechanism in common" (Science 119:623–626) [1], that postulated that the products of the partial reduction of oxygen, superoxide radical (O_2^-), hydrogen peroxide (H_2O_2) and hydroxyl radical ($HO•$), are the chemical species responsible for the toxicity of oxygen and of x-irradiation. The paper reported a synergic effect of hyperbaric oxygen and x-radiation in the survival of mice. The experiments were remarkable: the animals were in a solid iron chamber and were irradiated from outside. The paper proposed the concept that either an increase in oxygen pressure or in x-radiation or a decrease in cellular antioxidant defense (a new concept at that time) would equally lead to cellular damage. The products of the partial reduction of oxygen, O_2^-, H_2O_2 and $HO^•$ were early described by Michaelis in 1946 on theoretical basis [2]. Simultaneously, the toxic effects of oxygen and H_2O_2 on brain metabolism were recognized by Mann and Quastel [3]. The confirmation of the Gerschman view on the toxicity of O_2^- and H_2O_2 was established by the discovery of superoxide dismutase (SOD) in 1969 that implied the biological existence of O_2^- [4] and by the establishment of the "Fridovich dogma" in 1975, in which SOD and catalase were recognized as absolutely necessary for aerobic life in bacteria [5].

The concept of reactive oxygen species (ROS)) was successfully introduced by Seim in 1982 to describe the luminol chemiluminescence of activated human monocytes [6]. The concept was immediately adopted, although there was not a definition of the involved chemical species. However, since the biological effects of ROS were similar or undistinguished, the idea was accepted without hesitation. Originally, ROS meant extracellular O_2^-, H_2O_2 and $HO^•$ but the concept was later extended to the intracellular situation and the list of ROS grew with time. Today it is accepted that includes O_2^-, H_2O_2 and $HO^•$, some intermediates of the free radical-mediated lipid peroxidation, such as $ROO^•$ (peroxyl radical) and $^1O_2$ (singlet oxygen) and also organic peroxides (ROOH) and peroxynitrite ($ONOO^-$), this latter the product of the reaction between and O_2^- and NO.

2 The Mitochondrial Production of O_2^- and H_2O_2

The physiological generation of O_2^- and of H_2O_2 is quantitatively a mitochondrial phenomenon. There are other sources for these species but they are less important in quantitative terms and also there is a widespread and essential presence of mitochondria in aerobic tissues. About 1–2 % of the O_2 consumed in tissue respiration is converted to O_2^-. Then, O_2^- can be regarded as the precursor of all the reactive oxygen species. Early reports recognized successively the mitochondrial production of H_2O_2 and of O_2^-.

Table 3.1 Tissue O_2 uptake[a] and mitochondrial H_2O_2 production

Organ	1. O_2 uptake (μmol/min.g)	2. Mitochondrial H_2O_2 production			Ratio 2/1 (%)	References
		State 4	State 3	Organ rate[b]		
		(nmol/min.mg prot)		(nmol/min.g)		
Rat liver	1.36	0.60	0.10	13.1	0.96	[7, 8]
Rat heart	3.92	0.70	0.10	26.4	0.67	[8, 9, 10]
Rat brain	1.67[c]	0.19	0.01	4.1[d]	0.25	[11, 12]

[a]Lagendorff perfused liver and heart at 30 °C and brain in situ. Mitochondrial H_2O_2 production was determined at 30 °C

[b]Calculated for the physiological situation; with mitochondria 68 % of in state 4 and 32 % in state 3, and a mitochondrial content of 30 (liver), 52 (heart) and 16 (brain) mg of mitochondrial protein/g of tissue

[c]Brain in situ, 37 °C. Brain has other significant sources of H_2O_2

[d]For the H_2O_2/O_2 ratio the H_2O_2 rate was converted to 37 °C

Table 3.2 Ratios of O_2^- and H_2O_2 production in submitochondrial preparations

Source	Production rates (nmol/min.mg protein)		Ratio O_2^- / H_2O_2	References
	O_2^-	H_2O_2		
Rat heart	5.0	3.1	1.61	[14]
Rat heart	6.9	3.2	2.15	[14]
Beef heart	3.8	1.8	2.00	[15]
Ascites tumor	2.9	1.4	2.10	[16]
	2.2	1.2	1.83	[16]

Table 3.1 summarizes the mitochondrial rates of H_2O_2 production in states 4 and 3 in mitochondria isolated from liver, heart and brain, the H_2O_2 organ production and the ratio of H_2O_2 production and organ O_2 uptake. The very significant mitochondrial rate of H_2O_2 production accounts for 0.26–0.96 % of the rate of organ O_2 uptake, with a ratio of 0.96 and 0.67 % for liver and heart. Mitochondrial H_2O_2 production in states 4 and 3 accounts for 1.32 and 0.22 % of the respiration in rat liver, for 0.93 and 0.13 % in rat heart and for 0.18 and 0.01 % in rat brain (Table 3.1). Thus, H_2O_2 production is a physiological phenomenon of mitochondrial respiration and of oxidative metabolism. There was an immediate interest in the quantitative relationship of the mitochondrial production of H_2O_2 and the eventual participation of O_2^-. It was known that flavoprotein enzymes produce H_2O_2 by transfer of two reducing hydrogen atoms and that the dismutation of O_2^- generates H_2O_2 [4]. Loschen et al. reported the mitochondrial production of O_2^- [13] and shortly after an stoichiometric ratio of 2 O_2^- to 1 H_2O_2 was reported for the mitochondrial production of H_2O_2 [14–16]. Table 3.2 summarizes some reports on the ratios O_2^-/H_2O_2 in mitochondrial preparations. Submitochondrial particles were used, washed to remove the Mn-SOD and to be able to measure O_2^- production. There was agree-

ment among the experimental reports with ratios of 2 O_2^- to 1 H_2O_2. The result was interpreted as evidence confirming the roles of O_2^- and H_2O_2 and of Mn-SOD and catalase in bacteria [5] and also as evidence of the endo-symbiotic origin of modern aerobic eukaryotic cells from primitive rickettsia or thermoplasma-like organisms and from primitive eubacteria or spirochaeta-like nucleated cells [17]. The mitochondrial production of O_2^- and H_2O_2 is frequently treated as an electron leak from the main stream of electrons handled by the respiratory chain [18]. Theoretically, all the reduced components of the respiratory chain are auto-oxidable, however, most of the rates of auto-oxidation are minimal and negligible. Significant rates of electron leakage take place at complexes I and III.

In complex I, the steady state level of the reduced forms of the flavin, $FMNH_2$ and $FMNH^\bullet$, are able to reduce O_2 to O_2^- by simple collision. The totality of the O_2^- produced, about 0.5–0.7 nmol/min.mg protein, is released into the mitochondrial matrix [19]. In complex III, O_2^- is released both to the mitochondrial matrix and to the intermembrane space. The fraction of O_2^- released into the matrix (2.5 nmol O_2^-/min.mg protein) is about 75 % of the total produced O_2^- while the remaining 25 % is released to the intermembrane space [20]. It is considered that $HOO^\bullet$ is formed in the phospholipid phase and that, as a non-charged species, diffuses through the phospholipid bilayer of the inner mitochondrial membrane [20]. The O_2^- released to the matrix quickly undergoes spontaneous disproportionation yielding O_2 and H_2O_2. The reaction is strongly accelerated by the enzyme Mn-SOD reaching a very low steady-state concentration of O_2^-. The Cu,Zn-SOD of the intermembrane space disproportionates the O_2^- released in this compartment [21]. As a result, a very low steady state concentration of O_2^- is achieved (10^{-10} M) while H_2O_2 is kept at higher levels (10^{-7} M). The small and non-charged molecule H_2O_2 freely diffuses to the cytosol (Fig. 3.1). The cytosolic transition metals Fe^{2+} or Cu^{1+} are able to produce H_2O_2 homolysis with $HO^\bullet$ radical generation. This radical is a major hazard for cells due to its high reduction potential ($E^\circ = +2.87$ V) and reactivity, which determines

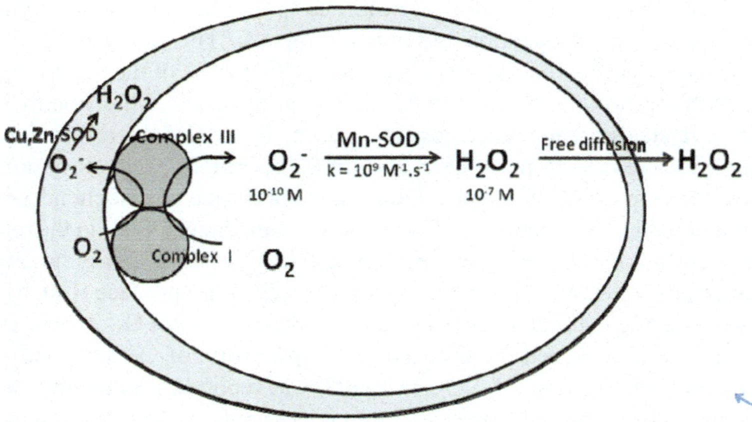

Fig. 3.1 Mitochondrial metabolism of O_2^- and H_2O_2. The former (O_2^-) is an intramitochondrial metabolite and the latter (H_2O_2) is a cellular metabolite

its brief half-life. Hydroxyl radical is extremely efficient at initiating phospholipid peroxidation and producing protein oxidation with formation of carbonyl groups (>C = O). The response to increased $HO^•$ radical generation is an almost instantaneous (seconds) increased lipid peroxidation and protein oxidation [22, 23].

3 Mitochondrial Function in Cell Physiology

All the chemical reactions that sustain life in cells and tissues constitute a very large group and are collectively called metabolism. Metabolism is constituted by anabolism and catabolism. Catabolic pathways generate chemical energy in the form of ATP, $NADPH_2$ and $FADH_2$. The anabolic pathways use precursor molecules (amino acids, sugars, fatty acids and nitrogenated bases) to build up cell macromolecules (proteins, polysaccharides, lipids and nucleic acids) using ATP to drive these synthesis. The distinction between catabolism and anabolism is illustrated by saying that in catabolism there is ATP production and in anabolism there is ATP utilization. Alternatively, acetyl-CoA formation and acetyl-CoA utilization can be considered as the characteristic of catabolism and anabolism, respectively [24, 25]. Both, ATP and acetyl-CoA are central and key mitochondrial metabolites. According to thermodynamics second law, chemical reactions only occur if there is a decrease in free energy (($\Delta G'$). ATP, with its large synthesis and hydrolysis energies to be formed and to be split, from and to ADP and Pi ($\Delta G^{\circ\prime} = 30.5$ kJ/mol), leads naturally to the idea that ATP is the cellular energy currency.

Lavoisier (1789) recognized for the first time that animals burn foods (chemical fuels) and release heath as inherent facts of life. Glucose oxidation is a strong exergonic process, described from left to right in (Eq. 3.1) [24, 26]:

$$C_6H_{12}O_6 + 6O_2 \leftrightarrow 6CO_2 + 6H_2O + 2840\ \text{kJ} \tag{3.1}$$

The process is reversible and goes from right to left in photosynthesis, where the energy provided by sunlight is used for glucose and O_2 synthesis in green plant thylakoids. Glucose utilization has two biochemical possibilities: the anaerobic pathway and the aerobic pathway. The first one, the anaerobic (about 2500 millions years old) provides 2 moles of ATP/mol of glucose. The second one, the aerobic (about 900 millions years old) provides 36 moles of ATP/mol of glucose. No wonder why biology evolution chose aerobic life. Utilization of O_2 provides 473 kJ/mol and evolved associated to the development of mitochondria. The organelle originated after the endo-symbiosis of primitive bacteria and eukaryotes [17]. Contemporaneous mitochondria are a double-membrane structure with an inner membrane that is extensively folded and contains the electron transfer chain with its four respiratory complexes (I, II, III and IV) and the F_1-ATP synthase (complex V). About 95 % of the total O_2 consumed in the tissues is utilized by mitochondrial cytochrome oxidase (complex IV). Mitochondrial function is to carry out a coupled process in which the oxidation of the coenzymes $NADH_2$ and $FADH_2$ and the reduction of cellular O_2 to H_2O are compulsorily linked to the synthesis of ATP. The energy needed for the transduction of chemical redox energy ($NADH_2$ oxidation to NAD) to the chemical energy of ATP synthesis (ADP + Pi → ATP) is provided by an electrochemical H^+ gradient [24]. The elucidation of the molecular mechanism of the process was a biochemical revolution that occurred in the last 50 years and produced two Nobel Prizes. Peter Mitchell received the Nobel Prize in Chemistry 1978 "for his contribution to the understanding of biological energy transfer through the formulation of the chemiosmotic theory" and John Walker and Paul Boyer received the Nobel Prize in Chemistry 1997 "for their elucidation of the enzymatic mechanism underlying the synthesis of adenosine triphosphate".

In the mitochondrial respiratory chain, complex I transfers electrons from $NADH_2$ to ubiquinone (UQ) and the energy released by this exergonic reaction is coupled to the extrusion of 4 H^+ into the intermembrane space. Complex II oxidizes succinate from the Krebs cycle without H^+ extrusion. The electrons received by complexes I and II are transferred to UQ, which becomes reduced to ubiquinol (QH_2). The quinol transfers its electrons to complex III, which in turn pumps 4 H^+ into the intermembrane space and reduces cytochrome c^{3+}. Cytochrome c transfers electrons to complex IV that reduces O_2 to H_2O and pumps another 2 H^+ into the intermembrane space. Due to the H^+ impermeability of the inner membrane a H^+ gradient is formed across the inner mitochondrial membrane. The gradient moves H^+ through the membrane F_0 channels in their way back to the matrix. A protein shaft with rotatory movements and conformational changes in F_1 synthase β-subunits, catalyzes ATP synthesis from ADP and Pi by complex V [24]. In cytochrome oxidase, O_2 is bound and reduced to 2 H_2O molecules, in 4 one electron steps that reduce the 2 Cu^{2+} and the 2 Fe^{3+} of the oxidase. The enzyme retains O_2 until it is fully reduced to H_2O. This mechanism dramatically lowers O_2^- release to almost nothing, because O_2^- entails a cellular risk for being a precursor of free-radical mediated reactions.

4 Steady States and Half Lives of Reactive Oxygen Species

The highly reactive chemical species constituted by the intermediates of the partial reduction of oxygen and of the free-radical mediated chain reactions in mammalian cells are kept at steady states. In this stable and non-equilibrium situation the rate of production of each species is equal to its rate of utilization. Table 3.3 provides a list with the steady states and half life values of these chemical species. They are kept at low or very low concentrations and exhibit a fast turnover with short half lives (the times needed under physiological conditions to decrease their steady state concentrations to 50 %).

O_2^- is not a good oxidant, rather it is a reductant for itself and for cytochrome c. However, as mentioned, O_2^- is considered the precursor of all the highly reactive molecules that participate in the free-radical mediated oxidation process (Table 3.3). The importance of O_2^- in free-radical mediated oxidations is that O_2^- dismutates to

Table 3.3 Steady state concentrations and half lives of the intermediates of the partial reduction of oxygen and of intermediates of the free-radical mediated chain reactions in mammalian cells

Chemical species	Steady state (M)	Half life ($t_{1/2}$) (s)
O_2^- (intramitochondrial)	10^{-10}	10^{-4}
O_2^- (cytosolic)	10^{-11}	10^{-3}
H_2O_2	10^{-7}	0.1
HO•	10^{-16}	10^{-9}
NO	10^{-7}	10^{-2}
$ONOO^-$	10^{-8}	0.5
UQH•	10^{-7}	0.2
R•	10^{-10}	10^{-7}
ROO•	10^{-9}	10^{-5}
ROOH	10^{-7-6}	0.1–0.3
$^1O_2$	10^{-15}	10^{-6}

Modified from Boveris et al. [27]

H_2O_2 and reduces the cytosolic transition metal ions, Fe^{3+} and Cu^{2+}. These ions, through the Haber-Weiss reaction, produce the homolysis of H_2O_2 and ROOH leading to the productions of HO• and RO• (Eqs. 3.2, 3.3, and 3.4):

$$O_2^- + Fe^{3+} \rightarrow O_2 + Fe^{2+} \tag{3.2}$$

$$\underline{H_2O_2 + Fe^{2+} \rightarrow HO^{\bullet} + HO^- + Fe^{3+}} \quad \text{Fenton reation} \tag{3.3}$$

$$O_2^- + H_2O_2 \rightarrow HO^{\bullet} + HO^- + O_2 \quad \text{Haber Weiss reaction} \tag{3.4}$$

The production of $HO^{\bullet}$ from H_2O_2 and Fe^{2+} by the Fenton reaction has been considered for a long time as the rate-limiting step for physiological lipid peroxidation [22, 27–29].

5 Oxidative Stress

The classic concept of oxidative stress was introduced by Sies in 1985 and defined as an unbalance situation in which oxidants predominate over antioxidants [30] (see Chap. 1 by H. Sies in this volume). The acceptation and success were instantaneous because the idea of oxidative stress made possible the rational interpretation of experimental data and the formulation of scientific hypothesis. The idea, popularized with the image of a scale with "oxidants" and "antioxidants" in the two arms, was immediately applied to a myriad of experimental and even clinical situations. Antioxidants were understood as a defense system constituted by antioxidant enzymes and by low molecular mass reductants. The group of the antioxidant enzymes is constituted by superoxide dismutases, catalase, glutathione peroxidases, and the thioredoxin system. The group of antioxidant substances is made by tocopherols, ascorbic acid, carotenoids and other minor antioxidants. The concept of oxidative stress was extrapolated to clinical medicine and extensively used [31]. As an example of the acceptation, there were more than 157000 entries in May 2016 under "oxidative stress" in PubMed. In terms of the free radical-mediated biochemical chain reactions, oxidative stress implies an increased rate of the endogenous free radical reactions. The increase in reaction rates is accompanied by increases in the steady-state concentrations of oxidative intermediates, in the rate of O_2 incorporation in the free radical reactions and in the content of final oxidation products [32]. The central involvement of the GSH/GSSG ratio was established by the redox hipothesis [33, 34] that potentiated the classic concept of oxidative stress. This hypothesis states that oxidative stress is an oxidative process that alters the redox balance of the thiol groups in low molecular weight molecules as GSH and small proteins that are involved in signal pathways and in the regulation of physiological functions. Today, the balance between reactive oxygen species (oxidants) and antioxidant defense (antioxidants) and the GSH/GSSG redox balance constitute the two main and differential characteristics of oxidative stress. The protective roles of GSH against oxidative stress and damage are multiple: scavenging $HO^{\bullet}$ and $RO^{\bullet}$ in the hydrophilic domain, reducing cofactor of glutathione peroxidases, part of the glutaredoxin system, and regeneration of the reduced forms of protein thiols [22, 35–38].

The redox potential of the couple GSH/GSSG is determined by the concentration of both chemical species, according to $E' = E^{\circ\prime} + RT \ln [GSH]^2/[GSSG]$ with $E^{\circ\prime} = -220$ mV. The ratio GSH/GSSG is a good estimation of the whole cell -SH/-SS- redox potential. In normal conditions, intracellular GSH steady state levels are physiologically maintained by the constant reduction of GSSG by $NADPH_2$ and glutathione reductase. The calculated normal cellular redox potential is about 340 mV. In oxidative stress and damage, GSH is oxidized, deeply changing the cellular redox potential. The transition metal ions Fe and Cu are essential for all living organisms because they participate in a wide variety of metabolic processes. However, their concentrations in the body tissues must be strictly regulated because these metal ions are toxic when their levels are increased [38–40]. In conditions of Fe and Cu overloads, rat brain dramatically changes GSH/GSSG redox potentials. From the −340 mV of normal conditions it is shifted to −170 mV and −64 mV for maximal Fe and Cu effects, respectively [41]. The indiscriminate use of the term oxidative stress without relation to redox biochemistry should be avoided. There are many examples where a decreased content of a supposed antioxidant is interpreted as due to oxidative stress.

6 Physiology of the Products of the Partial Reduction of Oxygen

Considering the intracellular properties of the products of the partial reduction of O_2, $HO^•$ radical is the most reactive, with the shortest half life (10^{-9} s) (Table 3.3), and oxidizes any cell molecule that it collides with and is the primary and main cause of toxicity and oxidative damage. It is understood that $HO^•$ does not have any signaling role in cells. Concerning O_2^-, its anionic charge limits its diffusion through membranes. It is produced by specific reactions in mitochondria and cytosol and is kept at different steady state levels in the two compartments (Table 3.3). O_2^- is a reductant that reduces itself in the dismutation to O_2 and H_2O_2, and reduces cytochome c, Fe^{3+} in ferritin and iron-sulfur clusters in proteins. Hydrogen peroxide is a stable molecule in the absence of specific enzymes and Fe^{2+} ions. It has the relative long half life of 0.1 s and is kept as an intracellular constituent at levels of 10^{-8} to 10^{-6} M (Table 3.3). Hydrogen peroxide is no longer considered harmful by its property of producing $HO^•$, but necessary for cellular homeostasis and communication. Important physiological processes are regulated by H_2O_2, among them: regulation of transcription factors (NF-kB and Nrf2), growth factor signaling, hypoxic response (HIF 1α and 2α), inflammation process and the immune response. Activation or inhibition of the transcription factors NF-kB and Nrf2 is crucial for cell homeostatis and adaptation, processes that are affected in human diseases. NF-kB is associated with the reactive oxygen species generated during inflammation. Nrf2, activated by H_2O_2, transition metals, NO and 4-HNE, is the major transcriptional regulator in the response and defense against oxidative stress [42]. The reactive chemical species listed in Table 3.3 show an integrated mitochondrial metabolism with NO (Fig. 3.2). There are two main characteristics for this fact: first, mitochondria produce its own

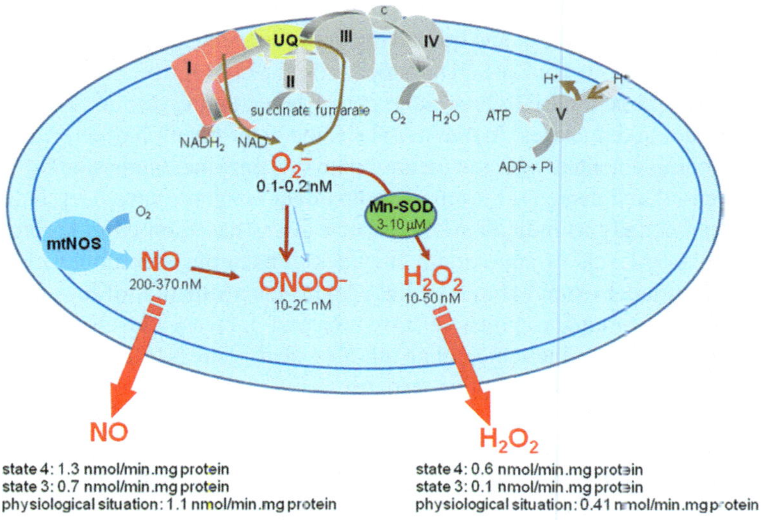

Fig. 3.2 Integrated mitochondrial metabolism of O_2-, H_2O_2, NO and $ONOO^-$.

NO by mitochondrial nitric oxide synthase (mtNOS)) [43], and second, O_2^- and NO are strongly linked by the diffusion-limited reaction of peroxynitrite formation ($O_2^- + NO \rightarrow ONOO^-$).

7 Experimental Determination of Oxidative Stress in Cells and Tissues

The experimental determination of oxidative stress in cells and tissues is of fundamental importance in studies of physiopathology with a clear extrapolation to evidence-based medicine. The most followed approach is the determination of the final oxidation products of phospholipids, proteins, and nucleic acids. Although there are no studies on the time course of the independent oxidation of the various cell components, it is considered that all the oxidative processes are simultaneous in the cell cytosol. A careful treatment is needed because the oxidation products are in dynamic steady states: they are produced, reduced, metabolized and eliminated. The most studied oxidation products are the ones of phospholipid peroxidation and of protein oxidation. Lipid peroxidation is the most studied and biologically relevant free radical chain reaction. The most determined products are malonaldehyde and 4-HO-nonenal. Malonaldehyde, usually determined as TBARS, accounts for about 10 % of the O_2 incorporated in lipid peroxidation. With advanced and sensitive equipments, minor products of the process, such as isoprostanes, pentane, $^1O_2$ and others, are determined [44, 45]. Singlet oxygen is determined by its monomol (1270 nm) and dimol (640–670 nm) emissions and by the formation of cholesterol endoperoxide. Only a small group of reactive oxygen species, H_2O_2, ROOH and $ONOO^-$, can be directly measured. Others species are at very low steady state concentrations (Table 3.3) and cannot be determined.

The determination of the activity of antioxidant enzymes as indicators of oxidative stress is not an easy task and trained operators are needed. The main antioxidant enzymes are: cytosolic Cu,Zn-SOD, mitochondrial Mn-SOD, peroxisomal catalase, cytosolic and mitochondrial glutathione peroxidases, and cytosolic and mitochondrial thioredoxin reductases. In the first place, the determination of enzyme activity has to be complemented with the determination of enzyme expression by Western blot. Increased and decreased levels of antioxidant enzymes have been indiscriminately taken equally as indication of oxidative stress based on two interpretations. First, decreased levels of antioxidant enzymes in bacteria, according to Fridovich [5], a concept latter extended to mammals, indicates a situation of low antioxidant protection and subsequent oxidative stress. Second, increased levels of antioxidant enzymes were taken as indication of an initial oxidative stress followed by increased enzyme biosynthesis in an adaptive response.

8 Aging

The concept that free radicals are responsible for the physiological decline in individuals with age was proposed by Harman in "the free radical theory of aging" in which aging results from HO$^\bullet$ initiated oxidations and damage to macromolecules [46, 47]. Organic antioxidants, 2-mercaptoetylamine and others, added to food were able to increase mouse lifespan. More recently, Navarro et al. [48] reported that mice refused food supplemented with 2-thioproline, that the animals detected as disagreeable, leading to caloric restriction and to increased lifespan. Other treatments, as increased neurologic activity, moderate exercise and vitamin E supplementation, were also able to increase average lifespan but not maximum lifespan [49]. Harman hypothesis gained momentum after the discovery of superoxide dismutases [4], the acceptation of the Gerschman theory of a common mechanism for oxygen and radiation toxicity [1], and the recognition of mitochondrial H_2O_2 production in aerobic tissues [7, 18]. At present, the free radical theory of aging, evolved from Gerschman and Harman theories [1, 47, 50], is preferred among the aging theories.

9 Conclusions

About 1 % of the O_2 consumed in normal respiration in the tissues is converted to O_2^- and H_2O_2. The chemical species of the partial reduction of O_2 (O_2^-, H_2O_2 and HO$^\bullet$) are physiologically produced as part of mitochondrial function and of aerobic life. Partially reduced O_2 species are in physiological situations at a non-zero level and at relatively low steady state concentrations. When these levels are increased, free-radical mediated oxidations take place in cells and tissues and produce biochemical and cellular damage. The situation is described as reversible oxidative stress, but if marked or sustained leads to oxidative damage. The cellular metabolism of O_2^-, H_2O_2, NO and $ONOO^-$ is integrated and constitutes the main source of cellular oxidative stress. Free radical mediated oxidative stress and damage have been extensively used to describe the molecular mechanisms of a long series of pathologies and of aging. Antioxidant enzymes and substances have proved to be effective against oxidative stress and damage in many cases. However, as cells and tissues have very different properties there is no effective treatment that can be used for all cases of oxidative stress and damage.

Acknowledgements The authors acknowledge to Mg. Rosario Mussaco Sebio and to Mg. Christian Saporito Magriña for their successful experimental work that produced four publications on the toxicity of Fe and Cu overloads.

References

1. Gerschman R, Gilbert D, Nye SW et al (1954) Oxygen poisoning and x-irradiation: a mechanism in common. Science 119:623–626
2. Michaelis L (1946) Fundamentals of oxidation and respiration. Am Sci 34:573–596
3. Mann PJG, Quastel JH (1946) Toxic effects of oxygen and of hydrogen peroxide on brain metabolism. Biochem J 40:139–144
4. McCord JM, Fridovich I (1969) Superoxide dismutase. An enzymic function for erythrocuprein (hemocuprein). J Biol Chem 244:6049–6055
5. McCord JM, Keele BB Jr, Fridovich I (1971) An enzyme-based theory of obligate anaerobiosis: the physiological function of superoxide dismutase. Proc Natl Acad Sci U S A 68:1024–1027
6. Seim S (1982) Production of reactive oxygen species and chemiluminescence by human monocytes during differentiation and lymphokine activation in vitro. Acta Pathol Microbiol Immunol Scand C 90:179–185
7. Boveris A, Oshino N, Chance B (1972) The cellular production of hydrogen peroxide. Biochem J 128:617–630
8. Boveris A, Costa LE, Cadenas E (1999) The mitochondrial production of oxygen radicals and cellular aging. In: Packer L, Cadenas E (eds) Understanding the process of aging. Marcel Dekker Inc, New York, pp 1–16
9. Loschen G, Flohé L, Chance B (1971) Respiratory chain linked H_2O_2 production in pigeon heart mitochondria. FEBS Lett 18:261–264
10. Cadenas E, Boveris A (1980) Enhancement of hydrogen peroxide formation by protophores and ionophores in antimycin-supplemented mitochondria. Biochem J 188:31–37
11. Gajavelli S, Kentaro S, Diaz J et al (2015) Glucose and oxygen metabolism after penetrating ballistic-like brain injury. J Cereb Blood Flow Metab 35:773–780
12. Adamo AM, Llesuy SF, Pasquini JM, Boveris A (1989) Brain chemiluminescence and oxidative stress in hyperthyroid rats. Biochem J 263:273–277
13. Loschen G, Azzi A, Richter C, Flohé L (1974) Superoxide radicals as precursors of mitochondrial hydrogen peroxide. FEBS Lett 42:68–72
14. Boveris A, Cadenas E (1975) Mitochondrial production of superoxide anions and its relationship to the antimycin insensitive respiration. FEBS Lett 54:311–314
15. Boveris A, Cadenas E, Stoppani AO (1976) Role of ubiquinone in the mitochondrial generation of hydrogen peroxide. Biochem J 156:435–444
16. Dionisi O, Galeotti T, Terranova T, Azzi A (1975) Superoxide radicals and hydrogen peroxide formation in mitochondria from normal and neoplastic tissues. Biochim Biophys Acta 403:292–300
17. Margulis L (1996) Archaeal-eubacterial mergers in the origin of eukarya: phylogenetic classification of life. Proc Natl Acad Sci U S A 93:1071–1076
18. Boveris A, Chance B (1973) The mitochondrial generation of hydrogen peroxide. General properties and the effect of hyperbaric oxygen. Biochem J 134:707–716
19. Turrens JF, Boveris A (1980) Generation of superoxide anion by the NADH dehydrogenase of bovine heart mitochondria. Biochem J 191:421–427
20. Muller FL, Liu Y, Van Remmen H (2004) Complex III releases superoxide to both sides of the inner mitochondrial membrane. J Biol Chem 279:49064–49073
21. Okado-Matsumoto A, Fridovich I (2001) Subcellular distribution of superoxide dismutases (SOD) in rat liver: Cu, Zn-SOD in mitochondria. J Biol Chem 276:38388–38393
22. Chance B, Sies H, Boveris A (1979) Hydroperoxide metabolism in mammalian organs. Physiol Rev 59:527–605
23. Boveris A, Cadenas E, Reiter R et al (1980) Organ chemiluminescence: noninvasive assay for oxidative radical reactions. Proc Natl Acad Sci U S A 77:347–351
24. Nelson DL, Cox MM (2000) Lehninger principles of biochemistry, 3rd edn. Worth Publishers, New York

25. Dawson AG (1979) Oxidation of cytosolic NADH formed during aerobic metabolism in mammalian cells. Trends Biochem Sci 4:171–176
26. Jequier E, Acheson K, Schutz Y (1987) Assessment of energy expenditure and fuel utilization in man. Ann Rev Nutr 7:187–208
27. Boveris A, Repetto M, Bustamante J et al (2008) The concept of oxidative stress in pathology. In: Alvarez S, Evelson P, Boveris A (eds) Free radical pathophysiology. Research Signpost, Kerala, pp 1–17
28. Cadenas E, Giulivi C, Ursini F et al (1994) Electronically-excited state formation during lipid peroxidation. Methods Toxicol 1B:384–399
29. Repetto M, Boveris A (2012) Transition metals: bioinorganic and redox reactions in biological systems. In: Mishra A (ed) Transition metals: characteristics, properties and uses. Nova Science Publihers, Hauppauge, pp 349–370
30. Sies H (1985) Introductory remarks. In: Sies H (ed) Oxidative stress. Academic, London
31. Sies H (1991) Oxidative stress: from basic research to clinical application. Am J Med 91:31–38
32. Gonzalez Flecha B, Cutrin JC, Boveris A (1993) Time course and mechanism of oxidative stress and tissue damage in rat liver subjected to *in vivo* ischemia-reperfusion. J Clin Invest 91:456–464
33. Sies H, Jones DP (2007) Oxidative Stress. In: Fink G (ed) Encyclopedia of stress, vol 3, 2nd edn. Elsevier, Amsterdam, pp 45–48
34. Kemp M, Go Y, Jones DP (2008) Non equilibrium thermodynamics of thiol/disulfide redox systems: a perspective on redox system biology. Free Radic Biol Med 44:921–937
35. Jaeschke H, Gores GJ, Cederbaum AI et al (2002) Mechanisms of hepatotoxicity. Toxicol Sci 65:166–176
36. Yuan L, Kaplowitz N (2009) Glutathione in liver diseases and hepatotoxicity. Mol Aspects Med 30:29–41
37. Orrenius S, Nicotera P, Zhivotovsky B (2011) Cell death mechanisms and their implications in toxicology. Toxicol Sci 119:3–19
38. Arnér ES, Holmgren A (2000) Physiological functions of thioredoxin and thioredoxin reductase. Eur J Biochem 267:6102–6109
39. Halliwell B, Gutteridge J (1989) Lipid peroxidation: a radical chain reaction. In: Free Radical Biology and Medicine, 2nd edn. Clarendon, Oxford, pp 188–276
40. Musacco-Sebio R, Ferrarotti N, Saporito-Magriñá C et al (2014) Rat brain oxidative damage in iron and copper overloads. Metallomics 6:1410–1416
41. Semprine J, Ferrarotti N, Musacco-Sebio R et al (2014) Brain antioxidant response to iron and copper acute intoxications in rats. Metallomics 6:2083–2089
42. Gorlach A, Dimova E, Petry A et al (2015) Reactive oxygen species, nutrition, hipoxia and diseases: problems solved? Redox Biol 6:372–385
43. Giulivi C, Poderoso JJ, Boveris A (1998) Production of nitric oxide by mitochondria. J Biol Chem 273:11038–11043
44. Niki E (2000) Oxidative stress and aging. Intern Med 39:324–326
45. Yoshida Y, Saito Y, Hayakawa M et al (2007) Levels of lipid peroxidation in human plasma and erythrocytes: comparison between fatty acids and cholesterol. Lipids 42:439–449
46. Harman D (1956) Aging: a theory based on free radical and radiation chemistry. J Gerontol 11:298–300
47. Harman D (1981) The aging process. Proc Natl Acad Sci U S A 78:7124–7128
48. Navarro A, Sanchez-Pino MJ, Gomez C et al (2007) Dietary thioproline decreases spontaneous food intake and increases survival and neurological function in mice. Antiox Redox Signal 9:131–141
49. Boveris A, Navarro A (2008) Brain mitochondrial dysfunction in aging. IUBMB Life 60:308–314
50. Jin K (2010) Modern biological theories of aging. Aging Dis 1:72–74

Chapter 4
Biochemistry and Physiology of Heart Mitochondrial Nitric Oxide Synthase

Tamara Zaobornyj, Darío E. Iglesias, Silvina S. Bombicino, Ivana A. Rukavina-Mikusic, and Laura B. Valdez

Abstract Heart mitochondria are the major source of reactive oxygen and nitrogen species and play a central role in cell energy provision and signaling. The NO produced by cardiac mtNOS is allowed to interact restrictedly with the co-localized effectors. NO exerts a high affinity, reversible and physiological inhibition of cytochrome *c* oxidase activity. A second effect of NO on the respiratory chain is accomplished through its interaction with ubiquinol-cytochrome *c* oxidoreductase. The ability of mtNOS to regulate mitochondrial O_2 uptake and O_2^- and H_2O_2 productions is named mtNOS functional activity. Several situations, including chronic hypoxia and ischemia-reperfusion, modify heart mtNOS activity or expression. The regulation of heart mtNOS by distinctive mitochondrial environments includes the effects of Ca^{2+}, O_2, L-arginine, NADPH, mitochondrial membrane potential ($\Delta\psi$) and the metabolic states. Together, this enzyme seems to be critical during the adaptation of heart mitochondria to changes in cellular bioenergetics.

Keywords Heart mitochondrial NOS • mtNOS functional activity • Nitric oxide • Membrane potential • Hydrogen peroxide • Mitochondrial complexes I, III and IV

1 Introduction

Mitochondria supply the cells with both the energy and the signals that coordinate cell death and survival [1]. Indeed, heart mitochondria produce the potentially toxic products of the partial reduction of oxygen: superoxide anion (O_2^-), hydrogen peroxide (H_2O_2), and hydroxyl radical ($HO^\bullet$) and signaling and regulatory molecules, as nitric oxide (NO) and H_2O_2 [2–5]. A fine control of mitochondrial respiration is critical to meet the energy demand of cardiac muscle [6–8]. During the progression

T. Zaobornyj (✉) • D.E. Iglesias • S.S. Bombicino • I.A. Rukavina-Mikusic • L.B. Valdez
Institute of Biochemistry and Molecular Medicine (IBIMOL; UBA-CONICET), Physical Chemistry Division, School of Pharmacy and Biochemistry, University of Buenos Aires, Buenos Aires, Argentina
e-mail: tamaraz@ffyb.uba.ar; lbvaldez@ffyb.uba.ar

© Springer International Publishing Switzerland 2016

R.J. Gelpi et al. (eds.), *Biochemistry of Oxidative Stress*, Advances in Biochemistry in Health and Disease 16, DOI 10.1007/978-3-319-45865-6_4

of heart disease, the loss of mitochondrial function leads to cell injury, cell dysfunction and death [8]. The signaling functions of NO in the heart depend on its production by the specific NO synthase (NOS) isoforms compartmentalized within cellular microdomains, and its interactions with several biomolecules involved in downstream signaling pathways. The strategic localization of heart mitochondrial nitric oxide synthase (mtNOS) in the organelles in charge of cellular energy metabolism allows for a tight control of different processes whereby mitochondria play pivotal roles [9, 10].

2 Nitric Oxide Regulates Mitochondrial Respiration and Superoxide Anion and Hydrogen Peroxide Productions

Even though NO is a highly diffusible molecule, the distances that this free radical reaches are short due to its high reactivity with several species which include heme groups, O_2^- and thiols [9]. As a result, the NO produced by cardiac mtNOS is allowed to interact restrictedly with the co-localized effectors. It is well known that NO activates soluble guanylatecyclase (sGC), which leads to the production of the second messenger 3,5′-cyclic guanosine monophosphate (cGMP) [8]. However, certain functions of NO in signaling and regulation of cardiac function are performed through cGMP-independent pathways including those that involve mitochondria (Fig. 4.1).

First, NO competes with O_2 for the binding site at the binuclear center of cytochrome *c* oxidoreductase or complex IV. This leads to a high affinity and reversible inhibition of the enzyme [11–13]. This effect was unraveled using a mathematical model based in experimental data and described different effects of NO in the mito-

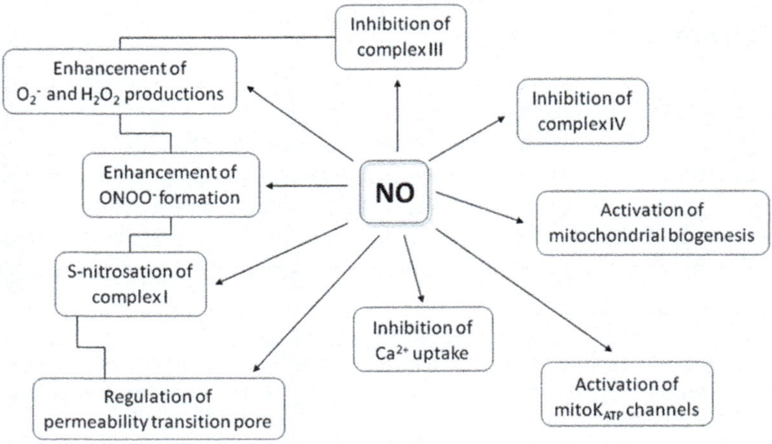

Fig. 4.1 Effects of NO in mitochondria

chondrial metabolic states [14, 15]. The mitochondrial NO steady-state concentrations in tissues, in the 50–350 nM range [6] are in the same range that produces an effective (50 %) inhibition of cytochrome oxidase activity [6, 7, 12–15]. The rate of respiration depends on O_2 concentration and on the O_2/NO ratio [6, 7]. In other words, the biological effect of a given NO concentration will depend on the simultaneous O_2 concentration; a lower PO_2 increases and extends the effects of NO [6, 7]. Additionally, another important effect of NO on mitochondrial respiratory chain is accomplished through the interaction of NO with respiratory complex III, ubiquinol-cytochrome *c* oxidoreductase [16, 17]. The activity of succinate-cytochrome *c* reductase (complexes II-III) was inhibited by about 50 % in the presence of about 1.3 μM NO released from the NO donors S-nitrosoglutathione (GSNO) or spermine-NONOate, while the activity of succinate-Q reductase (complex II) resulted unaffected [17]. These results indicate that NO specifically inhibits complex III activity. Interestingly, complexes II-III activity was also decreased (36 %) when submitochondrial particles (SMP) were incubated with mtNOS substrates and cofactors, suggesting that the inhibition is also produced by endogenous NO. Through its interaction with components of the electron-transfer chain, NO functions not only as a physiological inhibitor of cell respiration, but it also enhances the generation of reactive oxygen species [16, 17] and thereby triggers various mechanisms underlying the survival or death of the cells. A hyperbolic increase in O_2^- and H_2O_2 production rates by heart SMP was observed with a maximal effect at 500 μM GSNO [17]. Moreover, H_2O_2 production by heart coupled mitochondria was increased by 70 % when mitochondria were exposed to the NO donors [17].

The activity of heart mtNOS can be assessed indirectly through the two main effects that NO exert on the mitochondrial respiratory chain functions: the inhibition of O_2 consumption and the enhancement of H_2O_2 production. This mtNOS activity has been termed *mtNOS functional activity* [18] and it is determined by the difference in the rates of O_2 uptake or H_2O_2 production in isolated mitochondria in two different conditions. One of them is the condition where NO steady-state concentration is the highest, *i.e.* in the presence of sufficient L-arginine and SOD, condition in which active O_2 consumption is reduced and H_2O_2 production is enhanced. The second condition is by the contrary, when NO steady-state level is the lowest, condition in which active O_2 consumption is increased and H_2O_2 production is decreased. This latter situation is achieved in the absence of NOS substrates or cofactors and in the presence of a NOS inhibitor or a NO scavenger, *e.g.* oxyhemoglobin. Thus, supplementation of heart mitochondria with L-arginine and SOD decreases the respiration rate by 15–20 %, while supplementation of the mitochondrial preparation with a NOS inhibitor (L-NAME) and hemoglobin increases O_2 consumption by 10–15 %. Concerning H_2O_2 production, the addition of L-arginine and SOD enhances H_2O_2 production by 15–20 % in heart mitochondria, whereas the supplementation of the same preparation with L-NAME and hemoglobin declines H_2O_2 generation by 10–15 %. Thus, the changes in mtNOS functional activity clearly reveal variations in mitochondrial NO production rates and NO steady-state concentration [18]. For instance, the myocardial stunning observed in perfused rabbit heart after 15 min ischemia and 30 min reperfusion, is characterized by a decreased mtNOS activity and an impaired mtNOS functional activity [19].

3 Other Effects of NO and NO-Derived Species on Mitochondrial Functions

Other effects of NO in mitochondria-cytosol signaling are conveyed via nitrosation of proteins [20]. In nitrosation reactions, NO reversibly reacts with the nucleophilic centers in thiol residues contained within a specific consensus sequence of amino acids [21] of a broad array of low molecular weight compounds or proteins [21, 22]. When mitochondria are treated with NO donors, complex I is S-nitrosated resulting in a significant inhibition of this respiratory complex, an effect that can be reversed by light or reagents containing thiol groups [23]. It has been suggested that a transient reversible inhibition of the mitochondrial electron transfer minimizes ischemia-reperfusion injury, and that the blockade of electron transfer at complex I preserves respiration during reperfusion [24]. Studies have shown that complex I inhibition by nitrosation protects mitochondria during hypoxia and reoxygenation and cardiomyocytes during ischemia-reperfusion [25–27]. Nitric oxide also reacts with O_2^- that is formed by the mitochondrial respiratory chain during normal or pathological oxygen metabolism, yielding $ONOO^-$ [28]. This reaction occurs with a second order rate constant of about $2 \times 10^{10}\ M^{-1}\ s^{-1}$ and its rate is controlled by the diffusion of the reactants [29]. Peroxynitrite influences cardiac contractility and, in some cases, produces effects that are markedly different from those of NO [30, 31]. This NO congener may hinder mitochondrial functions and cause cell death. The switch from reversible inhibition of cellular respiration by NO to the pathological inhibition of mitochondrial function by the NO-derived $ONOO^-$ has been observed in many physiopathological conditions, in which intramitochondrial NO and O_2^- steady-state concentrations are enhanced. Pathological concentrations of NO are likely to affect respiration by mechanisms that are qualitatively different from those observed during reversible physiological regulation. Increased $ONOO^-$ steady-state concentration together with mitochondrial malfunctioning are hallmarks of heart hypoxia-reperfusion injury [19, 30, 31].

In addition, NO may act on other mitochondrial targets. It has been reported that NO increases oxidative phosphorylation efficiency [32], activates mitoK_{ATP} channels [33], regulates mitochondrial matrix pH and Ca^{2+} buffering capacity [34], triggers mitochondrial biogenesis [35] and modulates the mitochondrial permeability transition pore (MPT)) formation [36]. The regulation of mitochondrial Ca^{2+} accumulation by NO may provide means by which mtNOS can influence mitochondrial metabolism as well as survival [37]. Nitric oxide triggers mitochondrial biogenesis in several cell types and tissues, including those obtained from heart, through a cGMP-dependent mechanism [35]. The cell injury and death observed after reoxygenation of the tissue are in accordance with several reports showing that NO can prevent or accelerate mitochondrial permeability transitions [38]. This depends on whether NO concentrations are physiological or supraphysiological and whether O_2^- is generated simultaneously [39, 40]. In this regard, heart mtNOS plays an important role suggesting relevance of mtNOS in yet another important mitochondrial function.

4 Heart mtNOS Activity and Identity

Different groups have used various experimental approaches and reported the presence of mtNOS in the heart. A summary of the publications is included in Table 4.1. The first study showing the presence of a NOS isoenzyme located in mitochondria used the silver enhanced gold immunolabelling method and showed that about 85 % of the heart mitochondria were positive for the eNOS label [41]. After that, the studies of cardiac mtNOS have used various methods: immunohistochemistry [41, 42], spectrophotometry [19, 39, 43–51], radiometry [52, 53], fluorometry [40, 54–56], chemiluminescence [53] and electrochemistry [57]. Heart mtNOS has been studied

Table 4.1 mtNOS activity or expression in heart mitochondria

Methodology	Sensitivity [NO] (M)	References
Espectrophotometric (HbO_2 oxidation)	$1 \times 10^{-11}/10^{-12}$	Costa et al. [43]
		Boveris et al. [39]
		Saavedra-Molina et al. [44]
		Valdez et al. [45]
		Gonzales et al. [46]
		Zaobornyj et al. [47]
		Feller et al. [48]
		Boveris et al. [49]
		La Padula et al. [50]
		Zaobornyj et al. [51]
		Valdez et al. [19]
Radiometric ([^{3}H]L-citrulline/[^{14}C] L-citrulline))	1×10^{-7}	Zanella et al. [52]
		Zenebe et al. [53]
Microsensor (porphyrinic and amperometric)	1×10^{-7}	Kanai et al. [57]
Fluorometric (DAF-2, DAF2-DA, DAF-FM)	10^{-9}	Lopez-Figueroa et al. [54]
		Zanella et al. [55]
		Zorov et al. [56]
		Dedkova and Blatter [40]
Chemiluminescence		Zenebe et al. [53]
Immunohistochemistry		Bates et al. [41]
		Hotta et al. [42]

using isolated cardiac mitochondria [19, 39, 43–51] and through functional experiments in which mitochondrial membrane potential-dependent NO production was measured in cardiomyocytes [40, 49]. Kanai and co-workers [57] showed elegantly the presence of a NOS activity in mouse cardiac mitochondria by measuring, in a single individual mitochondrion, the NO production that followed to Ca^{2+} addition to the reaction medium, by using a porphyrinic microsensor. Moreover, Boveris group [19, 39, 43, 45, 47, 49] reported a heart mtNOS activity of 0.8–1.5 nmol NO $\times$ min^{-1} $\times$ mg protein^{-1} using the oxyhemoglobin (HbO_2) assay. Heart mitochondrial NO production accounts for about 60 % of total cellular heart NO generation, suggesting a central role of the mitochondrially produced NO in cardiomyocytes [51, 58]. Confocal microscopy using fluorescent probes for coupled mitochondria and for NO indicated the presence of NO in mitochondria [40, 55]. Other studies have detected mitochondrial NO release using a direct spin trapping technique with electron paramagnetic resonance spectrometry [59]. The physiological NO release from heart mitochondria has been calculated as 1.8 nmol NO $\times$ min^{-1} $\times$ mg protein^{-1}, considering the simultaneous oxidation of 3 mol of NADH (malate-glutamate) and 1 mol of succinate within the matrix, and taking into account that heart mitochondria oscillate between a NO-inhibited state 3 (47 %) and state 4 (53 %) [58]. In these conditions, NO release accounts for about 3 % of heart O_2 consumption.

Up to date, the immunochemical nature of heart mtNOS isoform is a subject under debate. The 51–57 % homology reported for nNOS, iNOS, and eNOS; and the cross-reactivity of isoform-specific anti-NOS antibodies, as well as the possible overexpression of iNOS and eNOS in experimental models could provide an explanation for the conflicting results [60]. However, the most convincing data implicate nNOS as the primary candidate for the NOS isoenzyme targeted into mitochondria [57, 60, 61]. In 2002, the sequence of rat liver mtNOS was reported as the α splice variant of the nNOS isoform, with post-translational modifications: acylation with myristic acid at the N-terminal and phosphorylation at the C-terminal region [61]. In the study of Kanai et al. [57], the NO production of an individual mouse cardiac mitochondrion was measured with a microsensor placed at the cytoplasmic face of the mitochondrial outer membrane. The sensor detected a NO signal in mitochondria isolated from eNOS-/- or iNOS-/- animals, while no signal was detected in mitochondria from nNOS-/- mice. In our laboratory, we have detected an increase in heart mtNOS expression after the exposure of rats to high altitude, using anti-nNOS and anti-iNOS antibodies [51]. Moreover, a decrease in heart mtNOS expression during the regression of the cardioprotection conferred by hypoxia was reported using anti-nNOS and anti-iNOS [50].

5 Regulation of Heart Mitochondrial NOS

Several situations can regulate heart mtNOS activity or expression. As an example, treatment with enalapril increased the production of NO by heart mitochondria [39, 62]. Interestingly, a receptor for angiotensin was found in inner mitochondrial membrane and it was suggested that the renin-angiotensin system directly regulates mitochondrial NO production [63]. We have reported that rats submitted to chronic hypoxia showed about 60 % enhanced heart mtNOS activity [47]. This up-regulation was associated with a preservation of heart contractility upon aging or after an ischemic insult. Concordant results were observed when rats were exposed to natural high altitude [46, 51]. Again, heart mtNOS activity and expression were specifically increased after exposure [51] and this enhancement showed a similar pattern to the one observed for hematocrit. Another study performed by us showed a decline of about 30 % in mitochondrial NO production after ischemia-reperfusion, without modification of mtNOS expression, together with a decrease of 30 % in mitochondrial complex I activity [19]. This latter mitochondrial dysfunction was named by Boveris et al. "complex I syndrome" in which complex I and mtNOS are partially inactivated associated with protein nitration and oxidative damage to proteins and phospholipids [64]. Localization of NO production within mitochondria provides a reciprocal regulation between mtNOS and the intramitochondrial medium (Fig. 4.2). Indeed, heart mtNOS activity depends on intramitochondrial Ca^{2+} concentration [40, 65]; the blockade of mitochondrial Ca^{2+} uniporter inhibits mitochondrial NO production. In energized mitochondria, elevation of extramitochondrial Ca^{2+} stimulates mtNOS activity and decreases respiration. Interestingly, results from our laboratory showed that energized and coupled heart mitochondria produce NO without supplementation with Ca^{2+} [66], indicating that the concentration of Ca^{2+} in the preparation may be sufficient to sustain a basal mtNOS activity. Concerning the

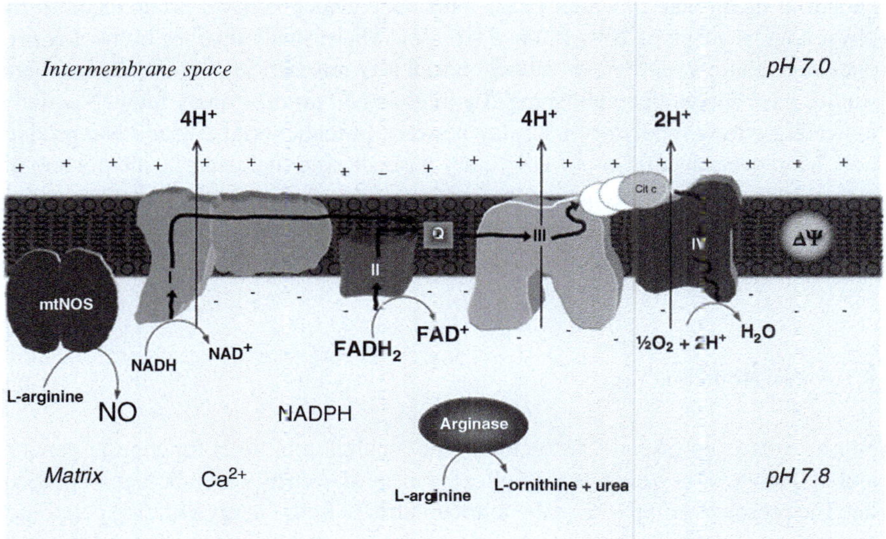

Fig. 4.2 Diagram showing the location of mtNOS within mitochondrion. Specific regulation of mitochondrial NO production occurs by the intramitochondrial levels of Ca^{2+}, O_2, L-arginine, arginases, and NADPH and by mitochondrial metabolic state and $\Delta\psi$

mtNOS activity determined in heart mitochondrial membranes, NO production was 80 % lower in the absence of Ca^{2+} in the reaction medium. Taking into account that heart mitochondrial matrix and cytosol Ca^{2+} concentrations are different [65–68], the existence of a Ca^{2+}-dependent NOS within mitochondria has significant consequences in terms of the differential regulation of this enzyme.

The mtNOS, as cytoplasmic NOS isoforms, requires O_2, L-arginine, and certain cofactors to produce NO. The study of NO production by heart mitochondrial membranes as a function of L-arginine concentration showed a hyperbolic response, with an apparent K_M value of about 35 μM [69]. Of note, rather lower intracellular levels of L-arginine and the presence of arginase-I and arginase-II activity in heart mitochondria have been reported [70]. The activity of mtNOS may be impaired under the conditions whereby L-arginine concentration within the heart mitochondria is diminished. As the content of NADP(H) in heart mitochondria is around 0.4 mM [71, 72], about 80 % of this nucleotide is in the reduced state (NADPH) and since NOS has a relatively low K_M value for NADPH (0.1–1 μM) [73], the intramitochondrial NADPH concentration is high enough to sustain mtNOS activity under physiological conditions.

The mitochondrial NO production is regulated by the metabolic state [49, 58, 66]. During the transition from resting (state 4) to active (state 3) respiration, heart mitochondrial NO release decreases about 60 %. The rate of NO release by mitochondria represents 10 % of the corresponding O_2 consumption in state 4 and only 1.5 % in state 3 [58]. This is consistent with the idea that mitochondrial respiration is more sensitive to exogenous NO in state 3 than in state 4 [12–15, 58, 66]. In addition, heart mitochondrial NO release shows an exponential dependence on the mitochondrial membrane potential ($\Delta\psi$). This dependence is more pronounced in the physiological range of $\Delta\psi$ (150 to 180 mV), where small changes in the $\Delta\psi$ produce noticeable variations of mitochondrial NO release [58, 60]. To date, several studies have shown that abolishing $\Delta\psi$ inhibits NO production by mtNOS activity, indicating a tight regulatory interplay between mitochondrial $\Delta\psi$ and NO production. Moreover, data from our laboratory have showed that heart inside out submitochondrial particles produce NO supported by succinate-dependent reversed electron flow in the respiratory chain, indicating a functional association between mtNOS and complex I proteins [74].

6 Conclusions

Nitric oxide is an essential molecule in the regulation of heart function in general, and of a range of key processes implicated in cardiac energy metabolism in particular. The presence of mtNOS in the mitochondria of high-energy utilizing cells, such as cardiomyocytes, indicates a precise regulation of metabolic pathways in which these organelles are involved. Heart mtNOS allows NO to optimize the balance between cardiac energy production and utilization, and to regulate oxygen and nitrogen free radical productions and Ca^{2+} homeostasis. Furthermore, heart mtNOS is regulated by physiological, pathological, and pharmacological situations.

Moreover, a spatially restricted localization of NO within mitochondria permits the regulation of NOS activity by the local environment within individual organelles. Thus, this enzyme seems to be critical during the adaptation of heart mitochondria to changes in cellular bioenergetics.

Acknowledgments This work was supported by research grants from the University of Buenos Aires (UBACYT 20020110100140, 20020130100731), Agencia Nacional de Promoción Científica y Tecnológica (PICT 2014-0964), and Consejo Nacional de Investigaciones Científicas y Técnicas (PIP 11220110100444).

References

1. Darley-Usmar V (2004) The powerhouse takes control of the cell: the role of mitochondria in signal transduction. Free Radic Biol Med 37:753–754
2. Boveris A, Oshino N, Chance B (1972) The cellular production of hydrogen peroxide. General properties and effect of hyperbaric oxygen. Biochem J 128:707–716
3. Boveris A, Cadenas E (1975) Mitochondrial production of superoxide anion and its relationship to the antimycin insensitive respiration. FEBS Lett 54:311–314
4. Boveris A, Cadenas E (1982) Production of superoxide radicals and hydrogen peroxide in mitochondria. In: Oberley LW (ed) Superoxide dismutase. CRC Press, Boca Raton, pp 15–30
5. Giulivi C, Poderoso JJ, Boveris A (1998) Production of nitric oxide by mitochondria. J Biol Chem 273:11038–11043
6. Boveris A, Costa LE, Cadenas E, Poderoso JJ (1999) Regulation of mitochondrial respiration by adenosine diphosphate, oxygen, and nitric oxide. Meth Enzymol 301:188–198
7. Boveris A, Costa LE, Poderoso JJ et al (2000) Regulation of mitochondrial respiration by oxygen and nitric oxide. Ann N Y Acad Sci 899:121–135
8. Aon MA, Cortassa S, Marbán E, O'Rourke B (2003) Synchronized whole cell oscillations in mitochondrial metabolism triggered by a local release of reactive oxygen species in cardiac myocytes. J Biol Chem 278:44735–44744
9. Wink DA, Mitchell JB (1998) Chemical biology of nitric oxide: insights into regulatory, cytotoxic, and cytoprotective mechanisms of nitric oxide. Free Radic Biol Med 25:434–456
10. Zaobornyj T, Ghafourifar P (2012) Strategic localization of heart mitochondrial NOS: a review of the evidence. Am J Physiol Heart Circ Physiol 303:1283–1293
11. Brown GC, Cooper CE (1994) Nanomolar concentrations of nitric oxide reversibly inhibit synaptosomal respiration by competing with oxygen at cytochrome oxidase. FEBS Lett 356:295–298
12. Cleeter MW, Cooper JM, Darley-Usmar VM et al (1994) Reversible inhibition of cytochrome c oxidase, the terminal enzyme of the mitochondrial respiratory chain, by nitric oxide. Implications for neurodegenerative diseases. FEBS Lett 345:50–54
13. Takehara Y, Nakahara H, Inai Y et al (1996) Oxygen-dependent reversible inhibition of mitochondrial respiration by nitric oxide. Cell Struct Funct 21:251–258
14. Antunes F, Boveris A, Cadenas E (2004) On the mechanism and biology of cytochrome oxidase inhibition by nitric oxide. Proc Natl Acad Sci U S A 101:16774–16779
15. Antunes F, Boveris A, Cadenas E (2007) On the biologic role of the reaction of NO with oxidized cytochrome c oxidase. Antioxid Redox Signal 9:1569–1579
16. Poderoso JJ, Carreras MC, Lisdero C et al (1996) Nitric oxide inhibits electron transfer and increases superoxide radical production in rat heart mitochondria and submitochondrial particles. Arch Biochem Biophys 328:85–92

17. Iglesias DE, Bombicino SS, Valdez LB, Boveris A (2015) Nitric oxide interacts with mitochondrial complex III producing antimycin-like effects. Free Radic Biol Med 89:602–613
18. Valdez LB, Zaobornyj T, Boveris A (2005) Functional activity of mitochondrial nitric oxide synthase. Meth Enzymol 396:444–455
19. Valdez LB, Zaobornyj T, Bombicino S et al (2011) Complex I syndrome in myocardial stunning and the effect of adenosine. Free Radic Biol Med 51:1203–1212
20. Stamler JS, Lamas S, Fang FC (2001) Nitrosylation, the prototypic redox-based signaling mechanism. Cell 106:675–683
21. Sun J, Xin C, Eu JP et al (2001) Cysteine-3635 is responsible for skeletal muscle ryanodine receptor modulation by NO. Proc Natl Acad Sci U S A 98:11158–11162
22. Scharfstein JS, Keaney JF Jr, Slivka A et al (1994) In vivo transfer of nitric oxide between a plasma protein-bound reservoir and low molecular weight thiols. J Clin Invest 94:1432–1439
23. Borutaite V, Budriunaite A, Brown GC (2000) Reversal of nitric oxide-, peroxynitrite- and S-nitrosothiol-induced inhibition of mitochondrial respiration or complex I activity by light and thiols. Biochim Biophys Acta 1459:405–412
24. Burwell LS, Nadtochiy SM, Tompkins AJ et al (2006) Direct evidence for S-nitrosation of mitochondrial complex I. Biochem J 394:627–634
25. Nadtochiy SM, Burwell LS, Ingraham CA et al (2009) In vivo cardioprotection by S-nitroso-2-mercaptopropionyl glycine. J Mol Cell Cardiol 46:960–968
26. Chen Q, Moghaddas S, Hoppel CL, Lesnefsky EJ (2006) Reversible blockade of electron transport during ischemia protects mitochondria and decreases myocardial injury following reperfusion. J Pharmacol ExpTher 319:1405–1412
27. Shiva S, Sack MN, Greer JJ et al (2007) Nitrite augments tolerance to ischemia/reperfusion injury via the modulation of mitochondrial electron transfer. J Exp Med 204:2089–2102
28. Radi R, Cassina A, Hodara R et al (2002) Peroxynitrite reactions and formation in mitochondria. Free Radic Biol Med 33:1451–1464
29. Kissner R, Nauser T, Bugnon P et al (1997) Formation and properties of peroxynitrite as studied by laser flash photolysis, high-pressure stopped-flow technique, and pulse radiolysis. Chem Res Toxicol 10:1285–1292
30. Mihm MJ, Bauer JA (2002) Peroxynitrite-induced inhibition and nitration of cardiac myofibrillar creatine kinase. Biochimie 84:1013–1019
31. Katori T, Donzelli S, Tocchetti CG et al (2006) Peroxynitrite and myocardial contractility: in vivo versus in vitro effects. Free Radic Biol Med 41:1606–1618
32. Clerc P, Rigoulet M, Leverve X, Fontaine E (2007) Nitric oxide increases oxidative phosphorylation efficiency. J Bioenerg Biomembr 39:158–166
33. Sasaki N, Sato T, Ohler A et al (2000) Activation of mitochondrial ATP-dependent potassium channels by nitric oxide. Circulation 101:439–445
34. Ghafourifar P, Richter C (1999) Mitochondrial nitric oxide synthase regulates mitochondrial matrix pH. Biol Chem 380:1025–1028
35. Nisoli E, Clementi E, Paolucci C et al (2003) Mitochondrial biogenesis in mammals: the role of endogenous nitric oxide. Science 299:896–899
36. Brookes PS, Salinas EP, Darley-Usmar K et al (2000) Concentration-dependent effects of nitric oxide on mitochondrial permeability transition and cytochrome c release. J Biol Chem 275:20474–20479
37. Dedkova EN, Blatter LA (2005) Modulation of mitochondrial Ca^{2+} by nitric oxide in cultured bovine vascular endothelial cells. Am J Physiol Cell Physiol 289:C836–C845
38. Crompton M (1999) The mitochondrial permeability transition pore and its role in cell death. Biochem J 341:233–249
39. Boveris A, Lores-Arnaiz S, Bustamante J et al (2002) Pharmacological regulation of mitochondrial nitric oxide synthase. Meth Enzymol 359:328–339
40. Dedkova EN, Blatter LA (2009) Characteristics and function of cardiac mitochondrial nitric oxide synthase. J Physiol 587:851–872

41. Bates TE, Loesch A, Burnstock G, Clark JB (1996) Mitochondrial nitric oxide synthase: an ubiquitous regulator of oxidative phosphorylation? Biochem Biophys Res Commun 218:40–44
42. Hotta Y, Otsuka-Murakami H, Fujita M et al (1999) Protective role of nitric oxide synthase against ischemia-reperfusion injury in guinea pig myocardial mitochondria. Eur J Pharmacol 38:37–48
43. Costa LE, La-Padula P, Lores-Arnaiz S et al (2002) Long-term angiotensin II inhibition increases mitochondrial nitric oxide synthase and not antioxidant enzyme activities in rat heart. J Hypertens 20:2487–2494
44. Saavedra-Molina A, Ramirez-Emiliano J, Clemente-Guerrero M et al (2003) Mitochondrial nitric oxide inhibits ATP synthesis. Effect of free calcium in rat heart. Amino Acids 24:95–102
45. Valdez LB, Zaobornyj T, Alvarez S et al (2004) Heart mitochondrial nitric oxide synthase. Effects of hypoxia and aging. Mol Aspects Med 25:49–59
46. Gonzales GF, Chung FA, Miranda S et al (2005) Heart mitochondrial nitric oxide synthase is upregulated in male rats exposed to high altitude (4,340 m). Am J Physiol Heart Circ Physiol 288:H2568–H2573
47. Zaobornyj T, Valdez LB, La Padula P et al (2005) Effect of sustained hypobaric hypoxia during maturation and aging on rat myocardium. II. mtNOS activity. J Appl Physiol 98:2370–2375
48. Fellet AL, Balaszczuk AM, Arranz C et al (2006) Autonomic regulation of pacemaker activity: role of heart nitric oxide synthases. Am J Physiol Heart Circ Physiol 291 H1246–H1254
49. Boveris A, Valdez LB, Zaobornyj T, Bustamante J (2006) Mitochondrial metabolic states regulate nitric oxide and hydrogen peroxide diffusion to the cytosol. Biochim Biophys Acta 1757:535–542
50. La Padula P, Bustamante J, Czerniczyniec A, Costa LE (2008) Time course of regression of the protection conferred by simulated high altitude to rat myocardium: correlation with mtNOS. J Appl Physiol 105:951–957
51. Zaobornyj T, Valdez LB, Iglesias DE et al (2009) Mitochondrial nitric oxide metabolism during rat heart adaptation to high altitude: effect of sildenafil, L-NAME, and L-arginine treatments. Am J Physiol Heart Circ Physiol 296:H1741–H1747
52. Zanella B, Giordano E, Muscari C et al (2004) Nitric oxide synthase activity in rat cardiac mitochondria. Basic Res Cardiol 99:159–164
53. Zenebe WJ, Nazarewicz RR, Parihar MS, Ghafourifar P (2007) Hypoxia-reoxygenation of isolated rat heart mitochondria causes cytochrome c release and oxidative stress: evidence for involvement of mitochondrial nitric oxide synthase. J Mol Cell Cardiol 43:411–419
54. Lopez-Figueroa MO, Caamano C, Morano MI et al (2000) Direct evidence of nitric oxide presence within mitochondria. Biochem Biophys Res Commun 272:129–133
55. Zanella B, Calonghi N, Pagnotta E et al (2002) Mitochondrial nitric oxide localization in H9c2 cells revealed by confocal microscopy. Biochem Biophys Res Commun 290:1010–1014
56. Zorov DB, Kobrinsky E, Juhaszova M, Sollott SJ (2004) Examining intracellular organelle function using fluorescent probes: from animalcules to quantum dots. Circ Res 95:239–252
57. Kanai AJ, Pearce LL, Clemens PR et al (2001) Identification of a neuronal nitric oxide synthase in isolated cardiac mitochondria using electrochemical detection. Proc Natl Acad Sci U S A 98:14126–14131
58. Valdez LB, Boveris A (2007) Mitochondrial nitric oxide synthase, a voltage-dependent enzyme, is responsible for nitric oxide diffusion to cytosol. Front Biosci 12:1210–1219
59. Boveris A, Valdez LB, Alvarez S et al (2003) Kidney mitochondrial nitric oxide synthase. Antioxid Redox Signal 5:265–271
60. Valdez LB, Zaobornyj T, Bombicino SS et al (2011) Regulation of heart mitochondrial nitric oxide synthase (mtNOS) by oxygen. In: Cadenas S, Palau F (eds) Mitochondrial pathophysiology. Transworld Research Network, Kerala, pp 29–42
61. Elfering SL, Sarkela TM, Giulivi C (2002) Biochemistry of mitochondrial nitric oxide synthase. J Biol Chem 277:38079–38086

62. Boveris A, D'Amico G, Lores-Arnaiz S, Costa LE (2003) Enalapril increases mitochondrial nitric oxide synthase activity in heart and liver. Antioxid Redox Signal 5:691–697
63. Abadir PM, Foster DB, Crow M et al (2011) Identification and characterization of a functional mitochondrial angiotensin system. Proc Natl Acad Sci U S A 108:14849–14854
64. Boveris A, Carreras MC, Poderoso JJ (2010) The regulation of cell energetics and mitochondrial signaling by nitric oxide. In: Ignarro L (ed) Nitric oxide. Elsevier Academic Press, London, pp 441–482
65. Dedkova EN, Ji X, Lipsius SL, Blatter LA (2004) Mitochondrial calcium uptake stimulates nitric oxide production in mitochondria of bovine vascular endothelial cells. Am J Physiol Cell Physiol 286:C406–C415
66. Valdez LB, Zaobornyj T, Boveris A (2006) Mitochondrial metabolic states and membrane potential modulate mtNOS activity. Biochim Biophys Acta 1757:166–172
67. Dedkova EN, Wang YG, Ji X et al (2007) Signalling mechanisms in contraction-mediated stimulation of intracellular NO production in cat ventricular myocytes. J Physiol 580:327–345
68. Dedkova EN, Blatter LA (2008) Mitochondrial Ca^{2+} and the heart. Cell Calcium 44:77–91
69. Alvarez S, Boveris A (2004) Mitochondrial nitric oxide metabolism in rat muscle during endotoxemia. Free Radic Biol Med 37:1472–1478
70. Steppan J, Ryoo S, Schuleri KH et al (2006) Arginase modulates myocardial contractility by a nitric oxide synthase 1-dependent mechanism. Proc Natl Acad Sci U S A 103:4759–4764
71. Chance B, Williams GR (1956) The respiratory chain and oxidative phosphorylation. Adv Enzymol 17:65–134
72. Boveris A, Stoppani AO (1970) Inhibition of electron and energy transfer in mitochondria by 19-nor-ethynyltestosterone acetate. Arch Biochem Biophys 141:641–655
73. Yui Y, Hattori R, Kosuga K et al (1991) Purification of nitric oxide synthase from rat macrophages. J Biol Chem 266:12544–12547
74. Bombicino SS, Iglesias DE, Zaobornyj T, Boveris A, Valdez LB. (2016) Mitochondrial nitric oxide production supported by reverse electron transfer. Arch Biochem Biophys. 607: 8–19

Chapter 5
Biochemistry of Nitric Oxide and Peroxynitrite: Sources, Targets and Biological Implications

Adrián Aicardo, Débora M. Martinez, Nicolás Campolo, Silvina Bartesaghi, and Rafael Radi

Abstract Nitric oxide (•NO) is a relatively stable free radical generated biologically that participates in a series of signal transducing and physiological processes in human biology. Enzymatic as well as non-enzymatic sources of •NO have been reported *in vivo*. In circumstances where •NO is overproduced and/or in the context of a pro-oxidant environment, it can turn into a toxic molecule and converts into a pathogenic mediator in human diseases. In particular, the diffusion-controlled reaction of •NO with superoxide radical anion ($O_2^{\bullet -}$) leads to the formation of peroxynitrite ($ONOO^-$), a strong oxidant and nucleophile that mediates much of the toxicity associated to •NO. Peroxynitrite promotes one- and two-electron oxidations and nitration reactions *via* a series of mechanisms several of which involve free radical intermediates. In this chapter we summarize key biochemical aspects concerning the mechanisms of formation of •NO, peroxynitrite and other reactive nitrogen species (RNS), their reaction with biomolecular targets and participation in the development of pathologies. In addition, we critically analyze current redox-based therapeutic strategies including the direct action of compounds on •NO-and peroxynitrite-derived reactive species as well as the induction or endogenous antioxidant mechanisms, with the overall goal to cope against nitroxidative stress conditions.

Keywords Nitric oxide • Peroxynitrite • Nitroxidative stress • Redox-active drugs • Tyrosine nitration • 3-Nitrotyrosine • Free radicals

---

A. Aicardo • D.M. Martinez • N. Campolo • R. Radi (✉)
Departamento de Bioquímica and Center for Free Radical and Biomedical Research, Facultad de Medicina, Universidad de la República, Montevideo, Uruguay
e-mail: rradi@fmed.edu.uy

S. Bartesaghi
Departamento de Bioquímica and Center for Free Radical and Biomedical Research, Facultad de Medicina, Universidad de la República, Montevideo, Uruguay

Departamento de Educacion Médica, Facultad de Medicina, Universidad de la República, Montevideo, Uruguay

© Springer International Publishing Switzerland 2016
R.J. Gelpi et al. (eds.), *Biochemistry of Oxidative Stress*, Advances in Biochemistry in Health and Disease 16, DOI 10.1007/978-3-319-45865-6_5

1 Introduction

At cellular and tissue level, the biological effects of reactive nitrogen species (RNS) are a consequence of the combination of the biochemical characteristics of their metabolism: sites and rates of formation, diffusion rates, presence of scavenger mechanisms, or concentration of targets. Thus, in this chapter we synthesize the current knowledge about the metabolism of two of the most biologically relevant RNS, nitric oxide ($^{\bullet}NO$) and peroxynitrite ($ONOO^-$).

2 Nitric Oxide: Sources and Destination of an Ubiquitous Reactive Nitrogen Species

2.1 *Biological Formation of* $^{\bullet}NO$

Nitric oxide is a nitrogen-derived radical defined three decades ago as an agent that can mediate different physiological processes such as the biochemical regulation of vascular smooth muscle [1, 2]. In addition, alterations in $^{\bullet}NO$ homeostasis is associated to endothelial dysfunction and vascular oxidative stress [3]. Thus, key aspects of the biochemistry of $^{\bullet}NO$ have been focused on understanding the ambivalent role that this relatively stable free radical species has under protective or harmful conditions.

In biological systems $^{\bullet}NO$ is ubiquitous, and may arise from different sources. These sources can be classified as nitric oxide synthase (NOS)-dependent and NOS-independent. Nitric oxide synthases are a family of heme-containing enzymes that synthesize $^{\bullet}NO$ using L-arginine, NADPH, and O_2 as substrates, and FAD, FMN, tetrahydrobiopterin (BH_4) and calmodulin as cofactors. In its native conformation these enzymes dispose as a homodimer with the formation of a zinc-thiolate cluster involving two cysteine residues of each monomer [4]. The presence of this cluster is essential for the correct folding and activity of the enzyme, because the electrons provided by NADPH are transferred from one monomer to the other during the catalytic cycle. Moreover, structural modifications may lead to a leakage of electrons facilitating the reduction of O_2 to produce superoxide radical anion ($O_2^{\bullet -}$) in a process called NOS uncoupling. Three NOS isoforms have been isolated, cloned, and characterized: neuronal NOS (nNOS), inducible NOS (iNOS) and endothelial NOS (eNOS). Although some reports have described the presence of a mitochondrial NOS, there is yet no total consensus over the presence of this isoform [5, 6]. The three isoforms of NOS are expressed in a wide variety of tissues but, differential regulation of their activities determines differences in the $^{\bullet}NO$ production under certain conditions. On one hand, eNOS and nNOS are constitutively expressed and are stimulated by intracellular calcium levels increments to produce low fluxes of $^{\bullet}NO$ (nanomolar range). On the other hand, iNOS is expressed in response to the presence of proinflammatory cytokines (interleukin-1, tumor necrosis factor α, interferon γ), particularly in immune system cells, to produce higher fluxes of $^{•}NO$ (more than 100-fold the $^{•}NO$ production of eNOS). The activity of iNOS is not regulated by calcium levels, thus it may produce $^{•}NO$ for hours even in the presence of low concentrations of calcium [7]. Moreover, the regulation of NOS activity is also affected by co- and post-translational modifications (acylation, myristoylation, phosphorylation, S-nitrosylation), substrate and cofactors availability, and subcellular localization [8].

In addition to NOS-derived $^{•}NO$, others chemical and enzymatic reactions can mediate the production of $^{•}NO$ by a NOS-independent mechanism. Nitrite (NO_2^-) originated from $^{•}NO$ autoxidation or from the diet vegetables are a NOS-independent source of $^{•}NO$. At the acidic pH of gastric lumen, dietary nitrite is protonated to nitrous acid (HNO_2) which decomposes to $^{•}NO$ and $^{•}NO_2$. Moreover, in the intravascular compartment deoxyhemoglobin ($HbFe^{2+}$) reduces NO_2^- to $^{•}NO$ and methemoglobin, whereas oxyhemoglobin oxidizes NO_2^- to NO_3^-. This system provides $^{•}NO$ to induce vasodilation when serum O_2 concentration is low [9]. In addition, the catalytic reduction of NO_2^- by xanthine dehydrogenase/xanthine oxidase has been suggested. These molybdenum-containing hydroxylases involved in the oxidative metabolism of purines, are capable of catalyzing the one electron reduction of NO_2^- to $^{•}NO$ *in vitro* and in cellular cultures [10]. Other NOS-independent sources of $^{•}NO$ are agents that deliver $^{•}NO$ as a product of a chemical reaction. These $^{•}NO$ donors are nitrites and nitrates, inorganic nitroso compounds, or nitrosothiols, some of which are clinically relevant pharmacological agents to induce vasodilation as a therapy for ischaemic cardiomiopathy (*i.e.* nitroglycerine, sodium nitroprusside).

2.2 *Fates of* $^{•}$*NO*

Once produced, $^{•}NO$ arising from any system, has different fates depending on its localization (cytosol, membranes, intravascular compartment, etc.) or concentration of the surrounding targets (Fig. 5.1).

2.3 *Autoxidation Reactions*

Under aerobic conditions $^{•}NO$ reacts with O_2 to form nitrogen dioxide ($^{•}NO_2$), a strong oxidizing and nitrating agent ($^{•}NO_2/NO_2^-$ $E^{\circ\prime} = 0.99$ V), with a reaction rate constant of $k = 2.8 \times 10^6$ $M^{-2}s^{-1}$ [11]. In turn, NO_2 may react with $^{•}NO$ to form dinitrogen trioxide (N_2O_3), an unstable species that decomposes into two molecules of nitrite. Thus, nitrite represents the main product of the $^{•}NO$ autoxidation pathway. Moreover, autoxidation reactions may occur faster in hydrophobic environments, such as membranes and lipoproteins, where the concentration of $^{•}NO$ and O_2 may be higher than in aqueous solutions.

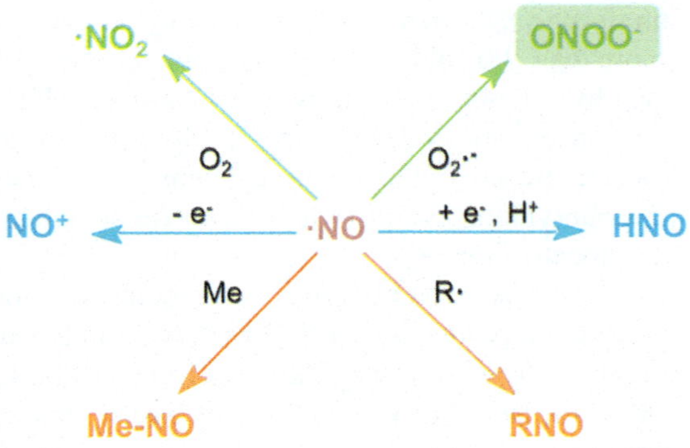

Fig. 5.1 Reactive nitrogen species derived from ${}^{\bullet}NO$ (Modified from Ref. [16])

2.4 Reactions with Radical Species

Nitric oxide reacts relatively slow with most biological molecules, although, it is highly reactive with other radical species such as: $O_2^{\bullet-}$, peroxyl radical ($ROO^{\bullet}$), tyrosyl radical ($Tyr^{\bullet}$) and thiyl radical ($RS^{\bullet}$). In the case of $O_2^{\bullet-}$, ${}^{\bullet}NO$ reacts in a close to diffuse-controlled reaction to yield $ONOO^-$, a reactive oxidizing and nitrating agent (it will be discussed below), or nitrite, respectively. Nitric oxide reacts with amino acid-derived radicals, such as $Tyr^{\bullet}$ yielding 3-nitrosotyrosine ($k = 10^9\ M^{-1}s^{-1}$), which in turn, may be further oxidized into 3-nitrotyrosine [12]. On one hand, the reaction of ${}^{\bullet}NO$ with organic peroxyl radicals, for example membrane phospholipids peroxyl radicals ($ROO^{\bullet}$), may yield nitrosated and nitrated lipids and may account for some of the "antioxidant" effects of ${}^{\bullet}NO$. The reaction of ${}^{\bullet}NO$ with lipid peroxyl radicals yields an organic peroxynitrite that may decompose through rearrangement to an organic nitrate, preventing the propagation of radicals by lipid peroxidation chains in biological membranes [13]. On the other hand, organic peroxynitrite may suffer homolytic cleavage to produce alkoxyl radical ($RO^{\bullet}$) and ${}^{\bullet}NO_2$, both stronger oxidants than peroxyl radicals. However, neither of them can further react with ${}^{\bullet}NO$ preventing propagation. Furthermore, the exposure of thiol groups (R-SH), such as glutathione or protein thiols, to ${}^{\bullet}NO$ derives in the nitrosylation of the thiol to give a nitrosothiol (R-SNO), as nitrosoglutathione or nitrosoproteins. In a fast reaction ($k = 2–3 \times 10^9\ M^{-1}s^{-1}$), ${}^{\bullet}NO$ reacts with thiyl radicals, but not significantly with reduced thiols, particularly at sites of inflammation and/or hydrophobic compartments where the concentration of ${}^{\bullet}NO$ is high enough to outcompete with other routes of thiyl radical decay ($R\text{-}S^{\bullet} + {}^{\bullet}NO \rightarrow R\text{-}SNO$) [14].

2.5 Reaction with Metal Centers

Another example of nitrosylation reactions promoted by •NO is the one that occurs when •NO reaches enzymes metal centers. In addition to the protection against lipid peroxidation chains, •NO also decreases the pro-oxidant effects of ferryl intermediates and amino acid radicals of heme proteins exposed to peroxides. It also reduces the reactivity of Fe^{2+} with H_2O_2 in the Fenton reaction. An example of protein nitrosylation reactions is the reaction of •NO with the heme a_3 group of cytochrome *c* oxidase to form nitrosyl-cytochrome a_3 complex. This reversible reaction inhibits oxygen binding, reducing electron transport and ATP production, however, if inhibition is prolonged in time mitochondrial $O_2^{\bullet-}$ formation increases and, in consequence, the $ONOO^-$ fluxes arising from the mitochondrion which may irreversible damage electron transport chain [15]. As a consequence of its ability to easily diffuse through membranes, •NO may react with different heme proteins in the vascular system depending on its localization. In smooth muscle cells it reacts with the heme of guanylate cyclase leading to its activation to increase cGMP levels and reduce intracellular calcium concentration to cause smooth vascular muscle relaxation. In contrast, in the intravascular compartment, •NO diffuses into erythrocytes where it may binds to the metal center of deoxyhemoglobin in a nitrosylation reaction, or it may react with oxyhemoglobin oxidizing it to methemoglobin and producing nitrate. Thus, erythrocytes significantly decrease the half-life of •NO in the intravascular compartment [9].

3 Peroxynitrite: A Potential Harm for Multiple Biomolecules

3.1 Biological Formation of $ONOO^-$

Among the different targets of •NO, $O_2^{\bullet-}$, is to be distinguished because of: a) its reaction rate with •NO is close to diffusion-controlled; and b) the biochemical importance of its product, $ONOO^-$, a strong oxidizing and nucleophilic agent. As an oxidant, it promotes one and two-electrons oxidation reactions by direct reaction with biomolecular targets. Peroxynitrite formation may arise from the reaction of the precursor radical species coming from independent cellular sources (*e.g.* mitochondrial $O_2^{\bullet-}$ reacting with NOS-derived •NO) but other sources such as uncoupled NOS can simultaneously yield $O_2^{\bullet-}$ and •NO and produce $ONOO^-$.

3.2 Fates of $ONOO^-$

Peroxynitrite may act as a one-electron oxidizing agent either directly (*e.g.* oxidation of cytochrome c^{2+}) or secondarily through the products of its homolysis. The protonation of $ONOO^-$ to form its conjugate acid (peroxynitrous acid, ONOOH, pKa = 6,8), weakens the O-O bond producing its homolysis and giving rise to $^{\bullet}OH$ and $^{\bullet}NO_2$, two strongly oxidizing/hydroxylating and nitrating species, respectively. Proton-catalyzed homolytic cleavage occurs with a first order reaction rate constant of 0.9 s^{-1} at 37 °C [17]. Thus, the quantitative relevance of $ONOO^-$ homolysis at cellular level is limited considering the higher rate constant of other bimolecular reactions of $ONOO^-$. However, the small percentage of $ONOO^-$ that decomposes through homolytic cleavage produce $^{\bullet}OH$ and $^{\bullet}NO_2$ (with a 30 % yield), both one-electron oxidants that may initiate radical chain reactions amplifying oxidative damage. Indeed, proton-catalyzed decomposition of $ONOO^-$ may become more relevant in hydrophobic phases (*e.g.* cellular membranes) resulting in the initiation of lipid peroxidation processes. However, most of the $ONOO^-$ formed (70 %) will isomerize to nitrate, and some proteins such as oxyhemoglobin may catalyze $ONOO^-$ isomerization [18]. In spite of its short half-life (~10 ms approximately at physiological pH), $ONOO^-$ can diffuse through membranes and it can affect surrounding cells to a range of one to two cellular diameters (5–20 μm).

Many biomolecules may be oxidized or nitrated by $ONOO^-$-derived radicals including tyrosine residues, thiols, DNA, and unsaturated membrane fatty acid [16]. However, in biological systems where these targets coexist, the preferential targets will depend on reaction rate constants and concentrations. Thus, in cells and tissues an important percentage of the $ONOO^-$ formed will react with carbon dioxide, certain thiol groups, seleno-containing proteins and some metal centers.

3.3 Carbon Dioxide

Despite its relatively low reaction rate, compared to the reaction with other targets ($k = 5.8 \times 10^4$ $M^{-1}s^{-1}$) [19], the nucleophilic addition of $ONOO^-$ to carbon dioxide (CO_2) is very important in biological systems due to physiological concentration of this gas (1–2 mM, in equilibrium with bicarbonate). As a result of the reaction, a transient intermediate is formed, nitroso-peroxo-carboxylate ($ONOOCO_2^-$), which rapidly homolyzes to carbonate radical ($CO_3^{\bullet-}$) and $^{\bullet}NO_2$ in a 35 % yield. In consequence, the reaction of $ONOO^-$ with CO_2 leads to the formation of secondary one-electron oxidants that can expand nitroxidative damage.

3.4 Thiols

Peroxynitrite can oxidize thiol groups by one- or two-electron reactions. Peroxynitrous acid (ONOOH) reacts directly with thiolates (RS^-) promoting a two-electron oxidation reaction to give sulfenic acid (RSOH) and nitrite as products. Moreover, these oxidized thiols are usually unstable, and biologically relevant low molecular weight (LMW) sulfenic acid derivatives react with thiols to form disulfides as main products. In the case of protein sulfenic acids, they can also react with

LMW thiols or another protein thiols to form disulfide bonds. In the first case, the formation of this adduct represents the main route of protein-S-glutathionylation in biological systems, a reversible post-translational modification that can either change protein function or protect cysteine over oxidation to non-reversible oxidized forms (sulfinic and sulfonic acids). Moreover, the reactivity of $ONOO^-$ towards thiol groups varies widely in terms of reaction rate constant. On one hand, free thiol group of Cys^{34} of bovine serum albumin has a second-order rate constant of $\sim 10^3\ M^{-1}s^{-1}$, and on the other hand the peroxidatic thiol of peroxiredoxins reacts with a second-order rate constant of $\sim 10^6$–$10^7\ M^{-1}s^{-1}$, allowing them to act as first line antioxidant defense against $ONOO^-$ [20].

Low molecular weight thiols, as well as protein thiols, can be oxidized by one-electron reaction leading to thiyl radical ($RS^•$) formation. As a result of $ONOO^-$ decay, either through homolysis or its reaction with carbon dioxide, different one-electron oxidants are produced ($^•OH$, $^•NO_2$, $CO_3^{•-}$) which are capable of oxidizing thiol groups to $RS^•$ radical. This sulfur radical is also unstable, and in the presence of oxygen reacts with it in a reversible diffusion-controlled rate reaction resulting in thioperoxyl radical ($RSOO^•$), which may propagate damage through radical chain reactions [20].

3.5 *Metal Centers*

Direct one-electron oxidations of transition metal centerss yields nitrogen dioxide and oxo-metal complexes, both one-electron oxidants capable of giving rise to secondary radical species, an example is the reaction of $ONOO^-$ with oxyhemoglobin. The reaction between $ONOO^-$ and the ferrous heme, mainly yields methemoglobin and nitrate (90 %), however a minor fraction of this reaction may result in a two-electron oxidation of the ferrous heme yielding ferrylhemoglobin and $^•NO_2$ (10 %) [18].

3.6 *Tyrosine Residues*

Among the multiple oxidation reactions that are mediated by $ONOO^-$ in biological systems, one of the most representative and studied modification is the nitration of protein tyrosine residues to 3-nitrotyrosine (NO_2Tyr). Nitration of tyrosine residues in proteins represents an *in vivo* oxidative post-translational modification that can affect protein structure and function and implies the substitution of a hydrogen atom by a nitro group ($-NO_2$) in one of the carbon atoms in the *ortho* position (with respect to the –OH group) of the phenolic ring of the tyrosine residue. Importantly, $ONOO^-$ does not cause the nitration of tyrosine by a direct bimolecular reaction, but rather *via* $ONOO^-$-derived radicals or one-electron oxidants. To note, under biological conditions, tyrosine nitration is mediated by free radical reactions in a

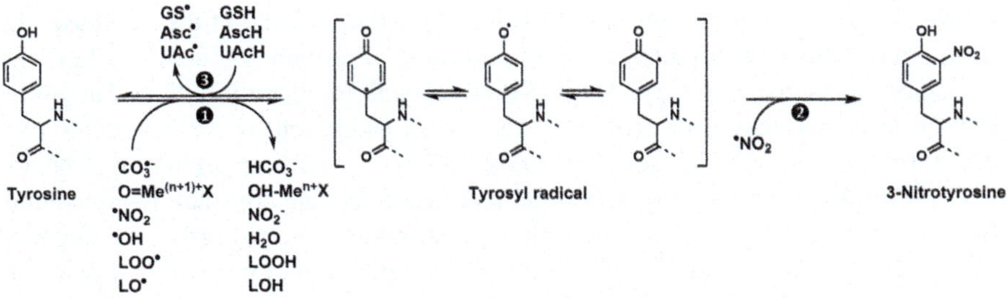

Fig. 5.2 Main reactions involved in the formation of 3-nitrotyrosine. Tyrosine nitration is achieved through two reaction steps that involve free radical chemistry. Initially, tyrosine is oxidized by several one-electron oxidants to tyrosyl radical (1); then, the tyrosyl radical reacts with nitrogen dioxide, producing the non-radical product 3-nitrotyrosine (2). Reducing agents such as glutathione (GSH), ascorbic acid (AscH) and uric acid (UAcH) can reduce tyrosyl radicals back to tyrosine (3), thus preventing the nitration of tyrosine residues

two-step process: first, the phenolic ring of tyrosine is oxidized to its one-electron oxidation product, the tyrosyl radical (Tyr•), and, then, the addition of $^{\bullet}NO_2$ to the Tyr• produces the non-radical product NO_2Tyr [21] (Fig. 5.2).

Many one-electron oxidants derived from $ONOO^-$ decomposition *in vivo* can produce the oxidation of tyrosine to tyrosyl radical: carbonate radical produced after the reaction of $ONOO^-$ with CO_2, several oxo-metal complexes ($O = Me^{(n+1)+}X$) that are produced from the one- or two-electron oxidation of certain transition metal centers (both from metalloproteins or low molecular weight complexes) by $ONOO^-$ [21], and $^{\bullet}OH$ and $^{\bullet}NO_2$ formed by peroxynitrous acid homolysis. Also, in hydrophobic environments like cell membranes and lipoproteins, certain intermediates of lipid peroxidation processes such as lipid peroxyl (ROO•) and alkoxyl (RO•) radicals can promote the one-electron oxidation of tyrosine residues, and so, act as a link between lipid oxidation and protein oxidation [22] (Fig. 5.3). Given that $ONOO^-$, through several of its decay pathways, produces both strong one-electron oxidants to oxidize tyrosine to the tyrosyl radical and $^{\bullet}NO_2$ to recombine with the produced Tyr•, it is one of the most relevant agents leading to protein tyrosine nitration in biological systems. Nevertheless, *in vivo* tyrosine nitration can also occur through $ONOO^-$ independent reactions. The most relevant of these alternative mechanisms of nitration is the one that depends on the hemeperoxidase catalyzed production of $^{\bullet}NO_2$ from hydrogen peroxide (H_2O_2) and nitrite (NO_2^-). Certain hemeperoxidases, mainly those released by leukocytes during inflammatory conditions, like myeloperoxidase (MPO) and eosinophilperoxidase (EPO), can actively mediate protein tyrosine nitration through a catalytic process that begins with the two-electron oxidation of the heme group of the active site of the peroxidase by H_2O_2. As a result of this oxidation step, an oxoferrylporphyrin π cation radical ($HP^{\bullet+}$-(Fe^{4+} = O)), known as compound I, is formed; this intermediate is a strong one- and two-electron oxidant which is able to oxidize tyrosine to Tyr•, as well as NO_2^- to $^{\bullet}NO_2$. Moreover, when compound I is reduced by one electron when it oxidizes, for example, tyrosine or NO_2^-, it converts into the respective oxo-ferryl intermediate (HP-(Fe^{4+} = O)), known as compound II. This species is also a strong

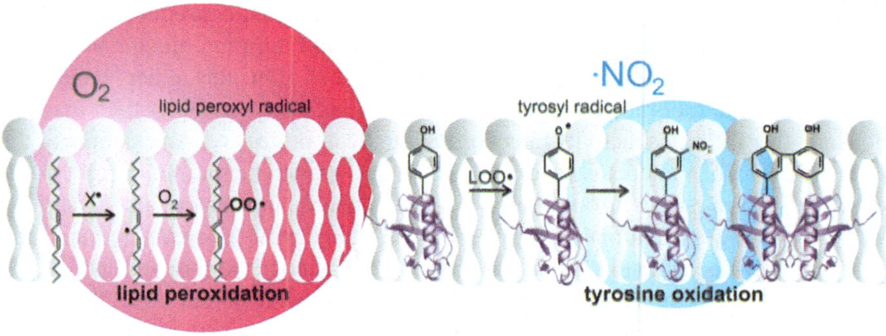

Fig. 5.3 Lipid peroxidation and tyrosine oxidation processes induced by $ONOO^-$ in hydrophobic biostructures. The *scheme* shows that $ONOO^-$-derived radicals such as $^{\bullet}OH$ and $^{\bullet}NO_2$ (indicated as $X^{\bullet}$) can initiate lipid peroxidation in structures such as biomembranes and lipoproteins. Lipid peroxyl radicals arising in the process are capable to promote the one-electron oxidation of amino acids such as tyrosine to yield tyrosyl radical that can subsequently react with $^{\bullet}NO_2$ and yield 3-nitrotyrosine in proteins. This connecting reaction is a prime example on how biomolecular oxidative processes can be intertwined and how $ONOO^-$ can lead to a variety of oxidation and nitration products via its secondary radicals. Moreover, in hydrophobic biostructures, peroxynitrite also causes the nitration of lipids (not shown) by the combination of lipid-derived radicals with $^{\bullet}NO_2$. These processes are relevant to explain chemical modifications in lipids and proteins during nitroxidative stress conditions (Reproduced from Ref. [22])

one-electron oxidant, able also to oxidize tyrosine to $Tyr^{\bullet}$ radical and to produce $^{\bullet}NO_2$ from nitrite; its one-electron reduction regenerates the hemeperoxidase to its ground state. It is by these reactions that, in the presence of H_2O_2 and NO_2^-, hemeperoxidases are able to produce both $Tyr^{\bullet}$ and $^{\bullet}NO_2$, leading thus to tyrosine nitration [23]. This mechanism of nitration is not limited to hemeperoxidases: free hemin and other low molecular weight metal complexes have also been shown that are able to catalyze tyrosine nitration from H_2O_2 and NO_2^- [24]. In close relationship with the hemeperoxidase-dependent tyrosine nitration, there is a $^{\bullet}NO_2$-independent mechanism of nitration that has been shown to occur on some heme proteins. This particular mechanism implies the addition of $^{\bullet}NO$ to a tyrosyl radical instead of $^{\bullet}NO_2$, to yield 3-nitrosotyrosine. This intermediate, when formed in the proximity of the heme group, can be further oxidized by two electrons to produce finally 3-nitrotyrosine in two steps of one-electron oxidation that are mediated by the heme group in the presence of additional H_2O_2 [25]. This mechanism may be of great relevance *in vivo* as it requires $^{\bullet}NO$ and not $^{\bullet}NO_2$, which has a much shorter biological half-life than $^{\bullet}NO$ [26]. Due to the fact that tyrosine nitration is mediated by free radical reactions that depend on the formation of the tyrosyl radical, several secondary products are usually produced along with 3-nitrotyrosine during tyrosine nitration and oxidation processes. The two most important and frequently formed of these products are 3,3'-dityrosine (DiTyr), which arises by the recombination of two $Tyr^{\bullet}$, and 3-hydroxytyrosine (DOPA), produced mainly as a product of $^{\bullet}OH$ addition to tyrosine and the subsequent loss of an electron of the adduct formed [26].

The nitration of tyrosine residues has different biological consequences depending on the protein affected. On one hand, tyrosine nitration inhibits the activity of manganese superoxide dismutase (Mn-SOD) [27]. On the other hand, this posttranslational modification may result in a gain of function rather than inactivation, and this additional activity may be either toxic or protective. An example of this is the neurotoxic outcome of the exposure of nerve growth factor (NGF) to $ONOO^-$, which transforms this neurotrophin factor into a neuronal proapoptotic signal [28]. However, as it was mentioned, nitration may also have protective effects, such is the case of nitrite-dependent intragastric nitration of pepsinogen and pepsin, which prevents from the gastric ulcer development under ulcerogenic conditions [29].

3.7 DNA

Purine nucleotides of DNA may be oxidized or nitrated in the presence of $ONOO^-$, and 8-oxo and 8-nitroguanine are two of the major products. In addition, $ONOO^-$ can also attack the deoxyribose backbone causing strand breaks [16].

3.8 Lipids

Membrane biomolecules such as unsaturated fatty acids, may be damaged by different reactive oxygen species (ROS) and RNS, including $ONOO^-$-derived radical species ($^{\bullet}OH$, $^{\bullet}NO_2$). Indeed, the one-electron oxidation of membrane fatty acids, mediated by ROS/RNS, yields very reactive alkyl radicals, which may abstract electrons from other fatty acids, initiating radical chain oxidation reactions. Moreover, alkyl radicals react fast with molecular oxygen producing an organic peroxyl radical ($ROO^{\bullet}$), a strong oxidizing agent, which may also propagate lipid peroxidation reactions in biological membranes [16] (Fig. 5.3). The reactions involved in the propagation of lipid peroxidation chains result in a variety of products including lipid hydroperoxides (*e.g.* 4-hydroxynonenal, 4-oxononenal), conjugated dienes, and malondialdehyde. In consequence, an increment in the detection of these products, either *in vitro* or *in vivo*, is considered as evidence of lipid oxidative damage [16].

Conversely to what occurs during propagation stage, the reaction of lipid radicals with other radical species results in termination reactions interrupting the propagation of lipid peroxidation. Indeed, the reaction of lipid radicals with RNS ($^{\bullet}NO_2$, $^{\bullet}NO$) yields nitrosated and nitrated lipids. However, some of the reactions between RNS and lipid radicals lead to intermediates that may give rise to reactive species. An example of this is the case of the reaction of peroxyl radicals with $^{\bullet}NO$ yielding an organic peroxynitrite (ROONO), a lipid derivative which may rearrange into an organic nitrate ($RONO_2$), or decompose into alkoxyl radical ($RO^{\bullet}$) and $^{\bullet}NO_2$. In the latter case, another molecule of $^{\bullet}NO$ is consumed to reduce the alkoxyl radical formed [30].

In general, nitrated unsaturated fatty acids represent a class of lipid-derived signaling mediators and represent the major bioactive oxides of nitrogen in the vasculature [31]. Nitrated fatty acids in human plasma and red blood cell membranes present the capacity to modulate signaling pathways linked to inflammation. The possibility of a pharmacological role of nitro fatty acids is an area under active investigation [32]. Nitrated linoleic acid, for example, displays cell-signaling activities associated to anti-inflammatory effects [32]. The derived nitroalkene obtained by arachidonic acid nitration has anti-inflammatory activity in activated macrophages by two mechanisms, inhibition of iNOS expression and a reduction in the production of proinflammatory cytokines [33]. Nitroarachidonic acid has been also found to serve as a potential inhibitor of cyclooxygenase [34]. Nitrated lipids also demonstrated pharmacological potential action by a relevant modulation of hypertension [35], vascular inflammation and hypertensive disorders [36], and cardiomyocytes protection in an isolated heart ischemia/reperfusion model [37]. Moreover, nitro fatty acids have also showed the capacity to reduce atherosclerotic lesion formation in animal model [38]. In summary, $ONOO^-$ and $ONOO^-$-derived radicals may be responsible for a wide range of nitroxidative modifications in a variety of biomolecules, reflecting the potential harm of its overproduction under pathological conditions (Fig. 5.4).

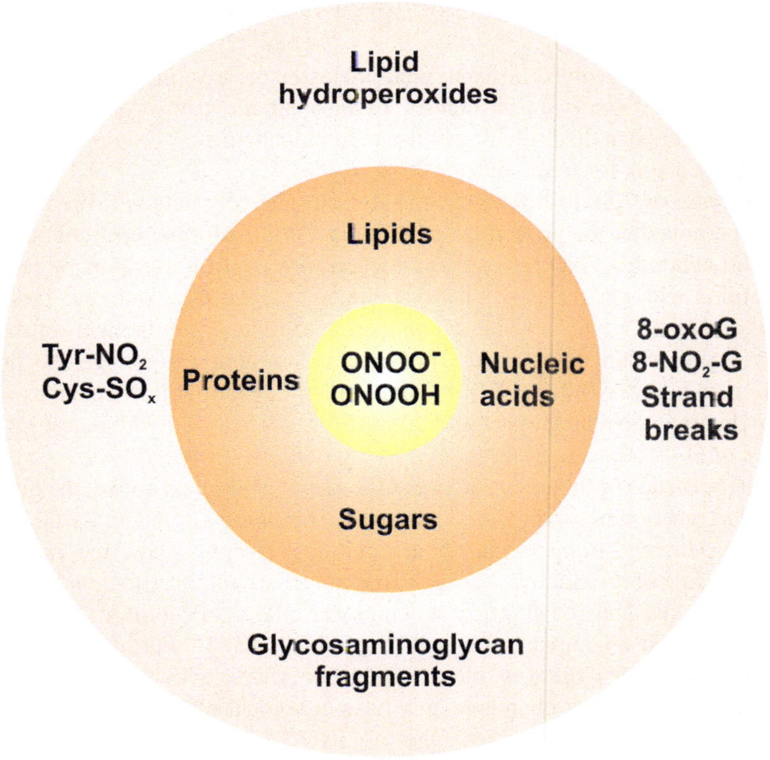

Fig. 5.4 Peroxynitrite induced nitroxidative damage to biomolecules

4 Physiology and Pathophysiology of $^{\bullet}NO$ and Peroxynitrite-Derived Oxidants

Nitric oxide formed under physiological conditions is not toxic agent, but excess $^{\bullet}NO$ and $^{\bullet}NO$-derived oxidants may exert biological effects depending on the concentration and time of exposure to these species, and the capacity of the cells to cope with nitroxidative challenge. Oxidant and free radical-induced damage has been thoroughly studied because it has been associated to prevalent human pathologies. Moreover, several of the most prevalent human pathologies currently, have been associated to an increase in the formation of free radicals and the induction of cellular oxidative damage. Indeed, reactive nitrogen species ($^{\bullet}NO$, $ONOO^-$, $^{\bullet}NO_2$) have been related to the development of cardiovascular, neurodegenerative, inflammatory, metabolic, and malignant neoplastic diseases in a process globally defined as nitroxidative damage [39]; and we will briefly discuss some of the data that supports the pathogenic role of reactive nitrogen species in in human pathology (see also Table 5.1).

4.1 Peroxynitrite and Mitochondrial Dysfunction

Mitochondria are involved in many vital processes of cells, including bioenergetic control, calcium homeostasis, control of biosynthetic pathway, and cell death. Mitochondria play a dual role in cellular redox biochemistry, because it may act as source or as a sink for free radicals.

In the case of $^{\bullet}NO$, its diffusion capacity and relatively long half-life of this free radical permits that the production of different NOS isoforms reach mitochondria. In the mitochondria, $^{\bullet}NO$ interrupts mitochondrial electron transport by reversible competition with oxygen for its binding site on cytochrome *c* oxidase. The inhibition of complex IV results in the accumulation of reduced components of the electron transport chain, favoring the one-electron reduction of oxygen to produce $O_2^{\bullet-}$ (*e.g.* in complexes I and III). In addition, mitochondrial flavoenzymes such as α-ketoglutarate dehydrogenase and glycerol phosphate dehydrogenase are also sources of $O_2^{\bullet-}$ [40].

In the presence of fluxes of $^{\bullet}NO$ and $O_2^{\bullet-}$, as we mentioned above, the diffusion-controlled reaction between them leads to the formation of $ONOO^-$, which in turn decays producing strong oxidizing and nitrating agents. Thus, the presence of $ONOO^-$ induces nitroxidative damage to biomolecules within mitochondria. Indeed, $ONOO^-$ induced tyrosine nitration of complexes I, II, and IV inhibits electron transport resulting in an impairment in ATP production [41]. Furthermore, $ONOO^-$ induces cytochrome *c* nitration increasing its peroxidatic activity and promoting its release to the cytosol, a phenomenon whose biological relevance is still not completely understood [42]. Another target of $ONOO^-$ are the metal centers of mitochondrial enzymes such as aconitase and MnSOD. The iron sulfur cluster of

Table 5.1 Evidence of the role of nitroxidative stress in relevant pathologies

Pathological condition	Biomarker	Redox therapy	References
Human atherosclerosis lesions	Protein tyrosine nitration at the vascular wall	Statins act as indirect antioxidants by inhibition of MPO and •NO-derived oxidants in patients	[62, 75]
Myocardial I/R injury	High 3-nitrotyrosine and malondialdehyde levels in coronary sinus	FP15 reduces myocardial infarct size in a porcine model of myocardial infarction	[66, 76, 77]
		Tocopherol reduced the incidence of non-fatal myocardial infarction in patients with coronary artery disease	
Stroke	Enhanced immunostaining for NOX4 in human brain with ischemic stroke	AEOL-10150 is effective in rodent stroke models.	[78, 79]
		FeTMPyP and FeTMPS reduce infarct size in a transient stroke model (rats)	
Hyper-cholesterolemia	Increased basal levels of $ONOO^-$ in circulating granulocytes	Vitamins C and E restore endothelial function in hyperlipidemic children	[30]
		Statins and AT1-receptor blockers inhibit NADPH oxidase activity and expression and re-couple NOS	
Diabetes	Increased plasma levels of oxidative stress and 3-nitrotyrosine in cardiomyocites and endothelial cells	Mercaptoalkylguanidines and FP15 (a metalloporphyrin) reduce the onset of type I diabetes in murine models	[81–83]
Arthritis	Increased 3-nitrotyrosine in synovial tissues	FP15 improves the outcome of autoimmune arthritis in murine models	[58, 59]
Systemic Lupus Erythematosus	Increased MDA/HNE protein adducts, 3-nitrotyrosine, iNOS, oxLDL, MDA-LDL adducts, versus controls.	N-acetylcysteine was effective in reversing GSH depletion and improving disease activity and fatigue in patients with SLE	[84–86]
	Lower serum levels of SOD		
Inflammatory Bowel Disease	High levels of •NO in intestinal lumen, an 3-nitrotyrosine in biopsies	N-acetylcysteine decreased •NO concentration and improved antioxidant defenses in mice.	[60, 61]

(continued)

Table 5.1 (continued)

Pathological condition	Biomarker	Redox therapy	References
Sepsis	Increased plasma and myocardial 3-nitrotyrosine, and urinary nitrite excretion	Treatment with metalloporphyrin (MnTE-2-PyP(5+)) reduced mitochondrial and diaphragmatic dysfunction in an animal model of sepsis	[87, 88]
	Decreased plasma levels of vitamins A and E, β-carotene and licopene	N-acetyl cysteine reduced NF-kB activation and IL-8 secretion in septic patients	
	Decreased ATP and GSH in skeletal muscle mitochondria		
Neuro-degenerative diseases	Increased 3-nitrotyrosine in Lewy bodies and polymorphonuclear cells in patients with Alzheimer's and Parkinson's diseases	NOS inhibitors or NOS gene deficiency improve the outcome in several models of neurodegeneration.	[68, 89–91]
	Nitrated α-synuclein in brains of patients with Parkinson's disease	Porphyrinic antioxidants improve the outcome of murine models of ALS	
	Increased plasma MDA and oxidized GSH, decrease in SOD activity, and in Se and Zn levels, in patients with amyotrophic lateral sclerosis		

aconitase [4Fe-4S], is readily oxidized [3Fe-4S] by $ONOO^-$ ($k = 1.1 \times 10^5\ M^{-1}s^{-1}$) releasing free iron and inactivating the enzyme [43]. In the case of MnSOD, $ONOO^-$ reacts with its metal center ($k = 1 \times 10^5\ M^{-1}s^{-1}$) leading to the inactivation of the enzyme through the nitration of tyrosine-34 in the proximity of the active site [44]. The inactivation of MnSOD results in an increment in $O_2^{\bullet-}$ steady-state concentration, which favours further $ONOO^-$ production closing a vicious cycle of nitroxidative damage. In addition, peroxiredoxins 3 (Prx-3) and 5 (Prx-5), mitochondrial isoforms, rapidly decompose $ONOO^-$ formed within mitochondria for Prx-5 $k = 7 \times 10^7\ M^{-1}s^{-1}$ [45]), preventing its reaction with other biomolecules. As a result, together with MnSOD, Prx-3 and -5 modulate mitochondrial fluxes of $ONOO^-$, regulating its signaling and toxic pathways.

4.2 Apoptosis and Necrosis

Under physiological conditions cells produce low levels of oxidants, as byproducts of basal metabolism, which are detoxified by cellular antioxidant defenses maintaining low steady state oxidants concentrations. Nevertheless, when the production of oxidants is increased (*e.g.* inflammation) antioxidant defenses may be overwhelmed and free radicals induce nitroxidative damage leading to cellular dysfunction and death. Indeed, depending on the level of cellular exposure to oxidants the mechanism of cellular death may vary. In the case of $ONOO^-$, exposure to high fluxes of this oxidant rapidly induces necrosis of the cell through ATP depletion, membrane disruption, and development of secondary inflammatory response. However, the exposure to low levels of $ONOO^-$ leads to an apoptotic cell death without the occurrence of inflammatory response.

The process that leads to necrotic cell death is a complex mechanism involving DNA damage and activation of the DNA repair enzyme poly(ADP-ribose) polymerase-1 (PARP-1)). PARP-1 is involved in the repair of DNA strand breaks induced by a variety of genotoxic insults including ROS/RNS (*i.e.* $ONOO^-$, H_2O_2, $CO_3^{\bullet -}$, $^{\bullet}OH$) [46]. In conditions associated with a low degree of DNA damage, PARP-1 binds to DNA strand breaks and transfers ADP-ribose units from NAD^+ to nuclear proteins. However, upon severe DNA injury, like in certain forms of ischemia reperfusion, overactivation of PARP-1 leads to a depletion of cellular NAD^+, impairing glycolysis, Krebs cycle and mitochondrial electron transport chain, resulting in a marked ATP depletion, cellular dysfunction and death through necrosis, as well as up-regulation of various pro-inflammatory pathways [47]. PARP-1 activation-mediated cell death has been consistently observed in disease processes associated with nitroxidative stress: unstable angina and myocardial infarction [48], stroke [49] and neurodegenerative disorders [50], among others [47].

The mechanism underlying $ONOO^-$ induced apoptosis vary among different cell types and experimental conditions. However, there are several common characteristics to all of them. First, it depends on caspase 3, 2, 8 and 9 activation. Secondly, another feature of $ONOO^-$ induced apoptosis is the formation of the permeability transition pore (PTP), a multiprotein complex on the inner mitochondrial membrane, composed by adenine nucleotide translocase (ANT), cyclophilin D, and the voltage-dependent anion channel (VDAC) [51]. The permeabilization of the inner mitochondrial membrane results in dissipation of mitochondrial membrane potential, interruption of electron transport, ATP synthesis, and formation of reactive oxygen species. Finally, the release of pro-apoptotic factors to cytosol promote caspase-dependent (cytochrome c, APAF-1, Smac/DIABLO) and caspase-independent (apoptosis inducing factor) cell death [52]. In addition, $ONOO^-$ also activates programmed cell death through mitochondrial independent mechanisms. It has been proposed that $ONOO^-$ activates p38 and JNK/MAPK pathways, revealing that apoptosis is not only a consequence of $ONOO^-$-dependent mitochondrial damage [53].

4.3 Peroxynitrite, Inflammation and Immune Responses

The pathophysiological actions of $^{\bullet}NO$-derived oxidants depends on their ability to switch to pro-inflammatory signaling cascades involving mitogen-activated protein kinases (MAPK), NF-kB and apoptotic pathways [51]. Indeed, $ONOO^-$ affects inflammatory response at multiple levels. Oxidation and nitration of membrane, cytosolic and nuclear receptors, as well as receptor ligands, had been linked to disturbances in downstream signaling [54, 55]. Peroxynitrite is generated by the activated immune system as an effector cytotoxic molecule. For instance, macrophages are able to fight *Trypanosoma cruzi* infection by the intraphagosomal formation of this cytotoxin [56]. Moreover, immunomodulation has recently emerged as a new function for tyrosine nitration, since the presence of a nitro group in a tyrosine residue can also elicit an immune response. Indeed, it was reported the detection of immunoglobulin against 3-nitrotyrosine in patients undergoing post-traumatic acute lung injury, indicating that nitroxidative stress is able to induce a immunological response even in the short-term [57]. These data suggest that autologous 3-nitrotyrosine-containing proteins may not be subjected to the constraints of immunological tolerance, raising the possibility that they may be able to generate a chronic inflammatory condition by acting as auto-antigens. Indeed, recent findings suggest that nitrated proteins may be involved in the development of autoimmune diseases such as systemic lupus erythematosus, arthritis and glomerulonephritis [58]. Many diseases are characterized by the occurrence of localized or systemic inflammation, and as we mentioned, nitroxidative stress plays an important role in the development and maintenance of inflammatory injury.

Rheumatoid arthritis and systemic lupus erythematosus show articular inflammatory damage as one of their characteristic features, and many studies in animal models and in humans support the role of RNS in the pathogenesis of arthritis. For example, increased levels of nitrite/nitrate and iNOS expression have been found in synovial tissue and fluid in animal models and humans with arthritis. Furthermore, increments in the detection of 3-nitrotyrosine has been observed in cartilage and subchondrial bone in an animal model of chronic arthritis that was reduced in iNOS knocked-out mice and by the treatment with agents that catalytically decompose $ONOO^-$ [58, 59]. Inflammatory bowel diseases are chronic disorders that undergo successive exacerbations and remissions episodes of intestinal inflammation. As an increased expression of iNOS and 3-nitrotyrosine levels in animal models and patients with these types of disorders has been observed, it was proposed that RNS play a defining role in the exacerbation process [61, 62].

4.4 Reactive Nitrogen Species in the Development of Cardiovascular Disease

The relationship between nitroxidative stress and the development of cardiovascular diseases have been widely studied, and nowadays there is a great deal of information about the pathogenic role of RNS. Endothelial dysfunction represent a situation in which the endothelium phenotype is characterized by low bioavailability of •NO, and represent an early step in the pathogenesis of most cardiovascular diseases [3]. The reduction in •NO availability, occurs as a result of many functional changes: decrease in •NO synthesis (due to the presence of the eNOS inhibitor asymmetric di-methyl-arginine (ADMA), eNOS uncoupling, or L-arginine depletion), or as a result of •NO scavenging by $O_2^{\bullet-}$. In addition, the latter reaction leads to the formation of $ONOO^-$, which promotes endothelial dysfunction through different mechanisms (up-regulation of adhesion molecules, vascular smooth muscle cell proliferation and migration, inhibition of ion channels, stimulation of platelets aggregation, inhibition of prostacyclin synthase, oxidation of BH_4). The modifications of endothelial cells evidenced in endothelial dysfunction are the starting point for the formation of atherosclerotic lesions in the vascular wall. Basically, atherosclerosis is a multifactorial progressive chronic inflammatory process comprising sub-intimal low density lipoprotein (LDL) accumulation, oxidation and nitration, foam cell formation, releasing of pro-inflammatory cytokines, proliferation and migration of smooth muscle cells, deposition of extracellular matrix, and platelet adhesion and aggregation In addition, RNS participation on atherosclerosis development has been supported by the demonstration of the increased levels of 3-nitrotyrosine, iNOS expression, oxidized and nitrated LDL in human atherosclerotic tissue [63, 64].

Myocardial ischemia represents a critical situation of nutrient and oxygen deprivation, where myocardial muscle is at risk of necrosis because of a reduced blood supply. Despite the fact that restitution of blood flow may prevent muscle cell death, it results in myocardial contractile dysfunction and the occurrence of arrhythmias in a phenomenon named ischemia reperfusion injury, which is thought to be related to ROS/RNS formation upon reperfusion [65]. In effect, *in vivo* studies using animal models of ischemia/reperfusion injury demonstrated that metalloporphyrins, which catalytically decompose $ONOO^-$, reduced myocardial infarction size and protein tyrosine nitration, and improved myocardial function [66]. Moreover, it has been found that patients undergoing cardiac surgery showed increased levels of 3-nitrotyrosine in the coronary sinus effluent after reperfusion, reflecting $ONOO^-$ formation during myocardial ischemia/reperfusion [67].

4.5 Neurodegenerative Diseases and Nitroxidative Damage

Nitroxidative stress has been related to the pathogenesis of different neurodegenerative diseases including Alzheimer's disease, Parkinson's disease, and amyotrophic lateral sclerosis (ALS). Indeed, $ONOO^-$ and other ROS/RNS have been implicated in the establishment of neuronal death and protein tyrosine nitration is a major component of $ONOO^-$ cytotoxicity. Nitration of tyrosine participates in neuronal death through different mechanisms: apoptosis induced by nitration of Hsp-90; Lewy bodies formation by oligomerization of α-synuclein [68–70]. In addition, in familial cases of ALS mutations of Cu,Zn-SOD result in a more efficient catalysis of $ONOO^-$-dependent tyrosine nitration of neuronal proteins inhibiting protein function, protein degradation and phosphorylation cascades [71].

4.6 Sepsis

Sepsis is a systemic condition characterized by systemic immune response and increased oxidative stress caused by microbial infection, which leads to severe multiple organ dysfunction and high risk of death. Electron paramagnetic resonance studies of tissue samples from animal models of sepsis evidenced the production of ROS/RNS during septic shock. In addition, increased 3-nitrotyrosine levels have been detected in blood and myocardial tissue in patients with sepsis [72].

4.7 Cancer

Different reactive nitrogen species may have antagonistic roles in cancer. On one hand, $^{\bullet}NO$ may act as a pro-carcinogenic agent by promoting angiogenesis, invasion of tumor cells, and DNA damage [55, 73]. Moreover, $ONOO^-$ can induce nitroxidative damage of DNA bases generating 8-oxoguanine and 8-nitroguanine which have the potential of triggering miscoding events during DNA replication [74]. Furthermore, $ONOO^-$ may promote damage by single-strand breaks, inactivation of repair enzymes, and p53 inhibition through tyrosine nitration- It has been reported that nitration of chemokine (C-C motif) ligand 2 (CCL2) results in a reduced recruitment of cytotoxic T lymphocytes to the core of tumors, trapping tumor-specific T cells in the stroma that surrounds cancer cells [75].

5 Pharmacology Directed to Prevent Nitroxidative Damage

As a result of the information accumulated over the past decades on the involvement of nitroxidative stress in the development of different pathologies, many pharmacological strategies have been proposed to prevent the damage mediated by

ROS/RNS. A wide variety of compounds, either natural or synthetic, have been utilized to reduce nitroxidative damage interfering at different levels on oxidants metabolism. In the case of $ONOO^-$ one rational possibility is to decrease its formation. Once formed, many agents with different redox activity (direct scavengers, promoters of catalytic decomposition or inducers of antioxidant defenses) have been used to prevent nitroxidative cellular damage.

5.1 *Inhibitors of $ONOO^-$ Formation*

The inhibition of $ONOO^-$ formation can be accomplished either by inhibiting the formation of its precursors ($O_2^{\bullet-}$ and $^{\bullet}NO_2$) or by scavenging them once formed. NADPH oxidases (NOXs) are membrane-associated multi-subunit complexes, which catalyze the univalent reduction of molecular oxygen to $O_2^{\bullet-}$. Several compounds are widely used in pre-clinical research as NOXs inhibitors (*i.e.*diphenyliodonium, DPI, and apocynin); however its pharmacological efficacy is limited due to their lack of specificity. As a result of their general ability to inhibit flavoenzymes they reduce the activity of other enzymes apart from NOXs, as mitochondrial respiratory complex I, xanthine oxidase and cytochrome P450 reductase [93]. Thus, results obtained from their use as NOXs inhibitors must be taken with caution. Instead of inhibiting its formation, the other strategy is elimination after its formation. SOD-mimics are a group of compounds able to catalyze the dismutation of $O_2^{\bullet-}$ to H_2O_2 and O_2, and the most promising of these are metalloporphyrins. Manganese porphyrins have been used in diverse animal models of conditions associated with nitroxidative stress, evidencing a protective cellular effect. Nevertheless, these compounds are not specific scavengers of $O_2^{\bullet-}$, and the protection afforded by SOD mimics seems more related to the detoxification of $ONOO^-$ and $ONOO^-$ derived radicals rather than the elimination of $O_2^{\bullet-}$ [94]. In the case of $^{\bullet}NO$, NOSs are the most relevant source of $^{\bullet}NO$ in cells and tissues. A wide range of inhibitors has been synthesized to reduce the activity of the three isoforms. However, studies using animal models of disease and data from clinical trials have produced confounding results, probably due to the lack of isoform specificity of these compounds. Many of the compounds used, are potent inhibitors of the three isoforms of NOS, including iNOS whose high $^{\bullet}NO$ production is related to nitroxidative damage. Although these may result in a reduction in the formation of RNS, it may also reduce eNOS and nNOS activity, whose low and maintained fluxes of $^{\bullet}NO$ have physiological and protective effects. In consequence, the universal inhibition of $^{\bullet}NO$ production may account for the negative results obtained in clinical trials [95]. Heme-binding to NOS monomers induce its dimerization, which results in the creation of binding sites for L-arginine and BH4, being a critical step for enzyme activity. Since dimerization involves a large interface in the oxygenase domain that contains the highest sequence variability among NOS isoform it is conceivable to develop isoform-selective allosteric NOS inhibitors that bind to the dimerization interface or that interfere with heme-induced dimerization. Indeed, imidazol-based compounds such as BBS-2 and BBS-4 have proven to be selective iNOS inhibitors in animal models [96].

5.2 Direct Antioxidants

Peroxynitrite scavengers include molecules that react with $ONOO^-$ itself, $ONOO^-$-derived radicals, or both. One of them is uric acid, the product of purine metabolism, which detoxifies $ONOO^-$ derived radicals rather than $ONOO^-$. As a result of the beneficial effects evidenced on animal models of disease, clinical trials have been designed to test the efficacy of uric acid in combination with recombinant tissue plasminogen activator in acute ischemic stroke patients [97]. Ascorbate (vitamin C) and tocopherol (vitamin E) are both biologically relevant antioxidants that have been used as pharmacological agents against oxidative damage. Ascorbate is a hydrophilic antioxidant important in cytosol and plasma, conversely tocopherol is a hydrophobic antioxidant of biomembranes and LDL particles. Neither of them reacts fast enough with $ONOO^-$ to outcompete its reaction with biomolecules or CO_2, nevertheless they can scavenge secondary radicals ($^{\bullet}OH$, $^{\bullet}NO_2$, $CO_3^{\bullet-}$, $ROO^{\bullet}$ producing ascorbyl and tocopheryl radicals, less reactive species that can be reduced back at the expense of glutathione or NADPH [98–100]. Spin traps agents are compounds which were classically utilized for the detection of radical species produced during chemical and enzymatic reactions. These agents react with radical species (*i.e.* free radicals, protein radicals, DNA radicals) generating a more stable radical adduct which is detected by different techniques (electron paramagnetic resonance, specific antibodies, mass spectroscopy), facilitating the detection of transient radical intermediates otherwise elusive. The scavenging of radical species by spin trap agents prevents the reaction of these harmful species with biomolecules, giving them the possibility to act as antioxidants. In consequence, the protective properties of these agents have been the focus of intensive research for years. Indeed, different types of spin trap agents have been utilized as pharmacological agents to prevent nitroxidative damage in different animal models of disease [101], and even in clinical trials, as is the case of 2,4-disulfophenyl-N-tert-butylnitrone (NXY-059) whose neuroprotective effect was tested in patients with ischemic stroke [102]. Tempol is a nitroxide radical that has been used to protect, animals and humans, from conditions associated with local or systemic nitroxidative damage. It has been shown that it reduces 3-nitrotyrosine levels, downregulates vascular NOX-2 in animal models of atherosclerosis, and protects from radiation induced damage in humans [103, 104]. Different groups have proposed the use peptides as antioxidant pharmacological agents, and more specifically tyrosine-containing peptides, in which the presence of tyrosine contributes substantially to their antioxidant activity. The phenolic lateral chain of tyrosine act as electron donors that allow the termination of radical chain reactions of lipid peroxidation process, as well as the reduction of one-electron oxidants ($OH^{\bullet}$, $CO_3^{\bullet-}$, $^{\bullet}NO_2$) [105]. Indeed, it was shown that the incubation of neurons with tyrosine-containing peptides protects them from $ONOO^-$ mediated induction of apoptosis [68]. Hydrophobic amino acids including proline, histidine, tryptophan and tyrosine are widely distributed in milk-derived proteins, caseins, soybean and gelatin and have shown antioxidant capacity. For example, tyrosine-based peptides like casein (Tyr-Phe-Tyr-Pro-Glu-Leu) and β-lactoglobulin may act as radical scavengers [106].

5.3 *Catalytic Antioxidants*

Another strategy that has been used to decrease $ONOO^-$-induced damage is based on the ability of certain compounds to accelerate $ONOO^-$ decomposition and decrease its steady-state levels. For example, metalloporphyrins (manganese (MnPs) and iron porphyrins (FePs)) take advantage of the fast reaction of $ONOO^-$ with transition metals to form an agent that has proved its efficacy in preventing nitroxidative damage *in vitro* and *in vivo*. Either of the mentioned porphyrins, MnPs and FePs, form rapid catalytic cycles with cellular reducing agents to outcompete with the reaction of $ONOO^-$ with biomolecules. In the case of 5,10,15,20-tetrakis(2,4',6'-trimethyl-3',5'-disulphonaphtophenyl) porphyrin iron III (7-) (FeTMPS), an iron metalloporphyrin (Fe^{III}Ps), it catalyzes the isomerization of $ONOO^-$ to NO_3^- with a reaction rate constant of 10^5–10^6 $M^{-1}s^{-1}$, with the formation of a oxo-Fe^{IV} complex that can be reduced by ascorbate [107]. Moreover, manganese porphyrins, Mn^{III}Ps, such as meso-tetrakis(4-carboxylatophenyl) porphyrins (3-) (MnTCPP), does not isomerize $ONOO^-$, but *in vivo*, it can be reduced to Mn^{II}Ps, by glutathione, ascorbate, and flavoenzymes (including the mitochondrial electron transport chain). As a result, this Mn^{II}Ps is able to promote the two-electron reduction of $ONOO^-$ to nitrite with a reaction rate constant of ~10^7 $M^{-1}s^{-1}$. Indeed, Mn^{II}Ps and Mn^{III}Ps also react fast with $CO_3^{\bullet-}$ (10^8–10^9 $M^{-1}s^{-1}$) enabling them to effectively scavenge $ONOO^-$ and $CO_3^{\bullet-}$ *in vivo*. Nevertheless, the reaction of $ONOO^-$ with Mn^{III}Ps or Fe^{III}Ps, yields the one-electron reduction product $^{\bullet}NO_2$. However, these may be neutralized by antioxidant defenses and $ONOO^-$ reactions with biomolecules are redirected.

As a result of the efficacy as "antioxidant agents" evidence in studies using different models of disease (I/R injury, radiation injury, neurodegenerative diseases, diabetes), phase I clinical trials have been designed to test a Mn^{II} Ps in ALS [108]. Other group of molecules that catalytically decompose $ONOO^-$ are organo-selenium compounds which act as glutathione peroxidase mimics. Ebselen (2-phenyl-1,2-benzoisoselenazol-3(2H)-one), is a prototypical lipid-soluble organic selenocompound which directly react with $ONOO^-$ to yield nitrite and a selenoxide, the latter being able to be reduced back to selenol either by GSH or NADPH-dependent thioredoxin reductase [109]. In accordance, many studies *in vitro* and *in vivo* of disease models treated with ebselen, showed that this compound prevented nitroxidative damage induced by $ONOO^-$ [110]. Nevertheless, ebselen double-blind clinical trials in patients with stroke and subarachnoid hemorrhage, gave non-conclusive results [111].

5.4 Antioxidant Inducers

Besides the use of synthetic scavengers of $ONOO^-$, an alternative strategy to combat the formation of RNS is the use of agents that enhance endogenous antioxidants defenses, and an example of this is another seleno-compound called diphenyldiselenide. This agent has proved to be effective in preventing mitochondrial dysfunction in endothelial cells treated with the exogenous fluxes of $ONOO^-$. Moreover, it was demonstrated that its protective role derived from the activation of Nrf2 pathway up-regulating the expression of enzymes related to the protection against oxidative stress: heme oxigenase-1, peroxiredoxin-1 and -3, and gamma-glutamyl cysteine synthetase [112]. Other types of compounds that have been employed as antioxidants are polyphenols. For example, resveratrol, in the past was used as a free radical scavenging agent, however after considering the low concentration reached in plasma and its important hepatic metabolism, other mechanisms of action have been proposed. Nowadays, flavonoids are considered to act indirectly through the inhibition of pro-oxidant enzymes (NADPH oxidases, MPO, lypoxigenases) and the activation of protective signaling cascades [113]. Thiol-containing molecules, such as N-acetylcysteine (NAC), together with others such as α-lipoic acid, have been used as "antioxidant agents" but their effects must be carefully interpreted. These agents do not react fast with $ONOO^-$, thus the do not have scavenging activity, rather they are able to enhance cellular antioxidant defenses (i.e. increasing glutathione and ascorbate concentrations, reduce activity/expression of iNOS, down-regulate pro-inflammatory genes such TNF α, and up-regulate phase II detoxification genes by Nrf2) [113].

6 Conclusions

In this chapter we reviewed different aspects of the nature of RNS, reflecting their biomedical importance. We have indicated biologically-relevant reactions and sites of formation of $^{\bullet}NO$, $ONOO^-$, and their corresponding derived radicals, together with cellular targets of this oxidizing and nitrating agents. In addition we listed different human pathologies in which RNS play an etiopathogenic role, and mentioned existing evidence on the efficacy of a variety of redox-active drugs used to treat these conditions. The better understanding of the biological chemistry of $^{\bullet}NO$ and other $^{\bullet}NO$-derived species such as $ONOO^-$, has helped to understand various aspects of redox processes in human physiology and pathology. Moreover, these developments are assisted for the identification of biomarkers and predictors of disease progression (*e.g.* 3-nitrotyrosine). Finally, the evolution of our understanding on the biological chemistry of $^{\bullet}NO$ and $ONOO^-$ has favoured the development of novel therapeutic strategies (some of them targeted to specific organelles such as mitochondria, see Fig. 5.5) based of redox-active drugs or the enhanced activity of endogenous and cytoprotective redox systems.

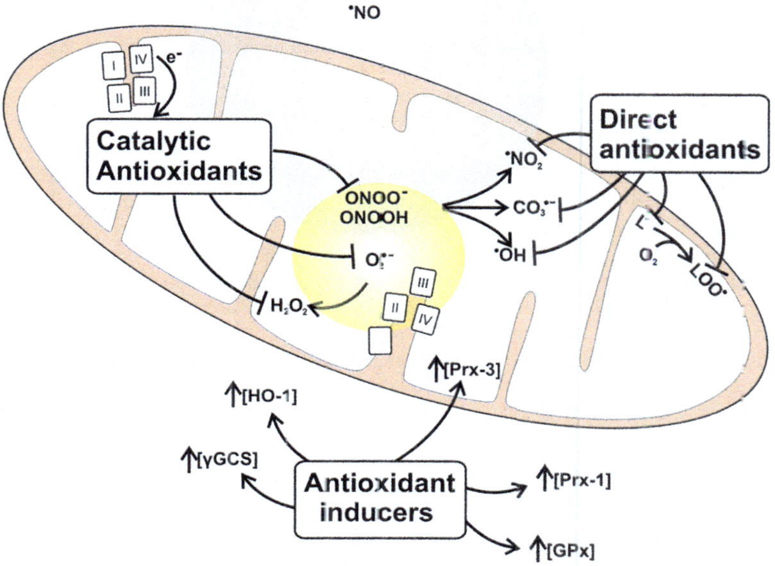

Fig. 5.5 Mechanisms of action of mitochondria-targeted redox active compounds that neutralize $ONOO^-$-dependent toxicity. See text for description of the different group of compounds (HO-1, hemoxygenase 1; Prx-1 and -3, peroxiredoxin 1 and 3; GPx, glutathione peroxidase; γGCS, gamma-glutamylcysteine synthetase)

Acknowledgements This work was supported by grants from Universidad de la República (CSIC and Espacio Interdisciplinario). Additional funding was obtained from PEDECIBA. A. Aicardo, D. M. Martinez and N. Campolo have been supported by fellowships of Agencia Nacional de Investigación e Innovación (ANII, Uruguay), Conselho Nacional de Desenvolvimento Científico e Tecnológico (CNPq), and Universidad de la República (CAP), respectively.

References

1. Ignarro LJ, Buga GM, Wood KS et al (1987) Endothelium-derived relaxing factor produced and released from artery and vein is nitric oxide. Proc Natl Acad Sci U S A 84:9265–9269
2. Palmer RM, Ferrige AG, Moncada S (1987) Nitric oxide release accounts for the biological activity of endothelium-derived relaxing factor. Nature 327:524–526
3. Feletou M, Vanhoutte PM (2006) Endothelial dysfunction: a multifaceted disorder. Am J Physiol Heart Circ Physiol 291:H985–H1002
4. Martasek P, Miller RT, Liu Q et al (1998) The C331A mutant of neuronal nitric-oxide synthase is defective in arginine binding. J Biol Chem 273:34799–34805
5. Lacza Z, Pankotai E, Busija DW (2009) Mitochondrial nitric oxide synthase: current concepts and controversies. Front Biosci 14:4436–4443

6. Boveris A, Valdez LB, Zaobornyj T, Bustamante J (2006) Mitochondrial metabolic states and membrane potential regulate nitric oxide and hydrogen peroxide diffusion to the cytosol. Biochim Biophys Acta 1757:535–542
7. Cho HJ, Xie QW, Calaycay J et al (1992) Calmodulin is a subunit of nitric oxide synthase from macrophages. J Exp Med 176:599–604
8. Qian J, Fulton D (2013) Post-translational regulation of endothelial nitric oxide synthase in vascular endothelium. Front Physiol 4:347–360
9. Rocha BS, Gago B, Pereira C et al (2011) Dietary nitrite in nitric oxide biology: a redox interplay with implications for pathophysiology and therapeutics. Curr Drug Targets 12:1351–1364
10. Maia LB, Pereira V, Mira L, Moura JJ (2015) Nitrite reductase activity of rat and human xanthine oxidase, xanthine dehydrogenase, and aldehyde oxidase: evaluation of their contribution to NO formation in vivo. Biochemistry 54:685–710
11. Nottingham WC, Sutter JR (1986) Kinetics of the oxidation of nitric oxide by chlorine and oxygen in nonaqueous media. Int J Chem Kinetics 18:1289–1302
12. Eiserich JP, Butler J, van der Vliet A et al (1995) Nitric oxide rapidly scavenges tyrosine and tryptophan radicals. Biochem J 310:745–749
13. Padmaja S, Huie RE (1993) The reaction of nitric oxide with organic peroxyl radicals. Biochem Biophys Res Commun 195:539–544
14. Madej E, Folkes LK, Wardman P et al (2008) Thiyl radicals react with nitric oxide to form S-nitrosothiols with rate constants near the diffusion-controlled limit. Free Radic Biol Med 44:2013–2018
15. Cassina A, Radi R (1996) Differential inhibitory action of nitric oxide and peroxynitrite on mitochondrial electron transport. Arch Biochem Biophys 328:309–316
16. Bartesaghi S, Romero N, Radi R (2012) Nitric oxide and derived oxidants. In: Pantopoulos K, Schipper HM (eds) Principles of free radical biomedicine. Nova, New York, pp 43–74
17. Beckman JS, Beckman TW, Chen J et al (1990) Apparent hydroxyl radical production by peroxynitrite: implications for endothelial injury from nitric oxide and superoxide. Proc Natl Acad Sci U S A 87:1620–1624
18. Romero N, Radi R, Linares E et al (2003) Reaction of human hemoglobin with peroxynitrite. Isomerization to nitrate and secondary formation of protein radicals. J Biol Chem 278:44049–44057
19. Denicola A, Freeman BA, Trujillo M, Radi R (1996) Peroxynitrite reaction with carbon dioxide/bicarbonate: kinetics and influence on peroxynitrite-mediated oxidations. Arch Biochem Biophys 333:49–58
20. Trujillo M, Alvarez B, Radi R (2016) One- and two-electron oxidation of thiols: mechanisms, kinetics and biological fates. Free Radic Res 50:150–171
21. Campolo N, Bartesaghi S, Radi R (2014) Metal-catalyzed protein tyrosine nitration in biological systems. Redox Rep 19:221–231
22. Bartesaghi S, Wenzel J, Trujillo M et al (2010) Lipid peroxyl radicals mediate tyrosine dimerization and nitration in membranes. Chem Res Toxicol 23:821–835
23. van der Vliet A, Eiserich JP, Halliwell B, Cross CE (1997) Formation of reactive nitrogen species during peroxidase-catalyzed oxidation of nitrite. A potential additional mechanism of nitric oxide-dependent toxicity. J Biol Chem 272:7617–7625
24. Thomas DD, Espey MG, Vitek MP et al (2002) Protein nitration is mediated by heme and free metals through Fenton-type chemistry: an alternative to the NO/O2- reaction. Proc Natl Acad Sci U S A 99:12691–12696
25. Gunther MR, Hsi LC, Curtis JF et al (1997) Nitric oxide trapping of the tyrosyl radical of prostaglandin H synthase-2 leads to tyrosine iminoxyl radical and nitrotyrosine formation. J Biol Chem 272:17086–17090
26. Radi R (2013) Protein tyrosine nitration: biochemical mechanisms and structural basis of functional effects. Acc Chem Res 46:550–559

27. MacMillan-Crow LA, Crow JP, Kerby JD et al (1996) Nitration and inactivation of manganese superoxide dismutase in chronic rejection of human renal allografts. Proc Natl Acad Sci U S A 93:11853–11858
28. Pehar M, Vargas MR, Robinson KM et al (2006) Peroxynitrite transforms nerve growth factor into an apoptotic factor for motor neurons. Free Radic Biol Med 41:1632–1644
29. Rocha BS, Gago B, Barbosa RM et al (2013) Pepsin is nitrated in the rat stomach, acquiring antiulcerogenic activity: a novel interaction between dietary nitrate and gut proteins. Free Radic Biol Med 58:26–34
30. O'Donnell VB, Chumley PH, Hogg N et al (1997) Nitric oxide inhibition of lipid peroxidation: kinetics of reaction with lipid peroxyl radicals and comparison with alpha-tocopherol. Biochemistry 36:15216–15223
31. Baker PR, Schopfer FJ, Sweeney S, Freeman BA (2004) Red cell membrane and plasma linoleic acid nitration products: synthesis, clinical identification, and quantitation. Proc Natl Acad Sci U S A 101:11577–11582
32. Baker PR, Schopfer FJ, O'Donnell VB, Freeman BA (2009) Convergence of nitric oxide and lipid signaling: anti-inflammatory nitro-fatty acids. Free Radic Biol Med 46:989–1003
33. Trostchansky A, Souza JM, Ferreira A et al (2007) Synthesis, isomer characterization, and anti-inflammatory properties of nitroarachidonate. Biochemistry 46:4645–4653
34. Trostchansky A, Bonilla L, Thomas CP et al (2011) Nitroarachidonic acid, a novel peroxidase inhibitor of prostaglandin endoperoxide H synthases 1 and 2. J Biol Chem 286:12891–12900
35. Zhang J, Villacorta L, Chang L et al (2010) Nitro-oleic acid inhibits angiotensin II-induced hypertension. Circ Res 107:540–548
36. Villacorta L, Chang L, Salvatore SR et al (2013) Electrophilic nitro-fatty acids inhibit vascular inflammation by disrupting LPS-dependent TLR4 signalling in lipid rafts. Cardiovasc Res 98:116–124
37. Koenitzer JR, Bonacci G, Woodcock SR et al (2016) Fatty acid nitroalkenes induce resistance to ischemic cardiac injury by modulating mitochondrial respiration at complex II. Redox Biol 8:1–10
38. Rudolph TK, Rudolph V, Edreira MM et al (2010) Nitro-fatty acids reduce atherosclerosis in apolipoprotein E-deficient mice. Arterioscler Thromb Vasc Biol 30:938–945
39. Peluffo G, Radi R (2007) Biochemistry of protein tyrosine nitration in cardiovascular pathology. Cardiovasc Res 75:291–302
40. Tahara EB, Navarete FD, Kowaltowski AJ (2009) Tissue-, substrate-, and site-specific characteristics of mitochondrial reactive oxygen species generation. Free Radic Biol Med 46:1283–1297
41. Radi R, Rodriguez M, Castro L, Telleri R (1994) Inhibition of mitochondrial electron transport by peroxynitrite. Arch Biochem Biophys 308:89–95
42. Abriata LA, Cassina A, Tortora V et al (2009) Nitration of solvent-exposed tyrosine 74 on cytochrome c triggers heme iron-methionine 80 bond disruption. Nuclear magnetic resonance and optical spectroscopy studies. J Biol Chem 284:17–26
43. Tortora V, Quijano C, Freeman B et al (2007) Mitochondrial aconitase reaction with nitric oxide, S-nitrosoglutathione, and peroxynitrite: mechanisms and relative contributions to aconitase inactivation. Free Radic Biol Med 42:1075–1088
44. Yamakura F, Taka H, Fujimura T, Murayama K (1998) Inactivation of human manganese-superoxide dismutase by peroxynitrite is caused by exclusive nitration of tyrosine 34 to 3-nitrotyrosine. J Biol Chem 273:14085–14089
45. Trujillo M, Clippe A, Manta B et al (2007) Pre-steady state kinetic characterization of human peroxiredoxin 5: taking advantage of Trp84 fluorescence increase upon oxidation. Arch Biochem Biophys 467:95–106
46. Koh DW, Dawson TM, Dawson VL (2005) Mediation of cell death by poly(ADP-ribose) polymerase-1. Pharmacol Res 52:5–14

47. Pacher P, Szabo C (2008) Role of the peroxynitrite-poly(ADP-ribose) polymerase pathway in human disease. Am J Pathol 173:2–13
48. Huang D, Yang CZ, Yao L et al (2008) Activation and overexpression of PARP-1 in circulating mononuclear cells promote TNF-alpha and IL-6 expression in patients with unstable angina. Arch Med Res 39:775–784
49. Love S, Barber R, Wilcock GK (1999) Neuronal accumulation of poly(ADP-ribose) after brain ischaemia. Neuropathol Appl Neurobiol 25:98–103
50. Love S, Barber R, Wilcock GK (1999) Increased poly(ADP-ribosyl)ation of nuclear proteins in Alzheimer's disease. Brain 122:247–253
51. Szabo C, Ischiropoulos H, Radi R (2007) Peroxynitrite: biochemistry, pathophysiology and development of therapeutics. Nat Rev Drug Discov 6:662–680
52. Tsujimoto Y, Shimizu S (2000) Bcl-2 family: life-or-death switch. FEBS Lett 466:6–10
53. Shacka JJ, Sahawneh MA, Gonzalez JD et al (2006) Two distinct signaling pathways regulate peroxynitrite-induced apoptosis in PC12 cells. Cell Death Differ 13:1506–1514.
54. Pacher P, Beckman JS, Liaudet L (2007) Nitric oxide and peroxynitrite in health and disease. Physiol Rev 87:315–424
55. Alvarez MN, Peluffo G, Piacenza L, Radi R (2011) Intraphagosomal peroxynitrite as a macrophage-derived cytotoxin against internalized Trypanosoma cruzi: consequences for oxidative killing and role of microbial peroxiredoxins in infectivity. J Biol Chem 286:6627–6640
56. Thomson L, Christie J, Vadseth C et al (2007) Identification of immunoglobulins that recognize 3-nitrotyrosine in patients with acute lung injury after major trauma. Am J Respir Cell Mol Biol 36:152–157
57. Hsu HC, Zhou T, Kim H et al (2006) Production of a novel class of polyreactive pathogenic autoantibodies in BXD2 mice causes glomerulonephritis and arthritis. Arthritis Rheum 54:343–355
58. Mapp PI, Klocke R, Walsh DA et al (2001) Localization of 3-nitrotyrosine to rheumatoid and normal synovium. Arthritis Rheum 44:1534–1539
59. Sandhu JK, Robertson S, Birnboim HC, Goldstein R (2003) Distribution of protein nitrotyrosine in synovial tissues of patients with rheumatoid arthritis and osteoarthritis. J Rheumatol 30:1173–1181
60. Lundberg JO, Hellstrom PM, Lundberg JM, Alving K (1994) Greatly increased luminal nitric oxide in ulcerative colitis. Lancet 344:1673–1674
61. Amrouche-Mekkioui I, Djerdjouri B (2012) N-acetylcysteine improves redox status, mitochondrial dysfunction, mucin-depleted crypts and epithelial hyperplasia in dextran sulfate sodium-induced oxidative colitis in mice. Eur J Pharmacol 691:209–217
62. Beckmann JS, Ye YZ, Anderson PG et al (1994) Extensive nitration of protein tyrosines in human atherosclerosis detected by immunohistochemistry. Biol Chem Hoppe Seyler 375:81–88
63. Depre C, Havaux X, Renkin J et al (1999) Expression of inducible nitric oxide synthase in human coronary atherosclerotic plaque. Cardiovasc Res 41:465–472
64. Ferdinandy P, Schulz R (2003) Nitric oxide, superoxide, and peroxynitrite in myocardial ischaemia-reperfusion injury and preconditioning. Br J Pharmacol 138:532–543
65. Bianchi C, Wakiyama H, Faro R et al (2002) A novel peroxynitrite decomposer catalyst (FP-15) reduces myocardial infarct size in an in vivo peroxynitrite decomposer and acute ischemia-reperfusion in pigs. Ann Thorac Surg 74:1201–1207
66. Hayashi Y, Sawa Y, Ohtake S et al (2001) Peroxynitrite formation from human myocardium after ischemia-reperfusion during open heart operation. Ann Thorac Surg 72:571–576
67. Ye Y, Quijano C, Robinson KM et al (2007) Prevention of peroxynitrite-induced apoptosis of motor neurons and PC12 cells by tyrosine-containing peptides. J Biol Chem 282:6324–6337
68. Giasson BI, Duda JE, Murray IV et al (2000) Oxidative damage linked to neurodegeneration by selective alpha-synuclein nitration in synucleinopathy lesions. Science 290:985–989

69. Hodara R, Norris EH, Giasson BI et al (2004) Functional consequences of alpha-synuclein tyrosine nitration: diminished binding to lipid vesicles and increased fibril formation. J Biol Chem 279:47746–47753
70. Casoni F, Basso M, Massignan T et al (2005) Protein nitration in a mouse model of familial amyotrophic lateral sclerosis: possible multifunctional role in the pathogenesis. J Biol Chem 280:16295–16304
71. Fukuyama N, Takebayashi Y, Hida M et al (1997) Clinical evidence of peroxynitrite formation in chronic renal failure patients with septic shock. Free Radic Biol Med 22:771–774
72. Fukumura D, Kashiwagi S, Jain RK (2006) The role of nitric oxide in tumour progression. Nat Rev Cancer 6:521–534
73. Suzuki N, Yasui M, Geacintov NE et al (2005) Miscoding events during DNA synthesis past the nitration-damaged base 8-nitroguanine. Biochemistry 44:9238–9245
74. Molon B, Ugel S, Del Pozzo F et al (2011) Chemokine nitration prevents intratumoral infiltration of antigen-specific T cells. J Exp Med 208:1949–1962
75. Shishehbor MH, Brennan ML, Aviles RJ et al (2003) Statins promote potent systemic antioxidant effects through specific inflammatory pathways. Circulation 108:426–431
76. Schewe T, Sies H (2005) Myeloperoxidase-induced lipid peroxidation of LDL in the presence of nitrite. Protection by cocoa flavanols. Biofactors 24:49–58
77. Stephens NG, Parsons A, Schofield PM et al (1996) Randomised controlled trial of vitamin E in patients with coronary disease. Lancet 347:781–786
78. Kleinschnitz C, Grund H, Wingler K et al (2010) Post-stroke inhibition of induced NADPH oxidase type 4 prevents oxidative stress and neurodegeneration. PLoS Biol 8:9
79. Mackensen GB, Patel M, Sheng H et al (2001) Neuroprotection from delayed postischemic administration of a metalloporphyrin catalytic antioxidant. J Neurosci 21:4582–4592
80. Eid HM, Lyberg T, Larsen J et al (2002) Reactive oxygen species generation by leukocytes in populations at risk for atherosclerotic disease. Scand J Clin Lab Invest 62:431–439
81. Martin-Gallan P, Carrascosa A, Gussinye M, Dominguez C (2003) Biomarkers of diabetes-associated oxidative stress and antioxidant status in young diabetic patients with or without subclinical complications. Free Radic Biol Med 34:1563–1574
82. Mabley JG, Southan GJ, Salzman AL, Szabo C (2004) The combined inducible nitric oxide synthase inhibitor and free radical scavenger guanidinoethyldisulfide prevents multiple low-dose streptozotocin-induced diabetes in vivo and interleukin-1beta-induced suppression of islet insulin secretion in vitro. Pancreas 28:E39–E44
83. Szabo C, Mabley JG, Moeller SM et al (2002) Pathogenetic role of peroxynitrite in the development of diabetes and diabetic vascular complications: studies with FP15, a novel potent peroxynitrite decomposition catalyst. Mol Med 8:571–580
84. Lai ZW, Hanczko R, Bonilla E et al (2012) N-acetylcysteine reduces disease activity by blocking mammalian target of rapamycin in T cells from systemic lupus erythematosus patients: a randomized, double-blind, placebo-controlled trial. Arthritis Rheum 64:2937–2946
85. Wang G, Pierangeli SS, Papalardo E et al (2010) Markers of oxidative and nitrosative stress in systemic lupus erythematosus: correlation with disease activity. Arthritis Rheum 62:2064–2072
86. Frostegard J, Svenungsson E, Wu R et al (2005) Lipid peroxidation is enhanced in patients with systemic lupus erythematosus and is associated with arterial and renal disease manifestations. Arthritis Rheum 52:192–200
87. Nin N, Cassina A, Boggia J et al (2004) Septic diaphragmatic dysfunction is prevented by Mn(III)porphyrin therapy and inducible nitric oxide synthase inhibition. Intensive Care Med 30:2271–2278
88. Paterson RL, Galley HF, Webster NR (2003) The effect of N-acetylcysteine on nuclear factor-kappa B activation, interleukin-6, interleukin-8, and intercellular adhesion molecule-1 expression in patients with sepsis. Crit Care Med 31:2574–2578

89. Good PF, Hsu A, Werner P et al (1998) Protein nitration in Parkinson's disease. J Neuropathol Exp Neurol 57:338–342
90. Baillet A, Chanteperdrix V, Trocme C et al (2010) The role of oxidative stress in amyotrophic lateral sclerosis and Parkinson's disease. Neurochem Res 35:1530–1537
91. Crow JP, Calingasan NY, Chen J et al (2005) Manganese porphyrin given at symptom onset markedly extends survival of ALS mice. Ann Neuro 58:258–265
92. Aldieri E, Riganti C, Polimeni M et al (2008) Classical inhibitors of NOX NAD(P)H oxidases are not specific. Curr Drug Metab 9:686–696
93. Ferrer-Sueta G, Radi R (2009) Chemical biology of peroxynitrite: kinetics, diffusion, and radicals. ACS Chem Biol 4:161–177
94. Paige JS, Jaffrey SR (2007) Pharmacologic manipulation of nitric oxide signaling: targeting NOS dimerization and protein-protein interactions. Curr Top Med Chem 7:97–114
95. Blasko E, Glaser CB, Devlin JJ et al (2002) Mechanistic studies with potent and selective inducible nitric-oxide synthase dimerization inhibitors. J Biol Chem 277:295–302
96. Amaro S, Canovas D, Castellanos M et al (2010) The URICO-ICTUS study, a phase 3 study of combined treatment with uric acid and rtPA administered intravenously in acute ischaemic stroke patients within the first 4.5 h of onset of symptoms. Int J Stroke 5:325–328
97. Rubbo H, Radi R, Anselmi D et al (2000) Nitric oxide reaction with lipid peroxyl radicals spares alpha-tocopherol during lipid peroxidation. Greater oxidant protection from the pair nitric oxide/alpha-tocopherol than alpha-tocopherol/ascorbate. J Biol Chem 275:10812–10818
98. Peluffo G, Calcerrada P, Piacenza L et al (2009) Superoxide-mediated inactivation of nitric oxide and peroxynitrite formation by tobacco smoke in vascular endothelium: studies in cultured cells and smokers. Am J Physiol Heart Circ Physiol 296:H1781–H1792
99. Bjelakovic G, Nikolova D, Gluud LL et al (2008) Antioxidant supplements for prevention of mortality in healthy participants and patients with various diseases. Cochrane Database Syst Rev 2008(2):CD007176
100. Villamena FA, Das A, Nash KM (2012) Potential implication of the chemical properties and bioactivity of nitrone spin traps for therapeutics. Future Med Chem 4:1171–1207
101. Shuaib A, Lees KR, Lyden P et al (2007) NXY-059 for the treatment of acute ischemic stroke. N Engl J Med 357:562–571
102. Cannizzo B, Quesada I, Militello R et al (2014) Tempol attenuates atherosclerosis associated with metabolic syndrome via decreased vascular inflammation and NADPH-2 oxidase expression. Free Radic Res 48:526–533
103. Metz JM, Smith D, Mick R et al (2004) A phase I study of topical Tempol for the prevention of alopecia induced by whole brain radiotherapy. Clin Cancer Res 10:6411–6417
104. Picot L, Ravallec R, Fouchereau-Peron M et al (2010) Impact of ultrafiltration and nanofiltration of an industrial fish protein hydrolysate on its bioactive properties. J Sci Food Agric 90:1819–1826
105. Rival SG, Boeriu CG, Wichers HJ (2001) Caseins and casein hydrolysates. Antioxidative properties and relevance to lipoxygenase inhibition. J Agric Food Chem 49:295–302
106. Jensen MP, Riley DP (2002) Peroxynitrite decomposition activity of iron porphyrin complexes. Inorg Chem 41:4788–4797
107. Orrell RW (1996) AEOL-10150 (Aeolus). Curr Opin Investig Drugs 7:70–80
108. Masumoto H, Sies H (1996) The reaction of ebselen with peroxynitrite. Chem Res Toxicol 9:262–267
109. Briviba K, Roussyn I, Sharov VS, Sies H (1996) Attenuation of oxidation and nitration reactions of peroxynitrite by selenomethionine, selenocystine and ebselen. Biochem J 319:13–15
110. Parnham M, Sies H (2000) Ebselen: prospective therapy for cerebral ischaemia. Expert Opin Investig Drugs 9:607–619

111. de Bem AF, Fiuza B, Calcerrada P et al (2013) Protective effect of diphenyl diselenide against peroxynitrite-mediated endothelial cell death: a comparison with ebselen. Nitric Oxide 31:20–30
112. Sies H, Schewe T, Heiss C, Kelm M (2005) Cocoa polyphenols and inflammatory mediators. Am J Clin Nutr 81S:304S–312S
113. Moldeus P, Cotgreave IA (1994) N-acetylcysteine. Meth Enzymol 234:482–492

Chapter 6
Nitro-arachidonic Acid: Downstream Signaling and Therapeutics

Homero Rubbo, Lucía González-Perilli, Mauricio Mastrogiovanni, Beatriz Sánchez-Calvo, and Andrés Trostchansky

Abstract Arachidonic acid (AA) represents a precursor of potent signaling molecules, i.e., prostaglandins and thromboxanes through enzymatic and non-enzymatic oxidative pathways. Arachidonic acid can be nitrated by reactive nitrogen species leading to the formation of nitro-arachidonic acid (NO_2-AA). In fact, nitration of AA by inflammatory stimulus could divert the "normal metabolic pathway" of AA to gain novel responses. In this prospective article, we describe the mechanisms of NO_2-AA formation in human cells as well as its key chemical and biological properties. This includes the ability of NO_2-AA to mediate unique redox signaling anti-inflammatory actions along with its therapeutic potential.

Keywords Nitro-arachidonic acid • Lipid nitration • Nitro-fatty acids • Inflammation

1 Introduction

Free as well as esterified fatty acids are important components of biological membranes that can be modified by reactive oxygen (ROS) and nitrogen (RNS) species. In particular, fatty acid nitration is expected to occur in hydrophobic compartments such as the lipid bilayer of cellular membranes or the lipophilic core of lipoproteins. Reactive nitrogen species both oxidize and nitrate unsaturated fatty acids yielding an array of hydroxyl, hydroperoxy, nitro and nitrohydroxy derivatives [1]. Although the detailed mechanisms for nitro-fatty acid (NO_2-FA)) formation in vivo still remain unknown, available knowledge suggests that nitrogen dioxide ($^\bullet NO_2$) is the rate limiting step. In hydrophobic compartments the most probable source for $^\bullet NO_2$ is the autooxidation of nitric oxide ($^\bullet NO$), which concentrates up to 20-fold in the membrane bilayer [2]. Also, $^\bullet NO_2$ can be generated from the decomposition of

H. Rubbo (✉) • L. González-Perilli • M. Mastrogiovanni • B. Sánchez-Calvo
A. Trostchansky
Departamento de Bioquímica and Center for Free Radical and Biomedical Research, Facultad de Medicina, Universidad de la República, General Flores 2125, Montevideo 11800, Uruguay
e-mail: hrubbo@fmed.edu.uy

© Springer International Publishing Switzerland 2016
R.J. Gelpi et al. (eds.), *Biochemistry of Oxidative Stress*, Advances in Biochemistry in Health and Disease 16, DOI 10.1007/978-3-319-45865-6_6

peroxynitrite, peroxidase-catalyzed oxidation of nitrite (NO_2^-) to $^{\bullet}NO_2$ or reduction of NO_2^- in acidic tissue environments (e.g., gastric compartment) [3]. In fact, $^{\bullet}NO$ and superoxide radical ($O_2^{\bullet -}$) are the precursors of peroxynitrite which under inflammatory conditions lead to an increase of oxidized as well as nitrated proteins and lipids [4]. Two different pathways have been suggested for peroxynitrite reactions with lipophilic targets: a) peroxynitrite-derived radicals are formed in aqueous media and then react with a substrate bound to the membrane surface, or b) peroxynitrous acid can undergo proton-catalyzed hemolysis to $^{\bullet}NO_2$ and $^{\bullet}OH$ with a ~30 % yield [5]. Peroxynitrous acid derived $^{\bullet}OH$ and $^{\bullet}NO_2$ initiate lipid oxidation processes, which are propagated by the action of peroxyl radicals (ROO•). In this regard, the relative lack of direct peroxynitrite targets in hydrophobic compartments [6] makes homolysis a more likely decay mechanism. Since both peroxynitrite and $^{\bullet}NO_2$ readily diffuse through membrane bilayers, reactions leading to $^{\bullet}NO_2$ generation may take place in the aqueous environment in proximity to the membrane or inside the lipid bilayer [7].

2 Mechanisms of Arachidonic Acid Nitration

Hydrophobic compartments represent a reservoir for NO_2-FA. Nitroarachidonic acid (NO_2-AA) can be formed in biological membranes from arachidonic acid (AA), the most abundant fatty acid present in the 2-carbon position of phospholipids to exert biological effects upon phospholipase A2 cleavage. A complex mixture of products has been identified after reaction of AA with peroxynitrite/$^{\bullet}NO_2$ including cis-trans isomerization and formation of nitro-hydroxyarachidonate (NO_2(OH)AA) [8]. The isomerization is likely to involve the reversible binding of $^{\bullet}NO_2$ and formation of a nitroarachidonyl radical followed by elimination of $^{\bullet}NO_2$ and generation of a trans bond. Hydrophobic membranes may facilitate reactions where $^{\bullet}NO_2$, arachidonyl (AA•) and arachidonylperoxyl (AAOO•) radicals are likely to be simultaneously present. The half-life of AAOO• is sufficiently long for the reaction with $^{\bullet}NO_2$ to occur, leading to arachidonylperoxynitrites (AAOONO)/nitrates that could rearrange into nitroarachidonic acid (NO_2-AA), NO_2 (OH)AA and nitroepoxyarachidonate [9]. The currently accepted mechanisms for $^{\bullet}NO_2$-mediated oxidation and nitration of polyunsaturated fatty acids (PUFAs) such as AA involve different routes including hydrogen atom abstraction and addition reactions as well as electrophilic substitution by nitronium ion (NO_2^+, [10]). Reaction of $^{\bullet}NO_2$ with PUFAs leads to the generation of isomerized, oxidized and/or nitro-allylic, nitroalkene, dinitro, or nitro-hydroxy lipid derivatives. The allylic hydrogen abstraction in PUFAs yields a carbon-centered lipid radical and nitrous acid (HONO), which rapidly decomposes [10]. In anaerobic conditions or when O_2 tension is low, a second molecule of $^{\bullet}NO_2$ reacts with the carbon-centered radical generating a nitro-alkane, where the $-NO_2$ moiety is bound to a saturated carbon center (Fig. 6.1). However, when the O_2 tension is higher, the carbon-centered radical reacts with O_2 to generate a nitro-peroxyl radical, an unstable and very reactive species that can react with a neighboring

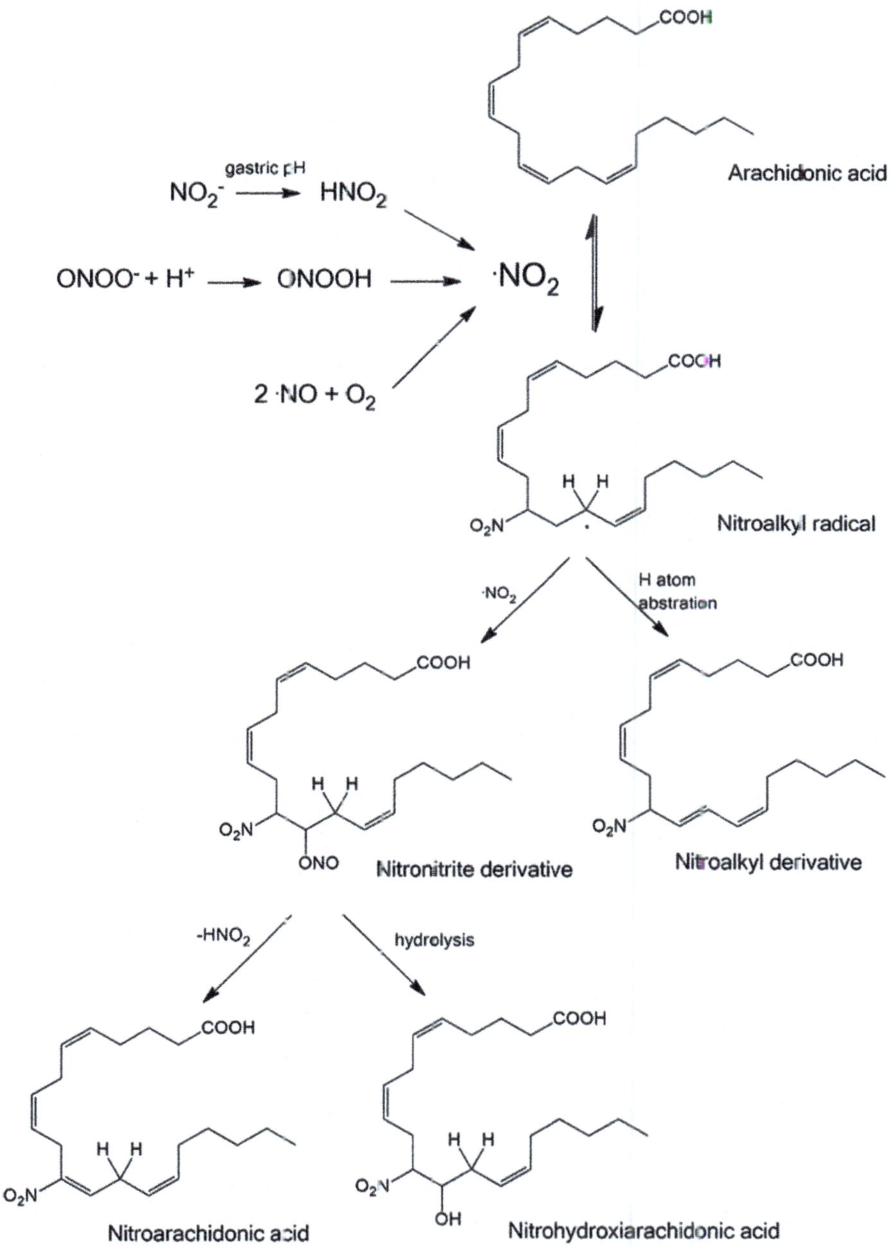

Fig. 6.1 Mechanisms of nitro-arachidonic acid (NO_2-AA) formation.

PUFA, thus initiating the propagation stage of lipoperoxidation, or with a second $^{\bullet}NO_2$ radical generating a nitro-nitrate alkane [10, 11]. A homolytic attack of $^{\bullet}NO_2$ on the double bond in AA yields a β-nitroalkyl radical (Fig. 6.1). In an aqueous environment, where the solubility of $^{\bullet}NO$ and its derived reactive species is rather low, the reaction will most probably revert, causing a cis-trans isomerization of AA. The generation of trans-AA along with NO_2-AA is unique to $^{\bullet}NO_2$-dependent mechanisms. Otherwise, when the O_2 levels are low, the β-nitroalkyl radical reacts with a second $^{\bullet}NO_2$ radical generating a dinitro-alkane acid and/or a nitro-nitrite alkane. Both these species are unstable and decompose to a nitroalkene, with the concurrent loss of HONO (Fig. 6.1). Alternatively, the loss of HONO leads to the generation of trans-nitroalkenes and/or trans-nitroalkanes. However, the nitro group preferentially binds the double bond in nitro-alkene derivatives in the cis configuration [10, 11].

3 Synthesis and Purification of NO_2-AA

The most successful method for NO_2-AA synthesis includes AA incubation with sodium nitrite ($NaNO_2$) following extraction of the reaction mixture and analysis by RP-HPLC and TLC [12]. HPLC fractions are then extracted using the Bligh and Dyer method and analyzed by ESI MS/MS. The main product of the reaction is trans-AA [8, 12]. Three NO_2-AA fractions have been identified exhibiting m/z of 348 that corresponds to the addition of the nitro ($-NO_2$) group to the fatty acid [12]. Their identity was confirmed by MRM scan mode following the m/z 348/301 transition, in addition to the MS/MS spectra. A nitrohydroxy-derived product was also present [8, 12]. Four major isomers of NO_2-AA have been identified and key biological properties have been determined, including their capacity to release $^{\bullet}NO$, induce endothelium-independent vasorelaxation and modulate macrophage activation [12]. Mass spectrometry analysis showed the presence of m/z 348 products corresponding to NO_2-AA; however it cannot distinguish between a $-NO_2$ or a $-ONO$ functional group. The IR spectrum of NO_2-AA, in contrast to AA, exhibited absorbance maxima at 1370 and 1555 cm^{-1} that corresponds to a $-NO_2$ group of a nitroalkene attached to the carbon chain [12]. Eight positional nitroalkene isomers can be formed from the addition of a nitro group to AA at the double bonds. A novel MS methodology using lithium was used in order to determine which nitrated isomers were obtained. The presence of lithium allows the fragmentation of NO_2-AA at the double bond yielding an oxime or an aldehyde, depending where the $-NO_2$ group was attached [12]. The theoretical daughter ions with specific m/z can be detected by MS: 9-, 12-, 14- and 15- NO_2-AA isomers. We clearly demonstrated the presence of mononitrated nitroalkenes: 9-nitroicosa-5,8,11,14-tetraenoic acid (9-NO_2-AA), 12-nitroicosa-5,8,11,14-tetraenoic acid (12- NO_2-AA), 14-nitroicosa-5,8,11,14-tetraenoic acid (14-NO_2-AA) and 15-nitroicosa-5,8,11,14-tetraenoic acid (15- NO_2-AA) [12]. Reaction batches contained in all cases a mixture of 23 % of 12- and 15- NO_2-AA, 55 % of 9- NO_2-AA and 22 % of 14- NO_2-AA [13].

4 Interaction of NO_2-AA with Lipid Metabolizing Enzymes

Prostaglandin endoperoxide H synthase (PGHS) is the key enzyme of AA metabolism as it final oxidation product, prostaglandin H2 (PGH2) is the precursor of various lipid mediators, e.g. prostaglandins and thromboxanes. The enzyme catalyzes the dioxygenation of AA to prostaglandin G2 and the subsequent reduction of PGG2 to PGH2 performed by the cyclooxygenase (COX) reaction, where two molecules of oxygen are added to AA, and the peroxidase (POX) reaction yielding PGH2, respectively [14]. Two isoforms of PGHS (PGHS-1 and -2) are found in mammalian tissues where both isoforms are of pharmacological importance because they are targets for nonsteroidal anti-inflammatory drugs (NSAIDs). Inflammatory processes and PUFAs are linked by eicosanoids, which represent mediators and regulators of inflammatory processes formed from 20 carbon length-PUFAs. Molecular cross-talk between $^{\bullet}NO$ and prostaglandin pathways are well documented. In fact, many of the reported signaling actions of $^{\bullet}NO$ are due to its interaction with iron-containing enzymes; in this sense, PGHS, an heme-containing enzyme, can be a potential target for $^{\bullet}NO$, modulating prostaglandin synthesis [14]. Moreover, $^{\bullet}NO$ can act as a reducing cosubstrate during PGHS-1-POX activity thus favoring the catalytic cycle of POX [15]. When analyzed with PGHS-1, $^{\bullet}NO$ spared POX while enhanced COX inactivation by peroxynitrite [14]. During catalysis, AA derived-radicals are formed being potential targets for $^{\bullet}NO$ reactivity thus influencing enzyme activity. We speculate that during PGHS-1-COX catalysis, AA-derived radicals can be "sequestered" by $^{\bullet}NO$ to form nitrogen-containing AA products that in turn could act as poor substrates or inhibitors of COX, thus diminishing enzyme activity [14]. As AA is the substrate for PGHS activity, we evaluated if nitration of the carbon chain of AA may divert the fatty acid from its normal metabolizing pathways. Nitroarachidonic acid was able to inhibit both POX and COX activities of PGHS-1 while only affected POX in PGHS-2 [13]. These inhibitory effects were only for NO_2-AA, since other nitroalkenes (i.e. nitro-oleic acid or nitro-linoleic acid) were unable to modify PGHS activity. Analysis indicates that NO_2-AA inhibits PGHS-1 and -2 being a poorly reversible POX inhibitor of PGHS-2, through a slow tight binding mechanism that leads to the formation of a stable binary complex where the dissociation rates of enzyme-inhibitor complexes are so slow that NO_2-AA appears to be practically irreversible. The most likely reaction involves the release of heme as a result of NO_2-AA reaction with the protein, being different from those reported for other well-known enzyme inhibitors [13]. Next, the capacity of NO_2-AA to modulate PGHS-1 activity was evaluated in human platelets [13, 16]. Platelets are key players during the first stage of hemostasis, undergoing a series of reactions such as adhesion to the endothelium, aggregation, release of granule content and morphological changes that lead to the formation of the platelet plug. Nitroarachidonic acid represents a potent inhibitor of thrombin-mediated platelet aggregation (IC50 of 1.3 μM, [13, 16]), inhibiting thromboxane A2 (TxA2) synthesis. In addition, NO_2-AA modulated platelet aggregation in response to activation of several membrane receptors i.e. protease activated receptors -PARs- for thrombin, P2Y receptors for ADP and TxA2 receptors for AA, following TxA2 synthesis or through direct stimulation of protein kinase C (PKC), indicating that NO_2-AA acts downstream membrane receptors to exert its antiplatelet effects [16]. The PKC family is centrally involved in platelet activation and aggregation. Stimulation of phospholipase C (PLC) causes generation of inositol 1,4,5-triphosphate (IP3) and diacylglycerol (DAG). The latter remains in the plasma membrane and activates the normally cytosolic conventional PKC isoforms (α/β/γ, cPKC), which act in a synergistic manner with Ca2+. Downstream consequences of PKC activation are granule secretion, integrin αIIbβ3 activation and irreversible platelet aggregation. WhileNO_2-AA did not mobilize Ca^{2+}, it was able to inhibit α-granule secretion [16]. PKCα was diffusively distributed in the cytosol of untreated platelets and migrated to the plasma membrane after thrombin-stimulation in a process that was abolished by NO_2-AA. Also, NO_2-AA prevented platelet shape change and cytoskeletal reorganization, which alters α tubulin distribution from a homogeneous pattern in the marginal band to a highly irregular one [16]. PKC activates mitogen-protein kinase (MEK)/extracellular-signal regulated kinase (Erk) and p38 MAPK. While NO_2-AA had no effect on Erk activation, it markedly reduced Erk2 phosphorylation in thrombin-activated platelets [16]. Nitroalkene bioactivities are primarily mediated through electrophilic reactivity leading to the formation of covalent adducts with reactive Cys or His residues in proteins. cPKC isoforms contain Cys rich motifs that are duplicated as a tandem domain which are critical for its interaction with membrane phospholipids, suggesting that PKC inhibition could be mediated by electrophilic modifications of the enzyme. In fact,NO_2-AA effects demonstrate that its electrophilic properties are needed for NO_2-AA-mediated inhibition of platelet activity [16]. Figure 6.2 summarize NO_2-AA effects on platelets: (1) NO_2-AA inhibits platelet activation through modulation of additional targets than PGHS-1; (2) NO_2-AA inhibits platelet responses downstream to PKC activation (α-granule secretion, Erk2 phosphorylation, PKC translocation to the membrane) while not affecting upstream responses (e.g. Ca^{2+} mobilization); (3) NO_2-AA inhibits PMA-induced aggregation and (4) NO_2-AA exerts unique platelet regulation compared to other nitroalkenes. These observations provide a possible novel mechanism for platelet regulation under conditions where AA acts as a mild agonist for hemostasis, but adopts potent anti-platelet properties at inflammatory environments associated with increased $^{\bullet}NO$ and RNS production when transformed into NO_2-AA.

5 Vasorelaxing and $^{\bullet}NO$ Releasing Properties of NO_2-AA

Balazy et al. demonstrated that AA(OH)NO_2 is a bioactive compound that induces vasorelaxation due to its capacity to release $^{\bullet}NO$ in a non-inhibitable indomethacin mechanism [8]. The release of $^{\bullet}NO$ from NO_2-AAwas demonstrated by using different experimental approaches [12], including oxyHb oxidation ($k = 9.8 \times 10^{-5}\ s^{-1}$) and EPR analysis of $^{\bullet}NO$-dependent cPTI formation. One of the first vascular

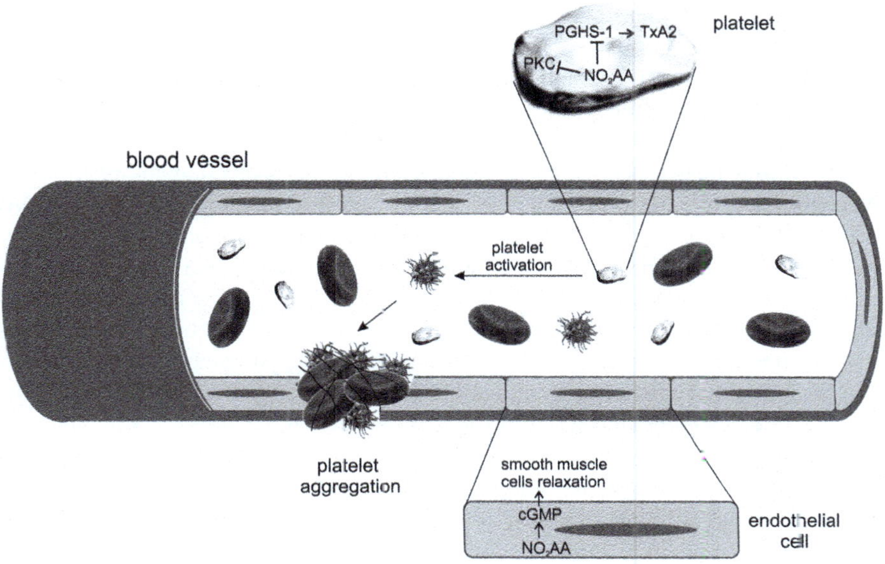

Fig. 6.2 Modulatory responses of NO_2-AA in vascular cells. NO_2-AA inhibits platelet activation through downstream of PGHS-1 and PKC signaling. NO_2-AA induces cGMP formation in endothelial cells, leading to smooth muscle cells relaxation.

protective actions attributed to NO_2-AA was its capacity to induce soluble guanylate cyclase (sGC)-mediated vasorelaxation (Fig. 6.2, [17]). We hypothesized that addition of a methyl group into the NO_2-AA structure could keep its biological properties and improve its bio-disposal for in vivo administration as a potential vascular protective drug. In fact, incorporation of the methyl moiety could increase the lipophilicity of NO_2-AA and presumably facilitate its storage in lipidic matrices while reduce its elimination. In this sense, a novel esterified nitroalkene, 6-nitro-methyl arachidonate was synthesized and characterized [17]. Similarly to the results achieved from AA nitration, the cis-alkene isomerization of AAMet was the main reaction (90 %) occurring during the incubation with $NaNO_2$, and the nitrated products were obtained in lower yields (between 1 and 5 %). Using AAMet as the fatty acid precursor, the main product obtained is the result of the substitution of the alkenic proton 6 by the $-NO_2$ moiety instead of alkenic protons 9-, 12-, 14- or 15-substitutions as occurred for NO_2-AA [17]. 6-nitro-methyl arachidonate promoted comparable extents of vasorelaxation on precontracted aortic rings through sGC activation. This was confirmed by the fact that both methylated and non-methylated NO_2-AA showed similar ability to induce cGMP formation in vascular smooth muscle cells. Moreover, vasorelaxation and cGMP formation were endothelium-independent, suggesting that its interaction with the endothelium did not contribute to generate •NO/•NO-like species responsible for sGC activation. In contrast with NO_2-AA, 6-nitro-methyl arachidonate was not able to release •NO in cell-free conditions but, in a cell environment, led to •NO generation [17]. So far, there have been reports on several and controversial mechanisms by which nitroalkenes release •NO spontaneously [18, 19]. Differences between the ability to release •NO in an aqueous milieu could be explained by the lower solubility of the ester compared with the free fatty acid. Alternatively, the position of the $-NO_2$ group (closer to the ester moiety) could decrease the alkene-capability to act as Michael acceptor preventing the formation of an intermediary nitronate for •NO release (Nef mechanism, [19]). It is known that, under neutral aqueous conditions, nitroalkene equilibrium with vicinal nitrohydroxy derivatives will facilitate formation of the nitronate anion and, following this event, the resulting nitroso intermediate provides a pathway that yields •NO via a reaction facilitated by reductants. This nitroso intermediate has a weak C-N bond that yields . •NO and a carbon-centered radical product stabilized by conjugation with both the alkene and the OH group. Nitric oxide detection during 6-nitro-methyl arachidonate incubation with endothelial cells [17] suggests that the methylated form could be converted to NO_2-AA upon hydrolytic release by cell lipases and esterases followed by •NO release. If this is true, 6-nitro-methyl arachidonate may be relevant as an esterified hydrolyzable reserve of NO_2-AA in hydrophobic compartments that could contribute to modulate the vascular tone through sGC activation. Alternatively, other cGMP-dependent signaling actions of NO_2-AA or 6-nitro-methylarachidonate may be involved in the modulation of vascular tone [8, 20–22]. Endothelium is not critical for mediating the vasoactive action of 6-nitro-methyl arachidonate since inhibition of nitric oxide synthase-1 or denudation of endothelium revealed similar extents of vessel relaxation [17]. In conclusion, 6-nitro-methylarachidonate could be considered as a potential novel vascular protective agent, where the presence of the methyl group serves to improve the bio-disposal of NO_2-AA.

6 Nitro-arachidonic Acid as an Inhibitor of Inflammatory Cell Function

Inflammation is characterized by the production of cytokines, AA-derived eicosanoids, RNS and ROS, i.e. through the activation of inducible nitric oxide synthase (NOS2) and NADPH oxidase (NOX). Indeed, one of the open questions in the field is related to the mechanisms involved in the observed anti-inflammatory actions of NO_2-AA at both molecular and cellular levels. Despite the well demonstrated role of NO_2-AA as stimulator of smooth muscle relaxation, the NO_2-AA-dependent attenuation of the inflammatory phenotype observed in several inflammatory cells follows •NO-independent mechanisms [16, 17, 23]. The anti-inflammatory activity of nitroalkenes is mainly mediated by rapid electrophilic and receptor-mediated reactions leading to nitroalkylation of proteins. These play critical roles in the regulation of inflammatory response. In fact, inflammatory cells, including blood neutrophils, monocytes, macrophages, platelets or endothelial cells, respond to pro-inflammatory signals by switching to an inflammatory phenotype, characterized by the expression of molecules for leukocyte and endothelium adhesion. This cause adhesion of blood cells to endothelium cells, penetration of blood cells across vascular walls and migration into the inflammation focus [16, 17, 23]. Secondary to the phenotype switch, inflammatory mediators are released, where cytokines play key roles in the initiation and regulation of the inflammatory response.

In macrophages, NO_2-AA down-regulates the lipopolysaccharide-induced secretion of proinflammatory cytokines (IL-6, TNFα) and NOS2 expression, contributing to the physiological shut down of the inflammatory response [12, 24]. One of the immediate responses upon macrophage activation involves the production of $O_2 \cdot^-$ due to the NADPH dependent univalent reduction of oxygen by the phagocytic NADPH-oxidase isoform (NOX2). NOX2 is an enzyme complex comprised of a membrane–bound flavor-cytochrome b558 (gp91phox and p22phox), three cytosolic subunits (p47phox, p40phox and p67phox) and GTPase Rac2. In resting macrophages, the NOX2 complex is unassembled. Upon activation, the cytosolic components associate with gp91phox and p22phox forming the active enzyme complex which at this time generates $O_2 \cdot^-$ [25]. It has been suggested a strong relationship between the AA pathway and NOX2 assembly/activation. The inhibition of phagocytic NOX2 activity by NO_2-AA (Fig. 6.3) has been demonstrated both in a macrophage cell line as well as in primary macrophages [23]. The mechanism includes the capacity of NO_2-AA to decrease $O_2 \cdot^-$ formation in the phagosome by

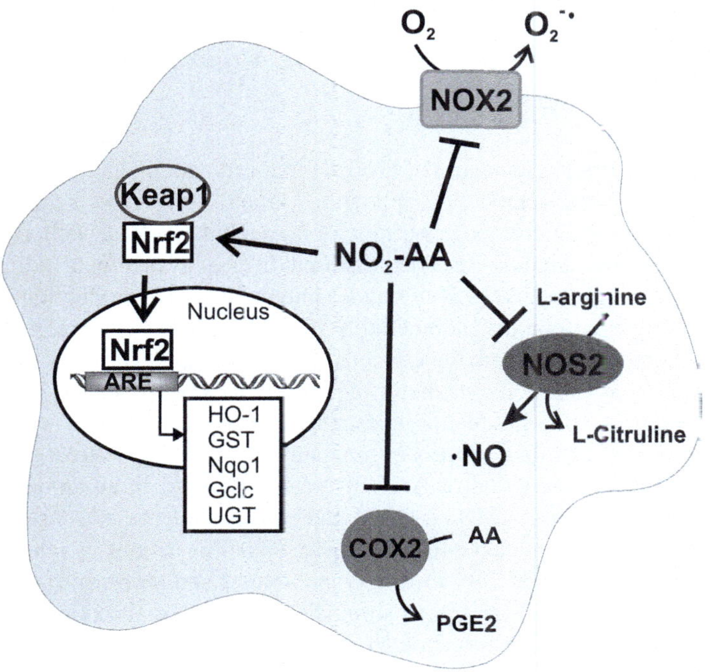

Fig. 6.3 Anti-inflammatory actions actions of NO_2-AA. NO_2-AA inhibits pro-inflammatory inducible enzymes in macrophages as well as activates the Nrf2 pathway, leading to the induction of antioxidant responsive elements (ARE)

inhibiting NOX2 activity/assembly. Phosphorylation of cytosolic subunits is not affected by NO_2-AA. After phosphorylation, the cytosolic subunits migrate to the membrane where they associate with gp91phox and p22phox achieving the active form of the enzyme. NO_2-AA affects the migration of the cytosolic subunits to the membrane, resulting in the inhibition of NOX2 activity. In fact, NO_2-AA modulates the translocation of p47phox avoiding the formationof the active enzyme complex, thus inhibiting NOX2 dependent $O_2\cdot^-$ formation [23]. This data is indicative that NO_2-AA inhibits NOX2 in activated macrophages through a mechanism that involves the prevention of the migration of the cytosolic subunits to the membrane, affecting the correct assembly of the active form of the enzyme. Data from an inflammatory model (thioglycolate-treated mice) also support this mechanism, where subcutaneous injection of NO_2-AA decreased macrophage NOX2 activity [23]. This suggests that NO_2-AA exerts beneficial anti-inflammatory effects when administered at pharmacological doses. In a chronic inflammatory processes, where nitration of AA may occur due to nitro-oxidative stress conditions, NO_2-AA may inhibit key inducible enzymes (e.g. NOX2, NOS2 or PGHS-2) aiding in the resolution of inflammation (Fig. 6.3).

7 Therapeutic Potential of NO_2-AA

7.1 Mitochondrial Protection During Angiotensin II-Mediated Kidney Injury

The role of ROS in angiotensin-II (ANG II)-induced endothelial dysfunction, cardiovascular and renal remodeling, inflammation and fibrosis has been well documented [26]. In fact, increased generation of intracellular ROS as well as activation of redox-sensitive signaling cascades represent critical events involved in ANG II actions [27]. In this way, ANG II stimulates intracellular $O_2\cdot^-$ production by activating NOX [28] and promotes endothelial NOS (NOS3) uncoupling [29]. When •NO production is also stimulated, the effects of ANG II could lead to the formation of peroxynitrite that can be the mediator of ANG II-mediated damage [25]. ANG II induces excessive ROS production and mitochondrial dysfunction, which eventually leads to apoptosis and necrosis of renal tubular cells [30]. Nitroalkenes administration result in a reduction of ANG II-induced damage in an animal model of hypertension by inhibiting ANG II type 1 receptor-dependent vasoconstriction [31]. In addition, nitroalkenes exert antihypertensive signaling actions by inhibiting soluble epoxide hydrolase [32]. Reactive oxygen species and peroxynitrite production are enhanced by ANG II as well as stimulates mitochondrial oxidants production depressing mitochondrial energy metabolism (Fig. 6.4).

In ANG II-treated kidney cells, NO_2-AA protects mitochondria through reduction of ROS and RNS production [33]. While stimulation with ANG II induced $O_2\cdot^-$ formation, pretreated cells with NO_2-AA diminished $O_2\cdot^-$ production and

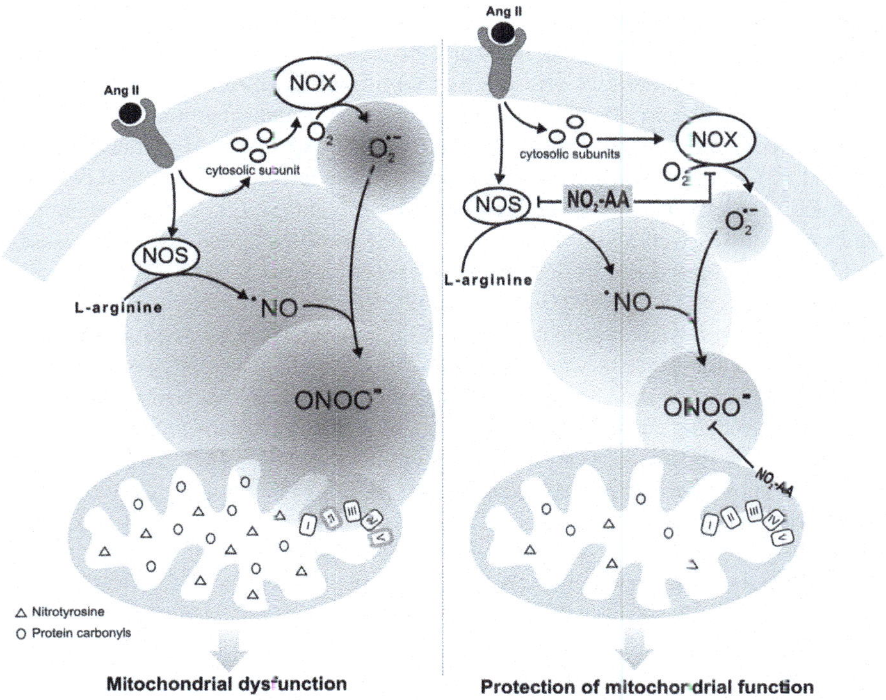

Fig. 6.4 NO_2-AA prevention of ANG II-mediated mitochondrial dysfunction. (Adapted from Ref. [33])

oxidative stress. Overall, NOX activity stimulated by ANG II is inhibited by NO_2-AA as a protective mechanism (Fig. 6.4). In addition,NO_2-AA improves mitochondrial coupling in renal cells treated with ANG II [33]. Moreover, NO_2-AA spares mitochondrial complexes from peroxynitrite-mediated oxidative damage. This data supports that NO_2-AA modulates ANG II-mediated inflammatory damage in kidney cells trough prevention of peroxynitrite formation and subsequent mitochondrial dysfunction (Fig. 6.4). The protection of mitochondrial function by NO_2-AA in a model of renal injury emphasizes its use as a novel mitochondrial-targeted antioxidant.

7.2 *Neuroprotection Through Nrf2 Activation*

Due to its electrophilic properties, nitroalkenes can activate Nuclear factor erythroid-2-related factor (Nrf2) through reversible nitroalkylation reactions [34–37]. Nrf2 activity is principally governed by Kelch-like ECH-associating protein 1 (Keap1)) a protein with elevated cysteine content, which renders it highly reactive to electrophiles. Villacorta et al. demonstrated a direct reaction of nitroalkenes with Keap1 impairing Keap1-mediated inhibition of Nrf2/Antioxidant Responsive Element (ARE) signaling (Fig. 6.3) [36]. It is currently unknown whether nitroalkenes exert actions in neurodegenerative diseases. For familial Amyothropic Lateral Esclerosis (ALS), approximately 10–20 % is caused by a toxic gain-of-function induced by mutations of superoxide dismutase 1 (SOD1). Over-expression of mutated forms of hSOD1 in rodents resulted in animal models of the disease, e.g. hSOD1G93A rats or mice. Toxicity to motor neurons requires mutant SOD1 expression in non-neuronal cells as well as in motor neurons [38]. Both Nrf2 and hemeoxygenase-1 (HO-1) are increased and co-localized with reactive astrocytes in the degenerating lumbar spinal cord of hSOD1G93A rats [39]. Recently, a potent protective role of NO_2-AA on astrocytes expressing the ALS-linked SOD1G93A mutation-mediated toxicity to motor neurons was described [40]. The effects of NO_2-AA on motor neuron degeneration induced by astrocytes revealed that NO_2-AA administration to cultured astrocytes caused (a) Nrf2 activation and antioxidant phase II enzymes induction and (b) an increase in total glutathione levels. Thus, the influence of NO_2-AA applied to astrocytes bearing the SOD1G93A mutation on astrocyte-mediated motor neurons death in co-culture conditions was explored. Pre-treatment of SOD1G93A astrocytes with NO_2-AA significantly reduced motor neurons loss. This effect was prevented by transfecting astrocytes with a Nrf2-siRNA before NO_2-AA treatment, further supporting that Nrf2 activation is mediating the protective effect of NO_2-AA. An increase in glutathione levels in astrocytes may account for the observed protection of motor neurons death via induction of the modulatory subunit of the glutamate-cysteine ligase which catalyzes glutathione synthesis [40]. Thus, the increase in antioxidant defenses induced by NO_2-AA treatment could potentially improve mitochondrial function in astrocytes and be partially responsible for the protection observed. These mechanisms may play an important role in NO_2-AA-triggered astrocyte-mediated increase in motor neuron survival. In summary, NO_2-AA induces ARE-driven gene expression which is dependent on Nrf2 activation as well as astrocytic glutathione production, preventing motor neurons death. Considering that the central nervous system is abundant in AA, it is possible that NO_2-AA is being generate as adaptive response during inflammatory conditions to protect motor neurons.

8 Conclusions

Lipid nitration represents a novel mechanism for RNS to transduce metabolic and inflammatory information. Since AA is the most abundant PUFA present in biological membranes, NO_2-AA formation is likely to occur under inflammatory conditions, supporting a role of lipid nitration in adaptive redox-sensitive signaling. Nitroarachidonic acid exerts potent biological actions that aid in the resolution of inflammation, including inhibition of inflammatory cell activation, pro-inflammatory cytokine and prostaglandin secretion. Moreover, recent data reveal that NO_2-AA mediate homeostatic signaling reactions in vivo. Further work is necessary to determine whether NO_2-AA supplementation exerts unique anti-inflammatory and cytoprotective actions in inflammatory diseases.

Acknowledgment This work was supported by grants from Universidad de la República (CSIC), Programa de Desarrollo de Ciencias Básicas (PEDECIBA) and Agencia Nacional de Investigación e Innovación (ANII, Fondo Innovagro), Uruguay. L. González and M. Mastrogiovanni have been supported by fellowships of ANII.

References

1. Rubbo H, Radi R, Trujillo M et al (1994) Nitric oxide regulation of superoxide and peroxynitrite-dependent lipid peroxidation. Formation of novel nitrogen-containing oxidized lipid derivatives. J Biol Chem 269:26066–26075
2. Liu X, Miller MJ, Joshi MS et al (1998) Diffusion-limited reaction of free nitric oxide with erythrocytes. J Biol Chem 273:18709–18713
3. Fazzari M, Trostchansky A, Schopfer FJ et al (2014) Olives and olive oil are sources of electrophilic fatty acid nitroalkenes. PLoS One 9:e84884
4. Rubbo H, Radi R (2008) Protein and lipid nitration: role in redox signaling and injury. Biochim Biophys Acta 1780:1318–1324
5. Radi R, Denicola A, Alvarez B et al (2000) The biological chemistry of peroxynitrite. In: Nitric oxide: biology and pathobiology. Academic, San Diego, pp 57–82
6. Botti H, Batthyany C, Trostchansky A et al (2004) Peroxynitrite-mediated alpha-tocopherol oxidation in low-density lipoprotein: a mechanistic approach. Free Radic Biol Med 36:152–162
7. Trostchansky A, Moller M, Barthesaghi S et al (2010) Nitric oxide redox biochemistry in lipid environments. In: Ignarro L (ed). San Diego: Elsevier, pp 61–102
8. Balazy M, Iesaki T, Park JL et al (2001) Vicinal nitrohydroxyeicosatrienoic acids: vasodilator lipids formed by reaction of nitrogen dioxide with arachidonic acid. J Pharmacol Exp Ther 299:611–619
9. Balazy M, Poff CD (2004) Biological nitration of arachidonic acid. Curr Vasc Pharmacol 2:81–93
10. O'Donnell VB, Eiserich JP, Chumley PH et al (1999) Nitration of unsaturated fatty acids by nitric oxide-derived reactive nitrogen species peroxynitrite, nitrous acid, nitrogen dioxide, and nitronium ion. Chem Res Toxicol 12 83–92
11. Trostchansky A, Rubbo H (2008) Nitrated fatty acids: mechanisms of formation, chemical characterization, and biological properties. Free Radic Biol Med 44:1887–1896
12. Trostchansky A, Souza JM, Ferreira A et al (2007) Synthesis, isomer characterization, and anti-inflammatory properties of nitroarachidonate. Biochemistry 46:4645–4653
13. Trostchansky A, Bonilla L, Thomas CP et al (2011) Nitroarachidonic acid, a novel peroxidase inhibitor of prostaglandin endoperoxide H synthases 1 and 2. J Biol Chem 286:12891–12900
14. Trostchansky A, O'Donnell VB, Goodwin DC et al (2007) Interactions between nitric oxide and peroxynitrite during prostaglandin endoperoxide H synthase-1 catalysis: a free radical mechanism of inactivation. Free Radic Biol Med 42:1029–1038
15. O'Donnell VB, Coles B, Lewis MJ et al (2000) Catalytic consumption of nitric oxide by prostaglandin H synthase-1 regulates platelet function. J Biol Chem 275:38239–38244
16. Bonilla L, O'Donnell VB, Clark SR et al (2013) Regulation of protein kinase C by nitroarachidonic acid: impact on human platelet activation. Arch Biochem Biophys 533:55–61
17. Blanco F, Ferreira AM, Lopez GV et al (2011) 6-Methylnitroarachidonate: a novel esterified nitroalkene that potently inhibits platelet aggregation and exerts cGMP-mediated vascular relaxation. Free Radic Biol Med 50:411–418

18. Lima ES, Bonini MG, Augusto O et al (2005) Nitrated lipids decompose to nitric oxide and lipid radicals and cause vasorelaxation. Free Radic Biol Med 39:532–539
19. Schopfer FJ, Baker PR, Giles G et al (2005) Fatty acid transduction of nitric oxide signaling. Nitrolinoleic acid is a hydrophobically stabilized nitric oxide donor. J Biol Chem 280:19289–19297
20. Cui T, Schopfer FJ, Zhang J et al (2006) Nitrated fatty acids: endogenous anti-inflammatory signaling mediators. J Biol Chem 281:35686–35698
21. Wright MM, Kim J, Hock TD et al (2009) Human haem oxygenase-1 induction by nitro-linoleic acid is mediated by cAMP, AP-1 and E-box response element interactions. Biochem J 422:353–361
22. Wright MM, Schopfer FJ, Baker PR et al (2006) Fatty acid transduction of nitric oxide signaling: nitrolinoleic acid potently activates endothelial heme oxygenase 1 expression. Proc Natl Acad Sci U S A 103:4299–4304
23. Gonzalez-Perilli L, Alvarez MN, Prolo C et al (2013) Nitroarachidonic acid prevents NADPH oxidase assembly and superoxide radical production in activated macrophages. Free Radic Biol Med 58:126–133
24. Ferreira AM, Ferrari MI, Trostchansky A et al (2009) Macrophage activation induces formation of the anti-inflammatory lipid cholesteryl-nitrolinoleate. Biochem J 417:223–234
25. Griendling KK, Sorescu D, Ushio-Fukai M (2000) NAD(P)H oxidase: role in cardiovascular biology and disease. Circ Res 86:494–501
26. Touyz RM (2005) Reactive oxygen species as mediators of calcium signaling by angiotensin II: implications in vascular physiology and pathophysiology. Antioxid Redox Signal 7:1302–1314
27. de Cavanagh EM, Inserra F, Ferder L (2011) Angiotensin II blockade: a strategy to slow ageing by protecting mitochondria? Cardiovasc Res 89:31–40
28. Kimura S, Zhang GX, Nishiyama A et al (2005) Role of NAD(P)H oxidase- and mitochondria-derived reactive oxygen species in cardioprotection of ischemic reperfusion injury by angiotensin II. Hypertension 45:860–866
29. Pueyo ME, Arnal JF, Rami J et al (1998) Angiotensin II stimulates the production of NO and peroxynitrite in endothelial cells. Am J Physiol 274:C214–C220
30. Sun L, Xiao L, Nie J et al (2010) p66Shc mediates high-glucose and angiotensin II-induced oxidative stress renal tubular injury via mitochondrial-dependent apoptotic pathway. Am J Physiol Renal Physiol 299:F1014–F1025
31. Zhang J, Villacorta L, Chang L et al (2010) Nitro-oleic acid inhibits angiotensin II-induced hypertension. Circ Res 107:540–548
32. Charles RL, Rudyk O, Prysyazhna O et al (2014) Protection from hypertension in mice by the Mediterranean diet is mediated by nitro fatty acid inhibition of soluble epoxide hydrolase. Proc Natl Acad Sci U S A 111:8167–8172
33. Sanchez-Calvo B, Cassina A, Rios N et al (2016) Nitro-arachidonic acid prevents angiotensin II-induced mitochondrial dysfunction in a cell line of kidney proximal tubular cells. PLoS One 11:e0150459
34. Kansanen E, Bonacci G, Schopfer FJ et al (2011) Electrophilic nitro-fatty acids activate NRF2 by a KEAP1 cysteine 151-independent mechanism. J Biol Chem 286:14019–14027
35. Kansanen E, Jyrkkanen HK, Volger OL et al (2009) Nrf2-dependent and -independent responses to nitro-fatty acids in human endothelial cells: identification of heat shock response as the major pathway activated by nitro-oleic acid. J Biol Chem 284:33233–33241
36. Villacorta L, Zhang J, Garcia-Barrio MT et al (2007) Nitro-linoleic acid inhibits vascular smooth muscle cell proliferation via the Keap1/Nrf2 signaling pathway. Am J Physiol Heart Circ Physiol 293:H770–H776
37. Tsujita T, Li L, Nakajima H et al (2011) Nitro-fatty acids and cyclopentenone prostaglandins share strategies to activate the Keap1-Nrf2 system: a study using green fluorescent protein transgenic zebrafish. Genes Cells 16:46–57

38. Clement AM, Nguyen MD, Roberts EA et al (2003) Wild-type nonneuronal cells extend survival of SOD1 mutant motor neurons in ALS mice. Science 302:113–117
39. Vargas MR, Pehar M, Cassina P et al (2005) Fibroblast growth factor-1 induces heme oxygenase-1 via nuclear factor erythroid 2-related factor 2 (Nrf2) in spinal cord astrocytes: consequences for motor neuron survival. J Biol Chem 280:25571–25579
40. Diaz-Amarilla P, Miquel E, Trostchansky A et al (2016) Electrophilic nitro-fatty acids prevent astrocyte-mediated toxicity to motor neurons in a cell model of familial amyotrophic lateral sclerosis via nuclear factor erythroid 2-related factor activation. Free Radic Biol Med 95:112–120

Chapter 7
Mitochondrial Transfer by Intercellular Nanotubes

Viviana Sanchez and Alicia Brusco

Abstract Cell-to-cell communication is a critical requirement for the coordination of cell behavior in tissue homeostasis and the conservation of multicellular organisms. Among several types of intercellular communication, tunneling nanotubes (TNTs) were discovered no more than a decade ago but are now known to constitute intercellular bridges connecting distant cells. Over the last decade, research has shown TNTs to have structural and functional properties which vary across cell types. TNTs permit cell-to-cell communication on the basis of membrane continuity between connected cells and are capable of transferring various types of intracellular components including calcium ions, cytoplasmic molecules and different types of organelles. In this chapter, we will describe the different mechanisms of TNT formation, their heterogeneous composition and their functional roles in physiological and pathological processes. In this context, we also discuss the importance of mitochondria transfer from stem cells to recipient cells with nonfunctional mitochondria, which results in a significant improvement in aerobic respiration. The transfer of healthy mitochondria through TNTs may rescue damaged cells and thus constitute an alternative therapeutic approach for pathologies involving oxidative stress.

Keywords Tunneling nanotubes • Intercellular transfer • Mitochondria • Mitochondrial transfer • Intercellular communication

1 Introduction

Cell behavior is the result of its interaction with the environment, which includes both the soluble and fibrous molecules of the extracellular matrix in contact with the plasma membrane and other near or distant cells. Intercellular communication plays a central role in the development and maintenance of multicellular organisms. Without cell-to-cell communication, processes such as tissue growth,

V. Sanchez (✉) • A. Brusco
Institute of Cell Biology and Neuroscience (UBA-CONICET), School of Medicine, University of Buenos Aires, Buenos Aires, Argentina
e-mail: vsanchez@fmed.uba.ar; hbrusco@fmed.uba.ar; aliciabrusco@gmail.com

© Springer International Publishing Switzerland 2016

R.J. Gelpi et al. (eds.), *Biochemistry of Oxidative Stress*, Advances in Biochemistry in Health and Disease 16, DOI 10.1007/978-3-319-45865-6_7

differentiation during development, remodeling of tissues and organs, cell division and the cell responses to environmental stimuli could not take place. The relationship between neighboring cells is highly developed in some tissues through molecular interchanges [1]. The exchange of molecular information between cells includes the secretion of molecules and their subsequent binding through receptor-ligand mechanisms, cell endocytosis or plasmatic membrane crossing, with or without energy utilization. Other types of cell-cell interaction include ionic exchange through gap junctions and the secretion of membranous carriers referred to as exosomes [2]. Large membrane structures such as desmosomes, zonula adherents and occludens allow efficient cohesion between neighboring cells, as observed in epithelial or cardiac cells. Moreover, cells are also capable of exploring the extracellular environment and of establishing direct contact over long distances, for example through long actin based extensions called cytonemes and first observed during Drosophila development [3]. A particularly interesting type of cell connection, called tunneling nanotubes (TNTs), has been recently described as thin cytoplasmic bridges with actin cytoskeleton which connect distant cells [4]. The discovery of this new type of communication highways has opened up new ways of viewing how cells interact with one another. The first reference to nanotubular connections between cells dates back to a study on pheochromocytoma PC12 cell cultures. TNTs were then shown to have a diameter of 50–200 nm and a length several times cell diameter, and to exhibit neither branches nor contact with the substrate. TNTs also permit the transference of organelles from one cell to the other [4]. These structures permit cell-to-cell communication on the basis of membrane continuity between TNT-connected cells. As an additional feature, these intercellular bridges have been found to be transient structures whose lifetime is no longer than 30–60 min (Fig. 7.1).

After the first description by Rustom et al. [4] the presence of TNTs has been found in numerous cell types both *in vitro* [5] and in tissue explants [6]. In turn, most *in vivo* reports are related to studies on embryos, which analyze the role of TNTs during the development of multicellular organisms [7]. Worth pointing out, TNTs have been found both in healthy [8] and tumoral tissues [9]. In addition, cultures of human microvascular endothelial cells have proven significantly long TNTs (≥100 μm) to be stable during the whole mitotic process, which extends over several hours and during which TNTs transport granules [10]. Finally, cellular compo-

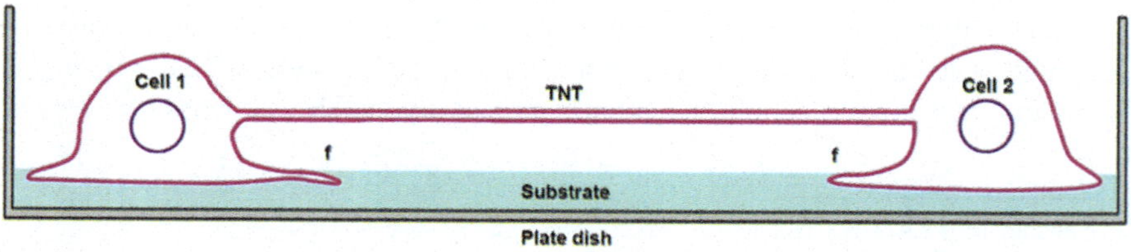

Fig. 7.1 TNTs as dynamic bridges connecting cells. Schematic representation of two cells in a culture dish. In contrast to filopodia (f), the TNT does not make contact with the substrate and shows a rectilinear trajectory and constant diameter for more than 5 times cell body length

nents transferred via TNTs include calcium ions, cytoplasmic molecules and different types of organelles.

2 TNT Morphology and Function

TNTs must be discriminated from filopodia. In contrast to filopodia, TNTs are not attached to the substratum and are about 5–10 times their length [11]. Also, TNTs and filopodia appear to serve different purposes. While filopodia act as important environmental sensors and play key roles in cell motility, TNTs appear to be a direct conduit for cell-to-cell communication, specifically in the transport of material from one cell to another. TNT *de novo* formation has not been fully elucidated yet, although time-lapse recording studies suggest two possible mechanisms. The first mechanism involves an outgrowth of a filopodia-like protrusion, rich in cytoskeletal filaments, toward a neighboring cell [12]. The second mechanism, called cell dislodgement, consists of attached cells departing from one another, after which a TNT is formed between them [11]. Further studies have shown that TNT formation and stabilization require anchoring junctions between the membranes of connected cells [13]. As mentioned above, initial TNT morphological and structural analyses revealed a diameter of 50–200 nm and a rectilinear trajectory several-cell-diameter long, confirming an F-actin cytoskeleton [4]. In other cell types such as immune cells and macrophages, the average length of TNTs has been estimated in 30 μm, and in some cases reaching over 140 μm. In addition, two different types of TNTs have been observed in these cell types, the first one composed only by F-actin (type I) and the second one exhibiting greater thickness (>700 nm in diameter) and both F-actin and microtubules (type II) [14]. In T cells, for example, only 20 % of TNTs present microtubules inside. It seems then possible that TNTs containing microtubules are derived from cell division, while only-actin TNTs are not. Worth mentioning, both types exhibit irregular diameter and membrane surface [11]. These heterogeneous TNT features probably represent cell specialized functions, although such categorization appears now elusive due to the wide range of TNT-like structures described both *in vitro* and *in vivo* (Table 7.1) [7] (Fig. 7.2).

2.1 TNT-Mediated Cargo Transfer

TNTs allow for the direct transfer of cellular components such as organelles, proteins, microRNAs and Ca^{2+}, and can in certain cases be highways for the transport of organelles belonging to the endoplasmic reticulum (ER), Golgi and the endosomal-lysosomal system and mitochondria. Moreover, different membrane protein and lipid components have been observed in different TNTs [15]. In connection with ion exchange, the major differences between gap junctions and TNTs are the distances reached and the sizes of the molecules transferred. Furthermore, TNTs

Table 7.1 Examples for cell-cell TNTs comunication and cargo transfer

Cell type	Cytoskeleton	Length/Thickness	Cargo	References
PC12	F-actin	10 mm/50–200 nm	Endosome/lysosome related proteins, lipid anchored proteins	[4]
Myeloid (dendritic) cells	Not determined	50–200 nm/>100 mm	Ca2+ signaling. Surface receptors, small molecules (Lucifer Yellow)	[17]
Co-culture of endothelial progenitor cells and MSCs	Not determined	5–120 mm/50–800 nm	Mitochondria, soluble components (GFP)	[18]
Human monocyte derived macrophages	F- actin and Microtubules	Not determined/ 700 nm	Golgi and RER, bacteria mitochondria, endosome/lysosome, membrane components,	[14, 43]
Myeloid cells in mouse cornea	F- actin	<0.8 mm/200–600 nm	Cx 43, YFP, CD11	[6, 8]
Junkat and T cells	F- actin	22–100 mm/<400 nm	HIV proteins, Mitochondria, Membrane components, death signals, virus	[11, 44]
Mouse CAD neuronal cells (variant of a CNS catecholaminergic cell line from a brain tumour)	F- actin	14–75 mm/50–80 nm	Exogenous and endogenous membrane components (GFP-Prion protein) Proteinaseous aggregates	[45]
Normal rat kidney cells	F- actin, Miosyn Va	70 mm/Not determined	Depolarization signals. Endosome-related organelles	[46]
Co-culture: vascular smooth muscle cells and MSCs	Not determined	Not determined	Mitochondria	[47]
Human umbilical vein endothelial cells	F- actin	21–30 mm/Not determined	Electrical signals	[46]
Co-culture: renal tubular cells and MSCs	Not determined	Not determined	Mitochondria and cytolpasmic components	[48]
Natural killer cells	F- actin and microtubules	21–100 mm/Not determined	Membrane components (MHC-1) Death signals	[26]
Primary human renal epithelial cells	F- actin	0.2/1 mm/>200 nm	Endosome-Lysosome	[40]

(continued)

Table 7.1 (continued)

Cell type	Cytoskeleton	Length/Thickness	Cargo	References
Human hepatocellular carcinoma cells	F- actin and microtubules	Not determined	Cadmium Telluride Quantum Dots (CdTe QDs)	[49]
Co-culture of H9C2 cardiomioblasts and MSCs	F- actin and microtubules	Not determined	Mitochondria, endosomal vesicles Ca2+ cytosolic GFP	[50]
Co-culture of human MSCs and adult myocandiocytes	F- actin	Not determined	Mitochondria	[51]
Co-cultures of human endothelial precursor cell and umbilical vein endothelial cells	F- actin	Not determined	Lysosomes	[52]
Human lung adenocarcinoma cells	F- actin, fascin, erzin	70 mm/ND	Vesicles, proteins, mitochondria	[9]
Human mesotelioma cells	F- actin, Miosyn Va	70 mm/Not determined	Vesicles, proteins, mitochondria	[53]
Co-culture of neurons and astrocytes	F- actin and Microtubules	30–140 mm/Not determined	Depolarization signals	[54]
Human Embryonic Kidney cells	F- actin	500 nm/ Not Determined	Calcium signaling (gap junctions) Fluorescent prion protein construct (GFP-Pr-PWt)	[12]
PC12 cells	TNT1: F- actin and microtubules TNT2 F-actin	TNT1 >40 mm/100–650 nm. TNT2 20 mm/70–200 nm	Alexa 488-Wheat Germ Agglutinin	[55]
Hematopoietic stem cells Co-culture with macrophages	Not determined	>40 mm/Not determined	Vesicular exchange (endosome/lysosome system)	[56]

appear to be more selective than gap junctions, allowing the transfer of electrical signals to the coupled acceptor cell only, thus preventing signal spreading to other cells [12].

Additional studies have revealed ER inside TNTs, which may be associated to the spreading of intercellular Ca^{2+} signals between nearby cells [16]. Also, a calcium flux has been reported through the thicker Type II TNTs in human cultured myeloid lineage dendritic cells and monocytes [17]. Studies on intercellular communication between human endothelial progenitor cells and rat cardiomyocytes have revealed nanotubular structures allowing the transport of organelles [18]. In

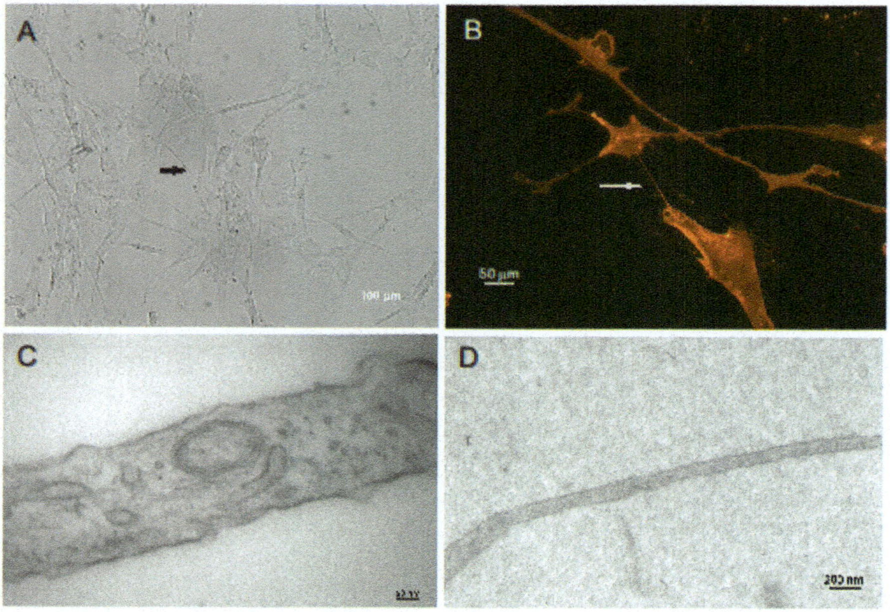

Fig. 7.2 TNTs in stem cells. (**a**) iPS cell culture. The *arrow* shows a bridge corresponding to a TNT between two neighboring cells. (**b**) MSC culture. The *arrow* shows a TNT between neighboring cells immunostained for the stem cell marker CD105. (**c**) Electron microscopy of thick TNT containing different organelles (**d**) Electron microscopy of thin TNT containing only microfilaments

cultures of human B cells, natural killer cells and macrophages, vesicular traffic in thick TNTs has been shown to include bacterial and viral particles surfing on the surface of thin TNTs [14]. T-cell nanotubes have been proven to transfer HIV-1 in a receptor-dependent way [11], while migrating endothelial cell TNTs appear to preserve intercellular communication through the transport of signals such as lipid mediators in the form of lipid droplets (Table 7.1) [10] (Fig. 7.3).

2.2 TNTs in Stem Cells

Embryonic stem (ES) cells, derived from the inner cell mass of mammalian blastocysts, have the ability to grow indefinitely while maintaining pluripotency. These properties have led to expectations that human ES cells might help to understand disease mechanisms and to design effective therapies. Several studies in this sense have shown different types of stem cells to develop TNTs as important transporters of cellular material. Mesenchymal stem cells (MSCs) are obtained from both fetal and adult tissues including bone marrow, adipose tissue and umbilical cord [19],

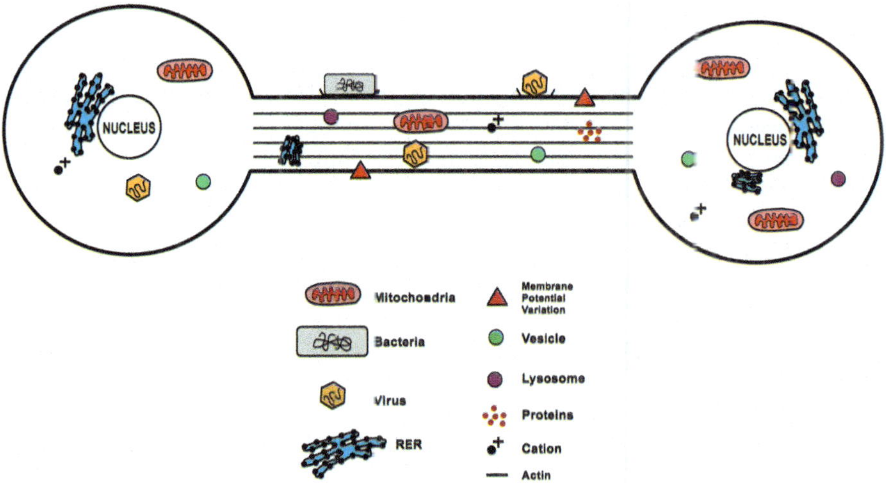

Fig. 7.3 TNT cargo transfer (Schematic representation of a TNT connecting two cells)

and are capable of multi-lineage differentiation into mesoderm-type cells. These cells are generally identified through a combination of physical, phenotypic and functional properties. MSCs must be plastic-adherent when maintained in standard culture conditions and must differentiate into osteoblasts, adipocytes and chondroblasts *in vitro*. New findings suggest that the ability of MSCs to alter tissue microenvironment via the secretion of soluble factors and the transfer of their own cellular components to neighboring cells may contribute significantly to tissue repair, although the underlying mechanisms remain uncertain.

The clinical use of human MSCs (hMSCs) requires the availability of a large number of functionally competent cells with a stable phenotype. However, *in vitro* expansion of hMSC in long-term culture is limited, and even ethical issues have been raised on the clinical use of this cell type. Inducing pluripotent status in somatic cells by direct reprogramming using transcription factors appears as a way to tackle these problems. Induced pluripotent stem (iPS) cells can be generated by the retrovirus-mediated transfection of four transcription factors, namely Oct3/4, Sox2, c-Myc and Klf4 [20]. These cells are indistinguishable from embryonic stem cells in morphology, proliferation, gene expression and teratoma formation. Furthermore, when transplanted into blastocysts, iPS cells can give rise to adult chimeras, which are competent for germ line transmission [21]. Several studies have shown that both cell types are capable of generating TNTs and transferring different cytoplasmic components from one cell to another. Analyses on the behavior of human hematopoietic progenitors when cultured with MSCs have revealed long and thin TNT-like processes to emerge from adjacent hematopoietic cells which are involved in intercellular communication [22].

2.3 TNTs in Nerve Tissue Cells

TNTs are also under active investigation for their role in maintaining nerve tissue homeostasis. Intense energy demands define the brain exceptional information storage and processing capacity, as well as its vulnerability. Astrocytes metabolize glucose to lactate, the major fuel for neurons. Glucose glycolytic pathways in neurons are redirected to the pentose phosphate shunt that serves to an efficient regeneration of NADPH and GSH, the latter a key free radical scavenger. Astrocytes have the capacity to deliver mitochondria to neurons in regions of high metabolic requirement, a process in which TNTs provide timely and critical support. Wang *et al.* [23] have reported TNT formation between astrocytes and neurons directly induced by H_2O_2 in the medium. Furthermore, both exosomes and TNTs are involved in the intercellular transfer of TDP-43 aggregates in human glioblastome U251 cells, a process similar to prion propagation [24].

2.4 TNTs in Immune Cells

Despite the number of functions ascribed to TNTs in various immune cells, overarching questions still remain regarding the biological significance of TNTs in immune cell function. Macrophages and dendritic cells have evidenced calcium flux propagated along TNTs independently of gap junctions or ATP release [17]. A study comparing cytokine production in co-cultures of primary B cell precursor acute lymphoblastic leukemia cells and primary mesenchymal stromal cells has shown an increase in several prosurvival cytokines such as IP10, IL-8 and MCP-1 when both cell types are connected by TNTs, which suggests that TNT communication modulates bone marrow microenvironment and enhances cell survival [25]. Also, natural killer cell activation has been proven to take place through receptor/ligand mechanisms in which target cells migrate toward natural killer cells using the TNT as a tether. This process generates tight cell contact leading to cell lysis and shows the biological relevance of TNTs [26].

2.5 TNTs and Cancer

Great interest has arisen in TNT cellular cargo in the context of human cancer. Studies on human prostate cancer DU145 cells have revealed filopodia and bridge structures facilitating vesicular traffic between cells. These structures were shown to have 100 nm to 5 μm diameter and 50–100 μm length, and a microtubular cytoskeleton and bead-like membrane bulges of 1–3 μm diameter [27]. Lou et al. [9] have demonstrated efficient transfer of cellular contents, including proteins, Golgi vesicles and mitochondria, between cells derived from several well established cancer cell lines. For the first time, they also demonstrated the *in vivo* relevance of these structures in humans, having effectively imaged nanotubes in intact solid tumors from patients. The existence of TNTs in solid tumors opens up a new avenue for targeted therapy and can be investigated as a means of delivering small molecule inhibitors or biological agents. Pasquier et al. [28] have described the spontaneous exchange of cytoplasmic material between endothelial cells or MSCs and multiple cancer cell lines. These authors have shown a preferential transfer of mitochondria from endothelial to cancer cells resulting in chemoresistance, which opens doors for direct intercellular exchange as a modulator of tumor phenotype by the stromal component. In addition, Thayanithy et al. [29] have reported direct cell-to-cell transfer of genetic material between tumor and stromal cells via TNT conduits. The work performed on osteosarcoma and ovarian cancers clearly showed intercellular tumor-tumor and tumor-stromal interaction via TNTs involving intercellular transfer of miRNAs, which could be of relevance in tumor formation, progression and recurrence.

3 Oxidative Stress

Oxidative stress (see Helmut Sies, in Chap. 1 in this volume) due to an increased content of reactive oxygen species (ROS) is one of the main etiological/pathological cellular mechanisms considered for various diseases, with the biochemical consequences of lipid peroxidation, protein oxidation and DNA damage [30]. For example, oxidative damage is involved in modifications of synaptic proteins as well in lipid peroxidation present in bipolar disorders [31]. As an example of the vicious cycle oxidative stress and cell damage constitute demyelinating lesions that lead to the generation of ROS, which contribute, among other factors, to the progression of oxidative stress and inflammation, both involved in a self-perpetuating circle [32]. In multiple sclerosis, a single prostaglandin has been postulated as a biomarker of oxidative stress in determining lipid peroxidation [33]. In turn, some degenerative brain disorders, such as Alzheimer's disease, are the product of energy dysregulation. During peak demand and as a consequence of insufficient energy supply, neurons promote a cascade of negative events that include an increase in free-radical mediated reactions and apoptosis, that is a typical consequence of oxidative stress, tau protein hyper phosphorylation and accumulation of extracellular β-amyloid plaques [34]. Cancer cells adapt to hypoxic conditions during progressive tumor cell growth by shifting the burden of energy metabolism from oxidative phosphorylation to glycolysis and continue to utilize glycolysis as the major energy source even in normoxic culture conditions [35]. In addition, Wang et al. [23] have shown that oxidative stress or serum starvation trigger selective TNT formation between stressed and unstressed astrocytes and postulated that a nanotube-mediated directional flow of healthy intracellular cargoes from unstressed to stressed cells exerts cell-sustaining effects.

3.1 Mitochondria

Mitochondria are essential organelles in plant and animal cells of prokaryotic origin and play a key role in the processes of oxidative phosphorylation and aerobic metabolism and also participate in glucose and fat metabolism, calcium signaling and triggering of apoptosis. Mitochondria are the major ROS producers and the main target of oxidative stress [36]; then mitochondria are considered involved in apoptotic cell death and in the aging process [37]. Mitochondria can transmit signals to the nucleus and initiate mitochondrial retrograde regulation leading to a bidirectional communication between these two subcellular structures. Studies employing a trans-mitochondrial hybrid system, mtDNA-depleted cancer cells as nuclear donors and non-cancerous cells as mitochondrial donors, have recently revealed that the oncogenic properties of an aggressive cancer cell line can be at least partially reversed by non-cancerous mitochondrial transfer and cross-talk through the suppression of several oncogenic pathways. This work constitutes an excellent tool to study specific effects of altered mitochondria under a defined nuclear background and reveals the importance of mitochondria in the development of certain diseases [38].

3.2 Mitochondrial Transfer Through TNTs

Mitochondrial transfer was first shown in 2006 to occur among cells *in vitro* and to have a physiological role in rescuing the respiration of deficient cells [39]. The authors reported that mitochondria were actively transferred from stem cells to recipient cells with nonfunctional mitochondria, and that the process resulted in a significant amelioration of aerobic respiration. Time lapse microscopy further demonstrated that MSCs in the co-cultures developed cytoplasmic projections directed toward the damaged cells and made contact with target cells, through long or thin cellular extensions. These experiments focused the attention of the research community on the possible role of mitochondrial transfer among cells and its possible physiological significance. Studies using MitoTrackerTM have further shown a dynamic transfer of mitochondria through TNTs in primary cultures of human renal tubular epithelial cells [40]. Work by our group on mitochondrial transport through MSC by TNTs has led us to estimate the speed of mitochondrial transport at 135–160 nm/min (Fig. 7.4).

Oxidative stress is closely related to mitochondrial dysfunction, as mitochondria are both generators of and targets for reactive species. Liu et al. [41] have determined that MSCs can repair post-ischemic endothelial cells with dysfunctional mitochondria by transferring functional mitochondria from healthy cells via TNT-like structures. The authors further prove that the formation of the TNT-like structures connecting the endothelial cells to the MSCs is dependent on F-actin

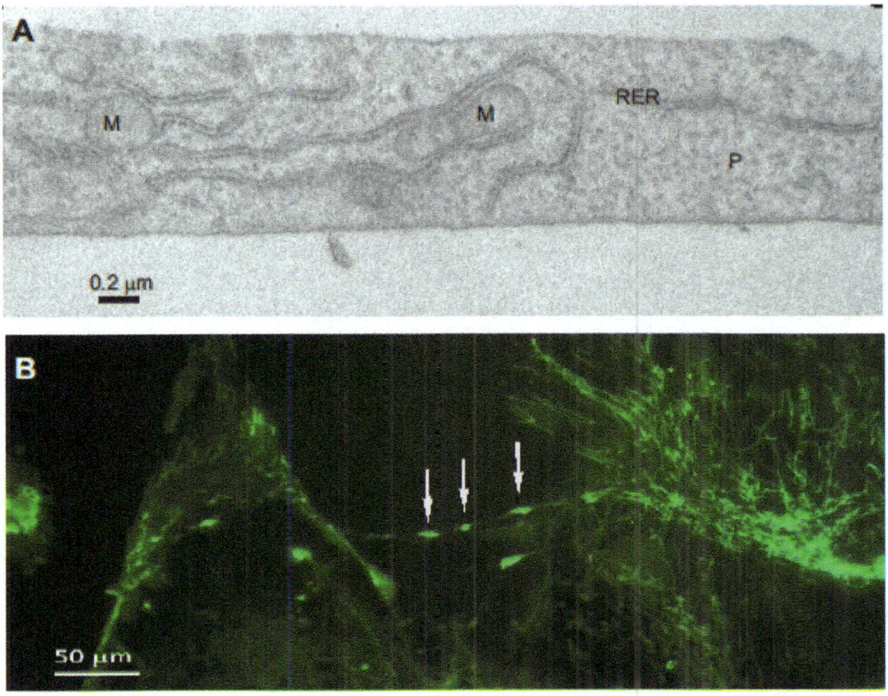

Fig. 7.4 Mitochondria inside TNTs. (**a**) Electron microscopy of thick TNT containing: *M* mitochondria, *RER* rough endoplasmic reticulum, *P* polyribosomes. (**b**) Mitochondria detection through Mito Tracker green. The *arrows* show mitochondria moving along a TNT

polymerization. MSCs and endothelial cells exchanged mitochondria by TNT-like structures. In particular, the stress induced on endothelial cells by O_2 and glucose deprivation, generates mitochondrial unidirectional transfer by the TNT-like structures, from the MSCs to the injured endothelial cells; mitochondrial transfer resulted in aerobic respiration recovery and in prevention of apoptosis. In addition, Han et al. [42] have recently employed an *in vitro* model of ischemia/reperfusion in H9c2 cardiomyocytes co-cultured with MSCs and were able to show that MSCs protect H9c2 cells against the apoptosis induced by ischemia/reperfusion and significantly restores impaired mitochondrial function. During this process, MSCs make contact with injured H9c2 cells via TNTs through which MSC mitochondria were transferred. Finally, in neurodegenerative diseases, when axonal transport alone cannot meet the dynamic supply and maintenance needs of terminal axons at great distances from neuron soma, as happens in Alzheimer's disease, astrocytes become efficient providers of mitochondria at critical locations through TNTs [34].

4 Conclusions

While widespread antioxidant therapies focus on reducing ROS damage through nutritional tools, the transfer of healthy mitochondria may rescue damaged cells and thus constitute an alternative approach. In this context, stress-induced TNT formation may be regarded as a defense mechanism of injured cells. In other words, under stress conditions such as oxidative stress, cells suffer mitochondrial damage which inevitably leads to cell death. And even if stressed cells are induced to form TNTs, these nanotubes fail to restore cell health in a stress-only environment. However, if stressed cells are in contact with healthy cells, TNTs can help transfer healthy mitochondria and thus rescue stress cell viability. In this context, MSCs become a promising therapeutic tool as donors of healthy mitochondria to injured cells. In summary, further research on the protective effects of stem cells through TNT-mediated mitochondrial transfer may provide novel insights into possible therapeutical approaches to different pathologies.

Acknowledgements The authors would like to thank Dr. N. Villalba, Dr. L. Fiore, Dr. R.J. Gelpi, Dr. A. Boveris and Dr. J.J. Poderoso for authorizing the use of photomicrographs obtained in ongoing collaborative work.

References

1. Kumar N, Gilula NB (1996) The gap junction communication channel. Cell 84:381–388
2. Denzer K, Kleijmeer MJ, Heijnen HF et al (2000) Exosome: from internal vesicle of the multivesicular body to intercellular signaling device. J Cell Sci 19:3365–3374
3. Ramirez-Weber FA, Kornberg TB (2000) Signaling reaches to new dimensions in Drosophila imaginal discs. Cell 103:189–192
4. Rustom A, Saffrich R, Markovic I et al (2004) Nanotubular highways for intercellular organelle transport. Science 303:1007–1010
5. Gerdes H-H, Bukoreshtliev NV, Barroso JF (2007) Tunneling nanotubes: a new route for the exchange of components between animal cells. FEBS Lett 581:2194–2201
6. Chinnery HR, Pearlman E, Mc Menamin PG (2008) Cutting edge: membrane nanotubes in vivo. A feature of MHC class II+ cells in the mouse cornea. J Immunol 180:5779–5783
7. Gerdes H-H, Rustom A, Wang X (2013) Tunneling nanotubes, an emerging intercellular communication route in development. Mech Develop 130:381–387
8. Seyed-Razavi Y, Hickey MJ, Zuffová L et al (2013) Membrane nanotubes in myeloid cells in adult mouse cornea represent a novel mode of immune cells interaction. Immunol Cell Biol 91:89–95
9. Lou E, Fujisawa S, Barlas A et al (2012) Tunneling nanotubes: a new paradigm for studying intercellular communication and therapeutics in cancer. Commun Integr Biol 5:399–403
10. Astanina K, Koch M, Jüngst C et al (2015) Lipid droplets as a novel cargo to tunneling nanotubes in endotelial cells. Sci Rep 5:11453. doi:10.1038/srep11453
11. Sowinski S, Jolly C, Berninghausen O et al (2008) Membrane nanotubes physically connect T cells over long distances presenting a novel route for HIV-1 transmission. Nat Cell Biol 10:211–219
12. Abounit S, Zurzolo C (2012) Wiring through tunneling nanotubes: from electrical signals to organelle transfer. J Cell Sci 125:1089–1098

13. Lokar M, Iglic A, Veranic P (2010) Protruding membrane nanotubes: attachment of tubular protrusions to adjacent cells by several anchoring junctions. Protoplasma 246:81–87
14. Önfelt B, Hedvetzki S, Benninger RKP et al (2006) Structurally distinct membrane nanotubes between human macrophages support long-distance vesicular traffic or surfing of bacteria. J Immunol 177:8476–8483
15. Schiller C, Diakopoulos KN, Rohwedder I et al (2013) LST1 promotes the assembly of a molecular machinery responsible for tunneling nanotube formation. J Cell Sci 126:767–777
16. Smith IF, Shuai J, Parker I (2011) Active generation and propagation of Ca^{2+} signals within tunneling membrane nanotubes. Biophys J 100:L37–L39
17. Watkins SC, Salter RD (2005) Functional connectivity between immune cells mediated by tunneling nanotubes. Immunity 23:309–318
18. Koyanagi M, Brandes RP, Haendeler J et al (2005) Cell-to-cell connection of endothelial progenitor cells with cardiac myocytes by nanotubes: a novel mechanism for cell fate changes? Circ Res 96:1039–1041
19. Bernardo ME, Locatelli F, Fibbe WE (2009) Mesenchymal stromal cells. Ann NY Acad Sci 1176:101–117
20. Takahashi K, Yamanaka S (2006) Induction of pluripotent stem cells from mouse embryonic and adult fibroblast cultures by defined factors. Cell 126:663–676
21. Maherali N, Sridharan R, Xie W et al (2007) Directly reprogrammed fibroblasts show global epigenetic remodeling and widespread tissue contribution. Cell Stem Cell 1:55–70
22. Freund D, Bauer N, Boxberger S et al (2006) Polarization of human hematopoietic progenitors during contact with multipotent mesenchymal stromal cells: effects on proliferation and clonogenicity. Stem Cell Dev 15:815–829
23. Wang Y, Cui J, Sun X, Zhang Y (2011) Tunneling-nanotube development in astrocytes depends on p53 activation. Cell Death Differ 18:732–742
24. Ding X, Ma M, Teng J et al (2015) Exposure to ALS-FTD-CSF generates TDP-43 aggregates in glioblastoma cells through exosomes and TNTs-like structure. Oncotarget 6:24178–24191
25. Polak R, de Rooij B, Pieters R et al (2015) B-cell precursor acute lymphoblastic leukemia cells use tunneling nanotubes to orchestrate their microenvironment. Blood 126:2404–2414
26. Chauveau A, Aucher A, Eissmann P et al (2010) Membrane nanotubes facilitate long-distance interactions between natural killer cells and target cells. Proc Natl Acad Sci U S A 107:5545–5550
27. Vidulescu C, Clejan S, O'Connor KC, (2004) Vesicle traffic through intercellular bridges in human prostate cancer cells. J Cell Mol Med 8:388–396
28. Pasquier J, Guerrouahen BS, Al Thawadi H et al (2013) Preferential transfer of mitochondria from endothelialto cancer cells through tunneling nanotubes modulates chemoresistance. J Transl Med 10:11:94
29. Thayanithy V, Dickson EL, Steer C et al (2014) Tumor-stromal cross talk: direct cell-to-cell transfer of oncogenic microRNAs via tunneling nanotubes. Transl Res 164:359–365
30. Dvash E, Rubistein M. (2016) A surprising mediator of oxidative DNA damage. Cell Cycle doi:10.1080/15384101.2016.1144989
31. Andreazza AC, Wang J-F, Salmasi F et al (2013) Specific subcellular changes in oxidative stress in prefrontal cortex from patients with bipolar disorder. J Neurochem 127:555–561
32. Ortiz GG, Pacheco-Moisés FP, Bitzer-Quintero OK et al (2013) Immnology and oxidative stress in multiple sclerosis: clinical and basic approach. Clin Devel Immunol 708659. http://dx.doi.org/10.1155/2013/708659
33. Mir F, Lee D, Ray H, Sadiq SA (2014) CSF isoprostane levels are a biomarker of oxidative stress in multiple sclerosis. Neurol Neuroimmunol Neuroinflamation 1:21. doi:10.1212/NXI.0000000000000021
34. Engel P (2014) Does metabolic failure at the synapse cause Alzheimer's disease? Med Hypoth 83:802–808
35. Ma Y, Bai RK, Trieu R, Wong LJ (2010) Mitochondrial dysfunction in human breast cancer cells and their transmitochondrial cybrids. Biochim Biophys Acta 1797:29–37

36. Ernster L (1993) Lipid peroxidation in biological membranes: Mechanisms and implications. In: Yagi K (ed) Active Oxygen, Lipid Peroxides and Antioxidants. CRC Press, Boca Raton, pp 1–38
37. Lenaz G, Bovian C, Formiggini G et al (1999) Mitochondria, oxidative stress, and Antioxidant defences. Acta Biochim Polonica 46:1–21
38. Kaipparettu BA, Ma Y, Park JH et al (2013) Crosstalk from non-cancerous mitochondria can inhibit tumor properties of metastatic cells by suppressing oncogenic pathways. PLoS One 8:e61747. doi:10.1371/journal.pone.0061747
39. Spees JL, Olson SD, Whitney MJ et al (2006) Mitochondrial transfer among cells can rescue aerobic respiration. Proc Natl Acad Sci U S A 103:1283–1288
40. Domhan S, Ma L, Tai A et al (2011) Intercellular communication by exchange of cytoplasamic material via tunneling nantotube like estructures in primary human renal epithelial cells. PLoS One 6:e21283. doi:10.1371/journal.pone.0021283
41. Liu K, Ji K, Guo L et al (2014) Mesenchymal stem cells rescue injured endothelial cells in an in vitro ischemia–reperfusion model via tunneling nanotube like structure mediated mitochondrial transfer. Microvasc Res 92:10–18
42. Han H, Hu J, Yan Q et al (2016) Bone marrow-derived mesenchymal stem cells rescue injured H9c2 cells via transferring intact mitochondria through tunneling nanotubes in an *in vitro* simulated ischemia/reperfusion model. Molec Med Rep 13:1517–1524
43. Kadiu I, Gendelman H (2011) Human immunodeficiency virus type 1 endocytic trafficking through macrophage bridging conduits facilitates spread of infection. J Neuroimmune Pharmacol 6:658–675
44. Arkwright P, Luchetti F, Tour J et al (2010) Fas stimulation of T-lymphocytes promotes rapid intercellular exchange of death signals via membrane nanotubes. Cell Res 20:72–88
45. Gousset K, Schiff E, Langevin C et al (2009) Prions hijack tunneling nanotubes for intercellular spread. Nat Cell Biol 11:328–336
46. Wang X, Veruki M, Bukoreshtliev N et al (2010) Animal cells connected by nanotubes can be electrically coupled through interposed gap-junction channels. Proc Natl Acad Sci U S A 107:17194–17199
47. Cselenyak A, Pankotai E, Horvath EM et al (2010) Mesenchymal stem cells rescue cardiomyoblasts from cell death in an in vitro ischemia model via direct cell-to-cell connections. BMC Cell Biol 20:11–29
48. Plotnikov EY, Khryapenkov TG, Galkina SI et al (2010) Cytoplasm and organelle transfer between mesenchymal multipotent stromal cells and renal tubular cells in coculture. Exp. Cell Res 316:2447–2455
49. Mi L, Xiong R, Zhang Y (2011) Microscopic observation of the intercellular transport of cdte quantum dot aggregates through tunneling-nanotubes. J Biomater Nanobiotechnol 2:173–180
50. He K, Shi X, Zhang X et al (2011) Long-distance intercellular connectivity between cardiomyocytes and cardiofibroblasts mediated by membrane nanotubes. Cardiovasc Res 92:39–47
51. Acquistapace A, Bru T, Lesault PF et al (2011) Human mesenchymal stem cells reprogram adult cardiomyocytes toward a progenitor-like state through partial cell fusion and mitochondrial transfer. Stem Cells 29:812–824
52. Yasuda K, Khandare A, Burianovsky L et al (2011) Tunneling nanotubes mediate rescue of prematurely senescent endothelial cells by endothelial progenitors: exchange of lysosomal pool. Aging 3:597–608
53. Lou E, Fujisawa S, Morozov A et al (2012) Tunneling nanotubes provide a unique conduit for intercellular transfer of cellular contents in human malignant pleural mesothelioma. PLoS One 7:e33093
54. Wang X, Bukoreshtliev NV, Gerdes HH (2012) Developing neurons form transient nanotubes facilitating electrical coupling and calcium signaling with distant astrocytes. PLoS One 7:e47429
55. Bénard M, Schapman D, Lebon A et al (2015) Structural and functional analysis of tunneling nanotubes using CW STED and gconfocal approaches. Biol Cell 107:419–425
56. Naphade S, Sharma J, Gaide-Chevronnay HP et al (2015) Brief reports: lysosomal cross-correction by hematopoietic stem cell-derived macrophages via tunneling nanotubes. Stem Cells 33:301–309

Chapter 8
Human Adaptation to Life at High Altitude

Gustavo F. Gonzales, Dulce E. Alarcón-Yaquetto, and Alisson Zevallos-Concha

Abstract Living at high altitude (HA) represents a daily challenge that over two hundred million people worldwide have to face. Populations living at HA are distributed mainly in Asia, Africa and America and they live in these settling with different periods of antiquity and evolution. Permanent life at HA is associated with a pathology unique in these places known as chronic mountain sickness or lack of adaptation to live at HA, which is characterized by excessive erythrocytosis. It has been reported that oxidative stress is increased at people acutely and chronically exposed at HA hypoxia. This may explain some features of the adaptation of highlanders. The present review summarizes findings related to different strategies of adaptation in populations living at HA. As features of human adaptation at HA we review data related to birth weight, gestational age, preeclampsia, hemoglobin and chronic mountain sickness in populations at HA and how gene evolution drives adaptation.

Keywords High altitude • Hypoxia • Chronic mountain sickness • Monge's disease • Himalayas • Andes

1 Introduction

Hypoxia tolerance seems to be a polygenic and mechanistically conserved trait in which selection for hypoxia tolerance acts on standing genetic variations in similar genes and pathways present in organisms that diverged by hundreds of millions of years [1]. First modern humans emerged in East Africa [2]. The most antique humans, Homo sapiens, dated 200000 years, was found in Ethiopia [3, 4]. These Homo sapiens from African origin migrated out 60000 years ago to spread worldwide with different environmental conditions including exposures to high altitude (HA) [5]. The main and more antique places in which humans migrated to HA are

G.F. Gonzales (✉) • D.E. Alarcón-Yaquetto • A. Zevallos-Concha
Instituto de Investigaciones de la Altura, and Department of Biological and Physiological Sciences, Faculty of Sciences and Philosophy, Universidad Peruana Cayetano Heredia, Distrito de Lima, Peru
e-mail: gustavo.gonzales@upch.pe

© Springer International Publishing Switzerland 2016

R.J. Gelpi et al. (eds.), *Biochemistry of Oxidative Stress*, Advances in Biochemistry in Health and Disease 16, DOI 10.1007/978-3-319-45865-6_8

the Himalayas (Asia), Ethiopia (Africa) and the Andes (South America). Genetic sequences (signatures) date to 30000 years ago in the ancient Tibet, during the Upper Paleolithic period, when the first human migration wave reached the Tibetan plateau. In 2013, the Tibetan population reached 5 million inhabitants living over 3500 m altitude [6].

Migration to America occurred later probably during last glaciation. Humans migrated from Asia to America about 16500 years ago through the Bering Strait [7]. In the late Pleistocene humans entered South America from a single migration wave [8] through the isthmus of Panama and split afterwards into two groups, one moving southward into the central and south Andes and a second group migrating eastwards populating the Amazonian region [9]. The earliest settlements in South America are dated 14000 years ago [10]. In the Pucuncho Basin at 4355 m in the Southern Peruvian Andes, where the highest-altitude Pleistocene archaeological sites were observed, have an antiquity of 12000 years [11]. During the 16th century, Spaniards conquered the Andean region and important admixture between Spaniards and native genes occurred [12]. It is unknown how this genetic admixture influenced the adaptation process to live at HA. In the Andes, the most antique population is that of Aymara origin, living at HA for more than 350 generations [13]. In Peru, these populations are mainly located in Puno at 3800 m [14]. At present, two hundred millions live permanently at HA, defined as altitudes >2500 m. From these, 55 million people live in Latin America [15]. Publications mainly deal with the effect of altitude >2000 m on physiological parameters. However, there are evidences that altitudes between 1500 and 2000 m affect human physiology [16, 17].

Populations living at HA are distributed mainly in Asia, Africa and America with different periods of antiquity in their places. The most antique populations are located in the Ethiopian highlands of the Semien mountains [18, 19], in the Tibetan plateau and the Himalayan valleys [20, 21], and in the Southern Andes in South America [13, 14, 22]. Intermediate level of antiquity are in the Peruvian Central Andes [14, 22], less antiquity in Rocky Mountains in United States and much less (<80 years) in the ethnic group Han in the Himalayas. Populations in which adaptation to live at HA is well documented are Tibetans, Ethiopians, and Andeans. These three populations with high multigenerational residence at HA show genes positively associated with adaptation [23–35].

It is important to highlight the contribution of the Peruvian school in the studies of life at HA. We recognize two big figures significantly contributing to research and medical training about life at HA: Carlos Monge-Medrano and Alberto Hurtado. Both are recognized worldwide for their contributions on human life adaptation to HA.

At present, it seems that in the ancestors of the modern Tibetans, there was a powerful selective pressure favoring variants in genes to allow adaptation to live at HA. The main manifestation of this selection is the limited erythropoiesis in Tibetans exposed to the hypoxia of HA [36]. In Tibetans, the hypoxia pathway gene EPAS1 seems associated with the low erythropoietic response at HA. This gene has an unusual haplotype structure and is explained by introgression of DNA from Denisovan or Denisovan-related hominids into humans. This suggests the mixture of Homo sapiens with other hominid species and that this admixture provided a genetic variation that allowed humans to adapt to the HA environments [5]. Tibetans are a mixture of two ancestral populations, one related to the East Asians and the other to the Sherpas [34]. However, evidence suggests that Tibetans are the ancestral populations of the Sherpas, whose adaptive traits for HA were inherited from their ancestors in Tibet [37]. Actually, evidence suggest that EPAS1 and EGLN1 genes are mostly likely responsible for HA adaptation and closely related to low hemoglobin concentration in Tibetans [38].

In the Himalayas as in the Andes there are populations co-habiting with differences antiquities at HA. Certainly, in the Tibet plateau there are populations of Tibetans residing there for 30000 years with the Chinese ethnic group Han with no more than 80 years living there. Interestingly, hemoglobin values are higher in Hans than in Tibetans [39]. In the Peruvian Southern Andes, the populations with highest antiquity at HA are at Cusco (Juliaca, Puno) and in the Peruvian Central Andes another group with less antiquity is at Huancayo (Cerro de Pasco) [14, 22]. In Peru, hemoglobin levels were lower in the Southern than in the Central Andes [40], as it happens in the Himalayas. The genes that allow lower hemoglobin concentration in the Southern Andes, compared with the Central Andes, have not been reported. Permanent life at HA is associated with a pathology unique in these places known as chronic mountain sickness (CMS) or Monge's disease, characterized by excessive erythrocytosis (EE; Hb ≥ 21 g/dl in men and ≥19 g/dl in women) [41]. Enhanced erythropoiesis is regulated by androgens and erythropoietin (EPO), by endocrine, autocrine and paracrine mechanisms [15, 42]. Increased ventilation and oxidative stress and hypoxemia have been proposed as the main CMS symptoms [43]. Sleep disordered breathing (SDB) and nocturnal hypoxemia are more severe in CMS patients than in lowlanders and are associated with systemic and pulmonary vascular dysfunction [44]. However, the fact that hemoglobin increases in EE without further changes in SpO2 (hemoglobin saturation) suggest that ventilation has a minor effect in EE (Fig. 8.1).

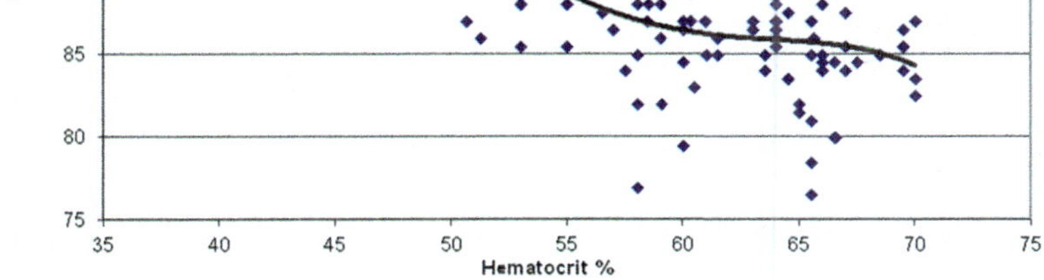

Fig. 8.1 Relationship between hematocrit and oxygen saturation in populations at low altitude (150 m) and high altitude (4340 m). Higher hematocrit (>63 %) is observed without further reduction of hemoglobin saturation (SpO_2)

This review summarizes the strategies of adaptation in populations living at HA, considering birth weight, gestational age, preeclampsia, hemoglobin and chronic mountain sickness in populations at HA and how gene may force adaptation.

2 The Challenge of Living at High Altitude

People living at HA have to endure a lower availability of oxygen for gas exchange in the lungs [45] because as altitude rises, air partial pressure of oxygen (pO_2) decreases in a proportional manner. There is a 40 % less pO_2 at 4500 m than at sea level [46]. The implications of the dwindled oxygen concentration are: headaches, dizziness, nausea, and sleeping disturbances. These are the common symptoms of people suddenly exposed to hypoxia, and constitutes the acute mountain sickness. Populations that lived in these extreme conditions for thousands of years have developed physiological and genetic adaptations to endure these harsh environments. An acute exposure to hypoxia produces increased heart rate and hyperventilation, that constitute accommodation mechanisms for a sudden exposure. A long term exposure to HA is accompanied by an increase in erythropoiesis (higher red blood cell production) and is called acclimatization. Millennial generational exposure to hypoxia produced genetic variants in Tibetans, Ethiopians and Aymaras populations.

At HA, the levels of reactive oxygen (ROS) and nitrogen species (RNOS) increase and the damage produced is as severe as the altitude exposed [47, 48]. In the last decades, the paradigm of oxidative stress at HA has changed. First, it was believed that the low availability of oxygen leads to a low production of reactive oxygen and nitrogen species (RNOS) but this was later considered not right when some markers of oxidative stress such as the urinary level of 8-iso-prostaglandin F2 α (8 iso $PGF_{2\alpha}$) were higher in people in Cerro de Pasco (4300 m) than in Lima (150 m) [47]. Apparently, the oxidative stress produced by long term exposure to HA hypoxia activates antioxidant defenses which contribute to the adaptation process [49]. Despite the low availability of oxygen, there are factors that might explain oxidative stress at HA. These are higher levels of angiotensin II [50] and of stimulation of the sympathetic nervous system. Also, UV light and its higher intensity at HA contributes to higher levels of oxidation [51, 52]. Alongside the increased oxidative stress and other challenges that habitants of the altitude have to bear, the most significant risk affects to the mother and child pair with low birth weight and preeclampsia.

3 Oxidative Stress at High Altitude

Life at HA occurs in an environment in which oxygen pressure is reduced and that results in hypoxemia. This situation is accompanied by oxidative stress [53] in acute [54] and chronic exposure to hypoxia [55]. The impact of hypoxia on oxidative stress is not the same under different hypoxic conditions. Healthy people living at HA have a moderate increase in oxidative-nitrosative stress [56]. Markers of reactive oxygen species (ROS) production, such as 8-iso-PGF2alpha, plasma TBARS and total glutathione are greater in people at HA compared with those at low altitude [47]. It has been proposed that at HA, ROS production is increased by decreased mitochondrial PO_2 [57]. However, the association between environmental hypoxia and oxidative stress is not free of experimental controversy. Rats from a hypobaric chamber at equivalent 4500 m, show decreased spontaneous *in situ* liver chemiluminescence, that means decreased singlet oxygen levels and decreased oxidative stress [58].

People with EE at HA had higher levels of oxidative stress than controls [59] correlated with hemoglobin concentrations [43]. The moderate increase in oxidative-nitrosative stress in normal highlanders does not affect vascular function, but the marked increase in oxidative stress in patients with EE seems to contribute to systemic vascular dysfunction [56]. Since hemoglobin is increased at HA, the World Health Organization (WHO) recommended the correction of the cut-off point of hemoglobin to define anemia at HA. The application of this correction produces an increase in the ratios of anemia at HA [60]. These people are treated with iron to reduce anemia. However, people with iron supplementation at HA showed reduced levels of endogenous antioxidants with increased lipid peroxidation products and protein carbonyl content in the lungs [61]. There are populations living in a same setting at HA but with different generational antiquity. At this time, there is a limited number of studies comparing oxidative stress in these populations. Interestingly, ox-LDL levels are higher among elderly Tibetan highlanders compared with those among elderly Han. As ox-LDL levels increase insulin resistance and arteriosclerosis [54], the more adapted subjects at HA are in a higher risk of diabetes and of metabolic syndrome.

It is uncertain how other markers of oxidative stress are affected in HA. An adaptation against oxidative stress may imply the existence of single nucleotide polymorphism (SNP) affecting gene function. For instance, Andean natives have the gene FAM213A with antioxidant properties and that could be useful in HA [62]. Deer mice from HA and human highlanders have genes involved in energy metabolism (Ckmt1, Ehhadh, Acaa1a) and angiogenesis (Notch4) that are more expressed [63]. The expression of regulators of the mitochondrial biogenesis, PGC-1α (PPARA) and mitochondrial transcription factor A (Tfam) positively correlate with muscle oxidative phenotype [63]. Tibetans have a PPARA haplotype positively associated with a decrease in fat oxidation [64]. Life at HA has been associated with low arterial blood pressure and low rate of hypertension, probably as effects of increased nitric oxide (NO) availability [65] and a lower carotid wall/lumen ratio in highlanders compared to lowlanders at sea level and HA [66]. Tibetans living at HA show elevated NO levels in the lung, plasma, and red blood cells that were at least double and up to an order of magnitude greater than in other populations regardless of altitude [65]. It is suggested that polymorphism of the gene NOS3 participates in this process [67]. CYBA (cytochrome b-245α) gene encodes a protein, which is a critical component of the $NADPH_2$-oxidase system and participates in the production of superoxide anion, which in turn causes oxidative stress [57]. CYBA risk alleles were positively correlated with markers of oxidative stress associated with hypertension without differences in HA respect to sea level [68].

In conclusion, one of the deleterious effects of HA induced hypoxia is oxidative stress. Effects of a higher magnitude are observed during acute exposure to hypoxia and more moderate effects in the situation of HA natives. Most of the studies deal with acute exposure to HA and a lower number of publications refer to the populations living at HA.

4 Epidemiology of Chronic Mountain Sickness (CMS)

Epidemiological studies in populations living at HA show that CMS varies with altitude, sex, age and generational time of residence [69–71]. Percentages of CMS tend to be higher in populations with lowest multigenerational residence at HA (the ethnic group Han in the Himalayas, inhabitants of the Peruvian Central Andes and of the Rocky Mountains in Colorado, USA) and lower in population with highest multigenerational residence at HA (Tibetans, Ethiopians, and Peruvians and Bolivian living in the Southern Andes). Most of the recent studies use the Qinghai score formulated by a scientific consensus to diagnose CMS [41], with EE as the main criterion. Using the Qinghai score, the rate of CMS in Tibetans is about 4 % [72]. In Himachal Pradesh (India) was 6.17 % at an altitude of 3281 m [70]. In Ladakh, in the Northern State of Jammu and Kashmir (India) at 4647 m was 4.87 % [73]. Chinese Han males that migrated and stayed at the Qinghai-Tibet plateau (3700–5000 m) for 2–96 months had a prevalence of CMS of 17.8 % [71]. The percentage of CMS among adult natives of Spiti Valley in the Indian Himalayas, at altitudes of 3000–4200 m free of cardiorespiratory diseases, was 28.7 %, and higher in women than in men (36.6 % vs. 15.7 %). The authors consider that the higher prevalence of CMS in women is due to Westernized life styles [69]. Although it is argued that CMS is almost absent in Ethiopians, no data is available. One report shows a lower CMS ratio in Ethiopians compared with people from the Peruvian Central Andes [74].

Differences in adaptation have been described for population living in the same altitude but with different generational times, as it happens between Hans and Tibetans in the Himalayas plateau [75], and between Peruvians of the Southern Andes and those in the Central Andes [14]. Apparently, strategies of adaptation to HA of populations of Tibetans are similar to the populations in the Southern Andes. In Peru, prevalence of EE in the Southern Andes was 4.5 % [76], lower to that observed in the Central Andes and similar to that observed in the Tibetans [72]. Moreover, percentage of pregnant women with hemoglobin >14.5 g/dL were significantly lower in Southern than in the Central Andes [40]. In Cerro de Pasco (4340 m, in the Central Andes) the ratio of EE was 15.6 % in the population aged 20–69 years [77]. In Junín (4100 m), in Peruvian Central Andes, the CMS ratios according to the Qinghai score were 28.6 % in men and 29.5 % in women [78]. The data suggest that population adaptation to HA depend on the generational time of residence

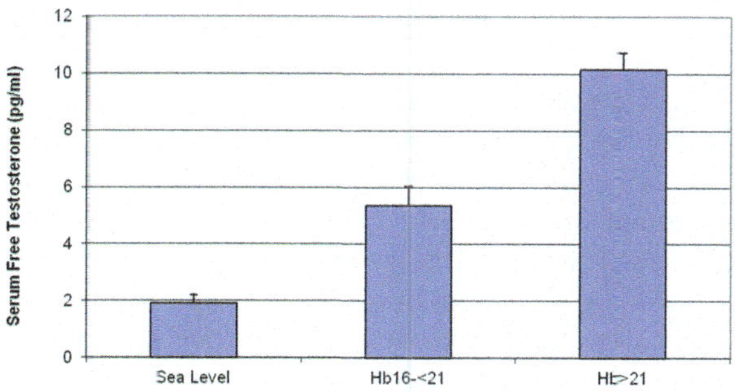

Fig. 8.2 Serum levels of free testosterone in men at sea level and at HA (with Hb 16- < 21 g/dl) and Hb > 21 g/dl) in Peru

at highlands. The ratio of CMS was similar in the Peruvian Southern Andes and in Tibetans. In the Peruvian Central Andes, there was an association between CMS and a SNP in the sentrin-specific protease 1 (SENP 1) gene with a different expression in Andean highlanders with and without CMS [79]. SENP 1 increases androgen receptor levels that would promote EE [15]. In the Tibetans, variants of the genes CYP17A1 and CYP2E1 are associated with EE [80]. CYP17A1 is responsible for testosterone synthesis. Higher androgen levels in natives of the Central Andes [81] and lower normal salivary testosterone values were reported in Aymara men in Bolivia in the Southern Andes, associated with high and low hemoglobin, respectively [82]. However, in the Central Andes, higher testosterone levels were associated with EE (Hb > 21 g/dl) (Fig. 8.2).

Men with African ancestry show higher serum testosterone [83], and less CAGn repeats, interpreted as higher androgen receptor levels-activity [84]. High androgen activity could be important for acclimatization; high serum testosterone increases with increasing hemoglobin values. Elevated testosterone has been observed during acute exposure to HA in rats [85] and humans [86]. In conclusion, adaptation to HA in terms of hemoglobin values or ratios of CMS are associated with genes regulating androgen activity.

5 Physiological Adaptations to Hypoxia: Birth Weight, Gestational Age and Neonatal Mortality

Important criteria for adaptation to HA are intrauterine fetal growth and survival of neonates. Increasing altitude was associated with reduced birth weight [87–89], that seems due to intrauterine growth restriction [90–92], rather than to pre-term delivery [93]. Reduction in fetal growth at HA goes parallel to high maternal hemoglobin [94]. In non-anemic pregnant women at HA, an increase of hemoglobin to values >14.5 g/dl decreased birth weight [95]. Normally, fetal growth is protected by higher uterine artery blood flow in Andeans at HA. When lower uterine blood flow occurs, as in early-onset pre-eclampsia or in descendants from European ancestry, fetal hypoxia restrict fetus growth [96], associated with high maternal hemoglobin levels [97].

In the Southern Andes populations with higher (>50 %) Aymara proportion have higher birthweight than those with less (<50 %) Aymara proportion (Fig. 8.3). Similarly, Tibetans have higher birth weight that Hans living in the same place [91]. Maternal PRKAA1 and EDNRA genotypes are associated with birth weight and PRKAA1 with uterine artery diameter and metabolic homeostasis at HA [98]. EPAS1, by regulating lipoxygenase expression, may play roles in the development of amnion and in birth weight in Tibetan newborns at HA [99]. HIF-targeted genes, ENDRA, PRKAA1 and NOS2A, in people of Southern Andes suggest that these genes have a role in adaptation to HA [24]. ENDRA is expressed in vascular smooth muscle and encodes a vasoconstrictor with action mediated through endothelin. PRKAA1 functions as an energy sensor under ATP-deprived conditions. NOS2 with other nitric oxide synthases, generates nitric oxide (NO) that is a potent vasodilator that reduces arterial blood pressure [24]. As discussed above, the process of adaptation to HA includes management of oxidative stress. Pregnancy and altitude increase catalase and superoxide dismutase activity in higher proportion in Andeans than in European women. SOD was lower in mothers of SGA age infants. It was suggested that an elevated antioxidant activity protects against altitude-associated illness in Andeans [100].

Prenatal and postnatal mortality rose with increasing elevation and were threefold higher across all altitudes in Hans than in Tibetans [92]. Similarly, in Peru stillbirth was 3 times higher at 4340 m than at sea level [101]. Inhabitants of the Southern Andes show lower stillbirths than of the Central Andes [102]. The lower

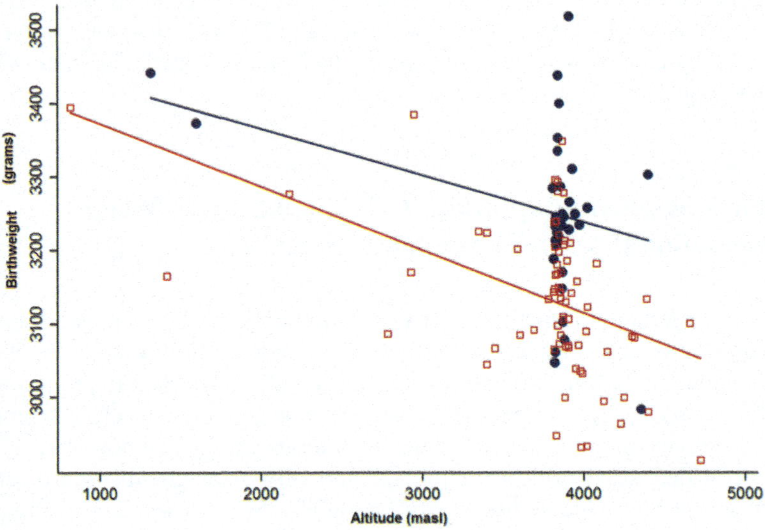

Fig. 8.3 Birth weight and altitude in two population from Puno: >50 % Aymaran (*blue*) and (2) <50 % Aymaran (*red*) component

Table 8.1 Hemoglobin changes during pregnancy and Odds Ratio (OR) crude or OR adjusted by maternal age, parity and altitude (n = 25,079 pregnant women)

Δ Hb	n	%	ORc	p	ORa	p	IC 95 %
−2	557	7.69	1.25	0	1.14	0.04	1.06 1.307
−2	385	6.08	1.04	0.51	0.99	0.96	0.865 1.147
0	450	5.73	1		1		
1	45	5.59	1.38	0.04	1.44	0.02	1.048 1.984
2	14	9.72	1.95	0.02	1.97	0.02	1.121 3.457

perinatal mortality in native mothers than in migrant mothers in Puno (3800 m) confirms the protective effect of multigenerational life at HA [103]. Mitochondrial DNA haplogroup B was associated with reduced neonatal death and decreased incidence of low birth weight in the Southern Andes [104]. The B mitochondrial DNA haplogroup rose in Asia 50000 years ago and is frequent in Chinese and Tibetan population but absent in Africans. Pre-eclampsia (PE) and gestational hypertension are more frequent at HA than at sea level [90, 105]. However, in Peru, incidence of pregnancy-induced hypertension was lower at HA than at sea level [97]; the discrepancy seems due to that PE is associated with increased hemoglobin levels in HA [97] (Table 8.1). Higher PE rates were higher in non-Tibetan than in Tibetans from Lhasa [106]; mothers at HA with normal or low hemoglobin values are protected against PE, whereas mothers with high hemoglobin (>14.5 g/dL) are in risk of PE. Another evidence of adaptation, is the lower rate of congenital malformations in the Southern Andes compared to those in the Central Andes [103, 107]. In conclusion, data related to maternal and perinatal health revealed that population with longer antiquity at HA have lower ratios of SGA, PE, malformations, and pre- and postnatal mortality.

6 Blood Hemoglobin in High Altitude Populations

Tibetans do not show an increasing effect of altitude on hemoglobin [38, 39] but Andean populations are distinct and show an increase in hemoglobin as altitude increases. However, this statement has been confronted. In fact, populations at the Southern Andes have less hemoglobin than those in the Central Andes [108]. Maternal hemoglobin in Puno at 3800 m in the Southern Andes are slightly lower than in Huancavelica at 3600 m in the Central Andes (Fig. 8.4). It seems that the lack of association of hemoglobin with altitude in Tibetans is not exclusive for that population. Multigenerational residence at HA is required to show an adaptive reduction in hemoglobin.

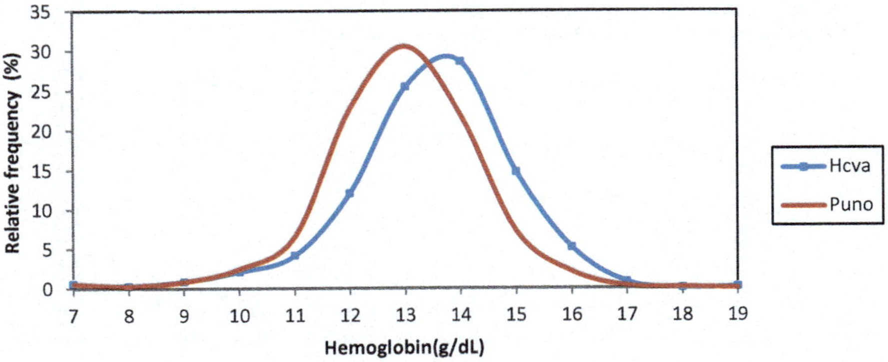

Fig. 8.4 Maternal hemoglobin distribution in Huancavelica (3600 m) and Puno (3800 m)

7 Responses to Hypoxia: Genetic Adaptations

There were attempts to identify genes associated with adaptation at HA and genes expressed in lack of adaptation in CMS. These genes are likely to be under selection in HA populations [29, 31–35]. However, identifying which human genes are involved in adaptation to HA is a challenge. Some of the genomic signals reported thus far among Tibetan, Andean, and Ethiopian are similar, while others appear unique in each population. The most commonly reported signatures of selection come from genomic segments containing genes of the hypoxia-inducible factor (HIF) pathway [109]. Hypoxia inducible factor(s) HIFs, including HIF1A and HIF2A, are key oxygen sensors and transcription factors that mediate the cell response to reduced oxygen. The regulatory cellular adaptation to hypoxia responds acutely and produces endogenous metabolites and proteins to regulate oxidative pathways [68].

7.1 Tibetans

In Tibetans, signatures of selection have been found in the HIF2A (EPAS1) gene, which encodes for HIF-2α, and in the prolyl hydroxylase domain protein 2 (PHD2, also known as EGLN1, which encodes for the key regulator, PHD2 [26–28]. Moreover, HIF2A variants were associated with different levels of HA hypoxia (2700 m, 3200 m, 3700 m and 4700 m) among native Tibetans [110]. Genes EPAS1, EGLN1 and PPARA are associated with decreased hemoglobin [64]. Other studies indicate hypoxia tolerance genes independent of HIF [23]. Hypoxia-specific selection seems stronger than genetic drift or hypoxia-independent natural selection [1]. EGLN1 in Tibetans shows a mutation that confers a higher affinity for oxygen. Preference for glycolysis and glucose oxidation at the expense of fatty acid oxidation may provide adaptation to decreased oxygen. The EPAS1 haplotype is associated with increased lactate levels and anaerobic metabolism, and the PPARA haplotype is associated with increased serum free fatty acids [64]. In the Himalayan populations, signals of selection were detected for EPAS1 but not in Andeans or Ethiopians residing at HA [29, 111]. EGLN1 gene seems to be under positive selection in Himalaya and Andes populations but not in Ethiopians [29, 111]. SNP of the CYP17A1 and CYP2E1 genes were associated with EE in Tibetans, with a correlation between these two SNPs and hemoglobin levels [80]. CYP17A1 is related with testosterone biosynthesis.

7.2 *Hans*

Carriers of this EPAS1 haplotype (G-G-G, rs13419896, rs4953354, and rs1868092) may have a higher risk for EE in the Han population [112].

7.3 *Ethiopians*

Using SNP genotype data, the strongest signal of selection to live at HA was in BHLHE41 (also known as DEC2 or SHARP1), a gene that regulates the same hypoxia response pathway in both adapted Tibetan and Andean populations [31]. Using the whole genome sequence of Ethiopian highlanders, three genes influence survival in low oxygen: cic, an ortholog of human CIC, Hsl, an ortholog of human LIPE, and Paf-AHα, an ortholog of human PAFAH1B3 [35]. Selection signals were observed in PPARA, a gene involved in energy metabolism, that produces food lower uptake and metabolic adaptation to HA. Selection signals were also observed in CDKAL1 and NEGR1 that are well-known genes increasing diabetes and obesity susceptibility [113]. Genomic analysis in two Ethiopian ethnic groups, Amhara and Oromo, were assessed in low and HA residents. The Oromo moved to HA only 500 years ago, and the Amhara group resides at HA at least for 5000 years, and possibly as far as 70000 years [29, 114]. Amhara and Oromo populations are very similar in the genome. However, hemoglobin was higher, SO2 % lower, EPO higher, with body iron store and hepcidin levels lower in Oromo than in Amhara subjects [114]. The difference is likely due to the different time at HA [29]. Several candidate genes for HA adaptation in Ethiopia have been considered, including CBARA1, VAV3, ARNT2 and THRB. Most of these genes have not been identified in Tibetan or Andean populations, with two of the genes (THRB and ARNT2) having a role in the HIF-1 pathway [30]. Ethiopians show increases in hemoglobin and reduction in oxygen saturation as altitude increases, independent of the generational antiquity at HA [30, 114]. As a summary, it seems that Ethiopian, Tibetan, and Andean populations living at HA have differently adapted to hypoxia with convergent evolution affecting different genes of the same pathway [30, 31].

7.4 *Andeans*

One gene known as important in cellular oxygen sensing: egl-9 homolog 1 (EGLN1), shows positive selection in Tibetans and Andeans; however, the pattern of variation differs in the two populations [111]. No associations between EPAS1 or EGLN1, SNP genotypes and hemoglobin concentration in Andeans were found [111]. The product of the TP53 gene is a transcription factor (p53) that activates or represses a large number of target genes that regulate a broad array of important cellular functions. The p53 factor is essential for genome integrity and a role in the adaptation to hypoxic environments has been proposed [115]. A diversity of five SNP located in genes of the TP53 pathway (TP53, rs1042522; MDM2, rs2279744; MDM4, rs1563828; USP7, rs1529916; and LIF, rs929271) was found in populations of native American or having a high level (>90 %) of native American ancestry [116]. Two genes involved in heart performance (VEGFB and ELTD1) have been implicated with the elevated hematocrit characteristic of the HA populations of Andean Aymaras [33]. Regulation of cerebral vascular flow also appears as part of the adaptive response [33]. The VEGFA tag SNP rs3025033 was associated with CMS in the population at Cerro de Pasco (4380 m) [117]. Aymaras and Quechuas in Bolivia show differences in chromosome 10 indicating la presence of genes for adaptation to HA. These genes are ANXA11, MAT1A, DYDC1, DYDC2, FAM213A, TSPAN14 and SH2D4B. The product of gene FAM213A and a related enhancer act as antioxidants to lower oxidative stress. The most significant signal in a extended haplotype homozygosity analysis was localized in the SFTPD gene, which encodes a pulmonary surfactant protein involved in normal respiration and innate host defense. These two genes and associated pathways seem involved in HA adaptation in Southern Andes populations [62].

8 Conclusions

Studies indicate that locale specific adaptation to HA have occurred due to several causes: founder effects, de novo mutations in each population, and introgression of DNA carrying beneficial mutations from nearby populations [1, 5, 34]. As populations at HA proceed from populations at low altitude, it seems that mutations improve tolerance, but the evidence shows the presence of adaptation genes even in populations at low altitude [118]. Life at HA in Peru provides an important model with its population to study human adaptation. Human beings are living in the Peruvian Andes since 12000 years BC, a period in which biological adaptation occurred. The Spaniard conquest may have altered the original pattern. It is clear that antiquity is an important component of adaptation to HA. The process of adaptation raises a question whether the pathology of sea level populations may start to change in the populations adapted to live at HA, as for example, diabetes mellitus.

Prevalence of hyperglycemia, obesity, and high blood pressure in Tibetan farmers and herdsmen are higher than in other HA communities [119, 120].

References

1. Jha AR, Zhou D, Brown CD et al (2015) Shared genetic signals of hypoxia adaptation in Drosophila and in high-altitude human populations. Mol Biol Evol 2015 pii:msv248
2. White TD, Asfaw B, DeGusta D et al (2003) Pleistocene Homo sapiens from middle Awash, Ethiopia. Nature 423:742–747
3. McDougall I, Brown FH, Fleagle JG (2005) Stratigraphic placement and age of modern humans from Kibish, Ethiopia. Nature 433:733–736
4. Fleagle JG, Assefa Z, Brown FH, Shea JJ (2008) Paleoanthropology of the Kibish formation, southern Ethiopia: introduction. J Hum Evol 55:360–365
5. Huerta-Sánchez E, Jin X, Bianba Z et al (2014) Altitude adaptation in Tibetans caused by introgression of Denisovan-like DNA. Nature 7513:194–197
6. Qi X, Cui C, Peng Y et al (2013) Genetic evidence of paleolithic colonization and neolithic expansion of modern humans on the Tibetan plateau. Mol Biol Evol 30:1761–1778
7. Goebel T, Waters MR, O'Rourke DH (2008) The late Pleistocene dispersal of modern humans in the Americas. Science 319:1497–1502
8. Reich D, Patterson N, Campbell D et al (2012) Reconstructing native American population history. Nature 488:370–374
9. Rothhammer F, Llop E, Carvallo P, Moraga M (2001) Origin and evolutionary relationships of native Andean populations. High Alt Med Biol 2:227–233
10. Rothhammer F, Dillehay TD (2009) The late Pleistocene colonization of South America: an interdisciplinary perspective. Ann Hum Genet 73:540–549
11. Rademaker K, Hodgins G, Moore K et al (2014) Paleoindian settlement of the high-altitude Peruvian Andes. Science 346:466–469
12. Homburger JR, Moreno-Estrada A, Gignoux CR et al (2015) Genomic insights into the ancestry and demographic history of South America. PLoS Genet 12:e1005602
13. Rothhammer F, Fuentes-Guajardo M, Chakraborty R et al (2015) Neonatal variables, altitude of residence and Aymara ancestry in northern Chile. PLoS One 10:e0121834
14. Hartinger S, Tapia V, Carrillo C et al (2006) Birth weight at high altitudes in Peru. Int J Gynaecol Obstet 93:275–281
15. Gonzales GF, Chaupis D (2015) Higher androgen bioactivity is associated with excessive erythrocytosis and chronic mountain sickness in Andean highlanders: a review. Andrologia 47:729–743
16. Aksoy AN, Batmaz G, Dane B et al (2015) Effects of altitude changes on Doppler flow parameters for uterine, umbilical, and mid-cerebral arteries in term pregnancy: a pilot study. J Turk Ger Gynecol Assoc 16:237–240
17. Galan HL, Rigano S, Chyu J et al (2000) Comparison of low- and high-altitude Doppler velocimetry in the peripheral and central circulations of normal fetuses. Am J Obstet Gynecol 183:1158–1661
18. Beall CM, Decker MJ, Brittenham GM et al (2002) An Ethiopian pattern of human adaptation to high-altitude hypoxia. Proc Natl Acad Sci U S A 99:17215–17218
19. Xing G, Qualls C, Huicho L et al (2008) Adaptation and mal-adaptation to ambient hypoxia; Andean, Ethiopian and Himalayan patterns. PLoS One 3:e2342
20. Aldenderfer MS (2003) Moving up in the world. Am Sci 91:542–550
21. Zhao M, Kong QP, Wang HW et al (2009) Mitochondrial genome evidence reveals successful late paleolithic settlement on the Tibetan plateau. Proc Natl Acad Sci U S A 106:21230–21235

22. Gonzales GF (2007) Peruvian contributions to the study on human reproduction at high altitude: from the chronicles of the Spanish conquest to the present. Resp Physiol Neurobiol 158:172–179
23. Wouters BG, Koritzinsky M (2008) Hypoxia signalling through mTOR and the unfolded protein response in cancer. Nat Rev Cancer 11:851–864
24. Bigham AW, Mao X, Mei R et al (2009) Identifying positive selection candidate loci for high-altitude adaptation in Andean populations. Hum Genomics 4:79–90
25. BighamAW KM, Leon-Velarde F et al (2010) Angiotensin-converting enzyme genotype and arterial oxygen saturation at high altitude in Peruvian Quechua. High Alt Med Biol 9:167–178
26. Beall CM, Cavalleri GL, Deng L et al (2010) Natural selection on EPAS1 (HIF2alpha) associated with low hemoglobin concentration in Tibetan highlanders. Proc Natl Acad Sci U S A 107:11459–11464
27. Simonson TS, Yang Y, Huff CD et al (2010) Genetic evidence for high-altitude adaptation in Tibet. Science 329:72–75
28. Yi X, Liang Y, Huerta-Sanchez E et al (2010) Sequencing of 50 human exomes reveals adaptation to high altitude. Science 329:75–78
29. Alkorta-Aranburu G, Beall CM, Witonsky DB et al (2012) The genetic architecture of adaptations to high altitude in Ethiopia. PLoS Genet 8:e1003110
30. Scheinfeldt LB, Soi S, Thompson S et al (2012) Genetic adaptation to high altitude in the Ethiopian highlands. Genome Biol 13:1–10
31. Huerta-Sánchez E, Degiorgio M, Pagani L et al (2013) Genetic signatures reveal high-altitude adaptation in a set of Ethiopian populations. Mol Biol Evol 30:1877–1888
32. Zhou D, Udpa N, Ronen R et al (2013) Whole-genome sequencing uncovers the genetic basis of chronic mountain sickness in Andean highlanders. Am J Hum Genet 93:452–462
33. Eichstaedt CA, Antão T, Pagani L et al (2014) The Andean adaptive toolkit to counteract high altitude maladaptation: genome-wide and phenotypic analysis of the Collas. PLoSOne 9:e93314
34. Jeong C, Alkorta-Aranburu G, Basnyat B et al (2014) Admixture facilitates genetic adaptations to high altitude in Tibet. Nat Commun 5:3281–3289
35. Udpa N, Ronen R, Zhou D et al (2014) Whole genome sequencing of Ethiopian highlanders reveals conserved hypoxia tolerance genes. Genome Biol 15:36–45
36. MacInnis MJ, Rupert JL (2011) Altitude adaptation, positive selection, and Himalayan genomics. High Alt Med Biol 12:133–139
37. Bhandari S, Zhang X, Cui C et al (2015) Genetic evidence of recent Tibetan ancestry to Sherpas in the Himalayan region. Sci Rep 5:16249–16256
38. Wu TY, Liu FY, Ouzhou L et al (2013) A genetic adaptive pattern: low hemoglobin concentration in the Himalayan highlanders. Zhongguo Ying Yong Sheng 29:481–493
39. Wu T, Wang X, Wei C et al (2005) Hemoglobin levels in Qinghai-Tibet: different effects of gender for Tibetans and Hans. J Appl Physiol 98:598–604
40. Gonzales GF, Tapia V, Gasco M, Carrillo C (2011) Maternal hemoglobin in Peru: regional differences and its association with adverse perinatal outcomes. Rev Peru Med Exp Salud Publica 28:484–491
41. León-Velarde F, Maggiorini M, Reeves JT et al (2005) Consensus statement on chronic and subacute high altitude diseases. High Alt Med Biol 6:147–157
42. Su J, Li Z, Cui S et al (2015) The local HIF-2α/EPO pathway in the bone marrow is associated with excessive erythrocytosis and microvessel density in chronic mountain sickness. High Alt Med Biol 16:318–330
43. Julian CG, Vargas E, Gonzales M et al (2013) Sleep-disordered breathing and oxidative stress in preclinical chronic mountain sickness (excessive erythrocytosis). Respir Physiol Neurobiol 186:188–196

44. Rexhaj E, Rimoldi SF, Pratali L et al (2015) Sleep disordered breathing and vascular function in patients with chronic mountain sickness and healthy high-altitude dwellers. Chest 10:1378/chest.15-1450
45. Peacock AJ (1998) ABC of oxygen: oxygen at high altitude. Brit Med J 317:1063–1066
46. Frisancho AR (2013) Developmental functional adaptation to high altitude: review. Am J Human Biol 25:151–168
47. Jefferson JA, Simoni J, Escudero E et al (2004) Increased oxidative stress following acute and chronic high altitude exposure. High Alt Med Biol 5:61–69
48. Dosek A, Ohno H, Acs Z et al (2007) High altitude and oxidative stress. Respir Physiol Neurobiol 158:128–131
49. Vij AG, Dutta R, Satija NK (2005) Acclimatization to oxidative stress at high altitude. High Alt Med Biol 6:301–310
50. Hakim IA, Harris R, Garland L et al (2012) Gender difference in systemic oxidative stress and antioxidant capacity in current and former heavy smokers. Cancer Epidemiol Biomark 21:2193–2200
51. Milledge JS (2013) Renin-aldosterone system. In: Lahiri S, West JB (eds) High altitude and man, Springer clinical physiology series. American Physiological Society, Bethesda, pp 53–56
52. Askew EW (2002) Work at high altitude and oxidative stress: antioxidant nutrients. Toxicology 180:107–119
53. Shi QH, Wei W, Ran JH et al (2015) Hydrogen therapy reduces oxidative stress-associated risks following acute and chronic exposure to high-altitude environment. Biomed Environ Sci 28:239–241
54. Sakamoto R, Okumiya K, Wang H et al (2015) Oxidized low density lipoprotein among the elderly in Qinghai-Tibet plateau. Wilderness Environ Med 26:343–349
55. Hu XQ, Huang X, Xiao D et al (2016) Direct effect of chronic hypoxia in suppressing large conductance Ca^{2+}-activated K+ channel activity in ovine uterine arteries via increasing oxidative stress. J Physiol 594:343–356
56. Bailey DM, Rimoldi SF, Rexhaj M et al (2013) Oxidative-nitrosative stress and systemic vascular function in highlanders with and without exaggerated hypoxemia. Chest 143:444–451
57. Cano I, Selivanov V, Gomez-Cabrero D et al (2014) Oxygen pathway modeling estimates high reactive oxygen species production above the highest permanent human habitation. PLoS One 9:e111068
58. Costa LE, Llesuy S, Boveris A (1988) Active oxygen species in the liver of rats submitted to chronic hypobaric hypoxia. Am J Physiol 255:C123–C129
59. Jefferson JA, Escudero E, Johnson RJ et al (2014) Increased oxidative stress at altitude. Chest 145:423–431
60. Gonzales GF, Tapia V, Gasco M (2014) Correcting haemoglobin cut-offs to define anaemia in high-altitude pregnant women in Peru reduces adverse perinatal outcomes. Arch Gynecol Obstet 290:65–74
61. Salama SA, Omar HA, Maghrabi IA et al (2014) Iron supplementation at high altitudes induces inflammation and oxidative injury to lung tissues in rats. Toxicol Appl Pharmacol 274:1–6
62. Valverde G, Zhou H, Lippold S et al (2015) A novel candidate region for genetic adaptation to high altitude in Andean populations. PLoS One 10:e0125444
63. Scott GR, Elogio TS, Lui MA et al (2015) Adaptive modifications of muscle phenotype in high-altitude deer mice are associated with evolved changes in gene regulation. Mol Biol Evol 32:1962–1976
64. Ge RL, Simonson TS, Gordeuk V et al (2015) Metabolic aspects of high-altitude adaptation in Tibetans. Exp Physiol 100:1247–1255
65. Beall CM, Laskowski D, Erzurum SC (2012) Nitric oxide in adaptation to altitude. Free Radic Biol Med 52:1123–1134

66. Lewis NC, Bailey DM, Dumanoir GR et al (2014) Conduit artery structure and function in lowlanders and highlanders: relationships with oxidative stress and role of sympathoexcitation. J Physiol 592:1009–1024
67. Pandey P, Qadar Pasha MA (2012) Oxidative stress at high altitude: genotype–phenotype correlations. Adv Genomics Gen 4:29–43
68. Kumar H, Choi DK (2015) Hypoxia inducible factor pathway and physiological adaptation: a cell survival pathway? Mediat Inflamm 2015:doi 584758
69. Negi PC, Asotra S, Vetri RK et al (2013) Epidemiological study of chronic mountain sickness in natives of Spiti Valley in the Greater Himalayas. High Alt Med Biol 14:220–229
70. Sahota IS, Panwar NS (2013) Prevalence of chronic mountain sickness in high altitude districts of Himachal Pradesh. Indian J Occup Environ Med 17:94–100
71. Jiang C, Chen J, Liu F et al (2014) Chronic mountain sickness in Chinese Han males who migrated to the Qinghai-Tibetan plateau: application and evaluation of diagnostic criteria for chronic mountain sickness. BMC Public Health 14:701–708
72. Niermeyer S, Andrade MP, Vargas E, Moore LG (2015) Neonatal oxygenation, pulmonary hypertension, and evolutionary adaptation to high altitude. Pulm Circ 5:48–62
73. Norboo T, Saiyed HN, Angchuk PT et al (2004) Mini review of high altitude health problems in Ladakh. Biomed Pharmacother 58:220–225
74. Appenzeller O, Claydon VE, Gulli G et al (2006) Cerebral vasodilatation to exogenous NO is a measure of fitness for life at altitude. Stroke 37:1754–1758
75. Weitz CA, Garruto RM, Chin CT (2015) Larger FVC and FEV1 among Tibetans compared to Han born and raised at high altitude. Am J Phys Anthropol doi. doi:10.1002/ajpa.22873
76. De Ferrari A, Miranda JJ, Gilman RH et al (2014) Prevalence, clinical profile, iron status, and subject-specific traits for excessive erythrocytosis in Andean adults living permanently at 3,825 meters above sea level. Chest 146:1327–1336
77. Monge CC, Arregui A, León-Velarde F (1992) Pathophysiology and epidemiology of chronic mountain sickness. Int J Sports Med 13:S79–S81
78. Gonzales GF, Rubio J, Gasco M (2013) Chronic mountain sickness is related with health status but not with hemoglobin at high altitudes. Respir Physiol Neurobiol 188:152–160
79. Cole AM, Petousi N, Cavalleri GL, Robbins PA (2014) Genetic variation in SENP1 and ANP32D as predictors of chronic mountain sickness. High Alt Med Biol 15:497–499
80. Xu J, Yang YZ, Tang F et al (2015) CYP17A1 and CYP2E1 variants associated with high altitude polycythemia in Tibetans at the Qinghai-Tibetan plateau. Gene 566:257–263
81. Gonzales GF, Tapia V, Gasco M et al (2011) High serum zinc and testosterone are associated with excessive erythrocytosis in men at high altitudes. Endocrine 40:472–480
82. Beall CM, Worthman CM, Stallings J et al (1992) Salivary testosterone concentration of Aymara men native to 3600 m. Ann Hum Biol 19:67–78
83. Richard A, Rohrmann S, Zhang L et al (2014) Racial variation in sex steroid hormone concentration in black and white men: a meta-analysis. Andrology 2:428–435
84. Ackerman CM, Lowe LP, Lee H et al (2012) Ethnic variation in allele distribution of the androgen receptor CAG. J Androl 33:210–215
85. Gonzales GF, Chung FA, Miranda S et al (2005) Heart mitochondrial nitric oxide synthase is upregulated in male rats exposed to high altitude (4340 m). Am J Physiol Heart Circ Physiol 288:H2568–H2573
86. He J, Cui J, Wang R et al (2015) Exposure to hypoxia at high altitude (5380 m) for one year induces reversible effects on semen quality and reproductive hormone levels in young male adults. High Alt Med Biol 16:216–222
87. Haas JD, Baker PT, Hunt EE (1977) The effects of high altitude on body size and composition of the newborn infant in southern Peru. Hum Biol 49:611–628
88. Ballew C, Haas JD (1986) Altitude differences in body composition among Bolivian newborns. Hum Biol 58:871–882
89. Khalid ME, Ali ME, Ali KZ (1997) Full-term birth weight and placental morphology at high and low altitude. Int J Gynecol Obstet 57:259–265

90. Keyes LE, Armaza JF, Niermeyer S et al (2003) Intrauterine growth restriction, preeclampsia, and intrauterine mortality at high altitude in Bolivia. Pediat Res 54:20–25
91. Moore LG, Young D, McCullough RE et al (2001) Tibetan protection from intrauterine growth restriction and reproductive loss at high altitude. Am J Hum Biol 13:635–644
92. Zamudio S, Droma T, Norkyel KY et al (1993) Protection from intrauterine growth retardation in Tibetans at high altitude. Am J Phys Anthropol 91:215–224
93. Levine LD, Gonzales GF, Tapia VL et al (2015) Preterm birth risk at high altitude in Peru. Am J Obstet Gynecol 91:210–218
94. Gonzales GF, Steenland K, Tapia VL (2009) Maternal hemoglobin level and fetal outcome at low and high altitudes. Am J Physiol Regul Integr Comp Physiol 297:R1477–R1485
95. Gonzales GF, Tapia V, Fort AL (2012) Maternal and perinatal outcomes in second hemoglobin measurement in nonanemic women at first booking: effect of altitude of residence in Peru. Obstet Gynecol 2012:368571
96. Browne VA, Julian CG, Toledo-Jalcin L et al (2015) Uterine artery blood flow, fetal hypoxia and fetal growth. Philos Trans R Soc Lond Biol Sci 370:20140068
97. Gonzales GF, Tapia V, Gasco M et al (2012) Association of hemoglobin values with adverse maternal outcomes among Peruvian populations living at different altitudes. Int J Gynaecol Obstet 117:134–139
98. Bigham AW, Julian CG, Wilson MJ et al (2014) Maternal PRKAA1 and EDNRA genotypes are associated with birth weight, and PRKAA1 with uterine artery diameter and metabolic homeostasis at high altitude. Physiol Genomics 46:687–697
99. Xu XH, Huang XW, Qun L et al (2014) Two functional loci in the promoter of EPAS1 gene involved in high-altitude adaptation of Tibetans. Sci Rep 4:7465–7474
100. Julian CG, Vargas E, Browne VA et al (2012) Potential role for elevated maternal enzymatic antioxidant status in Andean protection against altitude-associated SGA. J Matern Fetal Neonatal Med 25:1233–1240
101. Gonzales GF (1998) Demographic, reproductive, morbidity and mortality patterns at high altitude. Jap Soc Mountain Med 1998:174–179
102. Gonzales GF, Tapia V, Carrillo CE (2008) Stillbirth rates in Peruvian populations at high altitude. Int J Gynaecol Obstet 100:221–227
103. Passano S (1983) Características de las gestantes y de los recién nacidos en Puno (3812 m) Doctoral thesis, Universidad Peruana Cayetano Heredia, Lima, Peru
104. Myres JE, Malan M, Shumway JB et al (2000) Haplogroup-associated differences in neonatal death and incidence of low birth weight at elevation: a preliminary assessment. Am J Obstet Gynecol 182:1599–1605
105. Palmer SK, Moore LG, Young D et al (1999) Altered blood pressure during pregnancy and increased preeclampsia at high altitude (3100 meters) in Colorado. Am J Obstet Gynecol 180:1161–1168
106. Yangzom Y, Qian L, Shan M et al (2008) Outcome of hospital deliveries of women living at high altitude: a study from Lhasa in Tibet. Acta Paediatr 9:317–321
107. Orioli IM, Ribeiro MG, Castilla EE (2003) Clinical and epidemiological studies of ammniotic deformity, adhesion, and mutilation in a South American population. Am J Med Genet 118:135–145
108. Garruto RM, Dutt JS (1983) Lack of prominent compensatory polycythemia in traditional native Andeans living at 4200 meters. Am J Phys Anthropol 61:355–366
109. Simonson TS (2015) Altitude adaptation: a glimpse through various lenses. High Alt Med Biol 16:125–137
110. Basang Z, Wang B, Li L et al (2015) HIF2A variants are associated with different levels of high-altitude hypoxia among native Tibetans. PLoS One 10:e0137956
111. Bigham AW, Wilson MJ, Julian CG et al (2013) Andean and Tibetan patterns of adaptation to high altitude. Am J Hum Biol 25:190–197

112. Chen Y, Jiang C, Luo Y et al (2014) An EPAS1 haplotype is associated with high altitude polycythemia in male Han Chinese at the Qinghai-Tibetan plateau. Wilder Environ Med 25:392–400
113. Tekola-Ayele F, Adeyemo A, Chen G et al (2015) Novel genomic signals of recent selection in an Ethiopian population. Eur J Hum Genet 23:1085–1092
114. Lundgrin EL, Janocha AJ, Koch CD et al (2013) Plasma hepcidin of Ethiopian highlanders with steady-state hypoxia. Blood 122:1989–1991
115. Botcheva K (2014) p53 binding to human genome: crowd control navigation in chromatin context. Front Genet 5:447–453
116. Jacovas VC, Rovaris DL, Peréz O et al (2015) Genetic variations in the TP53 pathway in native Americans suggest adaptation to the high altitudes of the Andes. PLoS One 10:e0137823
117. Espinoza JR, Alvarez G, León-Velarde F et al (2014) Vascular endothelial growth factor-A is associated with chronic mountain sickness in the Andean population. High Alt Med Biol 15:146–154
118. Tomar A, Malhotra S, Sarkar S (2015) Polymorphism profiling of nine high altitude relevant candidate gene loci in acclimatized sojourners and adapted natives. BMC Genet 16:112–119
119. Mingji C, Onakpoya IJ, Perera R et al (2015) Relationship between altitude and prevalence of hypertension in Tibet: a systematic review. Heart 101:1054–1060
120. Sherpa LY, Deji SH et al (2013) Prevalence of metabolic syndrome and common metabolic components in high altitude farmers and herdsmen at 3700 m in Tibet. High Alt Med Biol 14:37–44

Part II
Cardiovascular Diseases

Chapter 9
Role of Oxidative Stress in Subcellular Defects in Ischemic Heart Disease

Monika Bartekova, Miroslav Barancik, and Naranjan S. Dhalla

Abstract Ischemic heart disease is caused by obstruction of the coronary arteries that reduces myocardial perfusion. The most efficient way to restore blood supply to the ischemic myocardium is reperfusion of the affected area. However, if reperfusion is initiated too late, it enhances ischemia-reperfusion damage to the heart. Oxidative stress is one of the major causes for the development of ischemia-reperfusion (I/R) injury to the heart, which is characterized by decreased functional performance, ultrastructural changes and metabolic alterations. Excessive formation of oxyradicals during the development of I/R injury has now been shown to produce a wide variety of abnormalities in cardiac subcellular organelles such as sarcolemma, sarcoplasmic reticulum, mitochondria and myofibrils. Different reactive oxygen species (ROS) have been shown to cause alterations in several proteins leading to their malfunction and changes in enzyme activities. Lipid peroxidation leads to depressed membrane fluidity and increased permeability, as well as changes in gene expression leading to impaired recovery of cardiac dysfunction due to I/R injury. Since the redox status of cardiomyocytes depends mainly on the balance between ROS production and availability of endogenous antioxidant defense systems, extensive efforts are made to maintain the redox status during I/R injury. Cardioprotection by ischemic conditioning as well as treatment of the heart with exogenous antioxidants have been observed to result in the increased availability of endogenous antioxidants and prevent oxidative damage to subcellular organelles. Thus antioxidant therapy is considered to be a promising strategy to decrease the I/R induced damage to the heart. Accordingly, this article is intended to describe the mechanisms of oxidative stress induced subcellular defects leading to cardiac dysfunction due to I/R injury and to outline the role of antioxidant defense mechanisms in the prevention as well as treatment of the ischemic heart disease.

M. Bartekova • M. Barancik
Institute for Heart Research, Slovak Academy of Sciences, Bratislava, Slovak Republic

N.S. Dhalla (✉)
Institute of Cardiovascular Sciences, St. Boniface Hospital Albrechtsen Research Centre, Department of Physiology and Pathophysiology, Max Rady College of Medicine, University of Manitoba, Winnipeg, Canada
e-mail: nsdhalla@sbrc.ca

© Springer International Publishing Switzerland 2016
R.J. Gelpi et al. (eds.), *Biochemistry of Oxidative Stress*, Advances in Biochemistry in Health and Disease 16, DOI 10.1007/978-3-319-45865-6_9

Keywords Oxyradicals • Iintracellular Ca^{2+}-overload • Antioxidants • Oxidative stress • Subcellular abnormalities • Cardiac dysfunction • Ischemia-reperfusion injury • Lipid peroxidation • Proteolysis

1 Introduction

Ischemic heart disease is the most common cause of death, especially in developed countries. It is characterized by reduced blood supply to the heart by obstruction of the coronary arteries leading to ischemic heart injury and cardiac heart dysfunction. Reduction of the artery lumen in ischemic heart disease is mostly caused by atherosclerosis of the coronary arteries leading to the low-flow ischemia with reduced supply of oxygen and metabolic substrates to the affected area. When the coronary arteries are affected by spasm or thrombosis, a full closure of the vascular lumen occurs leading to a no-flow ischemia [1]. In addition to the opening of collateral vessels, as an endogenous adaptive mechanism in response to ischemia, the restoration of blood supply to the ischemic heart is carried out by reperfusion via angioplasty, coronary by-pass or thrombolytic therapy. However, if the reperfusion is not initiated within a certain time, different pathological changes in the heart become irreversible and in fact the reperfusion even worsens the ischemic damage and accelerates the development of cardiac dysfunction. Both oxidative stress and intracellular Ca^{2+}-overload are considered to be the major mechanisms for the occurrence of the ischemia-reperfusion (I/R) injury to the heart [2]. It has been shown that there occurs an imbalance between the formation of oxidants and the availability of endogenous antioxidants in the heart [3]. Oxidative stress is associated with an increased formation of reactive oxygen species (ROS), which result in the modification of phospholipids and proteins, leading to lipid peroxidation and oxidation of thiol groups, respectively [4]. These changes alter the membrane permeability and produce functional modification of various cellular proteins. The consequences of I/R induced oxidative stress on heart function and cellular structure depend on the availability of endogenous antioxidants which may be enhanced by preventive interventions or treatment with different types of exogenous antioxidants.

2 Formation of ROS

Several molecular mechanisms and pathways are responsible for the production of free radicals in the cell. It should be mentioned that the production of one free radical can lead to the formation of other radicals through a sequential chain reaction. Accordingly, there are primary, secondary and tertiary sources of ROS [5]. The primary sources of ROS in the heart are mitochondrial cytochromes, as well as xanthine oxidoreductase, nicotinamide adenine dinucleotide phosphate (NADPH) oxidase, and nitric oxide synthase (NOS) [6]. In the vascular smooth muscle cells and endothelial cells, NADPH oxidase-induced production of ROS is triggered by vasoconstrictor agents such as angiotensin II and endothelin-1 through angiotensin type 1 receptors and endothelin receptors (ET_A, ET_B), respectively. In cardiomyocytes, the production of ROS is also stimulated by norepinephrine via α1-adrenergic receptors [5, 7]. On the other hand, NOS catalyzes the production of nitric oxide (NO) from the substrate L-arginine in presence of tetrahydrobiopterin (BH_4) as a co-factor. When BH_4 levels are reduced, the catalytic activity of NOS becomes functionally 'uncoupled' resulting in the generation of deleterious O_2^- rather than protective NO, thus acting as a secondary source of ROS which potentially increase their amounts in the cell [5, 8, 9]. It is pointed out that xanthine oxidoreductase (XOR) catalyzes the oxidation of hypoxanthine to xanthine and can further catalyze the oxidation of xanthine to uric acid. XOR requires molybdopterin, iron-sulphur centres, and FAD as cofactors. It appears in two forms, xanthine oxidase and xanthine dehydrogenase, which transfer electrons from xanthine to oxygen and to NAD+, yielding superoxide, hydrogen peroxide and NADH, respectively. Additionally, XOR can generate superoxide via NADH oxidase activity and can produce NO via nitrate and nitrite reductase activities. Interest in XOR has grown over the past decades because of its suspected role in the I/R injury [10].

3 Role of Mitochondria in the Oxidative Stress Due to I/R

Mitochondria play a crucial role in the energy metabolism of the cell where ATP is produced in the process of oxidative phosphorylation. This process takes place in the inner mitochondrial membrane with the aim to produce ATP via an electron transport from NADH and $FADH_2$ to oxygen, and where ROS are by-products of this process. Thus the oxidative phosphorylation in mitochondria contributes to the oxidative damage of the cell. In concurrence with this suggestion is the finding that uncoupling of mitochondrial respiration and oxidative phosphorylation can protect the myocardium from ischemic damage [11] suggesting the role of mitochondria-derived ROS in the pathophysiology of I/R injury and that the cellular protection is connected with the reduced capacity to produce energy during ischemia. On the other hand, the continuation of mitochondrial aerobic respiration in the absence of oxygen may result in excessive production of ROS, mitochondrial Ca^{2+}-overload, and activation of apoptotic cascades, all of which lead to cardiomyocyte death [11, 12]. Partial uncoupling of mitochondrial respiration during ischemia [13, 14] as well as blockade of electron transport [15] markedly decrease cardiac I/R injury. Moreover, uncoupling of mitochondrial respiration prior to an ischemic insult has been reported to be involved in ischemic preconditioning [14] suggesting that the down-regulation of mitochondrial function in anaerobic conditions is protective during myocardial ischemia. On the other hand, altered mitochondrial function may lead to depressed cardiac performance during I/R. A worsened heart function has been found after I/R associated with decreased mitochondrial respiration and oxidative phosphorylation. The observed alterations of cardiac performance and mitochondrial function in I/R hearts were prevented by pretreatment with radical scavenging agents or antioxidants [16]. All the above mentioned experimental studies support the view that oxidative stress plays an important role in inducing changes in cardiac performance and mitochondrial function due to I/R. The question is whether mitochondria contribute to oxidative stress via generating abundant amounts of ROS in I/R, or if oxidative stress caused by ROS generated in mitochondria alters mitochondrial function thus leading to a depressed cardiac function in I/R. Most likely both mechanisms participate in the oxidative stress induced injury of the myocardium associated with I/R.

4 Altered Ca^{2+} Levels Due to Oxidative Stress in I/R

Ca^{2+} ions are crucial in maintaining cardiac performance and therefore the regulation of intracellular concentration of Ca^{2+} is essential for proper heart function. Intracellular Ca^{2+} imbalance (intracellular Ca^{2+}-overload) and oxidative stress are the most prominent mechanisms of cardiac damage due to I/R injury in the heart [2]. It should be mentioned that these two factors may influence each other and lead to the subcellular defects and cardiac impairment not separately, but hand to hand. There is ample amount of evidence that oxidative stress induced by ischemia leads to altered Ca^{2+} handling in the I/R injury. It has been shown that global ischemia leads to depressed Ca^{2+}-pump and Na^{+}-Ca^{2+} exchange activities in the purified sarcolemmal fraction. These alterations were prevented when antioxidant enzymes (superoxide dismutase (SOD) plus catalase (CAT)) were added to the reperfusion medium [17]. Another study has reported that increased intracellular Ca^{2+} levels due to I/R were attenuated with antioxidants such as N-acetylcysteine (NAC) or N-(2-mercaptopropionyl)-glycine (MPG) in isolated rat heart [18]. Recent studies have also shown that mitochondria-derived ROS disturb Ca^{2+} cycling in cardiomyocytes due to the close proximity of mitochondria to the redox-sensitive sarcoplasmic reticulum Ca^{2+} release (ryanodine receptors) and Ca^{2+} uptake (SERCA) channels [19], and that leaky type 2 ryanodine receptors (RyR2) cause mitochondrial Ca^{2+}-overload and heart dysfunction in heart failure. Moreover, genetic enhancement of mitochondrial antioxidant activity was observed to improve mitochondrial function and reduce posttranslational modifications of RyR2 [20]. All these data support the view that I/R induced oxidative stress leads to the development of intracellular Ca^{2+}-overload and contribute to subcellular defects and cardiac dysfunction, most likely via mitochondria-derived ROS production leading to altered Ca^{2+} handling by both sarcolemma and sarcoplasmic reticulum.

5 Lipid Alterations Due to Oxidative Stress in I/R

Oxidation of membrane phospholipids, in other words of the polyunsaturated fatty acids that are located or stored in cardiomyocytes, is one of the most prominent manifestations of oxidative stress in the heart. ROS and reactive nitrogen species (RNS) are the electrophile substances most commonly involved in lipid peroxidation; their reaction results in the formation of lipid peroxidation products (LPPs) [21]. The heart contains a very high content of mitochondria relative to other organs and as it has been already mentioned, mitochondria are the largest source of intracellular ROS in cardiomyocytes [6, 22]. According to this fact, the most immediate targets of lipid peroxidation are the lipids present in the phospholipid bilayer of mitochondrial membranes [23]. Mitochondria have an enormous amount of unsaturated phospholipids, notably cardiolipin, a phospholipid exclusive of mitochondria, that is highly unsaturated and prone to peroxidation [24, 25]. Cardiolipin levels diminish dramatically in lipid peroxidation conditions, and its diminished levels have been observed in many pathological conditions including heart failure and I/R injury [26]. Thus, mitochondria are not only the largest endogenous source of LPPs, but are also a target of LPPs in the heart. Studies focused on the effects of lipid peroxidation on membrane physiology have demonstrated a marked reduction in membrane fluidity [27] and an increase in membrane permeability [28]. Furthermore, it has been shown that the increased resistance of rat hearts to I/R injury achieved by remote ischemic preconditioning induced by hind limb ischemia was associated with increased mitochondrial membrane fluidity [29, 30]. It can be concluded that oxidative stress induced phospholipid alterations and consequently reduced membrane fluidity play a critical role in mitochondrial impairment and depressed cardiac function due to I/R.

6 Protein Alterations Due to Oxidative Stress in I/R

It is evident that alterations in intracellular proteins due to oxidative stress are tightly connected with depressed cardiac function and tissue damage during I/R injury. One of the proposed mechanisms leading to protein alterations and degradation due to oxidative stress in I/R is the activation of endogenous proteinases such as matrix metalloproteinases (MMPs), cathepsins or calpain [31]. Targets of these proteolytic enzymes are different intracellular as well as extracellular proteins involved in the maintenance of cardiac function. For example, I/R induced depression in Na^+K^+-ATPase associated with attenuated cardiac function as a consequence of oxidative stress was suggested to be mediated via increased proteases activities in the heart [32]. It was also documented that the depression in levels of alpha2, alpha3, beta1, beta2, and beta3 isoforms of Na^+K^+-ATPase due to I/R was prevented by treatment of perfused hearts with antioxidants. Moreover, perfusion of the hearts with H_2O_2 was found to depress the levels of alpha and beta isoforms of Na^+K^+-ATPase, suggesting a significant role of oxidative stress in I/R-induced changes in the $Na^{+}K^{+}$-ATPase isoforms expression due to I/R [33]. Oxidative stress also leads to the cleavage and degradation of sarcomeric proteins. It has been shown that cardiac troponin I degradation is a prominent feature of the ischemic damage, and that its cleavage may contribute to ischemia-induced changes in force generation [34]. Myofilament protein degradation due to oxidative stress was attributed to μ-calpain, a myofibril-associated protease that is activated in ischemic cardiomyocytes [35, 36].

Oxidative stress may lead not only to protein cleavage via activation of proteases, but also to impaired protein function via depressing their phosphorylation. It has been shown that phosphorylation of the Ca^{2+} pump ATPase (SERCA) and phospholamban via sarcoplasmic reticulum (SR) of Ca^{2+}/calmodulin-dependent protein kinase (CAMK) were significantly decreased in I/R with significantly decreased sarcoplasmic CAMK activity. These changes were prevented by SOD plus CAT treatment suggesting the role of oxidative stress in this I/R-induced depression of phosphorylation [37]. It has been also found that mRNA levels and protein contents for SR Ca^{2+} pump ATPase and Ca^{2+} release channels were markedly depressed in the I/R hearts. Perfusion with SOD plus CAT partially prevented the I/R-induced SR Ca^{2+} transport activities and mRNA abundance, supporting the oxidative stress participation in these changes in calcium handling proteins [38]. Alterations in protein phosphorylation due to oxidative stress have been shown to affect sarcomeric proteins, essential for heart contractions. It has been shown that oxidative stress may lead to altered phosphorylation of troponins T and I, as well as cardiac myosin binding protein-C, a thick filament protein that is required for sarcomeric integrity, the regulation of cardiac contraction, and cardioprotection [39].

Oxidative stress and increased formation of ROS (or RNS) can result in oxidation or nitrosylation of contractile proteins leading to changes in their structural conformation and/or functionality [39]. It has been documented that H_2O_2 treatment of isolated rat hearts leads to the oxidation of both tropomyosin and actin [40]. Oxidative modifications of tropomyosin also have been detected in ischemic pig hearts and in the early post-myocardial infarction period in mouse hearts associated with the development of contractile dysfunction [41, 42]. Moreover, oxidative modifications of actin were detected during reperfusion of ischemic rat hearts [43]. In view of these observations, it is concluded that various ROS-induced post-translational modifications of different proteins, including direct oxidative modifications, protein phosphorylation, and protein cleavage by ROS-activated proteases significantly contribute to I/R-induced cardiac dysfunction.

7 DNA Alterations Due to Oxidative Stress in I/R

In addition to lipid and protein alterations, oxidative stress can change gene transcription and translation, as well as the integrity of the DNA repair systems [44]. One of the mechanisms regulating the gene expression is histone acetylation induced by the activation of the redox-sensitive transcription factors activator protein-1

(AP-1) and nuclear factor-kappa B (NF-kB). It was found that oxidative stress can enhance the expression of certain genes by stimulating AP-1- and NF-kB via altered histone acetylation. In addition, different oxidants have been shown to attenuate the histone deacetylase activity and expression. Thus, oxidative stress may lead to changes in chromatin structure [45]. Many transcription factors are regulated in a redox-sensitive fashion. For example, the transcription factor Nrf2 is activated upon thiol oxidation of its regulator protein Keap1 [46]. AP-1 is activated upon thiol oxidation via thioredoxin, which activates NF-kB [47]. It has been reported that inhibition of NF-kB and AP-1 translocation by heme oxygenase-1 (HO-1) activation decreases the occurrence of cardiomyocytic apoptosis following cardiac global ischemia and reperfusion during cardiopulmonary by-pass [48]. In addition to redox impact on the classical gene regulation, oxidative stress may regulate DNA repair. It has been observed that oxidative stress negatively impacts the DNA repair mechanism by inhibiting a number of DNA repair enzymes [49]. Oxidative stress also increases the burden of DNA lesions such as single strand breaks [50]. Finally, redox signaling and oxidative stress can affect the epigenetic regulation of genes by changes in the function of histones and DNA modifying enzymes, thereby altering the phenotype of cells [51]. Additionally, prominent miRNAs are regulated by oxidative stress and vice versa can regulate the antioxidant pathways [52]. Growing number of studies indicating the involvement of different miRNAs are in the development of I/R injury [53–56] as well as in cardioprotection [56, 57]. All these data demonstrate that oxidative stress-induced DNA damage as well as redox regulation of DNA repair and epigenetic factors are intimately involved in the development of I/R injury, and might represent potential targets for the protection and treatment of cardiovascular diseases, including ischemic heart disease.

8 Changes in Molecular Signaling Pathways Due to Oxidative Stress in I/R

ROS are key players in normal cardiovascular physiology and signaling, but many redox-sensitive signaling pathways are also activated during the development of ischemic heart disease. The increased production of ROS leads to cellular injury (necrosis or apoptosis) through direct effects (lipid peroxidation) or indirect effects (activation of the redox signaling pathways). I/R injury is intimately associated with mitochondrial dysfunction and increased oxidative stress. The I/R induced oxidative stress in turn influences the signaling pathways that contribute to apoptosis, endoplasmic reticulum stress, altered cell migration and proliferation [58]. A rise in ROS generated by mitochondria during reperfusion is also an important step for the activation of intrinsic apoptosis pathway in cardiomyocytes. It is noteworthy that ROS-sensitive signaling that significantly contributes to the cell death is the pathway involving c-Jun N-terminal kinases (JNKs) [59]. JNKs are involved in the regulation of apoptosis and play a crucial important role in ventricular remodeling after oxidative stress [60]. In addition to the activation of pro-apoptotic signal pathways (JNKs), oxidative stress triggers several cardioprotective signaling pathways [61]. For example, phosphatidylinositol 3-kinase/protein kinase B (PI3K/Akt) and extracellular-signal regulated protein kinases (ERKs) pathways are activated in response to I/R and have been identified as the pro-survival mediators, making cardiac cells resistant to ischemic cell death [62, 63]. Akt kinase is activated by receptor ligands, and its phosphorylation preserves mitochondrial integrity thereby protecting the cardiac cells against necrosis and apoptosis [64]. There is evidence that shows ERKs are activated during I/R and protect impaired cardiac function and cardiac injury [65]. Changes in the activation of these protein kinase signaling pathways also influence the regulation of expression and function of antioxidant defense system [66]. Important is the interplay of PI3K/Akt and ERK signaling with transcription factors and activation of Nrf2-ARE pathway (pathway of NF-E2-related factor-2 and antioxidant response element) [67, 68]. It has been demonstrated that mechanisms of increased myocardial resistance against oxidative stress during I/R injury might be associated with the enhancement of antioxidant defense system by activating Akt/ERK1/2/Nrf2 signaling pathways [66]. In addition, modulation of Nrf2, as an upstream signaling molecule, activates HO-1 [69], an enzyme with crucial role in protection of the myocardium against oxidative stress and I/R injury [70, 71].

Changes in the activation of signaling protein kinase cascades due to oxidative stress during I/R injury are often connected with the modulation of MMPs, enzymes involved in the degradation of extracellular matrix [72]. These enzymes are also involved in the development and progression of cardiovascular diseases and metabolic syndrome and recent data show that MMPs play an important role in progression of vascular diseases and pathogenesis of endothelial dysfunction [73]. Expression and activation of MMPs can be regulated by increased production of ROS [74] and this process is associated with the activation of ERKs [75]. The overall view on the subcellular changes during the I/R-induced oxidative stress, and its consequences on the heart are summarized in the Fig. 9.1.

9 Changes in Endogenous Antioxidants Due to Oxidative Stress in I/R

Endogenous antioxidants are substances produced within the body which inhibit or delay the oxidative damage to subcellular molecules. They can act through several mechanisms such as scavenging ROS and their precursors, inhibiting the formation of ROS, enhancing endogenous antioxidant generation, or reducing apoptotic cell death by upregulating the anti-apoptotic Bcl-2 gene [3]. Many substances have been suggested to act as endogenous antioxidants and these include SOD, CAT, and glutathione peroxidase (GSH-PX), which have been studied extensively. It is therefore

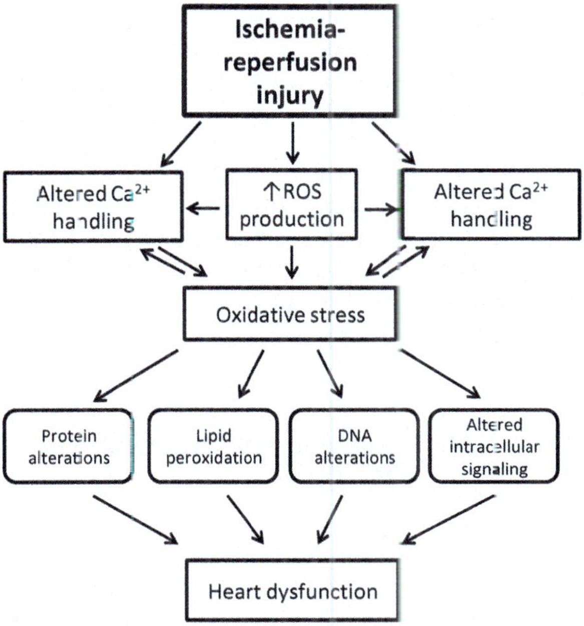

Fig. 9.1 Role of oxidative stress in ischemia-reperfusion induced subcellular alterations and cardiac dysfunction

intended to focus on the changes in activities and/or expression of these particular enzymes in the I/R conditions in the heart. SOD is an enzyme that catalyzes the dismutation of superoxide radical (O_2^-) into molecular oxygen (O_2) and hydrogen peroxide (H_2O_2). SOD is therefore an antioxidant enzyme which defends the cells exposed to conditions of oxidative stress. There are three SOD isoforms, namely the cytosolic copper-zinc SOD (CuZnSOD, or SOD1), mitochondrial manganese SOD (MnSOD, or SOD2), and the extracellular SOD (ecSOD, SOD3) [5]. The role of SOD in I/R injury is controversial. It has been reported that MnSOD activity was decreased due to I/R, but only after a long period of ischemia (60 min), and not after 30-min ischemia [76]. Another study has reported that CuZnSOD did not change due to I/R injury [77].

It should be noted that CAT is a common enzyme that catalyzes the decomposition of H_2O_2 to H_2O and O_2 and is also a very important enzyme in protecting the cell from oxidative damage. CAT is a membrane bound enzyme present in peroxisomes, but its activity has also been detected in the mitochondria [3]. The protective role of CAT in I/R injury has been suggested due to the observations of its up-regulation in ischemic preconditioning [78] as well as the protective effects of its treatment, which enhanced the intracellular CAT activity in I/R injury [79]. However, data on changes in its activity due to I/R are lacking, and thus the role CAT in oxidative stress due to I/R is unclear. GSH-PX catalyzes the peroxidation of H_2O_2 in the presence of reduced glutathione to form H_2O and oxidized glutathione.

In the heart, GSH-PX plays an important role as a scavenger of H_2O_2 since its activity is much higher than the CAT activity [3]. Like SOD and CAT, GSH-PX has been proposed to have a protective role in the oxidative stress-induced damage due to I/R injury; however, its role in the development of ischemia-induced oxidative stress has not been documented. Taken together, it seems that changes in endogenous antioxidant enzymes are not significantly involved in oxidative stress-induced damage to myocardium due to I/R injury, suggesting that the redox imbalance in I/R is more likely caused by the enhanced production of oxidants than by the decreased availability of endogenous antioxidants. On the other hand, there is evidence for the up-regulation and positive effects of endogenous antioxidant enzymes in cardioprotection against I/R injury [3].

10 Role of Endogenous Antioxidants in Cardioprotection Against I/R Injury

Numerous studies have shown that the activation of endogenous antioxidant systems has a potential to improve cardiac performance and to reduce the negative consequences of I/R injury. It has been found in a transgenic mice model that overexpression of MnSOD protects against myocardial I/R injury induced by global ischemia and subsequent reperfusion [80]. Similar protective effects were documented in transgenic mice with overexpression of SOD or CAT [81]. Increased activity of SOD associated with a protective effect against I/R injury has also been documented in rats treated with quercetin [82], resveratrol [83], atorvastatin [84] or astragalin [85]. Furthermore, endogenous antioxidants have been observed to be increased in ischemic preconditioning, an efficient endogenous cardioprotective mechanism achieved by one or more cycles of short ischemia prior to long-term ischemia. It has been shown that four brief episodes of regional ischemia and reperfusion significantly increased the activities of both mitochondrial MnSOD and GSH-PX [86]. Increase in the activity of MnSOD has been demonstrated in other studies using different animal models as well as different preconditioning protocols [78, 87]. On the other hand, data on changes in the activities of Cu-ZnSOD, GSH-PX and CAT due to ischemic preconditioning are controversial [78, 87]. Finally, there are also data pointing to no changes in any antioxidant enzymes due to preconditioning [88]. Despite some discrepancies in the existing literature, the antioxidant hypothesis for mediating ischemic preconditioning-induced cardioprotection cannot be ruled out due to a large body of evidence regarding the upregulation of the antioxidant enzymes in different preconditioning protocols.

11 Effectiveness of Exogenous Antioxidants in Prevention and Treatment of I/R

It has become evident that oxidative stress is caused by either increased production of oxidants or decreased availability of endogenous antioxidants within the cell, or both factors together and that oxidative stress markedly contributes to the subcellular defects and myocardial damage during I/R. These observations raise the question regarding the possibility and effectiveness of exogenous antioxidants administration to prevent and/or treat the negative consequences of cardiac ischemia. Numerous studies have been performed to explore the effects of various exogenous antioxidants in I/R injury. A large group of widely studied antioxidants is the family of polyphenols, which became popular due to observation of the "French paradox", showing decreased morbidity and mortality from cardiovascular diseases of people living in the Mediterranean Sea region. This is primarily due to a polyphenol-rich diet with high consumption of vegetables and fruits, as well as drinking red wine. It has been documented that quercetin, a polyphenol naturally occurring in the apple skin, onions and other natural sources, can prevent I/R injury upon acute as well as chronic administration to rats [89, 90]. Moreover, quercetin increased heart resistance to I/R in rats treated with doxorubicin [82].

Another polyphenol, resveratrol, present in the red wine, has been shown to improve heart function and reduce infarcted area in *in vivo* model of myocardial I/R in rats. This cardioprotection was associated with decreased myocardial myeloperoxidase levels, as well as serum creatinine kinase and lactate dehydrogenase levels. Moreover, resveratrol also markedly enhanced the activities of endogenous antioxidants SOD and GSH-PX, and reduced the level of malondialdehyde (MDA) [83]. Since the protective effect of resveratrol in I/R injury has been documented in many experimental studies, numerous mechanisms of its protective action such as inhibition of inflammation [91], modulation of antioxidant enzymes activity [92] or restoration of I/R altered microRNA expression [93] have been suggested to be involved in the resveratrol-induced cardioprotection. It should be mentioned that red wine probably exerts its positive effects in I/R by both antioxidant effects of polyphenols and protection of endothelial function via alcohol-induced vasodilatation [94]. Other antioxidant-rich substances shown to be protective in I/R are various natural plant mixtures, especially those used in the traditional eastern medicine such as *Hemidesmus hindicus* and *Hibiscus rosa* extracts [95], or *Bauhinia championii* flavone [96]. I/R-induced damage has been reported to be decreased also using curcumin [97, 98] and its analogues [99].

The oxidative stress-induced I/R injury has been shown to be attenuated by selenium, an essential trace element which is a component of the active center of GSH-PX. Selenium, added to the perfusion medium, caused a dose-dependent improvement in cardiac performance of isolated hearts and attenuated the decrease in the reduced/oxidized glutathione ratio, as well as the increased level of MDA due to I/R. It was suggested that selenium protects the heart against I/R injury due to its action on the redox state and deactivation of NF-kB [100]. Another antioxidant,

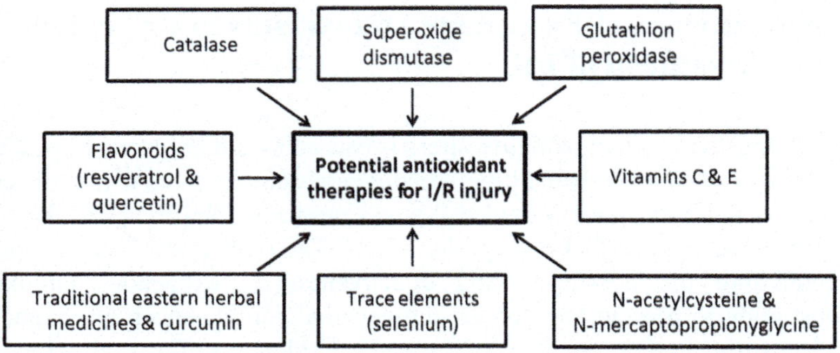

Fig. 9.2 Some antioxidant interventions commonly used for attenuating the ischemia-reperfusion induced injury in the heart

vitamin E, has been also documented to protect the oxidative stress induced damage in acute myocardial infarction (MI)model *in vivo* [101]. In addition to reducing the infarct size, vitamin E treatment prevented depression in left ventricular function as well as elevation of MDA content and conjugated diene formation in the heart. Total mortality was also reduced due to vitamin E treatment in the same study [101]. One of the important exogenous antioxidant shown to be protective in I/R injury of the heart is NAC, which is known to reduce oxidative stress by increasing the glutathione redox status. Treatment with NAC has been shown to protect isolated rat hearts against I/R injury induced by 30 min global ischemia. Similar effect was obtained using MPG, which scavenges both peroxynitrite and hydroxyl radicals. The cardioprotective effects of NAC and MPG were associated with attenuation of changes in Ca^{2+} handling by cardiomyocytes, supporting the view that the oxidative stress due to ROS generation and/or peroxynitrite formation plays a role in the development of intracellular Ca^{2+}-overload as a consequence of I/R injury in the heart [18]. NAC treatment, as well as the treatment with a mixture of SOD and CAT, has also been shown to attenuate the depression of cardiac function and reduction in the myofibrillar Ca^{2+}-stimulated ATPase activity as a consequence of I/R [102]. It has been also shown that intravenous NAC administration decreases oxidative stress (observed as reduced serum MDA levels after NAC treatment), infarct area and apoptotic activity in an *in vivo* rat model of I/R injury [103]. Moreover, administration of NAC in combination with streptokinase was shown to significantly diminish the oxidative stress and improved LV function in patients with acute MI [104]. Some of the potential therapies which are likely to produce beneficial effects in attenuating the I/R-induced injury are given in Fig. 9.2. It should be emphasized that despite numerous experimental studies documenting the positive effects of exogenous antioxidants in the prevention of I/R injury in animal models, the effectiveness of antioxidant supplementation in the prevention and/or treatment in patients with ischemic heart disease is controversial [105]. Nonetheless, new clinical studies are being carried out which are expected to help in making meaningful conclusions [106]

12 Conclusions

It has become evident that oxidative stress is one of the major causes of subcellular defects due to I/R injury of the heart. In this regard, several biochemical and molecular changes such as lipid peroxidation, protein alterations, and DNA damage during I/R lead to alterations in subcellular organelles, remodeling of extracellular matrix, cell apoptosis and necrosis, as well as heart dysfunction. Oxidative stress due to I/R injury is associated with enhanced production of ROS, mostly within the mitochondria, leading to imbalance between their production and degradation by endogenous antioxidant systems. Thus, interventions such as ischemic preconditioning which lead to enhanced activity and/or expression of endogenous antioxidants as well as treatment with exogenous antioxidant seem to be a promising strategy to reverse the redox imbalance in ischemic heart disease.

References

1. Dhalla NS, Saini-Chohan HK, Duhamel TA (2008) Strategies for the regulation of intracellular calcium in ischemic heart disease. Future Cardiol 4:339–345
2. Dhalla NS, Saini HK, Tappia PS et al (2007) Potential role and mechanisms of subcellular remodeling in cardiac dysfunction due to ischemic heart disease. J Cardiovasc Med 8:238–250
3. Dhalla NS, Elmoselhi AB, Hata T, Makino N (2000) Status of myocardial antioxidants in ischemia-reperfusion injury. Cardiovasc Res 47:446–456
4. Ceconi C, Cargnoni A, Pasini E et al (1991) Evaluation of phospholipid peroxidation as malondialdehyde during myocardial ischemia and reperfusion injury. Am J Physiol 260:H1057–H1061
5. Majzunova M, Dovinova I, Barancik M, Chan JY (2013) Redox signaling in pathophysiology of hypertension. Biomed Sci 20:69–74
6. Müller BA, Dhalla NS (2010) Mechanisms of the beneficial actions of ischemic preconditioning on subcellular remodeling in ischemic-reperfused heart. Curr Cardiol Rev 6:255–264
7. Balakumar P, Jagadeesh G (2010) Multifarious molecular signaling cascades of cardiac hypertrophy: can the muddy waters be cleared? Pharmacol Res 62:365–383
8. Montezano AC, Touyz RM (2012) Reactive oxygen species and endothelial function--role of nitric oxide synthase uncoupling and Nox family nicotinamide adenine dinucleotide phosphate oxidases. Basic Clin Pharmacol Toxicol 110:87–94
9. Zweier JL, Chen CA, Druhan LJ (2011) S-glutathionylation reshapes our understanding of endothelial nitric oxide synthase uncoupling and nitric oxide/reactive oxygen species-mediated signaling. Antioxid Redox Signal 14:1769–1775
10. Berry CE, Hare JM (2004) Xanthine oxidoreductase and cardiovascular disease: molecular mechanisms and pathophysiological implications. J Physiol 555:589–606
11. Chen Q, Camara AK, Stowe DF et al (2007) Modulation of electron transport protects cardiac mitochondria and decreases myocardial injury during ischemia and reperfusion. Am J Physiol Cell Physiol 292:C137–C147
12. Yoshioka J, Lee RT (2014) Thioredoxin-interacting protein and myocardial mitochondrial function in ischemia-reperfusion injury. Trends Cardiovasc Med 24:75–80

13. McLeod CJ, Aziz A, Hoyt RF Jr et al (2005) Uncoupling proteins 2 and 3 function in concert to augment tolerance to cardiac ischemia. J Biol Chem 280:33470–33476
14. Nadtochiy SM, Tompkins AJ, Brookes PS (2006) Different mechanisms of mitochondrial proton leak in ischaemia/reperfusion injury and preconditioning: implications for pathology and cardioprotection. Biochem J 395:611–618
15. Chen Q, Hoppel CL, Lesnefsky EJ (2006) Blockade of electron transport before cardiac ischemia with the reversible inhibitor amobarbital protects rat heart mitochondria. J Pharmacol Exp Ther 316:200–207
16. Makazan Z, Saini HK, Dhalla NS (2007) Role of oxidative stress in alterations of mitochondrial function in ischemic-reperfused hearts. Am J Physiol Heart Circ Physiol 292:H1986–H1994
17. Dixon IM, Kaneko M, Hata T et al (1990) Alterations in cardiac membrane Ca^{2+} transport during oxidative stress. Mol Cell Biochem 99:125–133
18. Saini-Chohan HK, Dhalla NS (2009) Attenuation of ischemia-reperfusion-induced alterations in intracellular Ca^{2+} in cardiomyocytes from hearts treated with N-acetylcysteine and N-mercaptopropionylglycine. Can J Physiol Pharmacol 87:1110–1119
19. Li Q, Su D, O'Rourke B et al (2015) Mitochondria-derived ROS bursts disturb Ca^{2+} cycling and induce abnormal automaticity in guinea pig cardiomyocytes: a theoretical study. Am J Physiol Heart Circ Physiol 308:H623–H636
20. Santulli G, Xie W, Reiken SR, Marks AR (2015) Mitochondrial calcium overload is a key determinant in heart failure. Proc Natl Acad Sci U S A 112:11389–11394
21. Anderson EJ, Katunga LA, Willis MS (2012) Mitochondria as a source and target of lipid peroxidation products in healthy and diseased heart. Clin Exp Pharmacol Physiol 39:179–193
22. Jezek P, Hlavatá L (2005) Mitochondria in homeostasis of reactive oxygen species in cell, tissues, and organism. Int J Biochem Cell Biol 37:2478–2503
23. Bacot S, Bernoud-Hubac N, Baddas N et al (2003) Covalent binding of hydroxy-alkenals 4-HDDE, 4-HHE, and 4-HNE to ethanolamine phospholipid subclasses. J Lipid Res 44:917–926
24. Paradies G, Petrosillo G, Paradies V, Ruggiero FM (2009) Role of cardiolipin peroxidation and Ca^{2+} in mitochondrial dysfunction and disease. Cell Calcium 45:643–650
25. Paradies G, Petrosillo G, Paradies V, Ruggiero FM (2010) Oxidative stress, mitochondrial bioenergetics, and cardiolipin in aging. Free Radic Biol Med 48:1286–1295
26. Chicco AJ, Sparagna GC (2007) Role of cardiolipin alterations in mitochondrial dysfunction and disease. Am J Physiol Cell Physiol 292:C33–C44
27. Chen JJ, Yu BP (1994) Alterations in mitochondrial membrane fluidity by lipid peroxidation products. Free Radic Biol Med 17:411–418
28. Goldstein RM, Weissmann G (1977) Effects of the generation of superoxide anion on permeability of liposomes. Biochem Biophys Res Commun 75:604–609
29. Ferko M, Kancirová I, Jašová M et al (2014) Remote ischemic preconditioning of the heart: protective responses in functional and biophysical properties of cardiac mitochondria. Physiol Res 63:S469–S478
30. Ferko M, Kancirová I, Jašová M et al (2015) Participation of heart mitochondria in myocardial protection against ischemia/reperfusion injury: benefit effects of short-term adaptation processes. Physiol Res 64:S617–S625
31. Müller AL, Hryshko LV, Dhalla NS (2013) Extracellular and intracellular proteases in cardiac dysfunction due to ischemia-reperfusion injury. Int J Cardiol 164:39–47
32. Singh RB, Hryshko L, Freed D, Dhalla NS (2012) Activation of proteolytic enzymes and depression of the sarcolemmal Na^{+}/K^{+}-ATPase in ischemia-reperfused heart may be mediated through oxidative stress. Can J Physiol Pharmacol 90:249–260
33. Ostadal P, Elmoselhi AB, Zdobnicka I et al (2004) Role of oxidative stress in ischemia-reperfusion-induced changes in Na+, K+-ATPase isoform expression in rat heart. Antioxid Redox Signal 6:914–923

34. Van Eyk JE, Powers F, Law W et al (1998) Breakdown and release of myofilament proteins during ischemia and ischemia/reperfusion in rat hearts: identification of degradation products and effects on the pCa-force relation. Circ Res 82:261–271
35. Di Lisa F, De Tullio R, Salamino F et al (1995) Specific degradation of troponin T and I by mu-calpain and its modulation by substrate phosphorylation. Biochem J 308:57–61
36. Zhang Z, Biesiadecki BJ, Jin JP (2006) Selective deletion of the NH2-terminal variable region of cardiac troponin T in ischemia reperfusion by myofibril-associated mu-calpain cleavage. Biochemistry 45:11681–11694
37. Netticadan T, Temsah R, Osada M, Dhalla NS (1999) Status of Ca^{2+}/calmodulin protein kinase phosphorylation of cardiac SR proteins in ischemia-reperfusion. Am J Physiol 277:C384–C391
38. Temsah RM, Netticadan T, Chapman D et al (1999) Alterations in sarcoplasmic reticulum function and gene expression in ischemic-reperfused rat heart. Am J Physiol 277:H584–H594
39. Steinberg SF (2013) Oxidative stress and sarcomeric proteins. Circ Res 112:393–405
40. Canton M, Menazza S, Sheeran FL et al (2011) Oxidation of myofibrillar proteins in human heart failure. J Am Coll Cardiol 57:300–309
41. Canton M, Skyschally A, Menabò R et al (2006) Oxidative modification of tropomyosin and myocardial dysfunction following coronary microembolization. Eur Heart J 27:875–881
42. Avner BS, Shioura KM, Scruggs SB et al (2012) Myocardial infarction in mice alters sarcomeric function via post-translational protein modification. Mol Cell Biochem 363:203–215
43. Eaton P, Byers HL, Leeds N et al (2002) Detection, quantitation, purification, and identification of cardiac proteins S-thiolated during ischemia and reperfusion. J Biol Chem 277:9806–9811
44. Mikhed Y, Görlach A, Knaus UG, Daiber A (2015) Redox regulation of genome stability by effects on gene expression, epigenetic pathways and DNA damage/repair. Redox Biol 5:275–289
45. Adcock IM, Cosio B, Tsaprouni L et al (2005) Redox regulation of histone deacetylases and glucocorticoid-mediated inhibition of the inflammatory response. Antioxid Redox Signal 7:144–152
46. Fourquet S, Guerois R, Biard D, Toledano MB (2010) Activation of NRF2 by nitrosative agents and H_2O_2 involves KEAP1 disulfide formation. J Biol Chem 285:8463–8471
47. Schenk H, Klein M, Erdbrügger W et al (1994) Distinct effects of thioredoxin and antioxidants on the activation of transcription factors NF-kappa B and AP-1. Proc Natl Acad Sci U S A 91:1672–1676
48. Yeh CH, Chen TP, Wang YC et al (2009) HO-1 activation can attenuate cardiomyocytic apoptosis via inhibition of NF-kappa B and AP-1 translocation following cardiac global ischemia and reperfusion. J Surg Res 155:147–156
49. Liu RH, Hotchkiss JH (1995) Potential genotoxicity of chronically elevated nitric oxide: a review. Mutat Res 339:73–89
50. Guetens G, De Boeck G, Highley M et al (2002) Oxidative DNA damage: biological significance and methods of analysis. Crit Rev Clin Lab Sci 39:331–457
51. Cyr AR, Domann FE (2011) The redox basis of epigenetic modifications: from mechanisms to functional consequences. Antioxid Redox Signal 15:551–589
52. Baird AM, O'Byrne KJ, Gray SG (2014) Reactive oxygen species and reactive nitrogen species in epigenetic modifications. In: Laher I (ed) Systems biology of free radicals and antioxidants. Springer, Berlin/Heidelberg, pp 437–455
53. Cheng Y, Liu X, Zhang S et al (2009) MicroRNA-21 protects against the H_2O_2-induced injury on cardiac myocytes via its target gene PDCD4. J Mol Cell Cardiol 47:5–14
54. Ye Y, Perez-Polo JR, Qian J, Birnbaum Y (2011) The role of microRNA in modulating myocardial ischemia-reperfusion injury. Physiol Genomics 43:534–542
55. Zhai C, Tang G, Peng L et al (2015) Inhibition of microRNA-1 attenuates hypoxia/re-oxygenation-induced apoptosis of cardiomyocytes by directly targeting Bcl-2 but not GADD45Beta. Am J Transl Res 7:1952–1962

56. Varga ZV, Zvara A, Faragó N et al (2014) MicroRNAs associated with ischemia-reperfusion injury and cardioprotection by ischemic pre- and postconditioning: protectomics. Am J Physiol Heart Circ Physiol 307:H216–H222
57. Yang J, Chen L, Ding J et al (2016) Cardioprotective effect of miRNA-22 on hypoxia/reoxygenation induced cardiomyocyte injury in neonatal rats. Gene 579:17–22
58. Giordano FJ (2005) Oxygen, oxidative stress, hypoxia, and heart failure. J Clin Invest 115:500–508
59. Xu J, Qin X, Cai X et al (2015) Mitochondrial JNK activation triggers autophagy and apoptosis and aggravates myocardial injury following ischemia/reperfusion. Biochim Biophys Acta 1852:262–270
60. Yin H, Chao L, Chao J (2008) Nitric oxide mediates cardiac protection of tissue kallikrein by reducing inflammation and ventricular remodeling after myocardial ischemia/reperfusion. Life Sci 82:156–165
61. Heusch G, Boengler K, Schulz R (2008) Cardioprotection: nitric oxide, protein kinases, and mitochondria. Circulation 118:1915–1919
62. Fujio Y, Nguyen T, Wencker D et al (2000) Akt promotes survival of cardiomyocytes in vitro and protects against ischemia-reperfusion injury in mouse heart. Circulation 101:660–667
63. Lips DJ, Bueno OF, Wilkins BJ et al (2004) MEK1-ERK2 signaling pathway protects myocardium from ischemic injury in vivo. Circulation 109:1938–1941
64. Lee S, Chanoit G, McIntosh R et al (2009) Molecular mechanism underlying Akt activation in zinc-induced cardioprotection. Am J Physiol Heart Circ Physiol 297:H569–H575
65. Kovacs K, Hanto K, Bognar Z et al (2009) Prevalent role of Akt and ERK activation in cardioprotective effect of Ca^{2+} channel- and beta-adrenergic receptor blockers. Mol Cell Biochem 321:155–164
66. Yu J, Wang L, Akinyi M et al (2015) Danshensu protects isolated heart against ischemia reperfusion injury through activation of Akt/ERK1/2/Nrf2 signaling. Int J Clin Exp Med 8:14793–14804
67. Kensler TW, Wakabayashi N, Biswal S (2007) Cell survival responses to environmental stresses via the Keap1-Nrf2-ARE pathway. Annu Rev Pharmacol Toxicol 47:89–116
68. Sun Z, Huang Z, Zhang DD (2009) Phosphorylation of Nrf2 at multiple sites by MAP kinases has a limited contribution in modulating the Nrf2-dependent antioxidant response. Plos One 4:e6588
69. Martin D, Rojo AI, Salinas M et al (2004) Regulation of heme oxygenase-1 expression through the phosphatidylinositol 3-kinase/Akt pathway and the Nrf2 transcription factor in response to the antioxidant phytochemical carnosol. J Biol Chem 279:8919–8929
70. Yet SF, Tian R, Layne MD et al (2001) Cardiac-specific expression of heme oxygenase-1 protects against ischemia and reperfusion injury in transgenic mice. Circ Res 89:168–173
71. Collino M, Pini A, Mugelli N et al (2013) Beneficial effect of prolonged heme oxygenase 1 activation in a rat model of chronic heart failure. Dis Model Mech 6:1012–1020
72. Ying Z, Yue P, Xu X et al (2009) Air pollution and cardiac remodeling: a role for RhoA/Rho-kinase. Am J Physiol Heart Circ Physiol 296:H1540–H1550
73. Raffetto JD, Khalil RA (2008) Matrix metalloproteinases and their inhibitors in vascular remodeling and vascular disease. Biochem Pharmacol 75:346–359
74. Schiffrin EL (2004) Vascular stiffening and arterial compliance. Implications for systolic blood pressure. Am J Hypertens 17:39S–48S
75. Gurjar MV, DeLeon J, Sharma RV, Bhalla RC (2001) Mechanism of inhibition of matrix metalloproteinase-9 induction by NO in vascular smooth muscle cells. J Appl Physiol 91:1380–1386
76. Turrens JF, Thornton J, Barnard ML et al (1992) Protection from reperfusion injury by preconditioning hearts does not involve increased antioxidant defenses. Am J Physiol 262:H585–H589
77. Arduini A, Mezzetti A, Porreca E et al (1988) Effect of ischemia and reperfusion on antioxidant enzymes and mitochondrial inner membrane proteins in perfused rat heart. Biochim Biophys Acta 970:113–121

78. Das DK, Engelman RM, Kimura Y (1993) Molecular adaptation of cellular defences following preconditioning of the heart by repeated ischaemia. Cardiovasc Res 27:578–584
79. Undyala V, Terlecky SR, Van der Heide RS (2011) Targeted intracellular catalase delivery protects neonatal rat myocytes from hypoxia-reoxygenation and ischemia-reperfusion injury. Cardiovasc Pathol 20:272–280
80. Chen Z, Siu B, Ho YS et al (1998) Overexpression of MnSOD protects against myocardial ischemia/reperfusion injury in transgenic mice. J Mol Cell Cardiol 30:2281–2289
81. Woo YJ, Zhang JC, Vijayasarathy C et al (1998) Recombinant adenovirus-mediated cardiac gene transfer of superoxide dismutase and catalase attenuates postischemic contractile dysfunction. Circulation 98:255–261
82. Barteková M, Simončíková P, Fogarassyová M et al (2015) Quercetin improves postischemic recovery of heart function in doxorubicin-treated rats and prevents doxorubicin-induced matrix metalloproteinase-2 activation and apoptosis induction. Int J Mol Sci 16:8168–8185
83. Cheng L, Jin Z, Zhao R et al (2015) Resveratrol attenuates inflammation and oxidative stress induced by myocardial ischemia-reperfusion injury: role of Nrf2/ARE pathway. Int J Clin Exp Med 8:10420–10428
84. Sun G, Li Y, Ji Z (2015) Atorvastatin attenuates inflammation and oxidative stress induced by ischemia/reperfusion in rat heart via the Nrf2 transcription factor. Int J Clin Exp Med 8:14837–14845
85. Qu D, Han J, Ren H et al (2016) Cardioprotective effects of astragalin against myocardial ischemia/reperfusion injury in isolated rat heart. Oxid Med Cell Longev 2016:8194690
86. Hoshida S, Kuzuya T, Fuji H et al (1993) Sublethal ischemia alters myocardial antioxidant activity in canine heart. Am J Physiol 264:H33–H39
87. Zhou X, Zhai X, Ashraf M (1996) Direct evidence that initial oxidative stress triggered by preconditioning contributes to second window of protection by endogenous antioxidant enzyme in myocytes. Circulation 93:1177–1184
88. Tang XL, Qiu Y, Turrens JF et al (1997) Late preconditioning against stunning is not mediated by increased antioxidant defenses in conscious pigs. Am J Physiol 273:H1651–H1657
89. Ikizler M, Erkasap N, Dernek S et al (2007) Dietary polyphenol quercetin protects rat hearts during reperfusion: enhanced antioxidant capacity with chronic treatment. Anadolu Kardiyol Derg 7:404–410
90. Bartekova M, Carnicka S, Pancza D et al (2010) Acute treatment with polyphenol quercetin improves postischemic recovery of isolated perfused rat hearts after global ischemia. Can J Physiol Pharmacol 88:465–471
91. Li J, Xie C, Zhuang J et al (2015) Resveratrol attenuates inflammation in the rat heart subjected to ischemia-reperfusion: Role of the TLR4/NF-kB signaling pathway. Mol Med Rep 11:1120–1126
92. Mokni M, Hamlaoui S, Karkouch I et al (2013) Resveratrol provides cardioprotection after ischemia/reperfusion injury via modulation of antioxidant enzyme activities. Iran J Pharm Res 12:867–875
93. Mukhopadhyay P, Mukherjee S, Ahsan K et al (2010) Restoration of altered microRNA expression in the ischemic heart with resveratrol. PLoS One 5:e15705
94. Das S, Santani DD, Dhalla NS (2007) Experimental evidence for the cardioprotective effects of red wine. Exp Clin Cardiol 12:5–10
95. Khandelwal VK, Balaraman R, Pancza D, Ravingerová T (2011) Hemidesmus indicus and Hibiscus rosa-sinensis affect ischemia reperfusion injury in isolated rat hearts. Evid Based Complement Alternat Med 2011:802937
96. Jian J, Xuan F, Qin F, Huang R (2015) Bauhinia championii flavone inhibits apoptosis and autophagy via the PI3K/Akt pathway in myocardial ischemia/reperfusion injury in rats. Drug Des Devel Ther 9:5933–5945
97. Brosková Z, Drábiková K, Sotníková R et al (2013) Effect of plant polyphenols on ischemia-reperfusion injury of the isolated rat heart and vessels. Phytother Res 27:1018–1022

98. Yang Y, Duan W, Lin Y et al (2013) SIRT1 activation by curcumin pretreatment attenuates mitochondrial oxidative damage induced by myocardial ischemia reperfusion injury. Free Radic Biol Med 65:667–679
99. Li W, Wu M, Tang L et al (2015) Novel curcumin analogue 14p protects against myocardial ischemia reperfusion injury through Nrf2-activating anti-oxidative activity. Toxicol Appl Pharmacol 282:175–183
100. Turan B, Saini HK, Zhang M et al (2005) Selenium improves cardiac function by attenuating the activation of NF-kappaB due to ischemia-reperfusion injury. Antioxid Redox Signal 7:1388–1397
101. Sethi R, Takeda N, Nagano M, Dhalla NS (2000) Beneficial effects of vitamin E treatment in acute myocardial infarction. J Cardiovasc Pharmacol Ther 5:51–58
102. Maddika S, Elimban V, Chapman D, Dhalla NS (2009) Role of oxidative stress in ischemia-reperfusion-induced alterations in myofibrillar ATPase activities and gene expression in the heart. Can J Physiol Pharmacol 87:120–129
103. Senturk T, Çavun S, Avcı B et al (2014) Effective inhibition of cardiomyocyte apoptosis through the combination of trimetazidine and N-acetylcysteine in a rat model of myocardial ischemia and reperfusion injury. Atherosclerosis 237:760–766
104. Yesilbursa D, Serdar A, Senturk T et al (2006) Effect of N-acetylcysteine on oxidative stress and ventricular function in patients with myocardial infarction. Heart Vessels 21:33–37
105. Stone PH, Lloyd-Jones DM, Kinlay S et al (2005) Effect of intensive lipid lowering, with or without antioxidant vitamins, compared with moderate lipid lowering on myocardial ischemia in patients with stable coronary artery disease: the vascular basis for the treatment of myocardial ischemia study. Circulation 111:1747–1755
106. Rodrigo R, Hasson D, Prieto JC et al (2014) The effectiveness of antioxidant vitamins C and E in reducing myocardial infarct size in patients subjected to percutaneous coronary angioplasty (PREVEC Trial): study protocol for a pilot randomized double-blind controlled trial. Trials 15:192–201

Chapter 10
Reduced Oxidative Stress as a Mechanism for Increased Longevity, Exercise and Heart Failure Protection with Adenylyl Cyclase Type 5 Inhibition

Stephen F. Vatner, Jie Zhang, and Dorothy E. Vatner

Abstract One common mechanism mediating longevity and healthful aging is the protection against oxidative stress. Most serious diseases that limit aging, in general, and healthful aging, in particular, are linked to increased oxidative stress and increased sympathetic stimulation. Conversely, inhibition of sympathetic tone at any level in the β-adrenergic receptor/adenylyl cyclase/G-protein signaling pathway is a major mechanism protecting against oxidative stress. Disruption of adenylyl cyclase type 5 (AC5), one of the two major isoforms of AC in the heart, protects against oxidative stress resulting in enhanced longevity and more importantly healthful aging as exemplified by increased exercise performance, and protection against diabetes, obesity, cardiomyopathy and cancer, all related to oxidative stress. Despite the overwhelming evidence in animal models that increased oxidative stress is a major mechanism in limiting healthful aging, disappointing clinical trials have impaired the translation to patients. Inhibition of AC5 is a potential novel therapeutic modality, since it extends longevity and protects against diabetes, obesity, cardiomyopathy and cancer, while improving exercise tolerance, with all having an oxidative stress component.

Keywords Oxidative stress • Human longevity • Healthful human aging • Cardiomyopathy • Sympathetic tone • Adenylyl cyclase type 5

1 Introduction

One common mechanism mediating longevity and healthful aging is the protection against oxidative stress. Importantly, most serious diseases that limit aging, in general, and healthful aging, in particular, are linked to increased oxidative stress.

S.F. Vatner (✉) • J. Zhang • D.E. Vatner
Department of Cell Biology and Molecular Medicine, Rutgers, New Jersey Medical School, Newark, NJ 07103, USA
e-mail: vatnersf@njms.rutgers.edu

© Springer International Publishing Switzerland 2016
R.J. Gelpi et al. (eds.), *Biochemistry of Oxidative Stress*, Advances in Biochemistry in Health and Disease 16, DOI 10.1007/978-3-319-45865-6_10

Increased sympathetic stimulation is involved in many of the diseases that limit healthful aging and conversely, inhibition of sympathetic tone is a major mechanism protecting against oxidative stress. Reducing sympathetic tone may be achieved inhibiting any component of the β-adrenergic receptor/adenylyl cyclase/G-protein signaling pathway. Disruption of adenylyl cyclase type 5 (AC5), one of the two major isoforms of AC in the heart, protects against oxidative stress resulting in enhanced longevity and more importantly, healthful aging, as exemplified by increased exercise performance, protection against diabetes, obesity, cardiomyopathy and cancer, all diseases related to oxidative stress [1–11].

2 Aging and Oxidative Stress

One common mechanism mediating longevity and healthful aging is the protection against oxidative stress [1–9, 12–26]. Table 10.1 (and Fig. 10.1d) illustrate that the most commonly studied longevity models have as a common theme, that is protection against oxidative stress. Oxidative stress is defined as an imbalance between the oxidant and antioxidant compounds, favoring the first [27]. Superoxide radical (O_2^-) has a prominent role in oxidative stress and impacts the production of many other reactive species, such as H_2O_2, peroxynitrite ($ONOO^-$) and peroxynitrite degradation products ($HO•$, NO_2, $CO_3•^-$), and lipid peroxyl ($ROO•$) and alkoxyl ($RO•$) radicals [28].

Mammals have three superoxide dismutases (SOD); CuZnSOD (SOD1, EC 1.15.1.1) which is located in the mitochondrial intermembrane space, cytosol and extracellular space, MnSOD (SOD2, EC 1.15.1.1) which exists in the mitochondrial matrix, and EC-SOD (SOD3, EC 1.15.1.1) which is a copper and zinc-containing enzyme exclusively located in extracellular spaces [29]. The enzyme, MnSOD, is key among endogenous antioxidant mechanisms as it maintains reactive O_2 species at subnanomolar levels. Without this mechanism, oxidative stress becomes overwhelming making death inevitable [30]. As an increase in oxygen reactive species results in oxidative stress, MnSOD protects against oxidative stress.

A major factor mediating enhanced oxidative stress in limiting healthful aging is to recognize that most serious diseases that limit aging, in general, and healthful aging, in particular, are linked to increased oxidative stress, *e.g.*, heart disease [31], cancer [32], pulmonary disease [33], stroke [34], Alzheimer's disease [35], diabetes [36] and chronic kidney disease [37]. As these diseases reduce longevity and healthful aging, it clearly demonstrates the importance of oxidative stress in limiting longevity. The converse is also worth noting. Premature aging and early death are best exemplified by the Hutchinson-Gilford progeria syndrome. This is an extremely rare genetic disease, that occurs due to a single nucleotide mutation in the LMNA gene [38, 39] that leads to accumulation of progerin, the mutant form of lamin A, in different cells of the body specially in the endothelial and smooth muscle cells [40] resulting in increased oxidative stress [41]. Patients with progeria are afflicted with an accelerated aging process that affects the skin, and the musculoskeletal and cardiovascular systems. Given the accelerated atherosclerotic process on these patients, they usually die between the first and second decade of life from myocardial infarction and stroke [42].

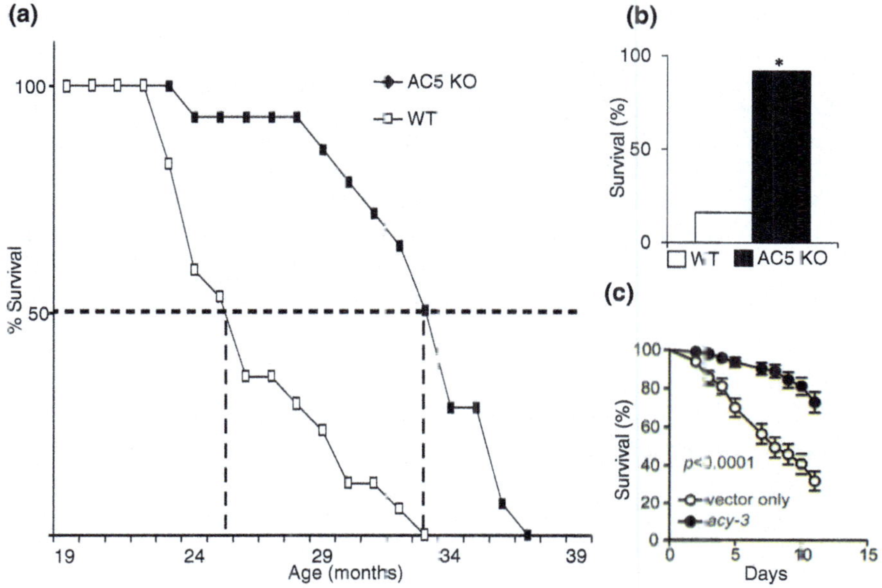

(d)

AC5KO [1-9]	Klotho OE [17]	Agtr1a KO [22]
Caloric Restriction [12]	p66shc KO [18]	Irs1 KO [23]
Snell Dwarf [13,14]	TRX OE [19]	R6/2•Irs2$^{+/-}$•IRSBtg [24]
GHR KO [15]	MCAT [20]	Surf1 KO [25]
Igf1r+/- [16]	MT OE [21]	Mclk1$^{+/-}$ [26]

Fig. 10.1 AC5 KO mice live longer and have better cardiac function and less remodeling with aging. (**a**) A cohort of female and male WT (n = 25) and AC5 KO mice (n = 13) demonstrated a significant survival difference between groups in the Kaplan-Meier curve analysis, $p < 0.01$. The *dotted line* indicates the time of 50 % survival. Roughly 50 % of WT mice died by 25 months, while 50 % of AC5 KO mice died by 33 months, $p < 0.01$. The maximum survival was also significantly different, $p < 0.02$, by the chi-square test. (**b**) At 30 months, only 16 % of WT mice survived, whereas 92 % of AC5 KO mice were still alive, $p < 0.05$ (**c**) C. elegans worms with down-regulated acy-3, the worm homologue to AC5, were exposed to oxidative stress with paraquat, and demonstrated significantly better survival upon exposure to paraquat, reflecting enhanced oxidative stress resistance, which was also observed in AC5 KO mouse. (**d**) Table of Longevity Models Related to Oxidative Stress Resistance (Redrawn from data in Yan et al. [2], Vatner et al. [3])

3 Sympathetic Tone and Oxidative Stress

Increased sympathetic stimulation is involved in many of the diseases that limit healthful aging, most particularly heart failure, and in turn is a major mechanism inducing increased oxidative stress [43–45]. Conversely, inhibition of sympathetic tone is a major mechanism protecting against oxidative stress. Reducing sympathetic tone may be mediated by blocking the effects of catecholamines and sympathetic nerve activity or at any level in the β-adrenergic receptor/adenylyl cyclase/G-protein signaling pathway [46, 47]. Adenylyl cyclase (AC) is a ubiquitous enzyme, which links β-adrenergic receptor (β-AR) stimulation to increased cAMP levels, and which regulates all organs and catalyzes the conversion of ATP to cAMP. There are two major AC isoforms in the heart, AC5 and AC6. Disruption of AC5 in mice, the AC5 knockout (KO) model protects against oxidative stress by reducing cAMP and protein kinase A (PKA), which in turn activates the Raf/MEK/ERK pathway, which increases MnSOD and protects against oxidative stress [2] (Fig. 10.2). As noted above, protection against oxidative stress is a common mechanism in longevity models (Fig. 10.1d), and is indeed a major mechanism mediating the increased longevity observed in the AC5 KO model (Figs. 10.1 and 10.2).

4 Mechanisms of AC5 KO Induced Longevity and Oxidative Stress Resistance

It is well recognized that oxidative stress is enhanced during the aging process [48]. The mechanism of oxidative stress induced aging seems to be multi-factorial, as noted by the variety of longevity models with protection against oxidative stress (Fig, 10.1d). Oxidative stress results in premature cell death with increased apoptosis and necrosis [49, 50]. The AC5 KO mouse increases longevity and stress resistance via activation of the Raf/MEK/ERK signaling pathway (Fig. 10.2) [2]. This signaling pathway is upregulated in the AC5 KO aging model [2] and is recognized as one of the main stress signaling pathways and central mediators activated in response to oxidative damage [51, 52]. Previous findings suggest that a decrease in the activation of the Raf/MEK/ERK pathway is associated with aging [53–56]. In addition, MnSOD, which is a major molecule protecting against oxidative stress, is upregulated in AC5 KO mice (Fig. 10.2) [2], but down-regulated, when AC5 is upregulated, as in the AC5 transgenic (Tg) heart (Fig. 10.2b) [7]. Thus, MnSOD is the downstream enzyme involved in the ERK signaling cascade [2, 7], protecting against oxidative stress and mediating longevity and stress resistance in AC5 KO mice (Fig. 10.2).

5 Oxidative Stress in Cardiomyopathy and Heart Failure

There is accumulating evidence that increased oxidative stress is involved in the pathogenesis of various types of cardiomyopathy [57–61] and affects cell types other than myocytes involved in mediating cardiomyopathy, *e.g.*, endothelial cells and fibroblasts. The AC5 KO model is protected against the cardiomyopathies of aging and heart failure, in part through protection against oxidative stress [7, 9] (Figs. 10.1, 10.2, and 10.3). Chronic β-AR stimulation induces cardiomyopathy and heart failure by increasing oxidative stress resulting in myocyte necrosis and apoptosis [1]. Transgenic mice with increased MnSOD are protected against the cardiomyopathy that develops with chronic catecholamine stimulation. Conversely, the AC5 KO model is protected against oxidative stress and catecholamine induced

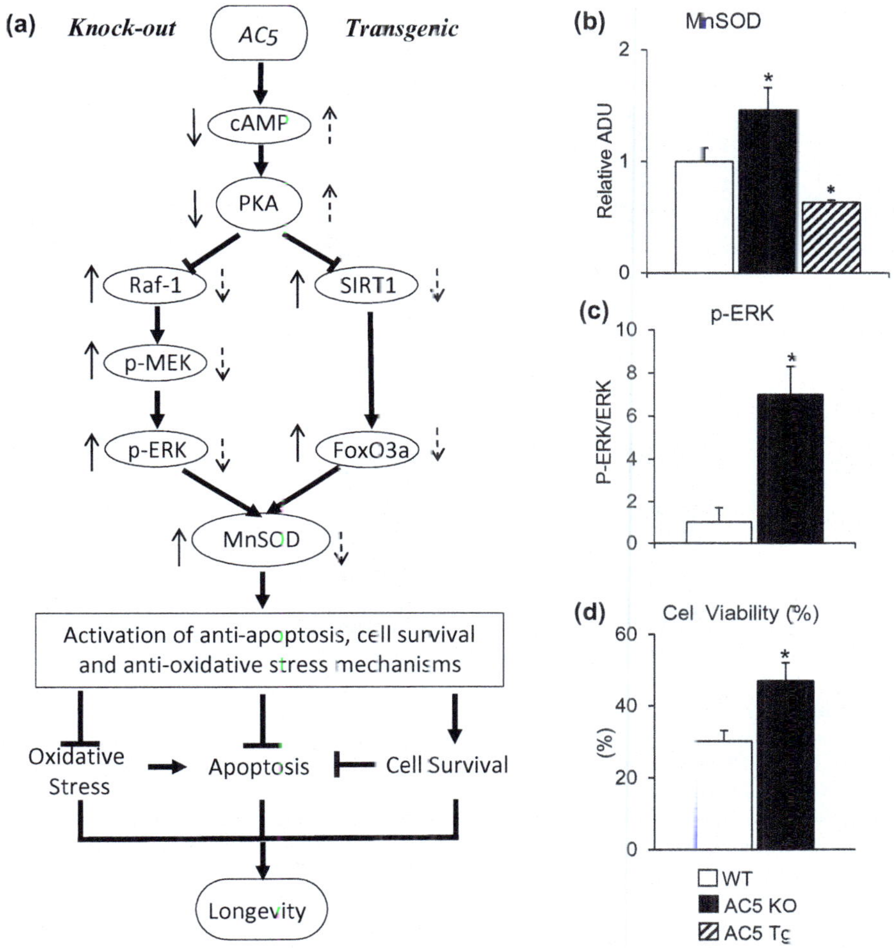

Fig. 10.2 ERK Pathway and Oxidative Stress in Aging AC5 KO. (**a**) The deletion of AC5 (*solid arrows*) reduces the intracellular cAMP production that leads to less activation of protein kinase PKA. PKA is one of the direct inhibitors of the cytoplasmic enzyme Raf-1. The increased activity of Raf-1 due to the reduced activity of PKA induces the complex formation between Raf-1 and MEK, and its subsequent activation. The activation of MEK leads to the translocation of ERK into the nucleus and the activation of a cascade that regulates cell proliferation and differentiation. On the right of the signaling diagram (*broken arrows*), the AC5 Tg model is shown to exert the opposite effect, ultimately reducing MnSOD, through a cAMP/PKA mechanism, SIRT1, which once activated interacts with FoxO3a. The deletion of AC5 eliminates the inhibition of SIRT1, inducing interaction between SIRT1 and FoxO3, translocation to the nucleus and enhancement of MnSOD expression. The overexpression of AC induces the opposite effect on the mediators. (**b**) The levels of MnSOD, determined by Western blotting, were significantly upregulated in AC5 KO mice compared to WT mice and down-regulated in AC5 Tg mouse hearts. (**c**) By western blotting, the level of ERK phosphorylation was significantly increased in AC5 KO mice compared with WT. (**d**) Cell viability was tested in response to oxidative stress in neonatal cardiomyocytes from AC5 KO and WT mice. Cardiomyocytes were treated with H_2O_2 (25 μM) and evaluated for cell viability using Cell Titer-Blue Cell viability assay. AC5 KO neonatal cardiomyocytes showed resistance to oxidative stress. *$p < 0.05$. Data are expressed as mean ± SEM (Redrawn from data in Yan et al. [2], Lai et al. [7])

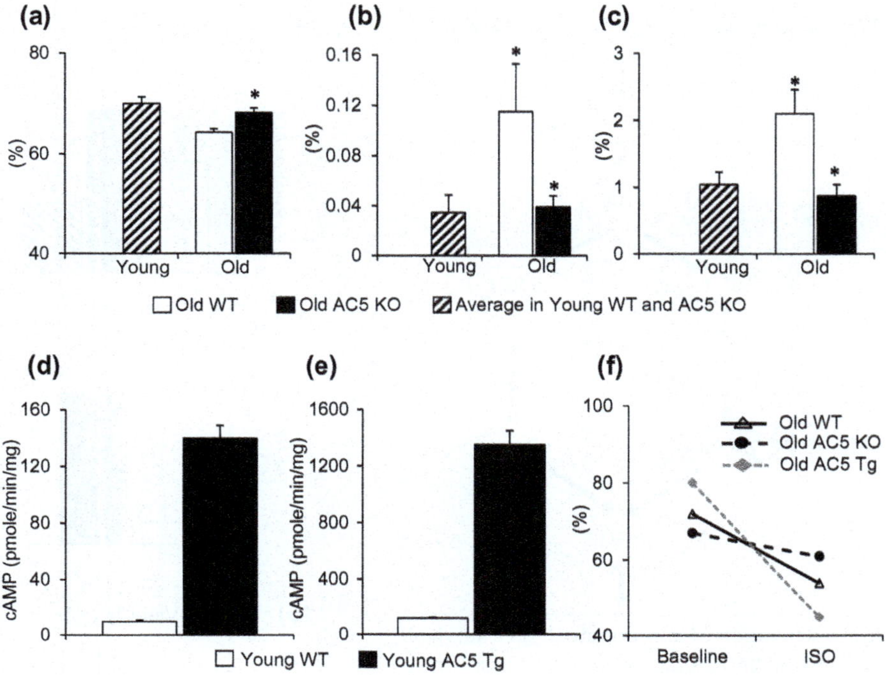

Fig. 10.3 Protection of Aging Cardiomyopathy (*Top*) and Catecholamine Cardiomyopathy (*Bottom*). Comparison of (**a**) LV ejection fraction, (**b**) LV apoptosis, and (**c**) LV fibrosis between AC5 KO and WT mice, in young (3–6 months, n = 4–9) and old mice (20–30 months, n = 4–9). Data from young WT and young AC5 KO mice were combined, since there were no significant differences (*shaded bars*). LV Ejection Fraction, Apoptosis and Fibrosis were significantly different in old WT (*open bars*) compared with young WT, characteristic of aging cardiomyopathy. In contrast, LV Ejection Fraction, Apoptosis and Fibrosis were different in old versus young AC5 KO, but were protected in aging AC5 KO and no longer significantly different from young WT. (**d**) and (**e**), AC activity at baseline and in response to forskolin stimulation was enhanced in AC5 Tg mice compared to WT. (**f**) Comparison of LV ejection fraction between AC5 KO, AC5 Tg and WT mice before and after chronic isoproterenol (ISO) infusion. Chronic ISO induced more severe cardiomyopathy in AC5 Tg, as reflected by the greater decline in LV ejection fraction, but less severe in AC5 KO mice, as reflected by the less severe decline in LV Ejection Fraction. Data are expressed as mean ± SEM, *$p < 0.05$ (Redrawn from data in Yan et al. [2], Okumura et al. [62], Lai et al. [7])

cardiomyopathy, but when oxidative stress protection was reduced by mating the AC5 KO mice with MnSOD +/− mice, there was a loss of the protection against the decreased cardiac function and increased cardiac fibrosis in response to chronic catecholamine stimulation in the double knockouts stress [7, 9]. Furthermore, AC5 KO mice are prevented from the cardiomyopathy induced by chronically enhanced β-AR signaling through enhancing resistance to oxidative stress [1]. These findings confirm the importance of oxidative stress in the pathogenesis of heart failure in general and in the protection afforded by the AC5 KO model in particular

6 Oxidative Stress Mediates the Augmented Cardiomyopathy in AC5 Transgenic Mice in Response to Chronic Catecholamine Stimulation

As noted above, we found that oxidative stress is an important mechanism mediating the enhanced longevity and protection against cardiomyopathy of aging in the AC5 KO mouse, where life span is increased through protection against oxidative stress by up-regulating the anti-oxidant, MnSOD [2] (Figs. 10.2 and 10.3). MnSOD also protects against the cardiomyopathy induced by increased AC5, as in the AC5 TG mouse, and eliminating the protective action of MnSOD abolishes the protection in the AC5 KO mouse against chronic catecholamine cardiomyopathy (Figs. 10.2 and 10.3) [7, 9]. In our experiments we have shown that the expression of MnSOD is tightly regulated not only through the MEK/ERK pathway but also by FoxO3a/SIRT1 mechanisms [7, 9, 63–65].

7 Reduced Oxidative Stress Mediates Enhanced Exercise in AC5 KO Mice

In contrast to the pathological mechanisms of cardiomyopathy and heart failure, exercise is the most physiological mediator of increased sympathetic activity, and exercise tolerance is enhanced in the AC5 KO (Fig. 10.4). The role of exercise and oxidative stress is a double-edged sword. On the one hand, it is well recognized that sympathetic activity increases oxidative stress [43–45], making it logical that exercise should increase oxidative stress. On the other hand increased exercise is known to increase longevity and protect against many diseases, including heart disease and cancer [66–68] and insufficient physical activity is the cause of many conditions that limit longevity. There are multiple cellular mechanisms mediating resistance to oxidative stress, by regular moderate exercise. These include reduction of basal formation of oxidants, improvement of the antioxidant defense system, and increased resistance of tissues against ROS damage [69]. Increased mitochondrial biogenesis, which is known to improve exercise performance is reduced in aging [70, 71] and provides a biological mechanism mediating the enhanced exercise tolerance in the AC5 KO model. The AC5 KO mice have increased exercise capacity mediated by higher oxidative stress resistance, mitochondrial biogenesis and enhancement in the mitochondrial activity as reflected by higher mitochondrial DNA content, increased ATP content, citrate synthase activity, and complex IV activity (Fig. 10.5). Furthermore, these in vivo observations were recapitulated in skeletal muscle myoblasts after knocking down their AC5. Altogether, our results indicate that improved mitochondrial number and function are involved in mediating the enhanced exercise capacity of AC5 KO mice.

The SIRT1/FoxO3a and MEK pathways, which are upregulated in AC5 KO, have also been shown to mediate mitochondrial biogenesis and oxidative stress [63,

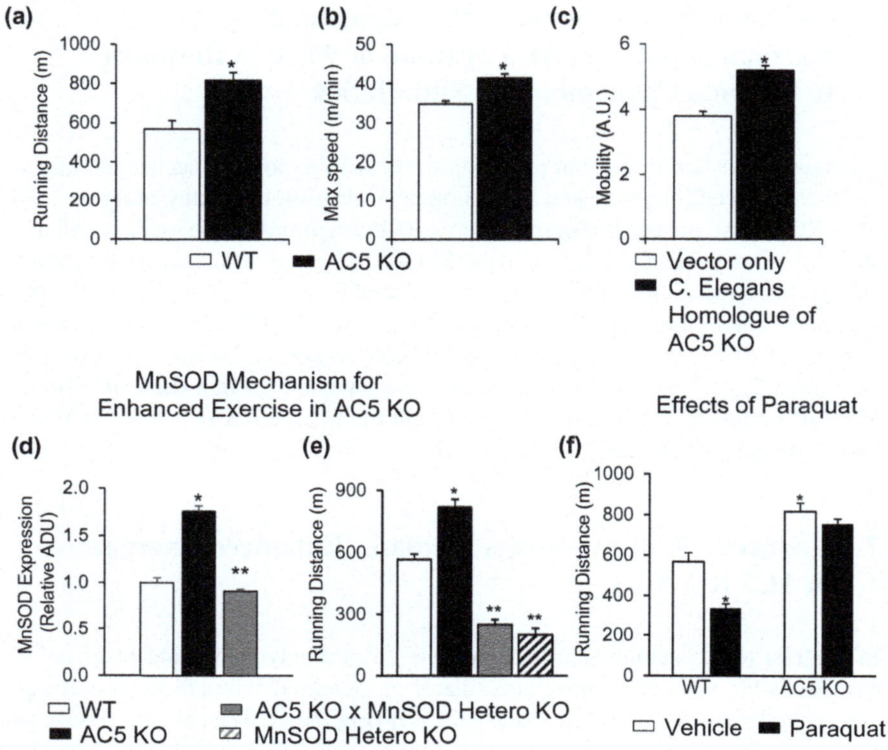

Fig. 10.4 AC5 KO mice have greater exercise capacity though increased oxidative stress resistance. Exercise capacity is enhanced in AC5 KO mice, as reflected by (**a**) running longer distances and (**b**) achieving higher speed. (**c**) The mobility, the mouse homologue to exercise, was enhanced in the C. elegans homologue of AC5 KO. (**d**) MnSOD expression was increased in the skeletal muscle from AC5 KO compared with WT mice; Mating AC5 KO mice with heterozygous MnSOD abolished increased MnSOD seen in the AC5 KO n = 4. (**e**) AC5 KO x MnSOD +/− had less exercise tolerance than the AC5 KO and similar to heterozygous MnSOD littermates. (**f**) Paraquat, which increases oxidative stress, significantly reduced exercise capacity in WT, but not in AC5 KO, indicating that the AC5 KO mice have better exercise tolerance at least partly due to higher oxidative stress tolerance. *p < 0.01 vs. WT and **p <0.05 vs. AC5 KO using one-way ANOVA; n = 5. Results are expressed as the mean ± SEM (Redrawn from data in Vatner et al. [3])

72], and mediate increased exercise performance [73, 74]. We confirmed that these pathways were also involved in mediating the enhanced exercise performance in the AC5 KO mouse by blocking SIRT1 activity using the inhibitor EX527 which abolished not only the enhanced exercise capacity, but also mitochondrial biogenesis in the AC5 KO mice (Fig. 10.5). The involvement of the Raf/MEK/ERK signaling pathway was confirmed by blocking the enhanced exercise performance in the AC5 KO by using the MEK inhibitor U0126 (Fig. 10.5) [3]. It is paradoxical that inhibition of the sympathetic tone, *e.g.* in the AC5 KO mouse results in reduced oxidative stress and improved exercise performance, although recognizing that exercise can

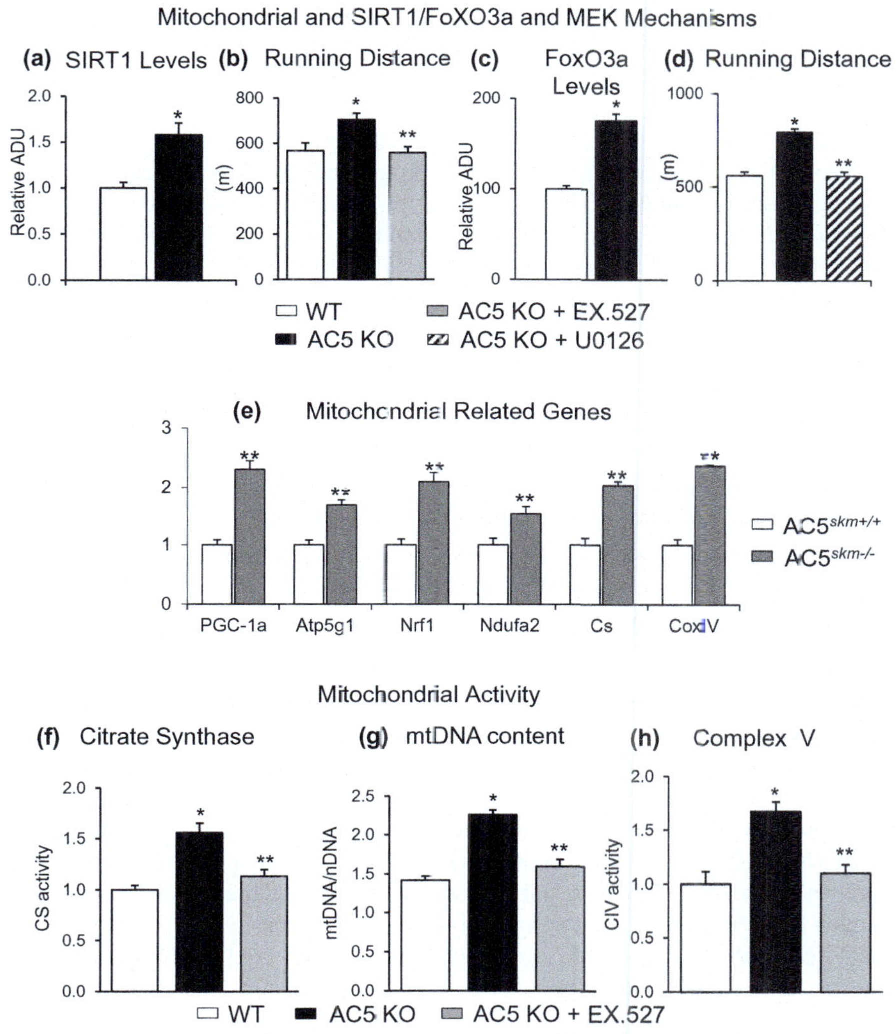

Fig. 10.5 Enhanced exercise tolerance in AC5 KO mice is mediated by mitochondrial biogenesis and the SIRT1 and MEK pathways. Western blot shows that (**a**) SIRT1 and (**c**) FoxO3a expression is increased in skeletal muscles of AC5 KO compared to WT mice, n = 4–6. (**b**) The enhanced exercise capacity in the AC5 KO mice was abolished with the SIRT1 inhibitor, EX 527 or (**d**) the MEK inhibitor, U0126, also abolished the enhanced exercise in AC5 KO mice. (**e**) Relative mRNA expression of mitochondria-related genes in the gastrocnemius muscle of AC5 SKM KO is shown: n = 6. (**f–h**) Increased mitochondrial biogenesis, as reflected by mitochondrial DNA content, as well as increased citrate synthase and complex IV activities in AC5 KO mice, were abolished after treatment with the SIRT1 inhibitor, EX 527. * $p < 0.01$ vs. WT and ** $p < 0.05$ vs. AC5 KO using one-way ANOVA. There were no significant differences between AC5 KO treated with either EX 527or U0126 and WT mice. Results are expressed as the mean ± SEM (Redrawn from data in Vatner et al. [3])

reduce oxidative stress [3]. The common denominator in that the AC5 KO mouse has an upregulation of MnSOD. To examine this mechanism, AC5 KO mice were treated with paraquat, which increases oxidative stress and reduces exercise performance in wild type mice but not in AC5 KO mice (Fig. 10.4f). Further studies were carried out with the AC5 KO mated with MnSOD heterozygous mice, resulting in a double KO model where MnSOD could no longer induce protection against oxidative stress. These double KO mice lost their ability for enhanced exercise, pinpointing the mechanism on increased MnSOD and reduced oxidative stress in AC5 KO mice. There is also a phylogenetic support for the role of AC5 as a key gene involved in healthy aging. The *C. elegans* worm acy-3 KO (the worm homolog of AC5) is a far more basic model than the AC5 KO mouse, but also shows greater longevity (Fig. 10.1c) and enhanced mobility (Fig. 10.4c) [3], akin to the AC5 KO model on increased longevity and exercise. The worm mutant showed improved mitochondrial metabolism and enhanced oxidative stress defense, via the induction of the expression of SOD-3, the worm homolog of MnSOD. The evolutionary conservation of the pathway involving decreased AC5 and increased MnSOD supports the importance and essential nature of this signaling mechanism.

8 Inhibition of AC5 Is Unique to All the AC Isoforms

It is noteworthy that disruption of the other 8 major mammalian isoforms of AC, which share regulation of many tissues or organs in the body [74, 75], have been reported to have the combined salutary features of enhanced longevity, exercise and protection against diabetes, obesity, cancer and heart disease. Potentially the best example relates to the two major isoforms of AC in the heart, AC5 and AC6. Inhibiting AC6 shares none of the salutary features of inhibiting AC5, as summarized in this review, with the majority of AC6 transgenic studies showing cardioprotection [76–79]. In addition, the AC5 KO is resistant to obesity [8], but AC6 KO did not affect body weight and AC3 KO actually induced obesity [80]. More relevant to this review, almost none of the other 8 AC isoforms have been shown to regulate oxidative stress. It could well be that other AC isoforms do regulate oxidative stress, but that this just has not been studied yet and would be an important future direction for AC research. It is also conceivable that the protection observed in the AC5 KO model, is not simply due to its ability to reduce AC activity, but rather to a related action, *e.g.*, its ability to upregulate MnSOD and protect against oxidative stress.

9 Clinical Translation

Despite the compelling evidence in animal models that reduction of oxidative stress is beneficial to treat cardiovascular diseases, the results in clinical trials and the translation to patients with cardiovascular disease have been controversial. Perhaps the clinical trial Q-SYMBIO, which aimed to test the use of the antioxidant coenzyme Q10 in patients with moderate to severe heart failure, showed the most promising results. In that study patients treated with coenzyme Q10 had 50 % less major cardiovascular events than the placebo treated, however given the underpowered nature of the study to assess mortality it still needs to be examined in a larger cohort. Vitamin E is another antioxidant compound that has been tested in multiple large and long clinical trials with contradictory results. For instance, the Cambridge Heart Antioxidant Study (CHAOS), showed a reduced rate of myocardial infarction with no mortality benefit [82]. Further studies on vitamin E, for example "The Physicians' Health Study II" [83] and "The Women's Antioxidant Cardiovascular Study" [84] did not show any beneficial effect of this vitamin supplementation. There are even some studies that question the safety of this vitamin, since there have been claims that this vitamin may increase mortality [85, 86]. Other compounds with antioxidant properties studied in clinical trials, such as vitamin C [83, 84] and the precursor of vitamin A, β carotene [84], have not shown any beneficial effect in terms of protecting cardiovascular events [87]. In view of the contradiction between the overwhelming evidence in animal models of cardiovascular disease demonstrating that increased oxidative stress is a major mechanism in the pathogenesis on the one hand and the disappointing results from clinical trials of agents that reduce oxidative stress, it seems likely that new therapeutic modalities will be discovered in the future. One candidate would be inhibition of AC5, since it extends longevity and protects against diabetes, obesity and cardiomyopathy, while improving exercise tolerance, with all having an oxidative stress component. Unfortunately disrupting the AC5 gene in patients is not feasible and therefore it becomes necessary to identify a pharmacological inhibitor of AC5. One example of a pharmacological compound that replicates many of the features of AC5 inhibition is an FDA approved anti-viral drug, AraA, which protects against the development of cardiomyopathy in mice [11], and more recently has been shown to be one of the rare interventions that can reduce infarct size when administered after coronary artery reperfusion [10], which is important clinically, since patients with myocardial infarction must be reperfused immediately and cannot wait for a drug to be administered. Accordingly, further work is required to develop a non-toxic AC5 inhibitor that can be given to patients orally.

10 Conclusions

Protection against oxidative stress is a common mechanism mediating longevity, and more importantly mediating healthful aging, since most serious diseases that limit healthful aging are linked to increased oxidative stress. Increased sympathetic stimulation, which increases oxidative stress, is involved in many of the diseases that limit healthful aging and conversely, inhibition of sympathetic tone is a major mechanism protecting against oxidative stress. Reducing sympathetic tone at the level of AC, in the AC5 KO mouse, not only extends longevity but also increases exercise performance, and protects against diabetes, obesity, cardiomyopathy and cancer, all related to oxidative stress [1–9], making AC5 a potential novel therapeutic modality for protecting against oxidative stress.

References

1. Yan L, Vatner SF, Vatner DE (2014) Disruption of type 5 adenylyl cyclase prevents beta adrenergic receptor cardiomyopathy: a novel approach to beta-adrenergic receptor blockade. Am J Physiol Heart Circ Physiol 307:H1521–H1528
2. Yan L, Vatner DE, O'Connor JP et al (2007) Type 5 adenylyl cyclase disruption increases longevity and protects against stress. Cell 130:247–258
3. Vatner DE, Yan L, Lai L et al (2015) Type 5 adenylyl cyclase disruption leads to enhanced exercise performance. Aging Cell 14:1075–1084
4. Okumura S, Takagi G, Kawabe J et al (2003) Disruption of type 5 adenylyl cyclase gene preserves cardiac function against pressure overload. Proc Natl Acad Sci U S A 100:9986–9990
5. Ho D, Zhao X, Yan L et al (2015) Adenylyl cyclase type 5 deficiency protects against diet-induced obesity and insulin resistance. Diabetes 64:2636–2645
6. De Lorenzo MS, Chen W, Baljinnyam E (2014) Reduced malignancy as a mechanism for longevity in mice with adenylyl cyclase type 5 disruption. Aging Cell 13:102–110
7. Lai L, Yan L, Gao S et al (2013) Type 5 adenylyl cyclase increases oxidative stress by transcriptional regulation of manganese superoxide dismutase via the SIRT1/FoxO3a pathway. Circulation 127:1692–1701
8. Yan L, Park JY, Dillinger JG et al (2012) Common mechanisms for calorie restriction and adenylyl cyclase type 5 knockout models of longevity. Aging Cell 11:1110–1120
9. Vatner SF, Park M, Yan L et al (2013) Adenylyl cyclase type 5 in cardiac disease, metabolism, and aging. Am J Physiol Heart Circ Physiol 305:H1–H8
10. Bravo C, Vatner DE, Pachon R et al (2016) An FDA approved anti-viral agent which inhibits adenylyl cyclase type 5 protects the ischemic heart even when administered after reperfusion. J Pharmacol Exp Ther 357:331–336
11. Iwatsubo K, Bravo C, Uechi M et al (2012) Prevention of heart failure in mice by an antiviral agent that inhibits type 5 cardiac adenylyl cyclase. Am J Physiol Heart Circ Physiol 302:H2622–H2628
12. Qiu X, Brown K, Hirschey MD et al (2010) Calorie restriction reduces oxidative stress by SIRT3-mediated SOD2 activation. Cell Metab 12:662–667
13. Flurkey K, Papaconstantinou J, Harrison DE (2002) The snell dwarf mutation pit1(dw) can increase life span in mice. Mech Ageing Dev 123:121–130
14. Salmon AB, Murakami S, Bartke A et al (2005) Fibroblast cell lines from young adult mice of long-lived mutant strains are resistant to multiple forms of stress. Am J Physiol Endocrinol Metab 289:E23–E29
15. Flurkey K, Papaconstantinou J, Miller RA et al (2001) Lifespan extension and delayed immune and collagen aging in mutant mice with defects in growth hormone production. Proc Natl Acad Sci U S A 98:6736–6741
16. Holzenberger M, Dupont J, Ducos B et al (2003) IGF-1 receptor regulates lifespan and resistance to oxidative stress in mice. Nature 421:182–187
17. Kurosu H, Yamamoto M, Clark JD et al (2005) Suppression of aging in mice by the hormone klotho. Science 309:1829–1833
18. Migliaccio E, Giorgio M, Mele S et al (1999) The p66shc adaptor protein controls oxidative stress response and life span in mammals. Nature 402:309–313
19. Mitsui A, Hamuro J, Nakamura H et al (2002) Overexpression of human thioredoxin in transgenic mice controls oxidative stress and life span. Antioxid Redox Signal 4:693–696

20. Schriner SE, Linford NJ, Martin GM et al (2005) Extension of murine life span by overexpression of catalase targeted to mitochondria. Science 308:1909–1911
21. Yang X, Doser TA, Fang CX et al (2006) Metallothionein prolongs survival and antagonizes senescence-associated cardiomyocyte diastolic dysfunction: role of oxidative stress. FASEB J 20:1024–1026
22. Benigni A, Corna D, Zoja C et al (2009) Disruption of the Ang II type 1 receptor promotes longevity in mice. J Clin Invest 119:524–530
23. Selman C, Lingard S, Choudhury AI et al (2008) Evidence for lifespan extension and delayed age-related biomarkers in insulin receptor substrate 1 null mice. FASEB J 22:807–818
24. Sadagurski M, Cheng Z, Rozzo A et al (2011) IRS2 increases mitochondrial dysfunction and oxidative stress in a mouse model of Huntington disease. J Clin Invest 121:4070–4081
25. Dell'agnello C, Leo S, Agostino A et al (2007) Increased longevity and refractoriness to Ca(2+)-dependent neurodegeneration in surf1 knockout mice. Hum Mol Genet 16:431–444
26. Lapointe J, Stepanyan Z, Bigras E et al (2009) Reversal of the mitochondrial phenotype and slow development of oxidative biomarkers of aging in long-lived Mclk1+'− mice. J Biol Chem 284:20364–20374
27. Gerschman R, Gilbert DL, Nye SW et al (1954) Oxygen poisoning and x-irradiation: a mechanism in common. Science 119:623–626
28. Forman HJ, Maiorino M, Ursini F (2010) Signaling functions of reactive oxygen species. Biochemistry 49:835–842
29. ZelkoIN MTJ, Folz RJ (2002) Superoxide dismutase multigene family: a comparison of the CuZn-SOD (SOD1), Mn-SOD (SOD2), and EC-SOD (SOD3) gene structures, evolution, and expression. Free Radic Biol Med 33:337–349
30. Miriyala S, Spasojevic I, Tovmasyan A et al (2012) Manganese superoxide dismutase and its mimics. Biochim Biophys Acta 1822:794–814
31. Dhalla NS, Temsah RM, Netticadan T (2000) Role of oxidative stress in cardiovascular diseases. J Hypertens 18:655–673
32. Valko M, Rhodes CJ, Moncol J et al (2006) Free radicals, metals and antioxidants in oxidative stress-induced cancer. Chem Biol Interact 160:1–40
33. Rahman I (2005) The role of oxidative stress in the pathogenesis of COPD: implications for therapy. Treat Respir Med 4:175–200
34. El Kossi MM, Zakhary MM (2000) Oxidative stress in the context of acute cerebrovascular stroke. Stroke 31:1889–1892
35. Markesbery WR (1997) Oxidative stress hypothesis in Alzheimer's disease. Free Radic Biol Med 23:134–147
36. Karunakaran U, Park KG (2013) A systematic review of oxidative stress and safety of antioxidants in diabetes: focus on islets and their defense. Diabetes Metab J 37:106–112
37. Oberg BP, McMenamin E, Lucas FL et al (2004) Increased prevalence of oxidant stress and inflammation in patients with moderate to severe chronic kidney disease. Kidney Int 65:1009–1016
38. De Sandre-Giovannoli A, Bernard R, Cau P et al (2003) Lamin a truncation in Hutchinson-Gilford progeria. Science 300:2055–2056
39. Eriksson M, Brown WT, Gordon LB et al (2003) Recurrent de novo point mutations in lamin A cause Hutchinson-Gilford progeria syndrome. Nature 423:293–298
40. McClintock D, Gordon LB, Djabali K (2006) Hutchinson-Gilford progeria mutant lamin A primarily targets human vascular cells as detected by an anti-lamin A G608G antibody. Proc Natl Acad Sci U S A 103:2154–2159
41. Viteri G, Chung YW, Stadtman ER (2010) Effect of progerin on the accumulation of oxidized proteins in fibroblasts from Hutchinson-Gilford progeria patients. Mech Ageing Dev 131:2–8
42. Hennekam RC (2006) Hutchinson-Gilford progeria syndrome: review of the phenotype. Am J Med Genet A 140:2603–2624
43. Singal PK, Yates JC, Beamish RE et al (1981) Influence of reducing agents on adrenochrome-induced changes in the heart. Arch Pathol Lab Med 105:664–669

44. Obata T, Hosokawa H, Yamanaka Y (1994) In vivo monitoring of norepinephrine and HO•. generation on myocardial ischemic injury by dialysis technique. Am J Physiol 266: H903–H908
45. Obata T, Yamanaka Y (1996) Cardiac microdialysis of salicylic acid to detect hydroxyl radical generation associated with sympathetic nerve stimulation. Neurosci Lett 211:216–318
46. Kukin ML, Kalman J, Charney RH et al (1999) Prospective, randomized comparison of effect of long-term treatment with metoprolol or carvedilol on symptoms, exercise, ejection fraction, and oxidative stress in heart failure. Circulation 99:2645–2651
47. Nakamura K, Kusano K, Nakamura Y et al (2002) Carvedilol decreases elevated oxidative stress in human failing myocardium. Circulation 105:2867–2871
48. Mehdi MM, Rizvi SI (2013) N, N-Dimethyl-p-phenylenediamine dihydrochloride-based method for the measurement of plasma oxidative capacity during human aging. Anal Biochem 436:165–167
49. Sastre J, Pallardo FV, Vina J (2003) The role of mitochondrial oxidative stress in aging. Free Radic Biol Med 35:1–8
50. Slater AF, Stefan C, Nobel I et al (1995) Signalling mechanisms and oxidative stress in apoptosis. Toxicol Lett 82–83:149–153
51. Finkel T, Holbrook NJ (2000) Oxidants, oxidative stress and the biology of ageing. Nature 408:239–247
52. Yoon SO, Yun CH, Chung AS (2002) Dose effect of oxidative stress on signal transduction in aging. Mech Ageing Develop 123:1597–1604
53. Lorenzini A, Tresini M, Mawal-Dewan M et al (2002) Role of the Raf/MEK/ERK and the PI3K/Akt(PKB) pathways in fibroblast senescence. Exp Gerontol 37:1149–1156
54. Hutter D, Yo Y, Chen W et al (2000) Age-related decline in Ras/ERK mitogen-activated protein kinase cascade is linked to a reduced association between Shc and EGF receptor. J Gerontol A Biol Sci Med Sci 55:B125–B134
55. Torres C, Francis MK, Lorenzini A et al (2003) Metabolic stabilization of MAP kinase phosphatase-2 in senescence of human fibroblasts. Exp Cell Res 290:195–206
56. Meloche S, Gopalbhai K, Beatty BG et al (2000) Chromosome mapping of the human genes encoding the MAP kinase kinase MEK1 (MAP2K1) to 15q21 and MEK2 (MAP2K2) to 7q32. Cytogenet Cell Genet 88:249–252
57. Romero-Alvira D, Roche E, Placer L (1996) Cardiomyopathies and oxidative stress. Med Hypotheses 47:137–144
58. Cesselli D, Jakoniuk I, Barlucchi L et al (2001) Oxidative stress-mediated cardiac cell death is a major determinant of ventricular dysfunction and failure in dog dilated cardiomyopathy. Circ Res 89:279–286
59. Wold LE, Ceylan-Isik AF, Ren J (2005) Oxidative stress and stress signaling: menace of diabetic cardiomyopathy. Acta Pharmacol Sin 26:908–917
60. Bandyopadhyay D, Chattopadhyay A, Ghosh G et al (2004) Oxidative stress-induced ischemic heart disease: protection by antioxidants. Curr Med Chem 11:369–387
61. Jacob MH, Pontes MR, Araujo AS et al (2006) Aortic-banding induces myocardial oxidative stress and changes in concentration and activity of antioxidants in male Wistar rats. Life Sci 79:2187–2193
62. Okumura S, Vatner DE, Kurotani R et al (2007) Disruption of type 5 adenylyl cyclase enhances desensitization of cyclic adenosine monophosphate signal and increases Akt signal with chronic catecholamine stress. Circulation 116:1776–1783
63. Brunet A, Sweeney LB, Sturgill JF et al (2004) Stress-dependent regulation of FoxO transcription factors by the SIRT1 deacetylase. Science 303:2011–2015
64. Salih DA, Brunet A (2008) FoxO transcription factors in the maintenance of cellular homeostasis during aging. Curr Opin Cell Biol 20:126–136
65. Sengupta A, Molkentin JD, Yutzey KE (2009) FoxO transcription factors promote autophagy in cardiomyocytes. J Biol Chem 284:28319–28331
66. Eaton CB (1992) Relation of physical activity and cardiovascular fitness to coronary heart disease: a meta-analysis of the independent relation of physical activity and coronary heart disease. J Am Board Fam Pract 5:31–42

67. Eaton CB (1992) Relation of physical activity and cardiovascular fitness to coronary heart disease: cardiovascular fitness and the safety and efficacy of physical activity prescription. J Am Board Fam Pract 5:157–165
68. Agress CM (1978) Energetics. Grosset & Dunlap, New York, 166 p
69. Kruk J, Duchnik E (2014) Oxidative stress and skin diseases: possible role of physical activity. Asian Pac J Cancer Prev 15:561–568
70. Rockl KS, Hirshman MF, Brandauer J et al (2007) Skeletal muscle adaptation to exercise training: AMP-activated protein kinase mediates muscle fiber type shift. Diabetes 56:2062–2069
71. Yamamoto H, Williams EG, Mouchiroud L et al (2011) NCoR1 is a conserved physiological modulator of muscle mass and oxidative function. Cell 147:827–839
72. Kops GJ, Dansen TB, Polderman PE et al (2002) Forkhead transcription factor FOXO3a protects quiescent cells from oxidative stress. Nature 419:316–321
73. Dufresne SD, Bjorbaek C, El-Haschimi K et al (2001) Altered extracellular signal-regulated kinase signaling and glycogen metabolism in skeletal muscle from p90 ribosomal S6 kinase 2 knockout mice. Mol Cell Biol 21:81–87
74. Chalkiadaki A, Igarashi M, Nasamu AS et al (2014) Muscle-specific SIRT1 gain-of-function increases slow-twitch fibers and ameliorates pathophysiology in a mouse model of Duchenne muscular dystrophy. PLoS Genet 10(7):e1004490
75. Onda T, Hashimoto Y, Nagai M et al (2001) Type-specific regulation of adenylyl cyclase. selective pharmacological stimulation and inhibition of adenylyl cyclase isoforms. J Biol Chem 276:47785–47793
76. Defer N, Best-Belpomme M, Hanoune J (2000) Tissue specificity and physiological relevance of various isoforms of adenylyl cyclase. Am J Physiol Renal Physiol 279:F400–F416
77. Takahashi T, Tang T, Lai NC et al (2006) Increased cardiac adenylyl cyclase expression is associated with increased survival after myocardial infarction. Circulation 114:388–396
78. Lai NC, Tang T, Gao MH et al (2008) Activation of cardiac adenylyl cyclase expression increases function of the failing ischemic heart in mice. J Am Coll Cardiol 51:1490–1497
79. Roth DM, Bayat H, Drumm JD et al (2002) Adenylyl cyclase increases survival in cardiomyopathy. Circulation 105:1989–1994
80. Roth DM, Gao MH, Lai NC et al (1999) Cardiac-directed adenylyl cyclase expression improves heart function in murine cardiomyopathy. Circulation 99:3099–3102
81. Wang Z, Li V, Chan GC et al (2009) Adult type 3 adenylyl cyclase-deficient mice are obese. PLoS One 4(9):e6979
82. Stephens NG, Parsons A, Schofield PM et al (1996) Randomised controlled trial of vitamin E in patients with coronary disease: Cambridge Heart Antioxidant Study (CHAOS). Lancet 347:781–786
83. Sesso HD, Buring JE, Christen WG et al (2008) Vitamins E and C in the prevention of cardiovascular disease in men: the Physicians' Health Study II randomized controlled trial. JAMA 300:2123–2133
84. Cook NR, Albert CM, Gaziano JM et al (2007) A randomized factorial trial of vitamins C and E and beta carotene in the secondary prevention of cardiovascular events in women: results from the Women's Antioxidant Cardiovascular Study. Arch Intern Med 167:1610–1618
85. Bjelakovic G, Nikolova D, Gluud LL et al (2007) Mortality in randomized trials of antioxidant supplements for primary and secondary prevention: systematic review and meta-analysis. JAMA 297:842–857
86. Lonn E, Bosch J, Yusuf S et al (2005) Effects of long-term vitamin E supplementation on cardiovascular events and cancer: a randomized controlled trial. JAMA 293:1338–1347
87. Myung SK, Ju W, Cho B et al (2013) Efficacy of vitamin and antioxidant supplements in prevention of cardiovascular disease: systematic review and meta-analysis of randomised controlled trials. BMJ 346:f10–f22

Chapter 11
Regulation of Protein Nitrosylation by Thioredoxin 1

Narayani Nagarajan and Junichi Sadoshima

Abstract Addition of an SNO group to a cysteine thiol in a protein is called protein S-nitrosylation. When the donor of the SNO group is another protein, then the transfer is referred to as trans-nitrosylation. One key protein that mediates trans-nitrosylation and denitrosylation is thioredoxin1 (Trx1). Trx1 is a 12 kDa redox protein with five cysteine residues: Cys32, 35, 62, 69 and 73. Factors such as pH conditions and cysteine residue redox status affect the S-nitrosylation of Trx1. Several studies have reported that Trx1 promotes trans-nitrosylation of caspase-3, ASK1, and peroxiredoxin-1. On the other hand, Trx1 can trigger denitrosylation of SNO-caspase-3, SNO-metallothionein, SNO-albumin, SNO-caspase-8, GAPDH, annexin-1, etc. By regulating the balance between S-nitrosylation and denitrosylation, Trx1 helps maintain the nitroso-redox balance in the cell. Trx1-mediated S-nitrosylation and denitrosylation of proteins affect several cellular functions, including apoptosis, heme protein maturation and insertion, and exocytosis as well as various pathophysiological conditions, such as inflammation, cancer and cardiovascular diseases. In this chapter, we shall discuss in detail the molecular mechanism of Trx1-mediated trans-nitrosylation and denitrosylation and its importance in the cellular milieu.

Keywords Thioredoxin 1 • Protein S-nitrosylation • Denitrosylation • Trans-nitrosylation • Cardiovascular diseases • Cancer • Inflammation

1 Introduction. Protein S-Nitrosylation

Protein S-nitrosylation is the covalent addition of a nitric oxide group (–NO) to a cysteine thiol in a protein [1]. S-nitrosothiol formation is selective to specific cysteine residues in a protein and, thus, is an important form of post-translational

N. Nagarajan • J. Sadoshima (✉)
Department of Cell Biology and Molecular Medicine, Cardiovascular Research Institute, Rutgers New Jersey Medical School, Newark, NJ 07103, USA
e-mail: sadoshju@njms.rutgers.edu

© Springer International Publishing Switzerland 2016
R.J. Gelpi et al. (eds.), *Biochemistry of Oxidative Stress*, Advances in Biochemistry in Health and Disease 16, DOI 10.1007/978-3-319-45865-6_11

modification. It can alter protein structure, protein-protein interaction, localization, and stability [2]. This significant second messenger system mediates several cellular signal transduction pathways [3–5].

1.1 Formation and Reduction of S-Nitrosothiols

The first question is where does this -NO come from? NO cannot directly oxidize amino acids and the kinetics of such reactions are too slow to be biologically possible. Hence, S-nitrosothiol formation occurs only as a secondary oxidation reaction [6, 7]. There are three mechanisms by which a protein can be S-nitrosylated, as follows:

Pathway 1. Oxygen-mediated pathway:

$$
\begin{aligned}
&NO + O_2 \rightarrow NO_3 \\
&NO_3 + NO \rightarrow 2\,NO_2 \\
&2\,NO_2 + RS^- \rightarrow RS\text{-}NO + NO_3^- \\
&(or) \\
&NO + NO_2 \rightarrow N_2O_3 \\
&N_2O_3 + RS^- \rightarrow RS\text{-}NO + NO_2^-
\end{aligned}
$$

Since NO cannot directly nitrosylate protein thiols, it mediates S-nitrosothiol formation indirectly. This occurs via a two-step mechanism, as described in the above equations. First, nitric oxide (NO) reacts with oxygen to form nitrogen trioxide (N_2O_3) or nitrogen dioxide (NO_2). These molecules then directly react with a protein thiol group to produce S-nitrosothiols [6, 7].

Pathway 2. Transition-metal ion mediated pathway:

$$
\begin{aligned}
&NO + \left[Cu^{2+}\right] - CP \rightarrow NO^+ + \left[Cu^+\right] - CP \\
&R\text{-}SH + NO^+ \rightarrow RSNO + H^+ \\
&R\text{-}SH + Cu^{2+} \rightarrow RS^- + Cu^+
\end{aligned}
$$

Metal ions such as Cu^{2+} and Fe^{3+} can catalyze S-nitrosothiol formation. This has been established by studies that showed that ceruloplasmin (CP) directly affects S-nitrosothiol formation. CP is a copper-containing plasma protein that mediates the oxidation of NO to generate a NO^+ ion. This can then nitrosylate protein thiols resulting in S-nitrosylation [8]. Alternatively, some reports suggest that the metal ions can mediate one-electron oxidation of thiols to thiyls radicals [6, 7] that can then interact with NO oxide to form S-nitrosothiols.

Pathway 3. Trans-nitrosylation

$$RSH + R'SNO \rightarrow RSNO + R'SH$$

Trans-nitrosylation is a mechanism of protein S-nitrosylation in which protein A, which is already S-nitrosylated, donates its NO group to another protein, protein B, resulting in the S-nitrosylation of protein B [3].

1.2 SNO-Reduction

S-nitrosylated protein thiols serve as signaling molecules. To regulate the signaling mechanism, S-nitrosylation is balanced by denitrosylation or SNO-reduction. Denitrosylation is the process of removal of a -NO group from a protein thiol, through either an enzymatic or non-enzymatic mechanism. Enzymatic denitrosylation is mediated by enzymes such as S-nitrosoglutathione reductase (GSNOR), thioredoxin reductase (TrxR), xanthine oxidase, etc. [9]. Non-enzymatic denitrosylation is mediated by free metal ions of copper or iron [10].

1.3 Importance of S-Nitrosylation in Biological Functions

The group of Stamler first reported that S-nitrosothiols are present in human circulation, suggesting that NO plays a significant role in various physiological and pathological conditions [11]. Following this discovery, various groups reported the significance of S-nitrosylation in cellular functions. S-nitrosylation signaling has been indicated in apoptosis through S-nitrosylation/denitrosylation of caspases, ASK1, JNK, and c-Jun, in the regulation of transcription factors like HIF, p53, and NF-KB, in calcium channel regulation by activation of ryanodine receptors [12], and of oncoproteins such as Src [5].

Any alteration in protein S-nitrosylation can, in turn, affect a variety of cellular functions, leading to various pathological conditions. Malfunctioning of this mechanism has been demonstrated in cardiovascular diseases, neurodegenerative disorders, and cancer [13, 14]. Sun et al. and Tong et al. have reported that there is an increased S-nitrosylation during ischemic preconditioning [15, 16]. One hypothesis is that S-nitrosylation of proteins protects the cysteine residues from irreversible oxidative damage and, hence, may be cardioprotective [2, 17]. S-nitrosylation of key proteins such as F_1-ATPase, SERCA2a, and PTEN are cardioprotective [16, 18]. On the other hand, aberrant S-nitrosylation of proteins like Bcl-2, p53, MKP-1, Fas, HIF-1 alpha, and Src tyrosine kinases plays a crucial role in the onset and progression of various cancers [13]. In addition, neurodegenerative disorders such as Alzheimer's and Parkinson's diseases are influenced by S-nitrosylation of parkin, PDI, Drp1, Cdk5, GAPDH, and UPS [19–21].

1.4 Regulation of S-Nitrosylation

At any given time, the balance between S-nitrosylation and denitrosylation of proteins is critical for the regulation of various cellular functions; this balance is governed by the redox milieu of the cell. Several mechanisms exist for regulating this balance, such as the Trx1 system, the GSNOR system, and xanthine oxidase [9]. In this chapter, we will limit our discussion to the Trx1-mediated S-nitrosylation and denitrosylation of proteins and the functional significance of the same.

2 Thioredoxin 1 and Its Relevance to Protein S-Nitrosylation

Thioredoxin1 (Trx1) is a redox protein that regulates oxidative stress in the cell. Several studies have shown that Trx1 controls the redox state of reactive cysteines in proteins, thus regulating its cellular function [22]. Trx1 is a 12 kDa protein that has five cysteine (Cys) residues: Cys32, 35, 62, 69, and 73 (Fig. 11.1). Cys32 and Cys35 are located in the active site of the protein and form a dithiol pair [22, 23]. These two sites are critical for the redox activity of Trx1. According to several reports, the other cysteine residues are involved in S-nitrosylation of Trx1, albeit under different conditions [24–27].

2.1 S-Nitrosylation of Trx1

Weichsel et al. used crystallized human Trx1 to study S-nitrosylation in Trx1 [24, 25]. Examination of the crystal structure of reduced Trx1 (rTrx1) indicated that Cys62 is buried deep within the tertiary structure of the protein. However, it is

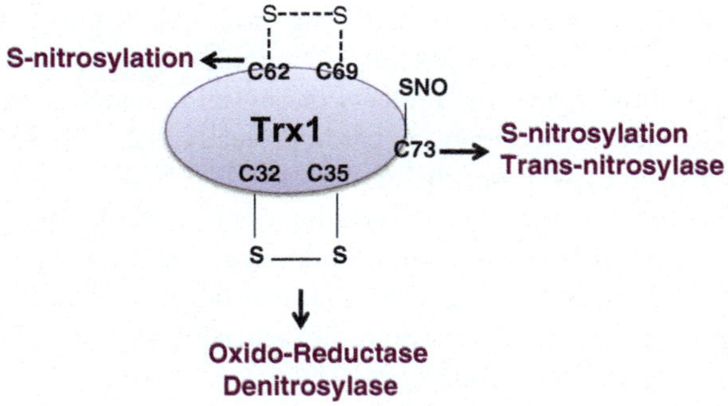

Fig. 11.1 Trx1 has five cysteine residues: Cys32,35,62,69 and 73. Cys32 and Cys35, the active sites, form a disulfide bond and are involved in oxido-reductase activity. They can also denitrosylate proteins. Cys62 and Cys69 have been reported to form a disulfide bond with each other and some reports suggests that they can be S-nitrosylated. Cys73 can be S-nitrosylated and functions as a trans-nitrosylase

readily S-nitrosylated at pH 7.0, possibly due to exposure of this site to the perimeter, as observed in the crystal structure of oxidized Trx1 (oTrx1). Cys69 is more exposed and can be S-nitrosylated under high pH conditions. On the other hand, Cys32 and Cys73 can react with an NO donor and undergo S-nitrosylation at pH 7.0, which is the physiological pH. In addition, these two cysteine residues are involved in disulfide bond formation. Cys32 forms an intramolecular disulfide bond with Cys35, while Cys73 is capable of intermolecular disulfide bond formation with Cys73 of another Trx1 molecule. Furthermore, it was observed that Cys73 forms the disulfide bond rapidly after S-nitrosylation.

Another factor that influences the S-nitrosylation of Trx1 is the redox state of the active cysteine residues. In a study conducted by Wu et al., it was demonstrated that only oTrx1 can be S-nitrosylated at Cys73 [27]. The oxidized form of recombinant Trx1 (with a disulfide bond between Cys32 and Cys35) was treated with GSNO; then, using MS/MS analysis, they found that Trx1 was S-nitrosylated at Cys73. In contrast, when rTrx1 was treated with GSNO, no SNO site was detected in Trx1. Using site-directed mutagenesis, they confirmed these results in HeLa cells overexpressing Trx1 C32S/C35S. According to their results, rTrx1 functions as a denitrosylase; however, formation of disulfide bonds leading to oTrx1 results in the denitrosylase property being lost and oTrx1 acting as a trans-nitrosylase via Cys73. In another report, Barglow et al. proposed that the redox state of the active Cys residues in Trx1 also affects the S-nitrosylation of non-active Cys residues [26]. Based on a reductive switch assay, they determined that rTrx1 favors S-nitrosylation of Cys62. This reaction occurs very rapidly, but is transient and of short duration. In contrast, oTrx1 promotes S-nitrosylation of Cys73 in a reaction that happens very slowly but is more stable. This data suggests that SNO-Cys62 is more of a signaling intermediate for trans-nitrosylation while SNO-Cys73 plays a dual role, as both a trans-nitrosylation agent and as a reservoir for NO storage. oTrx1 and rTrx1 are structurally very similar but exhibit different dynamics. Furthermore, Cys62 and Cys73 are located on opposite sides of the protein and have different binding partners. Thus, they can target different signaling pathways.

2.2 Importance of SNO-Trx1

Trx1 interacts with several proteins to maintain redox homeostasis. In addition, some of the above-mentioned studies propose that Trx1 may have several different downstream targets, depending upon its redox status and site of S-nitrosylation, and thus may play roles in several signaling pathways. This suggests that Trx1 could play a key role in maintaining the nitroso-redox balance in the cell. In the remainder of the chapter, we will discuss in detail the trans-nitrosylation and denitrosylation mediated by Trx1, SNO binding partners of Trx1, and their effect on cellular functions.

3 Regulation of Protein Trans-Nitrosylation by Trx1

Trx1 has been shown to increase cellular S-nitrosylation levels in endothelial cells. This was first demonstrated by Haendeler et al., who also showed that Trx1 is S-nitrosylated at Cys69 in endothelial cells [28] and that -SNO at Cys69 is important for the redox regulatory and ROS scavenging functions of Trx1. They further demonstrated that overexpression of Trx1 wild-type (WT) protein protected the endothelial cells from the reduction of S-nitrosylation induced by TNF-alpha, while expression of the Trx1 C69S (Cys mutated to Ser at the indicated site) mutant did not. In addition, the Trx1 C69S mutant only partially inhibited apoptosis compared to Trx1 WT, suggesting that overexpression of S-nitrosylable Trx1 decreases ROS formation and, thus, apoptosis. S-nitrosylable Trx1 could also trans-nitrosylate caspases and other apoptosis activators, inhibiting their activity to prevent the activation of apoptosis.

Following this report, Mitchell and Marletta demonstrated that Cys73 of Trx1 is capable of trans-nitrosylating caspase-3 in a reversible reaction to regulate apoptosis [29]. In that study, using the biotin-switch method and MALDI peptide mapping, they found that Trx1 was extensively S-nitrosylated at Cys73 while Cys69 S-nitrosylation was undetectable. This may be because Cys69 is only minimally exposed while Cys73, located on a ridge that is primed for protein-protein interaction, is the most surface-accessible residue. They also found that S-nitrosylation at Cys73 did not affect its activity. Rather, they determined that Trx1-Cys73-SNO specifically trans-nitrosylated caspase-3 at Cys163, leading to inactivation of caspase-3 activity and, thus, inhibiting apoptosis [30]. This mechanism was also observed in Jurkat cells, the human T cell lymphoma line. It was observed that, while Trx1 WT binds to and trans-nitrosylate caspase-3, and thus inhibit apoptosis, the Trx1 Cys73Ser mutant only binds to but not trans-nitrosylate caspase-3. Through co-immunoprecipitation studies, it was determined that the Trx1 Glu70 and Lys72 sites, two polar residues located adjacent to Cys73, are essential for the interaction of Trx1 with procaspase-3 to facilitate the trans-nitrosylation reaction.

Since Trx1 and S-nitrosylation have been known to play roles in the regulation of apoptosis, several other apoptotic proteins representing possible targets of trans-nitrosylation by Trx1 were investigated. Interaction of Trx1 with apoptosis signal-regulating kinase 1 (ASK1) has been known to inactivate and cause degradation of ASK1 to prevent apoptosis [31]. However, S-nitrosylation of Trx1 induces ASK1 activation [32]. In order for Trx1 to bind to ASK1, the -SH groups in Trx1 must be in their reduced form. Oxidation of these residues releases the bond between Trx1 and ASK1, resulting in activation of the kinase. Yasinska et al. reported that, in the presence of a two-fold excess of N_2O_3 compared to superoxide, Trx1 is S-nitrosylated, resulting in ASK1 release and activation [33]. However, a four-fold increase of the same, resulted in release of ASK1 and S-nitrosylation of ASK1 at Cys869, with a decreased kinase activity without affecting ASK1 protein stability. Thus, nitrosylation of Trx1 by N_2O_3-like compounds promotes ASK1 dissociation to stimulate apoptosis.

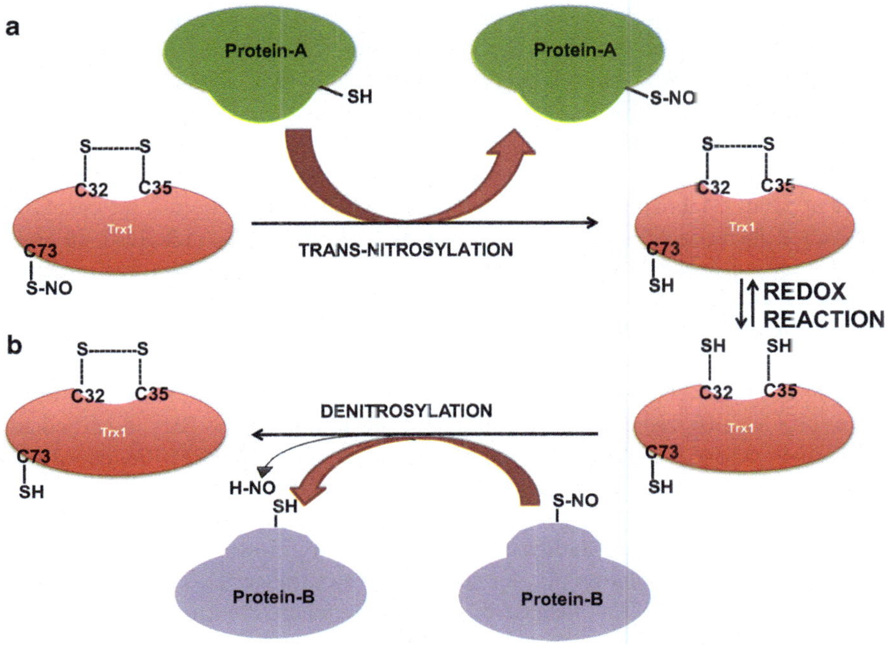

Fig. 11.2 Mechanism of trans-nitrosylation and denitrosylation by Trx1

In a global proteomics approach, Wu et al. used the biotin-switch technique together with MS/MS analysis to identify 47 novel trans-nitrosylation targets of Trx1 [27]. Using HeLa cells, they demonstrated the functional significance of Trx1-mediated trans-nitrosylation in peroxiredoxin-1 (Prx1), a trans-nitrosylation target identified in their global analysis. In HeLa cells overexpressing Trx1 C32S/C35S, Trx1 trans-nitrosylates Prx1 at Cys173 and Cys83 to protect Prx1 from H_2O_2-induced oxidative damage such as dimer disruption and sulfonylation. The group later developed an isotope-coded affinity tag (ICAT) method in conjunction with the biotin-switch technique to differentiate between trans-nitrosylation and denitrosylation targets of Trx1 in neuroblastoma cells [34, 35]. They identified 61 proteins that can be trans-nitrosylated by oTrx1 and 50 proteins that can be denitrosylated by rTrx1. However, not all of the trans-nitrosylation targets are capable of being denitrosylated by rTrx1, suggesting that trans-nitrosylation and denitrosylation targets of Trx1 are very specific (Fig. 11.2).

4 Regulation of Protein Denitrosylation by Trx1

In the cellular milieu, regulation of the S-nitrosylation and denitrosylation mechanisms is key to maintaining the nitroso-redox balance. Trx1 is one such key compound, playing an important role in regulating S-nitrosothiol homeostasis in the cell through its ability to specifically either trans-nitrosylate or denitrosylate protein molecules. Stoyanovsky et al. were the first to report that rTrx1 catalyzes the denitrosylation of GSNO and several S-nitrosoproteins, like SNO-caspase-3, SNO-metallothionein, and SNO-albumin, in chemical systems [36]. In the case of caspase-3, out of the many sites that were nitrosylated, only one site that is critical for the reactivation of caspase-3 activity was denitrosylated. In another study, Benhar et al. established Trx1 as the denitrosylase system required for caspase-3 activation in HeLa cells, RAW264.7 macrophages, and Jurkat cells [37]. In HeLa cells, SNO-caspase-3 was formed upon stimulation with S-nitrosocysteine. However, overexpression of Trx1 in HeLa cells triggered rapid denitrosylation of SNO-caspase-3 after 20 minutes. They further determined that the Trx1 Cys32 site forms mixed disulfides with caspase-3 to promote denitrosylation. In addition, Benhar et al. screened for S-nitrosylated substrates of Trx1 in human B lymphocytes (10C9 cells) treated with auranofin, a TrxR inhibitor [37]. They identified and validated caspase-9 and protein tyrosine phosphatase 1B (PTP1B) as denitrosylation targets of Trx1. Sengupta et al. demonstrated that the Trx1 system affects caspase-8 activity in HepG2 cells via S-nitrosylation [38]. In these cells, NO was able to nitrosylate caspase-8, thus inhibiting its activity. However, in TrxR-deficient HepG2 cells, denitrosylation of caspase-8 did not take place. This also directly correlated with the activity of caspase-8, such that regeneration of caspase-8 activity was not significant in Trx1-deficient cells. This suggests that Trx1 exerts its control of apoptosis via regulation of several key apoptotic proteins through S-nitrosylation.

To identify more denitrosylation substrates of Trx1 in a global profiling approach, Benhar et al. used stable isotope labeling by amino acids in cell culture (SILAC) in conjunction with the biotin-switch technique and mass spectrometry [39]. Using Jurkat cells and RAW264.7 cells, they identified 46 novel substrates for Trx1-mediated nitrosylation. They validated this approach by demonstrating increased S-nitrosylation of annexin-1 and 14-3-3θ following treatment with auranofin in Jurkat cells.

Trx1 also regulates apoptosis via trans-nitrosylation and denitrosylation of GAPDH [27, 35, 37]. S-nitrosylation of GAPDH promotes interaction of GAPDH with Siah1, an E3 ubiquitin ligase. This binding induces translocation of the complex to the nucleus, where GAPDH stabilizes Siah1 to promote degradation of nuclear proteins and, thus, trigger apoptotic cell death [40]. rTrx1 is responsible for the denitrosylation of GAPDH. Another important cellular function is that Trx1 regulates via S-nitrosylation heme insertion and, thus, heme protein maturation [41]. At a physiological pH level, NO inhibits heme protein maturation in cells via S-nitrosylation of GAPDH at Cys152. Chakravarti et al. manipulated the level of Trx1 expression in cells in order to understand the role of Trx1 in nitrosylation of GAPDH and the sensitivity of heme insertion to NO [41]. They found that overexpression of Trx1 prevented SNO-GAPDH accumulation and, thus, promoted heme maturation. However, knockdown of Trx1 protein in cells made heme insertion more sensitive to NO levels. Thus, Trx1-mediated regulation of protein S-nitrosylation is critical for regulating various cellular functions.

5 Therapeutic Implications and Functional Significance of Trx1-Mediated S-Nitrosylation

Trx1 acts as a central intermediary in the cellular system by mediating S-nitrosylation and denitrosylation, and thus regulating the redox balance in the cell. Several factors, including subcellular localization, the stability of SNO-Trx1, the redox status of Trx1, and the oxidative/nitrosative condition of the cellular environment, are all key to determining whether Trx1 functions as a trans-nitrosylase or denitrosylase and, thus, to determine its downstream targets. Trx1 has been shown to play a role in oxidative stress, cell death, gene transcription, and signal transduction [42]. The significance of S-nitrosylation by Trx1 has been extensively studied in apoptosis. Stimulation or prevention of apoptosis by Trx1-mediated S-nitrosylation occurs via regulation of the activities of key apoptotic proteins such as caspase-3, caspase-8, caspase-9, ASK, and GAPDH [29, 30, 33, 37, 38, 40, 43]. In addition, Trx1 regulates SNO-GAPDH levels to control heme protein insertion and maturation [41]. Furthermore, Trx1 reverses NO-mediated S-nitrosylation of N-ethylmaleimide sensitive factor (NSF) to denitrosylate and activate the enzyme and thus promote exocytosis in several conditions, including vascular inflammation, thrombosis, and insulin release [44]. Targets of Trx1 determined by global profiling of Trx1-mediated S-nitrosylation include substrates that play key roles in cytoskeletal organization, cellular metabolism, signal transduction, and redox balance [35, 39] (Table 11.1).

Trx1 has been shown to be important in pathophysiological conditions as well. For example, there is evidence that Trx1-mediated nitrosylation is significant in cancer [13]. Denitrosylation and, thus, activation of NF-kB by Trx1 have been shown to reduce inflammation in lung cancer cells [45]. In addition, other studies have shown that knockdown of the TrxR system in tumor cells sensitizes these cells to the cytotoxic effects of NO and S-nitrosylation [46]. In the cardiac system, Trx1 has been shown to be cardioprotective during ischemia-reperfusion [47]. Further, there is increased protein S-nitrosylation during ischemic preconditioning [48]. It is possible that Trx1 may S-nitrosylate proteins to protect them against irreversible oxidative damage. Though more studies are required to understand the precise role of Trx1-mediated S-nitrosylation, the studies conducted thus far, along with the global profiling data identified by several groups, point to the functional significance of Trx1-mediated regulation of the nitroso-redox balance in the cell.

Table 11.1 List of S-nitrosylation targets of Trx1

Study	Protein ID	Cell line	Trans-nitrosylation/ Denitrosylation	Trx1 site involved	Site of modification in target protein	Function	References
Haendeler et al.	Cellular proteins	Endothelial cells	Trans-nitrosylation	Cys69	–	Inhibit apoptosis	[28]
Mitchell et al.	caspase-3	Jurkat cells	Trans-nitrosylation	Cys73	Cys163	Inhibit apoptosis	[29, 30]
Yasinska et al.	ASK1	HEK293 cells	S-nitrosylation	–	Cys869	Activation of enzyme	[33]
Wu et al.	Prx1	HeLa cells	Trans-nitrosylation	Cys73	Cys173 and Cys83	Protect Prx1 from oxidative damage	[27]
Stoyanovsky et al.	S-nitrosoglutathione, caspase-3, albumin	recombinant proteins	Denitrosylation	–	–	Not studied	[36]
Benhar et al.	caspase-3	HeLa, RAW264.7, Jurkat cells	Denitrosylation	Cys32	–	Promote apoptosis	[37]
Benhar et al.	caspase-9, PTP1B	10C9 cells	Denitrosylation	Cys32	–	Not studied	[37]
Sengupta et al.	caspase-8	HepG2 cells	Denitrosylation	–	–	Regulation of extrinsic apoptosis	[38]
Benhar et al.	Annexin-1, 14-3-30	Jurkat cells	Denitrosylation	–	–	Not studied	[39]
Chakravarti et al.	GAPDH	RAW264.7 cells	Denitrosylation	–	Cys152	Promote heme maturation	[41]
Ito et al.	NSF	HUVEC cells	Denitrosylation	–	–	Exocytosis	[44]
Kelleher et al.	NF-KB	A549 cells	Denitrosylation	–	–	Inflammation	[45]

6 Conclusions

In summary, Trx1 is S-nitrosylated and can trans-nitrosylate or denitrosylate its protein targets. Which among the five cysteine residues of Trx1 is S-nitrosylated depends on several conditions, including the cell type, redox status of the active site cysteines, and pH. Several reports have been published that suggest that Trx1 plays an important role in trans-nitrosylation and denitrosylation of proteins such as caspase-3, caspase-8, ASK1, GAPDH, NEF, etc. Whether Trx1-mediated trans-nitrosylation or denitrosylation predominates is governed by various factors, including the protein target, cell type, nitroso-redox milieu of the cell, and redox status of Trx1. This mechanism plays a key role in the regulation of many cellular functions and, thus, affects several types of cancer, cardiovascular diseases and neuropathophysiological conditions.

References

1. Stamler JS, Simon DI, Osborne JA et al (1992) S-nitrosylation of proteins with nitric oxide: synthesis and characterization of biologically active compounds. Proc Natl Acad Sci U S A 89:444–448
2. Murphy E, Kohr M, Menazza S et al (2014) Signaling by S-nitrosylation in the heart. J Mol Cell Cardiol 73:18–25
3. Arnelle DR, Stamler JS (1995) NO+, NO, and NO- donation by S-nitrosothiols: implications for regulation of physiological functions by S-nitrosylation and acceleration of disulfide formation. Arch Biochem Biophys 318:279–285
4. Foster MW, McMahon TJ, Stamler JS (2003) S-nitrosylation in health and disease. Trends Mol Med 9:160–168
5. Hess DT, Matsumoto A, Kim SO et al (2005) Protein S-nitrosylation: purview and parameters. Nat Rev Mol Cell Biol 6:150–166
6. Broniowska KA, Hogg N (2012) The chemical biology of S-nitrosothiols. Antioxid Redox Signal 17:969–980
7. Li Q, Lancaster JR Jr (2012) A conspectus of cellular mechanisms of nitrosothiol formation from nitric oxide. Forum Immunopathol Dis Therap 3:183–191
8. Inoue K, Akaike T, Miyamoto Y et al (1999) Nitrosothiol formation catalyzed by ceruloplasmin. Implication for cytoprotective mechanism in vivo. J Biol Chem 274:27069–27075
9. Benhar M, Forrester MT, Stamler JS (2009) Protein denitrosylation: enzymatic mechanisms and cellular functions. Nat Rev Mol Cell Biol 10:721–732
10. Romero JM, Bizzozero OA (2009) Intracellular glutathione mediates the denitrosylation of protein nitrosothiols in the rat spinal cord. J Neurosci Res 87:701–709
11. Stamler JS, Jaraki O, Osborne J et al (1992) Nitric oxide circulates in mammalian plasma primarily as an S-nitroso adduct of serum albumin. Proc Natl Acad Sci U S A 89:7674–7677
12. Sun J, Xu L, Eu JP et al (2003) Nitric oxide, NOC-12, and S-nitrosoglutathione modulate the skeletal muscle calcium release channel/ryanodine receptor by different mechanisms. An allosteric function for O_2 in S-nitrosylation of the channel. J Biol Chem 278:8184–8189
13. Wang Z (2012) Protein S-nitrosylation and cancer. Cancer Lett 320:123–129
14. Foster MW, Hess DT, Stamler JS (2009) Protein S-nitrosylation in health and disease: a current perspective. Trends Mol Med 15:391–404

15. Tong G, Aponte AM, Kohr MJ et al (2014) Postconditioning leads to an increase in protein S-nitrosylation. Am J Physiol Heart Circ Physiol 306:H825–H832
16. Sun J, Murphy E (2010) Protein S-nitrosylation and cardioprotection. Circ Res 106:285–296
17. Murphy E, Kohr M, Sun J et al (2012) S-nitrosylation: a radical way to protect the heart. J Mol Cell Cardiol 52:568–577
18. Sun J, Morgan M, Shen RF et al (2007) Preconditioning results in S-nitrosylation of proteins involved in regulation of mitochondrial energetics and calcium transport. Circ Res 101:1155–1163
19. Nakamura T, Lipton SA (2015) Protein S-Nitrosylation as a therapeutic target for neurodegenerative diseases. Trend Pharmacol Sci 37:73–84
20. Nakamura T, Prikhodko OA, Pirie E et al (2015) Aberrant protein S-nitrosylation contributes to the pathophysiology of neurodegenerative diseases. Neurobiol Dis 84:99–108
21. Nakamura T, Tu S, Akhtar MW et al (2013) Aberrant protein s-nitrosylation in neurodegenerative diseases. Neuron 78:596–614
22. Holmgren A (1985) Thioredoxin. Annu Rev Biochem 54:237–271
23. Holmgren A, Lu J (2010) Thioredoxin and thioredoxin reductase: current research with special reference to human disease. Biochem Biophys Res Commun 396:120–124
24. Weichsel A, Brailey JL, Montfort WR (2007) Buried S-nitrosocysteine revealed in crystal structures of human thioredoxin. Biochemistry 46:1219–1227
25. Weichsel A, Kem M, Montfort WR (2010) Crystal structure of human thioredoxin revealing an unraveled helix and exposed S-nitrosation site. Protein Sci 19:1801–1806
26. Barglow KT, Knutson CG, Wishnok JS et al (2011) Site-specific and redox-controlled S-nitrosation of thioredoxin. Proc Natl Acad Sci U S A 108:600–606
27. Wu C, Liu T, Chen W et al (2010) Redox regulatory mechanism of transnitrosylation by thioredoxin. Mol Cell Proteomics 9:2262–2275
28. Haendeler J, Hoffmann J, Tischler V et al (2002) Redox regulatory and anti-apoptotic functions of thioredoxin depend on S-nitrosylation at cysteine 69. Nat Cell Biol 4:743–749
29. Mitchell DA, Marletta MA (2005) Thioredoxin catalyzes the S-nitrosation of the caspase-3 active site cysteine. Nat Chem Biol 1:154–158
30. Mitchell DA, Morton SU, Fernhoff NB, Marletta MA (2007) Thioredoxin is required for S-nitrosation of procaspase-3 and the inhibition of apoptosis in Jurkat cells. Proc Natl Acad Sci U S A 104:11609–11614
31. Saitoh M, Nishitoh H, Fujii M et al (1998) Mammalian thioredoxin is a direct inhibitor of apoptosis signal-regulating kinase (ASK) 1. EMBO J 17:2596–2606
32. Sumbayev VV (2003) S-nitrosylation of thioredoxin mediates activation of apoptosis signal-regulating kinase 1. Arch Biochem Biophys 415:133–137
33. Yasinska IM, Kozhukhar AV, Sumbayev VV (2004) S-nitrosation of thioredoxin in the nitrogen monoxide/superoxide system activates apoptosis signal-regulating kinase 1. Arch Biochem Biophys 428:198–203
34. Wu C, Parrott AM, Liu T et al (2013) Functional proteomics approaches for the identification of transnitrosylase and denitrosylase targets. Methods 62:151–160
35. Wu C, Parrott AM, Liu T et al (2011) Distinction of thioredoxin transnitrosylation and denitrosylation target proteins by the ICAT quantitative approach. J Proteomics 74:2498–2509
36. Stoyanovsky DA, Tyurina YY, Tyurin VA et al (2005) Thioredoxin and lipoic acid catalyze the denitrosation of low molecular weight and protein S-nitrosothiols. J Am Chem Soc 127:15815–15823
37. Benhar M, Forrester MT, Hess DT, Stamler JS (2008) Regulated protein denitrosylation by cytosolic and mitochondrial thioredoxins. Science 320:1050–1054
38. Sengupta R, Billiar TR, Kagan VE, Stoyanovsky DA (2009) Nitric oxide and thioredoxin type 1 modulate the activity of caspase 8 in HepG2 cells. Biochem Biophys Res Commun 391:1127–1130
39. Benhar M, Thompson JW, Moseley MA, Stamler JS (2010) Identification of S-nitrosylated targets of thioredoxin using a quantitative proteomic approach. Biochemistry 49:6963–6969

40. Hara MR, Agrawal N, Kim SF et al (2005) S-nitrosylated GAPDH initiates apoptotic cell death by nuclear translocation following Siah1 binding. Nat Cell Biol 7:665–674
41. Chakravarti R, Stuehr DJ (2012) Thioredoxin-1 regulates cellular heme insertion by controlling S-nitrosation of glyceraldehyde-3-phosphate dehydrogenase. J Biol Chem 287:16179–16186
42. Matsushima S, Zablocki D, Sadoshima J (2011) Application of recombinant thioredoxin1 for treatment of heart disease. J Mol Cell Cardiol 51:570–573
43. Li H, Wan A, Xu G, Ye D (2013) Small changes huge impact: the role of thioredoxin 1 in the regulation of apoptosis by S-nitrosylation. Acta Biochim Biophys Sin Shanghai 45:153–161
44. Ito T, Yamakuchi M, Lowenstein CJ (2011) Thioredoxin increases exocytosis by denitrosylating N-ethylmaleimide-sensitive factor. J Biol Chem 286:11179–11184
45. Kelleher ZT, Sha Y, Foster MW et al (2014) Thioredoxin-mediated denitrosylation regulates cytokine-induced nuclear factor kappaB (NF-kappaB) activation. J Biol Chem 289:3066–3072
46. Edes K, Cassidy P, Shami PJ, Moos PJ (2010) JS-K, a nitric oxide prodrug, has enhanced cytotoxicity in colon cancer cells with knockdown of thioredoxin reductase 1. PLoS One 5:e8786
47. Tao L, Gao E, Bryan NS et al (2004) Cardioprotective effects of thioredoxin in myocardial ischemia and reperfusion: role of S-nitrosation. Proc Natl Acad Sci U S A 101:11471–11476
48. Sun J, Nguyen T, Aponte AM et al (2015) Ischaemic preconditioning preferentially increases protein S-nitrosylation in subsarcolemmal mitochondria. Cardiovasc Res 106:227–236

Chapter 12
Thioredoxin Attenuates Post-ischemic Damage in Ventricular and Mitochondrial Function

Veronica D'Annunzio, Virginia Perez, Tamara Mazo, and Ricardo Jorge Gelpi

Abstract Thioredoxin (Trx) is an important antioxidant cellular system that plays an important role in cardioprotection against ischemia/reperfusion injury. The cardioprotective effects include a reduction of infarct size and an amelioration of ventricular and mitochondrial dysfunction that occurs in myocardial stunning. Particularly, Trx1 plays a protective role against the oxidative stress caused by an increase of reactive oxygen species concentration, as occurs during the reperfusion after an ischemic episode, and also could activate proteins related to pro-survival pathways such as MAP-kinases, Akt and GSK3-β. It has been also shown that, at least partially, Trx1 takes part of cardioprotective mechanisms such as ischemic preconditioning (PC) and postconditioning (PostC). However, comorbidities such as aging can modify this powerful cellular defense, leading to decrease cardioprotection, and even ischemic PC and PostC induced in aged animal models failed to decrease infarct size. Therefore, the lack of success of antioxidants therapies to treat ischemic heart disease could be solved avoiding the damage of Trx system.

Keywords Thioredoxin • Ischemia/reperfusion • Myocardial stunning • Infarct size • Aging • Mitochondrial function • Ischemic preconditioning • Ischemic postconditioning

1 Introduction

Thioredoxin is one of the most important cellular antioxidant systems known to present. In particular, thioredoxin-1 (Trx1) (12 kDa) is an important component in cellular defense against myocardial damage exerting anti-apoptotic [1], anti-inflammatory [2] and cardioprotective effects reducing infarct size and apoptosis in

V. D'Annunzio • V. Perez • T. Mazo • R.J. Gelpi (✉)
Institute of Biochemistry and Molecular Medicine (IBIMOL, UBA-CONICET)
and Institute of Cardiovascular Physiopathology, Faculty of Medicine,
University of Buenos Aires, Buenos Aires, Argentina
e-mail: rgelpi@fmed.uba.ar

© Springer International Publishing Switzerland 2016
R.J. Gelpi et al. (eds.), *Biochemistry of Oxidative Stress*, Advances in Biochemistry in Health and Disease 16, DOI 10.1007/978-3-319-45865-6_12

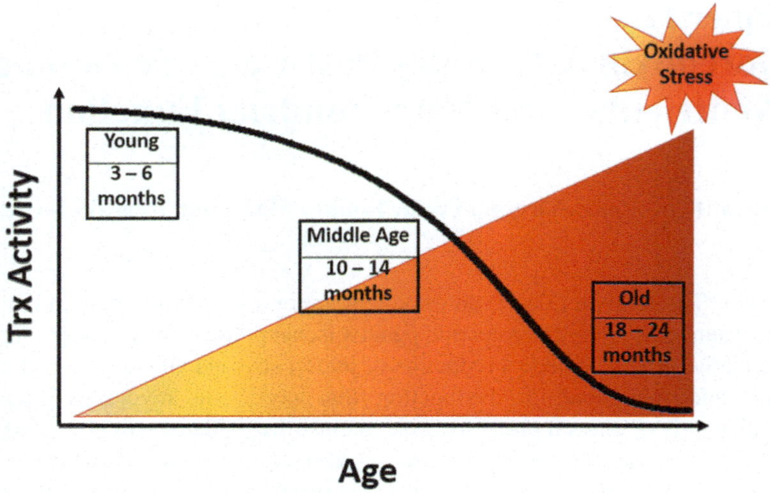

Fig. 12.1 Aging increases oxidative stress and decreases thioredoxin (Trx) activity

ischemia/reperfusion (I/R) processes [3]. Trx activity could be regulated by several mechanisms including expression levels, interaction with other molecules, and post-translational modifications. Regarding expression it has been shown that Trx transcript could be induced by several stimuli such as TNFα, Hydrogen peroxide, UV, thermal shock and also I/R [3]. Also, Trx1 can suffer post-translational modifications such as S-nitrosylation, oxidation and nitration, among others. S-nitrosylation has beneficial effects contributes to antiapoptotic effects through the inactivation of caspase-3 and BcL-2 [4]. Conversely, both oxidation and nitration reduced partial or total Trx1 activity, respectively [4]. In this sense, it has been shown that Trx inactivation, that occurs in aged and/or diabetic animals, could abolish the cardioprotective effects of this potent antioxidant due to nitrated Trx1 [5, 6] (Fig. 12.1).

Since that Trx1 over-expression in mice prolonged lifespan by 35 % compared to wild type mice, the finding could be related with an increased resistance against oxidative stress mainly in heart [7], Trx1 appears as a key protein to regulate oxidative stress involved in the majority of cardiovascular diseases. As we mentioned, Trx plays an important role in the cardioprotection against I/R injury, and several conditions could modify Trx expression and/or activity. It is important to gain knowledge concerning the intracellular mechanisms involved in the cardioprotective effects, the relationships among Trx1, myocardial stunning, aging and others physiological cardioprotective mechanisms such as ischemic preconditioning (PC) and postconditioning (PostC).

2 Thioredoxin Effects on the Stunned Myocardium

Heyndrickx et al. described for the first time the post-ischemic ventricular dysfunction (myocardial stunning) in 1976 [8]. They observed, in chronically instrumented dogs, that 15 min of coronary occlusion followed by reperfusion impaired the recovery of contractile state, which only was complete 24 h later. This finding was termed "post-ischemic systolic dysfunction", prioritizing the systolic impairment. A few years later, Przylenk et al. [9] observed an impaired diastolic function. The name "myocardial stunning" was termed in 1982 by Braunwald and Kloner [10] and was defined as a decrease in the contractile state accompanied by an alteration of the diastolic function with early relaxation impairment and later myocardial stiffness [11]. It is known that stunned myocardium involves an impairment of the calcium (Ca^{++}) homeostasis, accompanied by an increase of oxidative stress and damage [12, 13]. Several authors showed that during reperfusion a burst of superoxide radical anion (O_2^-) and H_2O_2 occurs, which leads to cell damage [13–16], and others noted that mitochondrial dysfunction is part of the post-ischemic deleterious mechanism [16, 17]. For these reasons, oxidative stress is a crucial factor in myocardial stunning physiopathology and due the aforementioned it is important to know the role that an antioxidant system, such as Trx1, may play in this pathophysiological entity.

Regarding Trx1 and myocardial stunning, Yoshioka et al. [18] reported that a deficiency in the thioredoxin interacting protein (TXNIP) improves the recovery of mitochondrial and ventricular function of the stunned myocardium, but they did not show a specific effect of Trx1 on ventricular function. Furthermore, they used a transgenic model, with TXNIP deficiency that has normal myocardial Trx1 activity and abnormal mitochondria morphology [19, 20]. Furthermore, Xiang et al. [21] also showed that targeted degradation of TXNIP mRNA by injection of a small nucleotide-based catalytic enzyme into ischemic myocardium leads to enhanced cardiomyocyte survival and reduced left ventricular remodeling. It is important to mention that adaptation to ischemia requires a reduction in cellular oxygen consumption. We and others have described this mitochondrial dysfunction in rabbits and mice [2, 22, 23]. Regarding Trx1, we reported that after 15 min of ischemia, the overexpression of Trx1 in transgenic mice reduces systolic and diastolic post-ischemic ventricular dysfunction [23]. Consistent with these results, we also observed that the beneficial effect was abolished in DN-Trx1 transgenic mice, a redox-inactive mutant of Trx1 that is dominant negative for endogenous Trx1 and in which the activity of Trx1 is highly reduced [24]. Even more, these mice show an exacerbation in myocardial stiffness and impairment of isovolumic relaxation rate compared to wild type mice at the end of reperfusion. Moreover, after I/R mitochondrial function was altered in wild type mice during late reperfusion, as shown by decreased state 3 mitochondrial O_2 consumption, accompanied by a lower complex I activity (Fig. 12.2). In DN-Trx1 mice, this alteration after I/R was exacerbated considering both O_2 consumption and complex I activity, in accordance with the exacerbation of diastolic dysfunction at 30 min of reperfusion. Conversely, mice

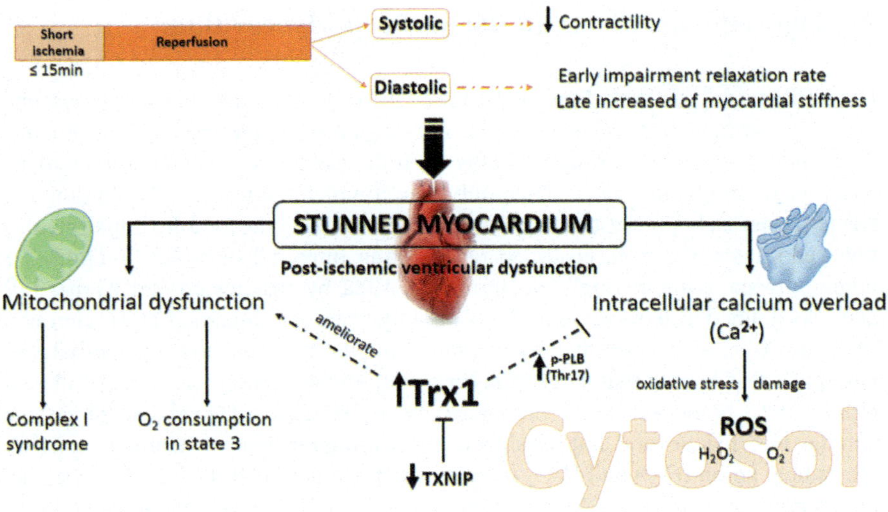

Fig. 12.2 Role of thioredoxin system in the physiopathology of myocardial stunning. *Trx1* Thioredoxin-1, *TXNIP* thioredoxin interacting protein, *p-PLB (Thr17)* phosphorylation of phospholamban at threonine-17, *ROS* reactive oxygen species, H_2O_2 hydrogen peroxide, O_2^- superoxide anion

hearts with Trx1 over-expression that were subjected to I/R showed a slighter decrease in O_2 consumption and in complex I activity. In relation to phospholamban (PLB) expression, we showed that transgenic Trx1 mice show a higher p-PLB^{Thr17} in comparison to wild type and DN-Trx1. Catalucci et al. [25] reported that at baseline conditions, a higher p-PLB^{Thr17} is accompanied by a greater Ca^{++} reuptake by the sarcoplasmic reticulum. Therefore, it is possible that Trx1 mice, which at baseline conditions present a higher p-PLB^{Thr17}, avoid at least partially, Ca^{++} overload during reperfusion when subjected to a stunning protocol (Fig. 12.2). This may justify the lack of isovolumic relaxation impairment at the onset of reperfusion. Due to the aforementioned, in this paper we show that an increase of antioxidant defenses in baseline conditions, in Trx1 mice, would avoid the relaxation impairment after ischemia, in early reperfusion in hearts subjected to a protocol of myocardial stunning.

Taken together, it is clear enough that Trx1 modulates the behavior of the ventricular dysfunction in the stunned myocardium, also Trx1 is able to modulate the activity of proteins related with intracellular calcium homeostasis. Finally, Trx1 over-expression can protect the mitochondrial function. As a consequence, Trx1 exerts beneficial effects in the stunned myocardium.

3 Myocardial Infarction

As we mentioned, Trx1 has cardioprotective effects against I/R injury in several animal models, both *in vivo* and *in vitro* [3, 26–33]. Turoczi et al. [26] revealed that Trx1 over-expression reduced infarct size and ameliorated the impairment in ventricular function recovery. In a similar manner, Shioji et al. [3] studied the role of Trx1 in myocytes cardiotoxicity induced by adriamycin (ADR) and showed that the formation of hydroxyl radicals in ADR-treated heart homogenates of Trx1-TG mice was decreased compared with normal mice. The finding indicate that Trx1 plays an important role in cardiomyocyte cellular defense against ROS damage. Furthermore, Kaga et al. [28] demonstrated that treatment with resveratrol induces Trx1, hemo-oxygenase and vascular endothelial growth factor, in infarcted hearts, resulting in infarct size reduction and in cardiac function improvement. All these data, suggest that Trx1 regulates gene transcription by interacting with several transcription factors, increasing antioxidant, anti-apoptotic and pro-angiogenic capacities, depending on cellular conditions. Similarly, Acta et al. [30] showed that the administration of recombinant human Trx reduced the incidence of reperfusion arrhythmias and Nakamura et al. [31] reported that Trx inactivation was a deleterious factor in the I/R injury in patients subjected to bypass surgery. Moreover, Tao et al. showed that administration of Trx1 *in vivo* exerts significant protective effects on myocardial apoptosis and decreases myocardial infarct size, likely by inhibiting p38-MAPK activation [32]. All these data indicates that Trx1 activity protects myocardial injury by I/R by regulation of gene transcription, by interacting with several transcription factors, and by increasing antioxidant, anti-apoptotic and pro-angiogenic capacities. We also demonstrated that cardiac Trx1 over-expression reduced infarct size in young mice, and for the first time we showed that Akt could be the probable target for Trx1 to provide protection against I/R injury [33]. Regarding interactions, it has been demonstrated that Trx1 is capable of indirectly phosphorylating Akt in other pathologies such as myocardial infarction and cancer [34–36]. This protein has been extensively studied and has a major role in physiological and pharmacological protective mechanisms; it is also part of the RISK-pathway that conferred cardioprotection [22, 35, 37]. Probably, the increased Akt phosphorylation in Trx1 transgenic mice involves the inactivation of PTEN, because it has been shown that Trx1 inhibits this molecule [22]. This mediator modulates activation of PI3K/Akt complex triggering intracellular events that confer protection against I/R injury [38]. A limitation is that PTEN modifications in the presence of Trx1 have not been studied in intact heart but only in cells in culture [37]. Also, the lack of positive results of this interaction in myocardial tissue could be related with Verrastro et al. [38] findings. The interactions appear to be dependent on the PTEN redox status, specifically oxidized PTEN avoid interactions with several molecules, such as Trx1. Nevertheless, when using DN-Trx1 mice we could not detect cardioprotection against infarct size, and neither an activation of cell survival proteins. Therefore, our data suggest that overexpression of Trx1 protects the myocardium against I/R injury by activating Akt and GSK3-β [32] (Fig. 12.3).

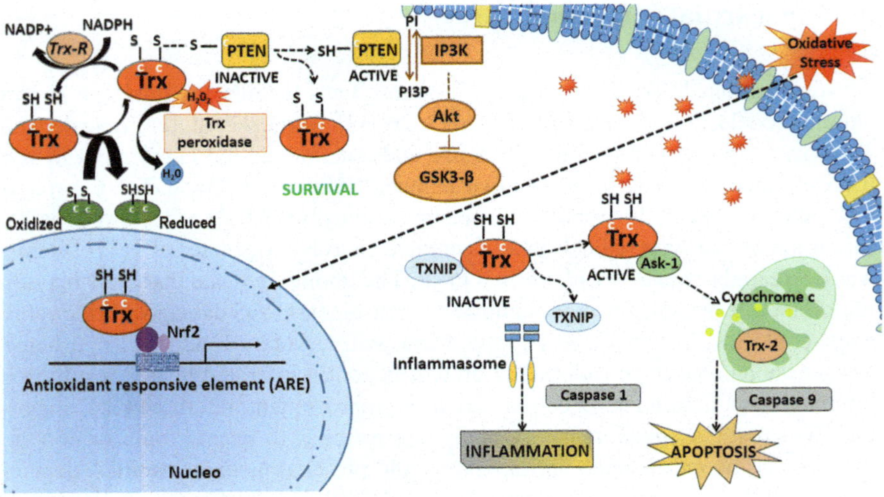

Fig. 12.3 Cardioprotective targets of thioredoxin system. *Trx* Thioredoxin, *Trx-R* Thioredoxin reductase, *NADP*$^+$ nicotinamide adenine dinucleotide phosphate, *NADPH* nicotinamide adenine dinucleotide phosphate reduced, H_2O_2 hydrogen peroxide, *PTEN* phosphatase and tensin homolog, *PI* phosphoinositol, *PI3P* phosphatidylinositol 3-phosphate, *IP3K* phosphor inositide 3-kinases, *Akt* protein kinase B, *GSK3-β* glycogen synthase kinase 3 beta, *TXNIP* thioredoxin interacting protein, *Ask1* apoptosis signal-regulating kinase 1, *Trx-2* Thioredoxin 2, *Nrf2* Nuclear factor 2 (erythroid-derived)

Due the aforementioned, is clear that Trx1 plays a key role in maintaining the redox balance in normal conditions. Also, in pathological conditions Trx1 is able to prevent the damage caused by oxidative stress when its intracellular concentration is increased, as in the case of over-expression models of Trx1 or in extracellular administration. Nevertheless, the inactivation that occurs in presence of comorbidities, such as aging, could abolish the protective effects of Trx1.

4 Thioredoxin Effects on Ischemic Preconditioning

The underlying mechanisms of ischemic preconditioning (PC) and ischemic postconditioning (PostC) increase cardiomyocyte survival pathways, at least in part by decreasing the oxidative stress damage that occurs during I/R injury [39–42]. Therefore, it is interesting to review the role that the Trx system plays in these endogenous protection mechanisms. The PC is a powerful physiological cardioprotective mechanism described for first time by Murry et al. in dogs [43], and later on in several species including humans [43–45]. Nowadays, it is known that during brief episodes of ischemia, many chemical mediators are released by the myocardium, including: adenosine, noradrenaline, bradikinine, opioids and endothelin [46–49]. All these agents could bind to myocyte membrane receptors and contribute to the activation of downstream kinases named as reperfusion ischemic salvage kinases (RISK-pathway). Different studies suggest that ROS are not only deleterious, but are also essential for the biology and physiology of myocytes [50, 51]. Therefore, depending on the level of myocyte antioxidant reserve, ROS can either be destroyed or persist. If they persist could play as a second messenger or signaling molecule after an I/R episode [50–53]. Taken together, the cardioprotective abilities of PC appear to be linked to redox signaling [52, 53] and it is in this sense that the antioxidant systems, such as Trx1, have an important role in conferred cardioprotection.

It has been shown that Trx1 inhibition abolished the PC cardioprotective effect [25]. Similarly, Chiueh et al. [54] showed that PC increases human Trx mRNA and protein for cytoprotection. Furthermore, cytosolic Trx1 and mitochondrial Trx-2 suppress free radical formation, lipid peroxidation, oxidative stress, and mitochondria-dependent apoptosis. Thus, to minimize or decrease the level and activity of either Trx1 or Trx-2 is detrimental to cell survival [54]. In conclusion, preconditioning adaptation and/or small and brief amounts of ROS induce a delayed nitric oxide-mediated compensatory mechanism for cell survival and vitality in the cardiovascular system.

The heart is effectively protected from reperfusion injury not only by preconditioning, but also by brief episodes of I/R during the early reperfusion period, in a process termed PostC, that follows to a pronounced ischemic insult [55]. PostC limits infarct size, reduces tissue edema and polymorphonuclear neutrophil accumulation and improves endothelial function in the myocardial risk area [55]. It has been suggested that PoctC is as powerful as PC in limiting infarct size and preserving post-ischemic endothelial function [55]. Given that PC has limited application in patients with acute myocardial infarction as a result of the clinical inability to predict the moment of coronary artery occlusion, the concept of ischemic PostC may be more useful than PC in the clinical setting. Besides endothelial preservation, inflammation and reduction of ROS, Yellon et al. [56] demonstrated that the inhibition of the phosphatidyl inositol 3 kinase (PI3K) and the MEK1/2, during the first 15 min of reperfusion, abolishes the PostC protective effect [56]; it is well known that these proteins are involved in the cardioprotection conferred by PC. Later, Tsang et al. [56] and Yang et al. [57] demonstrated the important role of the PI3K-Akt and MEK1/2/Erk complexes in the PostC protection mechanisms. It has also been shown that PostC decreased ROS release, showing a lower O_2^- and $ONOO^-$ productions and a lower heart injury [58, 59]. For several years our laboratory has been studying the cardioprotective mechanisms against I/R injury [58–63]. In previous studies we have demonstrated that PostC reduced infarct size though activation of A1 adenosine receptors and K^+_{ATP} sensitive channels [62]. Also ischemic PostC attenuates the MMP-2 activation during the first minutes of reperfusion and the administration of doxycycline reproduces the PostC effect on the infarct size by inhibiting MMP-2 [63]. In this way, PC and PostC share the intracellular signaling mechanisms and exerts cardioprotection, mainly by decreasing the oxidative stress that occurs during reperfusion [64]. We were the first to relate Trx1 to the PostC cardioprotection mechanism [65], showing that PostC reduced the infarct size in

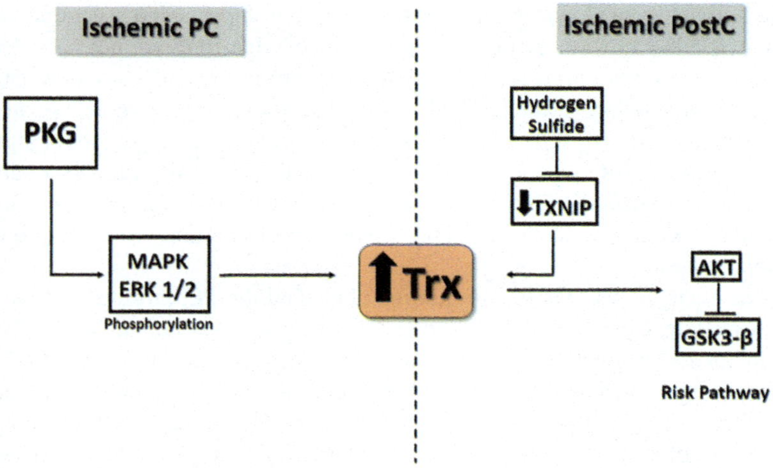

Fig. 12.4 Interactions between thioredoxin (Trx) and cardioprotective mechanisms such as ischemic preconditioning (PC) and ischemic postconditioning (PostC). *TXNIP* thioredoxin interacting protein, *Akt* protein kinase B, *GSK3-β* glycogen synthase kinase 3 beta, *PKG* protein kinase G, *MAPK* mitogen-activated protein kinases, *ERK 1/2* extracellular-signal-regulated kinases 1 and 2

young mice and that this cardioprotection was abolished in middle-aged and old mice. The reduction of infarct size in PostC is related to the Trx1 levels and with an improvement in the GSH/GSSG ratio (oxidative stress) in comparison to a control group. The cardioprotection was accompanied by Akt activation and by the phosphorylation and inhibition of GSK3-β, both proteins related to the RISK-pathway [65]. Therefore, it is possible that the preservation of Trx1 levels in young mice with the PostC protocol would reduce oxidative stress and allow activation of the RISK signaling pathway (Akt/GSK3-β) to reduce I/R injury (Fig. 12.4).

We can conclude that Trx1 plays an essential role in the adaptation to the process of I/R, given that Trx1 regulates the transcription of many genes that are regulated in PC [66]. During PostC the levels of Trx1 were preserved (66). In this way, Trx regulates the "redox switch" changing the I/R death signal into a survival cell signal mediated by PC [67] and triggering an adaptive response for the myocytes own defense.

5 Thioredoxin in Aging Hearts

Most of the studies that showed cardioprotective effects of Trx1 have been performed in young and healthy individuals. In aging, there is a progressive loss in adaptability although many biological functions remain normal on a basal level, but when they were subjected to stress (as I/R), the loss of functional reserve is revealed [68]. Oxidative stress is exacerbated during aging and it is also known that hearts of older individuals are more susceptible to I/R injury compared to younger animals [66, 69] and patients [70]. This increased susceptibility seems to occur due to modifications by protein oxidation [71], and Ca^{++} overload [72]. Moreover, increased levels of O_2^- by directly react with NO to produce $ONOO^-$. This seems evident in the increased formation of vascular $ONOO^-$ with age [73]. It is likely that the protein expression or activity of the antioxidant enzymes, such as is the case for Trx1, diminishes with aging. In the endothelial cells of elderly individual, an increase in the apoptosis induced by TNF release was observed. This deleterious effect was blocked in cells with Trx1 over-expression [74, 75]. The data suggest that Trx1 improves endothelial function and rescue endothelial cells from aging disorders. Therefore, it would be expected that in the heart an increased Trx1 attenuates structural and functional changes produced by the deleterious effects of aging. Liu et al. [76] demonstrated an increase of apoptosis in aged rat myocytes subjected to I/R as compared with young ones. Consistent with this research, Liu et al. [77] hypothesized that the increased vulnerability of aged myocardial tissue to I/R injury could be due to a decreased antioxidant capacity, rather than to increased oxidant production after an episode of I/R.

Regarding cardioprotection, it is known that age interfere in the cardioprotection mechanisms commonly used to reduce the infarct size in hearts subjected to I/R injury [78–80]. The increased oxidative stress in aging could, at least in part, explain the high mortality rate after I/R protocols [81, 82]. It has been demonstrated that aging is not only associated to an increased susceptibility to ischemia, but also to a decreased recovery of myocardial function following ischemia [83]. It seems that aging decreases the intrinsic tolerance to ischemia. This loss of tolerance to myocardial ischemia in the mouse starts during the middle-aged life (12 month old) and it becomes more evident during aging (18 month old and 24–28 month old) [84]. Zhang et al. [85] reported that the activity of Trx was significantly reduced in aging hearts, and indicated that for this reason, the infarct size was larger in aged C57/BL6 mice (20 month old) compared with young ones. In a similar manner, Azhar et al. [86] demonstrated an increase in the infarct size in C57/BL6 mice at 22–24 month old, subjected to 45 min of ischemia and 4 h of reperfusion. However, this reduction of tolerance to ischemia was not evidenced in the infarct size of C57/BL6 mice at 13 month old, subjected to 30 min of ischemia and 2 h of reperfusion *in vitro* [87]. We also evidenced an increase in the infarct size only in old mice (20 month old), but we did not observe changes in the infarct size in middle-aged mice (12 month old). We also demonstrated that PostC protocol do not confer protection in the middle-aged (12 month old) and in the old mice (>18 month old) [88]. The data agree with our finding that Trx1 levels are decreased in middle-aged and in old mice, as compared with young mice (Fig. 12.1). These results show that the protection conferred by PostC is abolished in middle-aged and old animal and suggests that the changes that occur with aging are able to modify the behavior of the endogenous antioxidant systems, modulating the results of the the physiological protection mechanisms, such as PostC, in an unfavorable manner.

Taken together, it is widely known that I/R injury is exacerbated in elderly populations and that many of the protective mechanisms lose their effectivity with advanced age [89, 90]. However, it is not clear whether this also occurs in middle-aged individuals when the deleterious effects of aging are starting to take place [89, 90]. It is known that although oxidation processes start when life begins, it is in middle-aged individuals that they reach sufficient levels to trigger deleterious mechanisms that affect different cell components [88], and that this ROS increase is able to modify expression and/or activity of several proteins [91–93]. We recently published that the cardioprotective effect in young mice of Trx1 overexpression, was abolished in middle-aged mice. The infarct size reduction was not observed, neither in Trx1 middle-aged nor in young and middle-aged DN-Trx1 mice. Although we found a significant increase in Trx1 expression in young and middle-aged transgenic mice, Trx1 activity was lower in middle-aged mice accompanied by an increase in the nitration of this protein. Finally, we also demonstrated that phosphorylation of Akt and GSK3-β was increased only in young Trx1 mice without any change in middle-aged mice and in DN-Trx1 mice [33].

6 Conclusions

A better understanding of the Trx1 system effects against I/R injury in general, and against myocardial stunning and infarction in particular, could prevent and ameliorate complications of ischemic heart disease. In this chapter we summarize data showing that there is a cardioprotective effect of Trx1 on myocardial infarction. The effect is age-dependent. PC and PostC were abolished in middle-aged and in old animals, at least in part by inactivation of the Trx system. Also, Trx1 ameliorates the systolic and diastolic dysfunction of myocardial stunning, including isovolumic relaxation and myocardial stiffness, by improving the free-radical mediated damage in ventricular and mitochondrial function. The description of these new regulatory mechanisms in myocardial stunning opens the possibility to new therapeutic strategies in I/R injury. Therefore, the lack of success of antioxidants therapies to treat ischemic heart disease could be solved avoiding the damage of the Trx system. This way the cardioprotective mechanisms can be useful and applied in the clinical setting in aged patients.

References

1. Ago T, Sadoshima J (2006) Thioredoxin and ventricular remodeling. J Mol Cell Cardiol 41:762–773
2. Nakamura H, Hoshino Y, Okuyama Y, Yodoi J (2009) Thioredoxin 1 as a new therapeutics. Adv Drug Rev 61:303–309
3. Shioji K, Nakamura H, Yodoi J (2003) Redox regulation by thioredoxin in cardiovascular disease. Antioxid Redox Signal 5:795–802
4. Rassaf T, Luedike P (2010) Between nitros(yl)ation and nitration: regulation of thioredoxin-1 in myocardial ischemia/reperfusion injury. J Mol Cell Cardiol 49:343–346

5. Wang K, Zhang J, Wang X et al (2013) Thioredoxin reductase was nitrated in the aging heart after myocardial ischemia/reperfusion. Rejuvenation Res 16:377–385
6. Liu Y, Qu Y, Wang R et al (2012) The alternative crosstalk between RAGE and nitrative thioredoxin inactivation during diabetic myocardial ischemia-reperfusion injury. Am J Physiol Endocrinol Metab 303:E841–E852
7. Mitsui A, Hamuro J, Nakamura H et al (2002) Overexpression of human thioredoxin in transgenic mice controls oxidative stress and life span. Antioxid Redox Signal 4:693–696
8. Heyndrickx GR, Millard RW, McRitchie RJ (1975) Regional myocardial functional and electrophysiological alterations after brief coronary artery occlusion in conscious dogs. J Clin Invest 56:978–985
9. Przyklenk K, Patel B, Kloner RA (1987) Diastolic abnormalities of postischemic "stunned" myocardium. Am J Cardiol 60:1211–1213
10. Braunwald E, Kloner RA (1982) The stunned myocardium: prolonged postischemic ventricular dysfunction. Circulation 66:1146–1149
11. Mosca SM, Gelpi RJ, Cingolani HE (1993) Dissociation between myocardial relaxation and diastolic stiffness in the stunned heart: its prevention by preconditioning. J Mol Cell Biochem 129:171–178
12. González GE, Rodríguez M, Donato M et al (2006) Effects of low-calcium reperfusion and adenosine on diastolic behavior during the transitory systolic overshoot of the stunned myocardium in the rabbit. Can J Physiol Pharmacol 84:265–272
13. Hess ML, Kukreja RC (1995) Free radicals, calcium homeostasis, heat shock proteins, and myocardial stunning. Ann Thorac Surg 60:760–766
14. Bolli R, Jeroudi MO, Patel BS et al (1989) Marked reduction of free radical generation and contractile dysfunction by antioxidant therapy begun at the time of reperfusion. Evidence that myocardial "stunning" is a manifestation of reperfusion injury. Circ Res 65:607–622
15. Valdez LB, ZaobornyjT BS et al (2011) Complex I syndrome in myocardial stunning and the effect of adenosine. Free Radic Biol Med 51:1203–1212
16. Boveris A, Lores-Arnaiz S, Bustamante J et al (2002) Pharmacological regulation of mitochondrial nitric oxide synthase. Methods Enzymol 359:328–339
17. Boveris A, Valdez LB, Zaobornyj T, Bustamante J (2006) Mitochondrial metabolic states regulate nitric oxide and hydrogen peroxide diffusion to the cytosol. Biochim Biophys 1757:535–542
18. Yoshioka J, Chutkow WA, Lee S (2012) Deletion of thioredoxin-interacting protein in mice impairs mitochondrial function but protects the myocardium from ischemia-reperfusion injury. J Clin Invest 122:267–279
19. Yoshioka J, Imahashi K, Gabel SA et al (2007) Targeted deletion of thioredoxin-interacting protein regulates cardiac dysfunction in response to pressure overload. Circ Res 101:1328–1338
20. Yoshioka J, Lee RT (2014) Thioredoxin-interacting protein and myocardial mitochondrial function in ischemia-reperfusion injury. Trends Cardiovasc Med 24:75–80
21. Xiang G, Seki T, Schuster MD et al (2005) Catalytic degradation of vitamin D up-regulated protein 1 mRNA enhances cardiomyocyte survival and prevents left ventricular remodeling after myocardial ischemia. J Biol Chem 280:39394–39402
22. Meuillet EJ, Mahadevan D, Berggren M et al (2004) Thioredoxin-1 binds to the C2 domain of PTEN inhibiting lipid phosphatase activity and membrane binding: a mechanism for the functional loss of PTEN's tumor suppressor activity. Arch Biochem Biophys 429:123–133
23. Perez V, D'Annunzio V, Valdez LB et al (2016) Thioredoxin-1 attenuates ventricular and mitochondrial post-ischemic dysfunction in the stunned myocardium of transgenic mice. Antioxid Redox Signal. doi:10.1089/ars.2015.6459
24. Yamamoto M, Yang G, Hong C et al (2003) Inhibition of endogenous thioredoxin in the heart increases oxidative stress and cardiac hypertrophy. J Clin Invest 112:1395–1406
25. Catalucci D, Latronico MV, Ceci M et al (2009) Akt increases sarcoplasmic reticulum Ca2+ cycling by direct phosphorylation of phospholamban at Thr17. J Biol Chem 284:28180–28187

26. Turoczi T, Chang VW, Engelman RM et al (2003) Thioredoxin redox signaling in the ischemic heart: an insight with transgenic mice overexpressing Trx1. J Mol Cell Cardiol 35:695–704
27. Matsushima S, Zablocki D, Sadoshima J (2011) Application of recombinant thioredoxin1 for treatment of heart desease. J Mol Cell Cardiol 51:570–573
28. Kaga S, Zhan L, Matsumoto M, Maulik N (2005) Resveratrol enhances neovascularization in the infarcted rat myocardium through the induction of thioredoxin-1, heme oxygenase-1 and vascular endothelial growth factor. J Mol Cell Cardiol 39:813–822
29. Altschmied J, Haendeler J (2009) Thioredoxin-1 and endothelial cell aging: role in cardiovascular diseases. Antioxid Redox Signal 11:1733–1740
30. Aota M, Matsuda K, Isowa N et al (1996) Protection against reperfusion-induced arrhythmias by human thioredoxin. J Cardiovasc Pharmacol 27:727–732
31. Nakamura H, Vaage J, Valen G et al (1998) Measurements of plasma glutaredoxin and thioredoxin in healthy volunteers and during open-heart surgery. Free Radic Biol Med 24:1176–1186
32. Tao L, Gao E, Bryan NS et al (2004) Cardioprotective effects of thioredoxin in myocardial ischemia and reperfusion: role of S-nitrosation. Proc Natl Acad Sci U S A 101:11471–11476
33. D Annunzio V, Perez V, Mazo T et al (2016) Loss of myocardial protection against myocardial infarction in middle-aged transgenic mice overexpressing cardiac thioredoxin-1. Oncotarget. doi:10.18632/oncotarget.7726
34. Dhalla NS, Elmoselhi AB, Hata T, Makino N (2000) Status of myocardial antioxidants in ischemia-reperfusion injury. Cardiovasc Res 47:446–456
35. Yang XM, Krieg T, Cui L et al (2004) NECA and bradykinin at reperfusion reduce infarction in rabbit hearts by signaling through PI3K, ERK, and NO. J Mol Cell Cardiol 36:411–421
36. Sartelet H, Rougemont AL, Fabre M (2011) Activation of the phosphatidylinositol 3′-kinase/AKT pathway in neuroblastoma and its regulation by thioredoxin-1. Hum Pathol 42:1727–1739
37. Hausenloy DJ, Tsang A, Yellon DM (2005) The reperfusion injury salvage kinase pathway: a common target for both ischemic preconditioning and postconditioning. Trends Cardiovasc Med 15:69–75
38. Verrastro I, Tveen-Jensen K, Woscholski R et al (2015) Reversible oxidation of phosphatase and tensin homolog (PTEN) alters its interactions with signaling and regulatory proteins. Free Radic Biol Med 90:24–34
39. Penna C, Perrelli MG, Pagliaro P (2013) Mitochondrial pathways, permeability transition pore, and redox signaling in cardioprotection: therapeutic implications. Antioxid Redox Signal 18:556–569
40. Sanada S, Komuro I, Kitakaze M (2011) Pathophysiology of myocardial reperfusion injury: preconditioning, postconditioning, and translational aspects of protective measures. Am J Physiol Heart Circ Physiol 301:H1723–H1741
41. Ge H, Zhao M, Lee S, Xu Z (2015) Mitochondrial Src tyrosine kinase plays a role in the cardioprotective effect of ischemic preconditioning by modulating complex I activity and mitochondrial ROS generation. Free Radic Res 49:1210–1217
42. Yu P, Zhang J, Yu S et al (2015) Protective effect of sevoflurane in postconditioning against cardiac ischemia/reperfusion injury via ameliorating mitochondrial impairment, oxidative stress and rescuing autophagic clearance. PLoS One 10:e0134666
43. Murry CE, Jennings R, Reimer KA (1986) Preconditioning with ischemia: a delay of lethal cell injury in ischemic myocardium. Circulation 74:1124–1136
44. Cave A, Horowitz G, Apstein C (1994) Can ischemic preconditioning protect against hypoxia-induced damage? Studies of contractile function in isolated rat hearts. J Mol Cell Cardiol 26:1471–1486
45. Ytrehus K, Liu Y, Downey J (1994) Preconditioning protects ischemic rabbit heart by protein kinase C activation. Am J Physiol 266:H1145–H1152
46. Liu GS, Thornton J, Van Winkle DM et al (1991) Protection against infarction afforded by preconditioning is mediated by A1 adenosine receptors in rabbit heart. Circulation 84:350–356

47. Banerjee A, Locke-Winter C, Rogers KB et al (1993) Preconditioning against myocardial dysfunction after ischemia and reperfusion by an alpha 1-adrenergic mechanism. Circ Res 73:656–670
48. Oldenburg O, Qin Q, Sharma AR et al (2002) Acetylcholine leads to free radical production dependent on K(ATP) channels, G(i) proteins, phosphatidylinositol 3-kinase and tyrosine kinase. Cardiovasc Res 55:544–552
49. Krieg T, Cui L, Qin Q et al (2004) Mitochondrial ROS generation following acetylcholine-induced EGF receptor transactivation requires metalloproteinase cleavage of proHB-EGF. J Mol Cell Cardiol 36:435–443
50. Maulik N, Watanabe M, Zu YL et al (1996) Ischemic preconditioning triggers the activation of MAP kinases and MAPKAP kinase-2 in rat hearts. FEBS Lett 396:233–237
51. Maulik N, Sato M, Price BD, Das DK (1998) An essential role of NFkappaB in tyrosine kinase signaling of p38 MAP kinase regulation of myocardial adaptation to ischemia. FEBS Lett 429:365–369
52. Yao Z, Tong J, Tan X et al (1999) Role of reactive oxygen species in acetylcholine-induced preconditioning in cardiomyocytes. Am J Physiol 277:H2504–H2509
53. Oldenburg O, Critz SD, Cohen MV, Downey JM (2003) Acetylcholine-induced production of reactive oxygen species in adult rabbit ventricular myocytes is dependent on phosphatidylinositol 3- and Src-kinase activation and mitochondrial K (ATP) channel opening. J Mol Cell Cardiol 35:653–660
54. Chiueh CC, Andoh T, Chock PB (2005) Induction of thioredoxin and mitochondrial survival proteins mediates preconditioning-induced cardioprotection and neuroprotection. Ann N Y Acad Sci 1042:403–418
55. Ago T, Yeh I, Yamamoto M et al (2006) Thioredoxin-1 upregulates mitochondrial proteins related to oxidative phosphorylation and TCA cycle in the heart. Antioxid Redox Signal 8:1635–1650
56. Tsang A, Hausenloy DJ, Mocanu MM, Yellon DM (2004) Postconditioning: a form of "modified reperfusion" protects the myocardium by activating the phosphatidylinositol 3-kinase-Akt pathway. Circ Res 95:230–232
57. Yang XM, Proctor JB, Cui L et al (2004) Multiple, brief coronary occlusions during early reperfusion protect rabbit hearts by targeting cell signaling pathways. J Am Coll Cardiol 44:1103–1110
58. Donato M, D'Annunzio V, Buchholz B et al (2010) Role of matrix metalloproteinase-2 in the cardioprotective effect of ischaemic postconditioning. Exp Physiol 95:274–281
59. D'Annunzio V, Donato M, Erni L et al (2009) Rosuvastatin given during reperfusion decreases infarct size and inhibits matrix metalloproteinase-2 activity in normocholesterolemic and hypercholesterolemic rabbits. J Cardiovasc Pharmacol 53:137–144
60. D'Annunzio V, Donato M, Buchholz B et al (2012) High cholesterol diet effects on ischemia-reperfusion injury of the heart. Can J Physiol Pharmacol 90:1185–1196
61. Buchholz B, D'Annunzio V, Giani JF et al (2014) Ischemic postconditioning reduces infarct size through the α1-adrenergic receptor pathway. J Cardiovasc Pharmacol 63:504–511
62. Donato M, D'Annunzio V, Berg G et al (2007) Ischemic postconditioning reduces infarct size by activation of A1 receptors and K+(ATP) channels in both normal and hypercholesterolemic rabbits. J Cardiovasc Pharmacol 49:287–292
63. Buchholz B, Perez V, Siachoque N et al (2014) Dystrophin proteolysis: a potential target for MMP-2 and its prevention by ischemic preconditioning. Am J Physiol Heart Circ Physiol 307:H88–H96
64. Skyschally A, Schulz R, Heusch G (2008) Pathophysiology of myocardial infarction: protection by ischemic pre- and postconditioning. Herz 33:88–100
65. Perez V, D'Annunzio V, Mazo T et al (2016) Ischemic postconditioning confers cardioprotection and prevents reduction of Trx1 in young mice, but not in middle-aged and old mice. Mol Cell Biochem 415:67–76

66. Frolkis VV, Frolkis RA, Mkhitarian LS, Fraifeld VE (1991) Age-dependent effects of ischemia and reperfusion on cardiac function and Ca2+ transport in myocardium. Gerontology 37:233–239
67. Das DK, Maulik N (2003) Preconditioning potentiates redox signaling and converts death signal into survival signal. Arch Biochem Biophys 420:305–311
68. Cusack BJ, Mushlin PS, Andrejuk T et al (1991) Aging alters the force-frequency relationship and toxicity of oxidative stress in rabbit heart. Life Sci 48:1769–1777
69. Lesnefsky EJ, Gallo DS, Ye J et al (1994) Lust aging increases ischemia-reperfusion injury in the isolated, buffer perfused heart. J Lab Clin Med 124:843–851
70. Lesnefsky EJ, Lundergan CF, Hodgson JM et al (1996) Increased left ventricular dysfunction in elderly patients despite successful thrombolysis: the GUSTO-I angiographic experience. J Am Coll Cardiol 28:331–337
71. Flitter WD (1993) Free radicals and myocardial reperfusion injury. Br Med Bull 49:545–555
72. Atake K, Chen D, Levitsky S (1992) Effect of aging on intracellular Ca2+, pHi, and contractility during ischemia and reperfusion. Circulation 86:371–376
73. Van der Loo B, Labugger R, Skepper JN (2000) Enhanced peroxynitrite formation is associated with vascular aging. J Exp Med 192:1731–1744
74. Asai K, Kudej RK, Shen YT et al (2000) Peripheral vascular endothelial dysfunction and apoptosis in old monkeys. Arterioscler Thromb Vasc Biol 20:1493–1499
75. Hoffmann J, Haendeler J, Aicher A et al (2001) Aging enhances the sensitivity of endothelial cells toward apoptotic stimuli: important role of nitric oxide. Circ Res 89:709–715
76. Liu P, Xu B, Cavalieri TA, Hock CE (2002) Age-related difference in myocardial function and inflammation in a rat model of myocardial ischemia-reperfusion. Cardiovasc Res 56:443–453
77. Liu P, Xu B, Cavalieri TA, Hock CE (2004) Attenuation of antioxidative capacity enhances reperfusion injury in aged rat myocardium after MI/R. Am J Physiol Heart Circ Physiol 287:2719–2727
78. Ferdinandy P, Hausenloy DJ, Heusch G (2014) Interaction of risk factors, comorbidities, and comedications with ischemia/reperfusion injury and cardioprotection by preconditioning, postconditioning, and remote conditioning. Pharmacol Rev 66:1142–1174
79. Przyklenk K (2011) Efficacy of cardioprotective 'conditioning' strategies in aging and diabetic cohorts: theco-morbidity conundrum. Drugs Aging 28:331–343
80. Boengler K, Schulz R, Heusch G (2009) Loss of cardioprotection with ageing. Cardiovasc Res 83:247–261
81. Abete P, Napoli C, Santoro G et al (1999) Age-related decrease in cardiac tolerance to oxidative stress. J Mol Cell Cardiol 31:227–236
82. Fenton RA, Dickson EW, Meyer TE, Dobson JG (2000) Aging reduces the cardioprotective effect of ischemic preconditioning in the rat heart. J Mol Cell Cardiol 32:1371–1375
83. Sumeray MS, Yellon DM (1998) Characterisation and validation of a murine model of global ischaemia reperfusion injury. Mol Cell Biochem 186:61–68
84. Willems L, Zatta A, Holgren K et al (2005) Age-related changes in ischemic tolerance in male and female mouse hearts. J Mol Cell Cardiol 38:245–256
85. Zhang H, Tao L, Jiao X et al (2007) Nitrative thioredoxin inactivation as a cause of enhanced myocardial ischemia/reperfusion injury in the aging heart. Free Radic Biol Med 43:39–47
86. Azhar G, Gao W, Liu L, Wei JY (1999) Ischemia-reperfusion in the adult mouse heart: influence of age. Exp Gerontol 34:699–714
87. Boengler K, Konietzka I, Buechert A et al (2007) Loss of ischemic preconditioning's cardioprotection in aged mouse hearts is associated with reduced gap junctional and mitochondrial levels of connexin-43. Am J Physiol Heart Circ Physiol 292:H1764–H1769
88. Fan Q, Chen M, Fang X et al (2013) Aging might augment reactive oxygen species (ROS) formation and affect reactive nitrogen species (RNS) level after myocardial ischemia/reperfusion in both humans and rats. Age (Dordr) 35:1017–1026
89. Uhl GS, Farrell PW (1983) Myocardial infarction in young adults: risk factors and natural history. Am Heart J 105:548–553

90. Cohen D, Manuel DG, Tugwell P et al (2013) Inequity in primary and secondary preventive care for acute myocardial infarction? Use by socioeconomic status across middle-aged and older patients. Can J Cardiol 29:1579–1585
91. Babušíková E, Lehotský J, Dobrota D et al (2012) Age-associated changes in Ca(2+)-ATPase and oxidative damage in sarcoplasmic reticulum of rat heart. Physiol Res 61:453–460
92. Choksi KB, Nuss JE, DeFord JH (2011) Mitochondrial electron transport chain functions in long lived Ames dwarf mice. Aging (Milano) 3:754–767
93. Kim JY, Kim OY, Paik JK et al (2013) Association of age-related changes in circulating intermediary lipid metabolites, inflammatory and oxidative stress markers, and arterial stiffness in middle-aged men. Age (Dordr) 35:1507–1519

Chapter 13
Antioxidant Supplementation in Cardiovascular Disease and Hypertension

José Milei, Susana Vila, Francisco Azzato, Giuseppe Ambrosio, and Matilde Otero-Losada

Abstract This chapter is organized in three topics. First, relevant aspects of oxidative stress and oxidative intermediaries that are essential for comprehension of the pathophysiological consequences of oxidative stress are briefly commented. Second, we share our experience concerning with antioxidant supplementation in elderly cardiovascular hypertensive patients. Supplementation with antioxidants and its benefit-risk relationship have been largely discussed in the elderly population. Numerous studies have focused on the utility of antioxidant supplementation in the treatment of cardiovascular diseases. Yet, whether antioxidant supplementation has any preventive and/or therapeutic value in cardiovascular pathology is still a matter of debate because the evidence is inconclusive. Our results on the benefits of antioxidants supplementation in elderly cardiovascular hypertensive patients undertaking periodical cardiovascular checking, are discussed focusing on potential improvement of the biochemical profile associated with oxidative metabolism. Third, our findings on biochemical indicators and histological analysis in the oxidative stress and myocardial injury, following cardioplegic arrest/reperfusion in human cardiac surgery, along with the role of the antioxidant therapy in cardiac surgery, are discussed. Our above mentioned findings are discussed in the light of updated information.

Keywords Antioxidants • Oxidative stress • Hypertension • Cardiovascular disease • Elderly patients • Ischemia • Reperfusion • Reactive oxygen species • Thioredoxins

If you don't know your blood pressure, it's like not knowing the value of your company
(Mehmet Oz)

J. Milei • S. Vila • F. Azzato • G. Ambrosio • M. Otero-Losada (✉)
Institute of Cardiological Research (ININCA, UBA-CONICET), School of Medicine, University of Buenos Aires and National Research Council, Buenos Aires, Argentina
e-mail: josemilei@gmail.com; molly1063@gmail.com

© Springer International Publishing Switzerland 2016
R.J. Gelpi et al. (eds.), *Biochemistry of Oxidative Stress*, Advances in Biochemistry in Health and Disease 16, DOI 10.1007/978-3-319-45865-6_13

1 Introduction. Reactive Oxygen Species

Several studies have proposed the essential role of reactive oxygen species (ROS) in the pathogenesis of myocardial ischemia–reperfusion injury [1]. In this context, ROS includes hydrogen peroxide (H_2O_2), superoxide radical (O_2^-), hydroxyl radical (HO) and peroxynitrite ($ONOO^-$), that have been shown to increase upon reperfusion of the heart following ischemia [2, 3]. ROS are also generated during mitochondrial oxidative metabolism as well as in cellular response to xenobiotics, cytokines, and bacterial invasion. Oxidative stress refers to an imbalance due to excess ROS or oxidants over the capability of the cell to have an effective antioxidant response. Oxidative stress results in macromolecular damage and is implicated in various disease states such as atherosclerosis, diabetes, cancer, neurodegeneration, and aging. Paradoxically, accumulating evidence indicates that ROS also serve as critical signaling molecules in cell proliferation and survival. Cellular ROS sensing and metabolism are tightly regulated by a variety of proteins involved in redox mechanisms. ROS directly interact with critical signaling molecules to initiate signaling in a broad variety of cellular processes, such as proliferation and survival (MAP kinases, PI3 kinase, PTEN, and protein tyrosine phosphatases), antioxidant gene regulation (thioredoxin, peroxiredoxin, Ref-1, and Nrf-2), mitochondrial oxidative stress, apoptosis, and aging (p66Shc), iron homeostasis through iron–sulfur cluster proteins (IRE–IRP), and ATM-regulated DNA damage response [4].

Cell redox homeostasis results from the critical balance between mitochondrial production of ROS and antioxidant defense systems. When the cellular antioxidant defense mechanism fails, ROS accumulate in the cytoplasm. Different proteins, including receptors, ionic channels, transporters or components of transduction pathways are substrates for oxidation by ROS [5]. ROS are both detrimental and protective for cardiac myocyte functions, electrophysiology and pharmacology, via different cellular pathways, mostly depending on the type and amount of ROS synthesized. While the literature clearly indicates ROS effects on cardiac contractility, their effects on cardiac excitability are relatively under appreciated. Cardiac excitability depends on the functions of various cardiac sarcolemmal or mitochondrial ion channels carrying various depolarizing or repolarizing currents that maintain cellular ionic homeostasis. ROS alter the functions of these ion channels to various degrees to determine excitability by affecting the cellular resting potential and the morphology of the cardiac action potential. Thus, the redox balance regulates cardiac excitability, and under pathological disregulation, may alter action potential propagation and cause arrhythmia. The understanding of how the redox state affects cellular excitability may lead to potential prophylaxis or treatment for various arrhythmias. A recent and interesting review focuses on the pathophysiological role of oxidative stress in cardiac excitation [6].

2 Angiotensin and the Renin-Angiotensin System

The cardiovascular effects of angiotensin II involve oxidative stress [7]. Clinical and experimental studies have demonstrated that activation of systemic and local renin-angiotensin system (RAS) contributes to the pathogenesis of ischemia/reperfusion (I/R) injury [8, 9]. Angiotensin II induces oxidative stress via NADPH oxidase activation and production of proinflammatory cytokines and growth factors leading to cardiovascular-renal remodeling. Angiotensin II also stimulates the RhoA/Rho kinase (ROCK) pathway, which is deeply involved in the development of cardiovascular and renal remodeling simultaneous with oxidative stress [10]. Actually, experimental evidence has provided a mechanistic rationale for olmesartan's antioxidant/anti-inflammatory properties and anti-atherosclerotic/anti-remodeling benefits reported by clinical trials [10]. Angiotensin II has been shown to activate multiple downstream pathways resulting in endothelial dysfunction and oxidative stress [11]. Angiotensin II is involved in ROS generation, in arrhythmias, cell death, heart failure, ischemia/reperfusion injury, cardiac hypertrophy and hypertension [11]. A growing body of evidence supports the crucial role of the renin-angiotensin aldosterone system in the pathophysiology of myocardial I/R damage [12, 13]. Clinical and experimental studies have demonstrated that the activation of systemic or local renin-angiotensin system is implicated in ischemia-induced cardiac injury [8]. Sirtuin-3 (silent information regulator of transcription 3) deacetylase, the major mitochondrial NAD-dependent lysine deacetylase, protects the heart against oxidative stress by survival factors upregulation. Interestingly the decrease in SIRT3 protein levels subsequent to I/R has been suggested as a novel signaling mechanism involved in I/R injury [8, 14]. SIRT3 regulates a variety of functions, and its inhibition may disrupt mitochondrial function to impact recovery from I/R injury [9]. Losartan, an angiotensin II type I receptor blocker, at non-hypotensive dose exerts anti-ischemic effects in part by normalizing the SIRT3 protein level and upregulating the survival factors encoding genes transcription in ischemic tissue of the heart [8, 9]. Losartan is an AT1 receptor blocker that was introduced as hypotensive medication, but studies showed that apart from hemodynamic effects, it might have cardioprotective influences by affecting cardiomyocytes directly. According to previous studies, even at non-hypotensive dose, losartan may increase the ventricular fibrilation threshold and decrease infarct size [8], premature ventricular beats, ventricular tachycardia episodes, and mortality rate [15–17]. Losartan reduces the production of reactive oxygen species (ROS) by inhibiting NADPH oxidase [16, 18] and increases the production of NO in endothelial cells by activating nitric oxide synthases [19].

2.1 Angiotensin

Angiotensin (Ang)-(1–7) exhibits cardioprotective effects in myocardial I/R injury. Myocardial I/R induces significant cardiac dysfunction, including ventricular arrhythmia and a reduction of left ventricular systolic pressure (LVSP), cardiomyocyte apoptosis and oxidative stress, which results in increased malondialdehyde (MDA) production and decreased superoxide dismutase (SOD) activity. Pretreatment of hearts with 1.0 nmol/L Ang-(1–7) for 30 min prior to ischemia considerably attenuated I/R induced ventricular arrhythmia, apoptosis and MDA production, and enhanced LVSP and SOD activity. These cardioprotective effects of Ang-(1–7) were antagonized by the COX inhibitor indomethacin, presented as an enhancement of VA, apoptosis and MDA production as well as a reduction of LVSP and SOD activity. In conclusion, Ang-(1–7) antioxidative cardioprotection involves COX activity [32].

2.2 Insulin

Insulin and the inhibition of the renin-angiotensin system have independent benefits for I/R injury [20]. During reperfusion of the ischemic heart, insulin activates the AKT and AMPK pathways and inhibits the deleterious effects of angiotensin II receptor type I on superoxide dismutase (SOD) expression and cardiac function [20]. Interest in the effects of insulin on the heart came with the recognition that hyperglycemia in the context of myocardial infarction is associated with increased risks of mortality, congestive heart failure, and cardiogenic shock [21–24]. More recently, instigated by research findings on stress hyperglycemia in critical illness, this interest has been extended to the influence of insulin on clinical outcome after cardiac surgery. Even in non-diabetic individuals, stress hyperglycemia commonly occurs as a metabolic response to critical illness, e.g., after surgical trauma. It is recognized as a major pathophysiological feature of organ dysfunction in the critically ill. The condition stems from the insulin resistance brought about by dysregulation of homeostatic processes, which implicates immune/inflammatory, endocrine, and metabolic pathways [25]. It has been associated with adverse clinical outcomes, including increased mortality, increased duration of mechanical ventilation, increased intensive care unit (ICU) and hospital stay, and increased risk of infection [26–29]. Insulin is a pleiotropic hormone with various effects on glucose metabolism, the central nervous system, immune system, and cardiovascular system. It signals via the phosphatidyl-inositol 3-kinase (PI3K) pathway, the E3 ubiquitin-protein ligase CBL (Cbl)/Cbl-associated protein (CAP) pathway, and the mitogen-activated protein kinase pathway. The primary cardioprotective action of insulin is thought to occur through upregulation of components of both the PI3K and mitogen-activated protein kinase pathways, which are sometimes collectively referred to as the reperfusion injury salvage kinase (RISK) [21, 30]. Studies on tight glycemic control with intensive insulin therapy specific to the setting of cardiac surgery are limited in number and greatly varied in methodology. A fairly recent meta-analysis of 7 randomized controlled trials concluded that tight glycemic control during or after cardiac surgery reduced mortality in the ICU, postsurgical atrial fibrillation, the use of epicardial pacing, the duration of mechanical ventilation and the length of stay in the ICU [31]. Actually insulin appears to sensitize cardiac muscles to the effects of intracellular Ca^{2+} [32].

2.3 *The Na^+-K^+ Pump*

The Na^+-K^+ pump is an essential heterodimeric membrane protein, which maintains electrochemical gradients for Na^+ and K^+ across cell membranes in all tissues. Reversible glutathionylation (postranslational redox modification of the Na^+-K^+ pump β1 subunit has been identified as a regulatory mechanism of the pump activity. Oxidative inhibitions of the Na^+-K^+ pump by angiotensin II and by the β1-adrenergic receptor-coupled signaling via NADPH oxidase activation, points out to the relevance of this regulatory mechanism in cardiovascular physiology and pathophysiology. This has implications for dysregulation of intracellular Na^+ and Ca^{2+} as well as for increased oxidative stress in heart failure, myocardial I/R, and regulation of vascular tone under conditions of elevated oxidative stress. Strategies that reverse oxidative inhibition of the Na^+-K^+ pump may likely provide cardiovascular protection [33].

2.4 *Ca^{2+} Transport*

Ca^{2+}/calmodulin-dependent protein kinase II (CaMKII) is an important molecule linking Ang II, ROS, inflammatory signals and cardiovascular pathological conditions [7]. Ca^{2+}/calmodulin-dependent protein kinase II δ mediates myocardial ischemia/reperfusion injury through NF-kB [33]. CaMKII inhibition protects against necrosis and apoptosis in irreversible ischemia-reperfusion injury [34]. Upregulation of CaMKIIδ during ischaemia-reperfusion involves sarcolemmal Ca^{2+} cycling proteins and is associated with reperfusion-induced arrhythmias and mechanical dysfunction of the rat heart [35].

2.5 *Immune Cells and Inflammation*

It is considered that I/R injury includes an inflammatory condition characterized by innate immunity and by an adaptive immune response. Myocardial I/R experiments suggested that immune cells may mediate reperfusion injury. Specifically, monocytes, macrophages, T-cells, mast cells, platelets and endothelial cells are involved in relation to the complement cascade, toll-like receptors, cytokines, oxidative stress, renin-angiotensin system, and in to the microvascular system in the signaling mechanisms of ischemia/reperfusion [36]. Atrial fibrillation (AF) is the most common arrhythmia associated with coronary artery surgery and is an important factor contributing to postoperative morbidity and mortality. There is growing evidence that dysregulation of the oxidant-antioxidant balance, inflammatory factors and discordant alteration of energy metabolites may play a significant role in its pathogenesis [37]. Both, oxidative stress and neutrophils have been suggested as the keystones in I/R injury [38]. Etanercept attenuated inflammation and oxidative stress in an I/R injured rat heart [39]. Etanercept treatment decreased neutrophil accumulation, oxidative stress, myocardial TNF-α induction, myocardial infarction area, myocardial myeloperoxidase (MPO) levels, malondialdehyde (MDA) levels, serum creatinine kinase (CK) and lactate dehydrogenase (LDH) levels, and myocardial, and serum TNF-α production while increasing the activities of the antioxidant enzymes superoxide dismutase (SOD) and glutathione peroxidase (GSH-PX) [39].

2.6 α-Tocopherol (Vitamin E)

Myocardial I/R constitutes a clinically relevant problem associated with thrombolysis, angioplasty and coronary bypass surgery. The myocardium injury to I/R includes cardiac contractile dysfunction, arrhythmias as well as irreversible myocyte damage. These changes are considered to be the consequence of an imbalance between the increased formation of oxidants and a decreased availability of endogenous antioxidants in the heart. Nearly 26 years ago, arterial and coronary sinus blood measurements revealed a delayed recovery of myocardial O_2 consumption and lactate utilization and the myocardial release of conjugated dienes (indicators of lipid peroxidation) at 3 and 60 min after reperfusion in 10 patients undergoing surgical revascularization. In addition, the myocardial concentrations of α-tocopherol decreased after reperfusion, indicating the consumption of this tissue antioxidant. These results supported the concept of a contribution of O_2-derived free radicals and lipid peroxidation to the myocardial injury after cardioplegic arrest [40]. Later on, a clinical study evaluated whether orally administered α – tocopherol acetate was effective in increasing myocardial α-tocopherol levels and to exert a protective effect on the cardioplegic arrest followed by reperfusion. Twenty stable angina patients having aortocoronary bypass surgery were evaluated in the study. The patients received preoperatively the natural stereoisomer of α-tocopherol acetate labeled with deuterium (D) at 100 mg ($n = 6$), 300 mg ($n = 8$) or 900 mg ($n = 6$) of D-α – tocopherol acetate for 14 consecutive preoperative days (control group: 6 non supplemented patients). Left ventricular deuterated and non-deuterated α-tocopherol levels were measured by gas chromatography/mass spectrometry. Although there was a decrease ($p < 0.05$) in myocardial α-tocopherol levels at the onset of reperfusion (cross-clamp removal), myocardial tocopherol levels were not statistically different from preoperative levels by 20 min of reperfusion. The conclusion was that 300 mg of α-tocopherol should be taken orally for 14 consecutive days to double myocardial alpha α-tocopherol heart content [41].

2.7 *Melatonin*

Melatonin has proved to exert protective effects against cardiac diseases induced by oxidative stress. The administration of melatonin, as a supposed result of its antioxidant properties, has been reported to reduce hypertension and cardiotoxicity induced by clinically used drugs [42–44]. Melatonin attenuates the molecular and cellular damages resulting from cardiac I/R in which destructive O_2 free radicals are involved. The anti-inflammatory and antioxidative properties of melatonin are considered being involved in the protection against chronic vascular disease and atherosclerosis [44].

3 Antioxidant Supplementation in Normotensive and Hypertensive Subjects

Atherosclerotic cardiovascular diseases are a major cause of mortality and morbidity in the general population [45]. Numerous studies have focused on the utility of antioxidant supplementation in the treatment of cardiovascular diseases [46]. Yet, whether antioxidant supplementation has any preventive and/or therapeutic value in cardiovascular pathology is still a matter of debate because the evidence is inconclusive [47–53]. Observational studies of vitamins C and E, the most prevalent natural antioxidant vitamins, suggest that supplemental use of these vitamins may lower the risk for coronary events [54]. High doses of antioxidants may pose a risk due to adverse effects [55]. Advertising and marketing encourage consumption of vitamins supplements regardless proper indication and supplements are ready available on-the-counter for self-medication. The estimated prevalence of dietary-supplement use among US adults was reported to be 73 % not long ago [53]. In some populations, supplements are consumed to enhance general wellbeing following the advice of friends and magazines [56]. Oxidative stress is considered involved in the pathogenesis of atherosclerotic cardiovascular disease and ROS participation in cardiovascular pathology has been reviewed [57]. It is then natural to consider the eventual role of antioxidants in the improvement of atherosclerotic cardiovascular disease. Antioxidants administration has proved to exert protection against damage in animal studies [58–61] and an imbalance of the oxidant/antioxidant relationship has been reported in experimental models of disease [62–65]. The aim of the present study is to evaluate whether the supplementation with antioxidants effectively modifies the biochemical profile associated with oxidative metabolism in elderly patients

Table 13.1 Plasma levels of antioxidants and oxidative stress parameters in the patients before the study. From Otero-Losada et al. [67]

	Smoker (n = 32)	Non-smoker (n = 80)	Hypertensive (n = 20)	Sedentary (n = 71)	ACVD (n = 26)
α-Tocopherol (α-TP)	19.2 ± 2.7	22.2 ± 2.3	21.3 ± 3.2	22.0 ± 2.3	20.4 ± 2.4
β-carotene	0.30 ± 0.03	0.34 ± 0.02	0.32 ± 0.0	0.33 ± 0.04	0.32 ± 0.03
Ubiquinol-10	0.24 ± 0.02	0.35 ± 0.07	0.32 ± 0.05	0.31 ± 0.04	0.32 ± 0.05
Glutathione	0.62 ± 0.06	0.75 ± 0.05	0.72 ± 0.04	0.74 ± 0.06	0.73 ± 0.06
TBARS	2.45 ± 0.23	1.97 ± 0.04	2.08 ± 0.09	2.14 ± 0.08	2.20 ± 0.10
TBARS/α-TP	0.13 ± 0.04#	0.08 ± 0.02	0.10 ± 0.02	0.10 ± 0.02	0.10 ± 0.03
Lipid oxidation (%)	18.7 ± 2.5*	12.6 ± 2.2	16.6 ± 2.0	15.3 ± 2.3	16.2 ± 2.1

undertaking periodical cardiovascular check. Vitamin C, vitamin E, and β-carotene [66] are considered important antioxidants in humans and were tested in this study. The plasma ratio of oxidants to antioxidants is higher in conditions associated with accelerated aging compared with expected physiological aging.

A few years ago, we reported a study on antioxidant supplementation in elderly cardiovascular patients [67]. We tested the hypothesis that antioxidants supplementation might improve the oxidative biochemical profile in elderly patients under periodical cardiovascular check. Patients received daily supplementation with α-tocopherol 400 mg, β-carotene 40 mg and ascorbic acid 1000 mg with dinner for 2 months. Plasma concentrations of α-tocopherol, β-carotene, ascorbic acid, ubiquinol, glutathione and TBARS (thiobarbituric acid reactive substances) were determined before and after antioxidant supplementation. A large number (n = 229) of patients participated in the study comprising 29 % smokers, 18 % hypertensives, 63 % sedentary subjects and 23 % patients with atherosclerotic cardiovascular disease (ACVD) [66]. Plasma levels of TBARS were determined before and after antioxidant supplementation. Antioxidant supplementation increased the plasma levels of α-tocopherol by an overall effect of about 35 % over pre-supplementation values from 21.0 μM (average of the groups in Table 13.1) to 25.8 μM and to 28.9 μM depending on the basal level (Tables 13.1 and 13.2). Similarly, supplementation increased the plasma levels of β-carotene by an overall effect of about 50 % over pre-supplementation values (Tables 13.1 and 13.2). Smoking status was strongly associated with ACVD and with high TBARS levels and TBARS/α-TP ratio (indication of lipid peroxidation). The pre-treatment levels of antioxidants or TBARS were not associated with age, diabetes condition or sedentarism. Smoking status was strongly associated with ACVD. Forty-two percent of smokers had ACVD compared with 16 % of ACVD cases in non-smokers [66]. Actually, ACDV were correlated with a correlation coefficient of 0.87; the 76 % of ACVD prevalence in the patients under study could be explained by smoking status [66]. Basal TBARS/α-TP ratio (pro-oxidant/antioxidant imbalance) was higher in smokers compared to non-smokers: 0.11 vs. 0.06 respectively (Table 13.1).

Table 13.2 Effect of antioxidant supplementation on the plasma levels of antioxidants in humans. Biochemical profile related to oxidative metabolism and oxidative stress. From Otero-Losada et al. [67]

	Plasma levels (μM)		Ratio (post/pre-treatments)	Statistical significance
	Pre-treatment	Post-treatment		
α-Tocopherol (α-TP)	>18	28.9 ± 2.6	1.3 ± 0.1	NS
α-Tocopherol (α-TP)	<18	25.8 ± 2.7	1.6 ± 0.1	$p < 0.01$
β-carotene	>0.3	0.48 ± 0.05	1.6 ± 0.2	$p < 0.02$
β-carotene	<0.3	0.40 ± 0.05	1.7 ± 0.1	$p < 0.02$
Ubiquinol	0.34 ± 0.06	0.46 ± 0.08	1.3 ± 0.1	NS
Glutathione	0.75 ± 0.09	0.96 ± 0.10	1.3 ± 0.1	NS
TBARS	1.5 ± 0.1	1.1 ± 0.1	0.7 ± 0.1	NS
Lipid oxidation(%)	14.6 ± 2.2	5.3 ± 0.4	0.4 ± 0.1	$p < 0.0003$

α-Tocopherol, β-carotene, glutathione, and ubiquinol levels were dissociated from the smoking condition (yet an overall trend to lower antioxidant levels was observed in smokers) [67]. The plasma levels of both α-tocopherol and β-carotene were increased by antioxidant supplementation, as said before. Moreover, cluster analysis split the sample into two categories of patients based on pre-treatment α-tocopherol and β-carotene (Table 13.2).

The value of 18 μM α-tocopherol, taken as the cutting point between lower and normal α-tocopherol plasma concentrations, was reported as a critical level by other authors [68]. Ubiquinol, glutathione and TBARS did not show significant changes following supplementation and did not show relation to α-tocopherol or β-carotene concentrations (Table 13.2). Supplementation with antioxidants reduced the percentage of oxidation, an indirect estimation of the amount of antioxidants based on the increase in TBARS by plasma incubation, from 14.6 to 5.2 %. The overall decrease in percentage of lipid oxidation following antioxidant supplementation was accounted by plasma from patients with low, <18 μM, α-tocopherol levels [68].

Overall, no objective beneficial effects were observed after antioxidant supplementation, except for subjective observations such as "feeling more vital" or "general wellbeing". A psychological, perceptual placebo effect was not ruled out at that time [67]. The strong association of smoking with a more pro-oxidative situation has been reported previously [69, 70]. It was concluded that antioxidant supplementation, at doses similar to those available in over the counter multivitamin supplements, might have benefits for patient health [66, 67].

4 Clinical Trials of Antioxidant Supplementation

Only a few of several meta-analyses of the clinical trials of antioxidant supplementation are mentioned below [71–75]. Several cohort studies suggested reduced cardiovascular risk in persons taking vitamin E supplements. However, randomized clinical trials of vitamin E did not show any benefit of vitamin E supplementation in terms of prevention of coronary heart disease and death [71]. Identical rates of cardiovascular death were found for the placebo and vitamin groups. Moreover, a small but significant increase in CVD was found associated with α-tocopherol and β-carotene supplementation in a meta-analysis of 7 trials using vitamin E in >81000 patients and 8 β-carotene in >138000 patients [72]. A meta-analysis of 19 clinical trials comprising a total of 135967 participants revealed that supplementation with high doses (16.5–2000 IU/d) of vitamin E may cause a slight increase in mortality [74]. A further meta-analysis of the same 19 clinical trials with the inclusion of 10 additional trials (2495 participants, vitamin E doses 136–5000 IU/d) was performed but yielded contradictory results. Results seem to indicate that the increased mortality ratio was not related to the supplementation with high doses of vitamin E [73]. As part of a European multicenter project, a study (400 healthy volunteers, 25–45 years) observed that supplementation with α-tocopherol and/or carotenoids increased the respective serum levels, without significant side effects (except carotenodermia) or changes in biochemical or hematological indices [74].

Oxidative stress has been implicated in aging, in the pathophysiology of aging and in age-associated diseases [75, 76]. Antioxidants supplementation has become a practice for delaying aging and for the prevention of atherosclerosis and cardiovascular disease. Clinical studies have not demonstrated a benefit of vitamin E in the primary and secondary prevention of cardiovascular disease and even vitamin E supplementation was associated with increased mortality, heart failure, and hemorrhagic stroke [45]. The American Heart Association does not support the use of vitamin E supplementation to prevent cardiovascular disease and recommends the consumption of foods rich in antioxidant vitamins and minerals [45]. Supplementation is usually recommended on the assumption that endogenous antioxidant levels are below the accepted normal values, underestimating the adverse effects that may appear. Supplementation adds an extra burden to the liver and kidneys particularly in elderly patients. A simple laboratory analysis provides information on antioxidant levels and may additionally help to redirect towards a more accurate diagnosis by ruling out (or not) hypothetical nutritional deficits. In our 2013 report [67] an increased oxidative stress was observed in smokers in agreement with a previous report [77]. Smoking was associated with higher TBARS/α-TP ratios suggesting an increase in plasma lipid peroxidation. Cigarette smoking is widely accepted to be a major cardiovascular risk factor. Low plasma levels of antioxidants have been associated with endothelial dysfunction, the first step towards atherosclerosis and increased cardiovascular risk [78, 79]. In this review we consider the relationship between ACVD and the plasma levels of antioxidants, and found no association between ACVD and low serum concentrations of any antioxidant. The small number of ACVD patients in the study (n = 26) may partly account for this discrepancy. It is possible that some of these patients had increased consumption of vegetables and other sources of antioxidants after they suffered a major event of coronary or peripheral vascular disease. Other authors did not find differences in plasma α-tocopherol but observed higher α/γ-tocopherol ratio in patients with coronary heart disease [80]. In our report we concluded that the effectiveness of antioxidant supplementation to modify plasma biochemistry mainly depended on basal antioxidants levels [67]. We also suggested that awareness of the plasma antioxidants levels might be advisable before starting supplementation with antioxidants in elderly cardiovascular patients, a population in which special precautions are recommended. Arguably, excess antioxidant levels in tissues may lead to deleterious consequences [81–84] However, no adverse effects were reported during the course of our study [57].

5 Oxidative Stress and Lipid Peroxidation in Post-ischemic Reperfusion

5.1 *Biochemical Findings in Human Cardiac Surgery*

In animal models, formation of oxidants during post I/R exerts deleterious effects on heart function that is mediated by oxidative stress. Cardioplegic arrest/reperfusion during cardiac surgery might similarly induce oxidative stress. However, the phenomenon had not been precisely characterized in patients until a few years ago. Then, we studied the role of antioxidant therapy in cardiac surgery which was a matter of debate in the cardiovascular surgery community and remains a challenging question for some surgeons even now. The purpose of our investigation was to ascertain whether the relationship between oxidant formation and development of myocardial injury also translate to patients subjected to cardioplegic arrest [85]. Twenty-four patients undergoing coronary artery bypass participated in our study, whose characteristics are shown in Table 13.3.

Trans-cardiac blood aliquots and myocardial biopsy samples were taken both before cardioplegic arrest and following reperfusion. The surgical procedure had only modest and transient effects on cardiac function. Hemodynamic parameters returned to normal levels within a few hours postoperatively. Ejection fraction and peak filling rate remained unchanged (Table 13.4), except for time-to-peak filling rate at discharge that was shorter than the baseline, but the difference was not significant.

Before ischemia, the concentrations of reduced glutathione (GSH) were similar in systemic arterial blood and in venous blood from the coronary sinus (0.7 and 0.8 μM, respectively). During reperfusion after cardioplegic arrest, the GSH concentration in arterial blood entering the heart remained stable (0.8 μM) but, in contrast, GSH in blood drawn from the coronary sinus significantly increased (to 1.8 μM); thus, a net trans-cardiac release of GSH, *i.e.*, a positive difference between venous

Table 13.3 Characteristics of the 24 patients under study

Gender (male/female) (n)	20/4
Age (y)	60.1 ± 9.4
Weight (kg)	80 ± 13
Smoking (%)	58
Drinking (%)	29
Angina episodes/week	4.7 ± 2.4
Previous myocardial infarction (%)	50
Hypertension (%)	67
Diabetes (%)	21
Hypercholesterolemia (%)	42
Drug therapy (% of treated patients)	
Atenolol	30 %
Statin	18 %
ACE inhibitors	9 %
Coronary lesions (occluded vessels with >70 % estenosis)	
Left anterior descending artery	3
Circumflex coronary artery	4
Right coronary artery	9
Left main coronary artery	8

Table 13.4 Radionuclide angiography data

	Enrollment	Pre-operative	Post-operative
LVEF (%)	53.1 ± 8.5	51.9 ± 9.0	52.5 ± 10.7
EDV/s	1.7 ± 0.7	1.7 ± 0.5	1.6 ± 0.7
Time-to-peak filling (ms)	195 ± 63	193 ± 56	166 ± 65

Data are mean ± SD. *LVEF* left ventricular ejection fraction, *EDV* end diastolic volume (From Milei et al. [84])

minus arterial concentrations, occurred at 5 min of reperfusion that persisted after 20 min (Fig. 13.1).

Prior to cardiac arrest, TBARS in the systemic arterial blood was 3.7 μM and in the venous blood from the coronary sinus was slightly lower, 3.3 μM, likely indicating a cardiac extraction of lipid peroxides (Fig. 13.3) consistent with metabolism by mitochondria aldehyde-dehydrogenase and aldose-reductase [35, 36]. Five min after reperfusion, TBARS concentrations were similar in arterial blood entering the heart (3.1 μM) and in coronary sinus blood (3.1 μM), therefore, no net cardiac release was observed, a condition that persisted at 20 min of reperfusion (Fig. 13.2).

In the living heart, TBARS were 0.99 μmol/g heart in biopsies taken before cardioplegic arrest, and 0.92 μmol/g heart in biopsies taken 10 min after reperfusion. Thus, no increase in membrane lipid peroxidation was documented in whole cardiac tissue. Previous animal studies of O_2 radical-mediated reperfusion injury had shown that peroxidation of membrane lipids is prominent in subcellular organelles [85–87]. To investigate this specific point, we analyzed cardiac biopsies that were

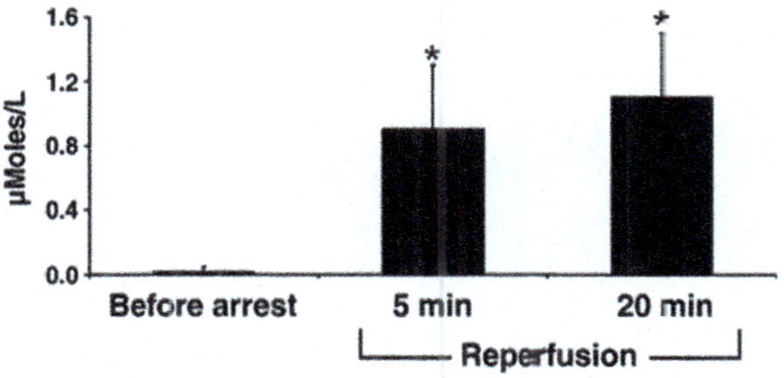

Fig. 13.1 Transcardiac glutathione release before cardioplegic arrest, and at 5 and 20 min reperfusion. Data are coronary sinus minus aorta concentrations. *$p < 0.05$ vs pre-ischemia (Taken from Milei et al. [85]

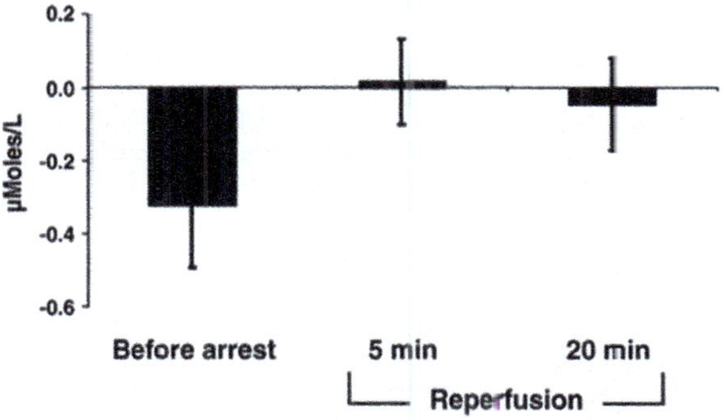

Fig. 13.2 Transcardiac release of lipid peroxides before cardioplegic arrest, and at 5 and 20 min reperfusion. Data are coronary sinus minus aorta concentrations. p = NS (Taken from Milei et al. [85]

processed to isolate subcellular fractions. Also in this case, there was no increase in TBARS after post ischemic reperfusion, neither in the whole homogenate, nor in mitochondrial or cytosol fractions (Table 13.5). It is worth noting that, in our hands, neither coronary sinus blood nor cardiac biopsies showed increased lipid peroxide concentrations.

5.2 *Histological Findings in Human Cardiac Surgery*

Intracellular myocyte architecture was normal in preischemic biopsies (Fig. 13.3). Sarcomere alignment was normal with intact sarcolemmal membrane, swelling tubules, intercalated disks, and basement membranes, Nuclei and cytoplasmic glycogen were well preserved (Fig. 13.3, left panel). Mild cytosolic and intermyofibrillar edema was occasionally seen and mitochondria showed tightly packed cristae, with no or minimal signs of edema (Fig. 13.3, right panel) (Fig. 13.4).

The overall injury score showed no differences between preischemic and reperfusion biopsies (1.7 vs 1.8). In addition, since mitochondria are a target of I/R injury, a systematic evaluation was performed scoring 150 mitochondria/section (*i.e.*, >2000 mitochondria/biopsy). Before cardiac arrest, the great majority of mitochondria showed minimal or no alterations (Fig. 13.5). After reperfusion, there was a decrease in the number of normal mitochondria with a concomitant increase in the

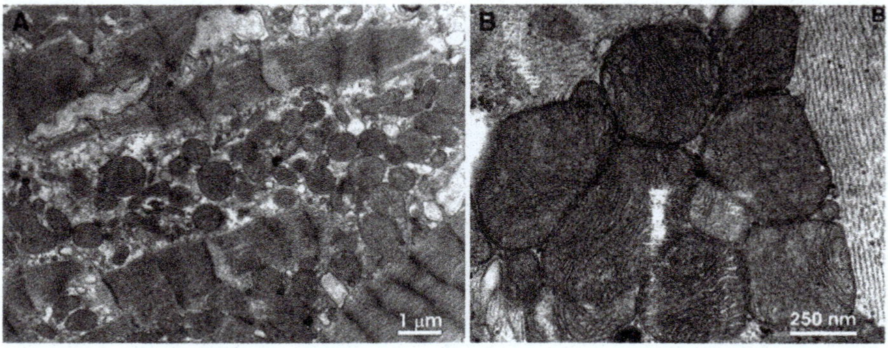

Fig. 13.3 Typical electron microscopy appearance of myocardial biopsies before cardioplegic arrest. *Left panel*: normal mitochondria, mild cytosolic and mild intermyofibrillar edema, and modest T-tubules dilation is observed (×8000). *Right panel*: cross-section at higher magnification, showing mitochondria with intact membranes and tightly packed cristae. (×10000) (From Milei et al. [84])

Table 13.5 Subcellular distribution of lipid peroxidation products

	Whole homogenate	Mitochondria	Cytosol
Before arrest	0.99 ± 0.42	1.04 ± 0.16	1.27 ± 0.30
Reperfusion	0.92 ± 0.37	0.70 ± 0.19	1.04 ± 0.28

From Milei et al. [84]

proportion of mitochondria showing ultrastructural alterations (Fig. 13.5); however, without statistical significance [84]. At the end, a full recovery of left ventricular systolic and diastolic functions was observed. In our study, the ultrastructural alterations that characterize I/R injury (contraction-band necrosis, massive mitochondrial swelling, sarcolemmal disruption) were mostly absent. The finding that myocyte ultrastructure was largely preserved helped explaining why recovery of systolic and diastolic function was unimpeded, and this was internally consistent with absence of biochemical "signatures" of irreversible oxidative injury in the patients under study [84].

Endogenous antioxidants are another defense mechanism, besides antioxidant enzymes. In our patients, the cardiac concentrations of ubiquinol decreased upon reperfusion [66], consistent with the notion that endogenous antioxidants are "consumed" in the process of oxidant detoxification [86, 87]. The finding supports the idea that hearts were exposed to ROS. However, since oxidation of glutathione and antioxidant vitamins occurs upstream in the chain of events following ROS formation, in order to establish whether oxidative injury has occurred, the measurements must be complemented by other indices of oxidant attack.

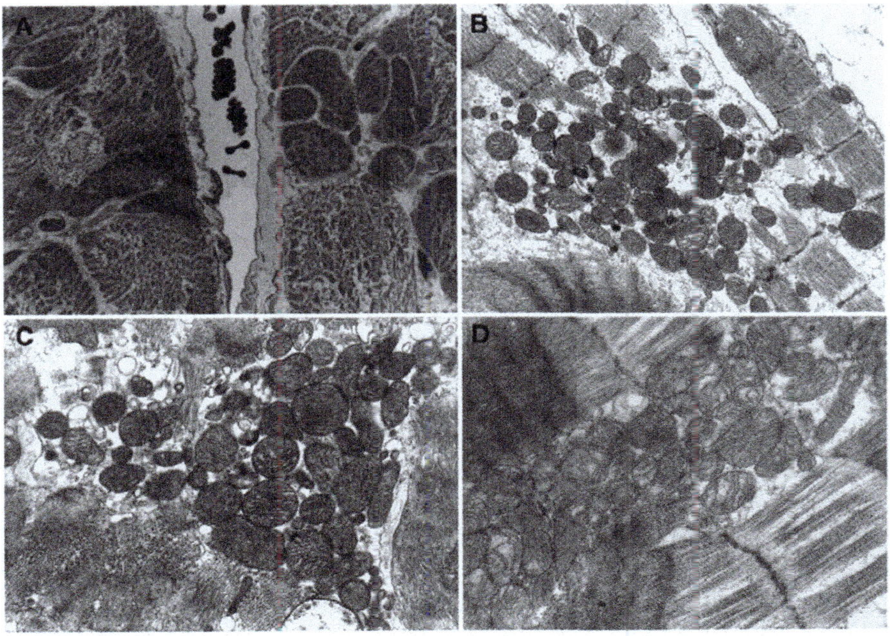

Fig. 13.4 Prototypical biopsies taken after 10 min reperfusion following cardioplegic arrest. *Panel A*. Large-field view by light microscopy of 1 μm-thick section. Myocardial architecture and vessel structure are well preserved. Mild intermyofibrillar edema is observed. 1 % Toluidine-borax staining; ×400. *Panel B*. Transmission electron micrograph of reperfusion biopsy. Intermyofibrillar edema with mild sarcomere hypercontraction, glycogen decrease, and mild cytosolic edema. Mild mitochondrial damage (degree 0–2); ×8000. *Panel C*. Sarcomere hypercontraction, mitochondrial swelling, damage degree 1–2, and intermyofibrillar edema; ×10,000. *Panel D*. Focal sarcomere fragmentation is shown. Mitochondrial damage degree 3, lipid vacuoles, T-tubules dilation and myofibrillar disorganization, with glycogen loss; ×8000 (Taken from Milei et al. [85])

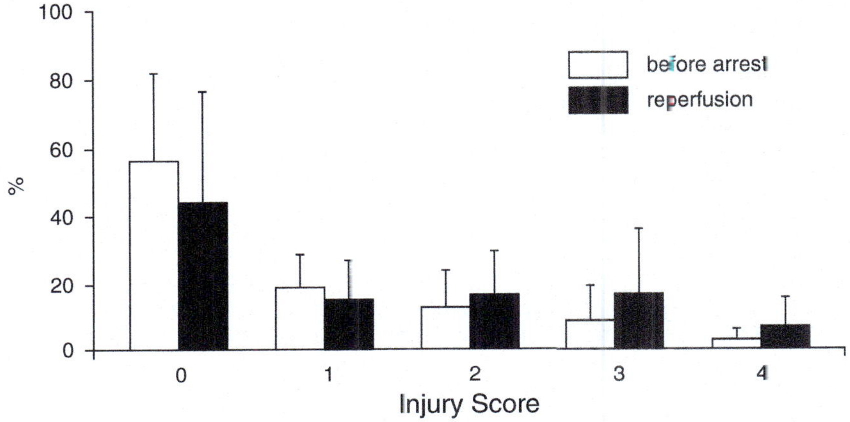

Fig. 13.5 Mitochondrial injury score in biopsies taken before arrest and 10 min after reperfusion. Data are the percentage of score distribution (From Milei et al. [85])

6 Conclusions

Antioxidant supplementation at doses similar to those routinely available in over the counter multivitamin supplements, was found to have some benefits in aged patients with a cardiac condition. The patients made subjective manifestations of general wellbeing after antioxidant supplementation. Supplementation increased the plasma levels of α-tocopherol by about 35 % over pre-supplentation values. Similarly, supplementation increased the plasma levels of β-carotene by about 50 % over pre-supplementation values. In both cases, the increased levels depended on the respective basal levels. Smoking status was strongly associated with atherosclerotic cardiovascular disease and with high TBARS levels and TBARS/α-TP ratio (indication of lipid peroxidation). The high prevalence of atherosclerotic cardiovascular disease in smokers agrees with the concept of an increased free-radical mediated oxidation in atherosclerotic pathology. Cardiac concentrations of ubiquinol decreased upon reperfusion in patients undergoing coronary artery bypass. This finding is consistent with the idea that endogenous antioxidants are "consumed" during oxidant detoxification and confirms heart exposition to ROS upon reperfusion.

Acknowledgements The authors acknowledge the patients who willingly participated in our studies and provided written consent.

References

1. Dhalla NS, Liu X, Panaglia V, Takeda N (1998) Subcellular remodeling and heart dysfunction in chronic diabetes. Cardiovasc Res 40:239–247
2. Dhalla NS, Elmoselhi AB, Hata T, Makino N (2000) Status of myocardial antioxidants in ischemia–reperfusion injury. Cardiovasc Res 47:446–456
3. Ferrari R, Alfieri O, Curello S et al (1990) Occurrence of oxidative stress during reperfusion of the human heart. Circulation 81:201–211
4. Grech ED, Baines M, Steyn R et al (1995) Evidence that continuous normothermic blood cardioplegia offers better myocardial protection than intermittent hypothermic cardioplegia. Br Heart J 74:517–521
5. Jain AK, Mehra NK, Swarnakar NK (2015) Role of Antioxidants for the Treatment of Cardiovascular Diseases: Challenges and Opportunities. Curr Pharm Des 21:4441–4455
6. Kaminski KA, Bonda TA, Korecki J, Musial WJ (2002) Oxidative stress and neutrophil activation--the two keystones of ischemia/reperfusion injury. Int J Cardiol 86:41–59
7. Aggarwal NT, Makielski JC (2013) Redox control of cardiac excitability. Antioxid Redox Signal 18:432–468
8. Wen H, Gwathmey JK, Xie LH (2012) Oxidative stress-mediated effects of angiotensin II in the cardiovascular system. World J Hypertens 2:34–44
9. Klishadi MS, Zarei F, Hejazian SH et al (2015) Losartan protects the heart against ischemia reperfusion injury: sirtuin3 involvement. J Pharm Pharm Sci 18:112–123
10. Porter GA, Urciuoli WR, Brookes PS, Nadtochiy SM (2014) SIRT3 deficiency exacerbates ischemia-reperfusion injury: implication for aged hearts. Am J Physiol Heart Circ Physiol 306:H1602–H1609

11. Ravarotto V, Pagnin E, Maiolino G et al (2015) The blocking of angiotensin II type 1 receptor and RhoA/Rho kinase activity in hypertensive patients: Effect of olmesartan medoxomil and implication with cardiovascular-renal remodeling. J Renin Angiotensin Aldosterone 16:1245–1250
12. Wei X, Zhu X, Hu N et al (2015) Baicalin attenuates angiotensin II-induced endothelial dysfunction. Biochem Biophys Res Commun 465:101–107
13. Yang BC, Phillips MI, Ambuehl PE et al (1997) Increase in angiotensin II type 1 receptor expression immediately after ischemia-reperfusion in isolated rat hearts. Circulation 96:922–926
14. Baker KM, Booz GW, Dostal DE (1992) Cardiac actions of angiotensin II: role of an intracardiac renin-angiotensin system. Annu Rev Physiol 54:227–241
15. Matsushima S, Sadoshima J (2015) The role of sirtuins in cardiac disease. Am J Physiol Heart Circ Physiol 309:H1375–H1389
16. Sladek T, Sladkova J, Kolar F et al (1996) The effect of AT1 receptor antagonist on chronic cardiac response to coronary artery ligation in rats. Cardiovasc Res 31:568–576
17. Safari F, Hajizadeh S, Shekarforoush S et al (2012) Influence of ramiprilat and losartan on ischemia reperfusion injury in rat hearts. J Renin-Angiotensin-Aldosterone Syst 13:29–35
18. Zhu B, Sun Y, Sievers RE et al (2000) Comparative effects of pretreatment with captopril and losartan on cardiovascular protection in a rat model of ischemia-reperfusion. J Am Coll Cardiol 35:787–795
19. Mollnau H, Wendt M, Szöcs K et al (2002) Effects of angiotensin II infusion on the expression and function of NAD (P) H oxidase and components of nitric oxide/cGMP signaling. Circ Res 90:E58–E65
20. Messadi-Laribi E, Griol-Charhbili V, Pizard A et al (2007) Tissue kallikrein is involved in the cardioprotective effect of AT1-receptor blockade in acute myocardial ischemia. J Pharmacol Exp Ther 323:210–216
21. de Oliveira UO, Belló-Klein A, de Oliveira ÁR et al (2012) Insulin alone or with captopril: effects on signaling pathways (AKT and AMPK) and oxidative balance after ischemia-reperfusion in isolated hearts. Fundam Clin Pharmacol 26:679–689
22. Ng KW, Allen ML, Desai A et al (2012) Cardioprotective effects of insulin: how intensive insulin therapy may benefit cardiac surgery patients. Circulation 125:721–728
23. Bellodi G, Manicardi V, Malavasi V et al (1989) Hyperglycemia and prognosis of acute myocardial infarction in patients without diabetes mellitus. Am J Cardiol 64:885–888
24. Zeller M, Cottin Y, Brindisi MC et al (2004) Impaired fasting glucose and cardiogenic shock in patients with acute myocardial infarction. Eur Heart J 25:308–312
25. Oswald GA, Smith CC, Betteridge DJ, Yudkin JS (1986) Determinants and importance of stress hyperglycaemia in non-diabetic patients with myocardial infarction. BMJ (Clin Res Ed) 293:917–922
26. Van Cromphaut SJ (2009) Hyperglycaemia as part of the stress response: the underlying mechanisms. Best Pract Res Clin Anaesthesiol 23:375–386
27. Whitcomb BW, Pradhan EK, Pittas AG et al (2005) Impact of admission hyperglycemia on hospital mortality in various intensive care unit populations. Crit Care Med 33:2772–2777
28. Sung J, Bochicchio GV, Joshi M, Bochicchio K, Tracy K, Scalea TM (2005) Admission hyperglycemia is predictive of outcome in critically ill trauma patients. J Trauma 59:80–83
29. Krinsley JS (2003) Association between hyperglycemia and increased hospital mortality in a heterogeneous population of critically ill patients. Mayo Clin Proc 78:1471–1478
30. Cheung NW, Napier B, Zaccaria C, Fletcher JP (2005) Hyperglycemia is associated with adverse outcomes in patients receiving total parenteral nutrition. Diabetes Care 28:2367–2371
31. Hausenloy DJ, Tsang A, Mocanu MM, Yellon DM (2005) Ischemic preconditioning protects by activating prosurvival kinases at reperfusion. Am J Physiol Heart Circ Physiol 288:H971–H976

32. Haga KK, McClymont KL, Clarke S et al (2011) The effect of tight glycaemic control, during and after cardiac surgery, on patient mortality and morbidity: A systematic review and meta-analysis. J Cardiothoracic Surgery 6:3. doi:10.1186/1749-8090-6-3
33. Liao X, Wang L, Yang C et al (2011) Cyclooxygenase mediates cardioprotection of angiotensin-(1–7) against ischemia/reperfusion-induced injury through the inhibition of oxidative stress. Mol Med Rep 4:1145–1150
34. Ling H, Gray CB, Zambon AC et al (2013) Ca2+/Calmodulin-dependent protein kinase II δ mediates myocardial ischemia/reperfusion injury through nuclear factor-kB. Circ Res 112:935–944
35. Vila-Petroff M, Salas MA, Said M et al (2007) CaMKII inhibition protects against necrosis and apoptosis in irreversible ischemia-reperfusion injury. Cardiovasc Res 73:689–698
36. Adameova A, Carnicka S, Rajtik T et al (2012) Upregulation of CaMKIIδ during ischaemia-reperfusion is associated with reperfusion-induced arrhythmias and mechanical dysfunction of the rat heart: involvement of sarcolemmal Ca2 + −cycling proteins. Can J Physiol Pharmacol 90:1127–1134
37. Zuidema MY, Zhang C (2010) Ischemia/reperfusion injury: The role of immune cells. World J Cardiol 2:325–332
38. Elahi MM, Flatman S, Matata BM (2008) Tracing the origins of postoperative atrial fibrillation: the concept of oxidative stress-mediated myocardial injury phenomenon. Eur J Cardiovasc Prev Rehabil 15:735–741
39. Kaminski KA, Bonda TA, Korecki J, Musial WJ (2002) Oxidative stress and neutrophil activation-the two keystones of ischemia/reperfusion injury. Int J Cardiol 86:41–59
40. Yang M, Chen J, Zhao J, Meng M (2014) Etanercept attenuates myocardial ischemia/reperfusion injury by decreasing inflammation and oxidative stress. PLoS One 9:e108024
41. Weisel RD, Mickle DA, Finkle CD et al (1989) Myocardial free-radical injury after cardioplegia. Circulation 80:14–18
42. Mickle DA, Weisel RD, Burton GW et al (1991) Effect of orally administered alpha-tocopheryl acetate on human myocardial alpha-tocopherol levels. Cardiovasc Drugs Ther 5(Suppl 2):309–312
43. Tengattini S, Reiter RJ, Tan DX et al (2008) Cardiovascular diseases: protective effects of melatonin. J Pineal Res 44:16–25
44. Reiter RJ, Tan DX (2003) Melatonin: a novel protective agent against oxidative injury of the ischemic/reperfused heart. Cardiovasc Res 58:10–19
45. Domínguez-Rodríguez A, Abreu-Gonzalez P (2010) Myocardial ischemia–reperfusion injury: possible role of melatonin. World J Cardiol 2:233–236
46. Saremi A, Arora R (2010) Vitamin E and cardiovascular disease. Am J Ther 17:e56–e65
47. Katsiki N, Manes C (2009) Is there a role for supplemented antioxidants in the prevention of atherosclerosis? Clin Nutr 28:3–9
48. Kailash P (2009) Vitamin E does not slow the progression of hypercholesterolemic atherosclerosis. Int J Angiol 18:89–95
49. Vivekananthan DP, Penn MS, Sapp SK et al (2003) Use of antioxidant vitamins for the prevention of cardiovascular disease: meta-analysis of randomised trials. Lancet 361:2017–2023
50. Lonn E, Yusuf S, Hoogwerf B et al (2002) Effects of vitamin E on cardiovascular and microvascular outcomes in high-risk patients with diabetes: results of the HOPE study and MICRO-HOPE substudy. Diabetes Care 25:1919–1927
51. Hodis HN, Mack WJ, LaBree L et al VEAPS Research Group (2002) Alpha-tocopherol supplementation in healthy individuals reduces low-density lipoprotein oxidation but not atherosclerosis: the Vitamin E Atherosclerosis Prevention Study (VEAPS). Circulation 106:1453–1459
52. Salonen JT, Nyyssonen K, Salonen R et al (2000) Antioxidant Supplementation in Atherosclerosis Prevention (ASAP) study: a randomized trial of the effect of vitamins E and C on 3-year progression of carotid atherosclerosis. J Intern Med 248:377–386

53. Kris-Etherton PM, Lichtenstein AH, Howard BV et al (2004) Antioxidant vitamin supplements and cardiovascular disease. Circulation 110:637–641
54. Sadovsky R, Collins N, Tighe AP et al (2008) Patient use of dietary supplements: a clinician's perspective. Curr Med Res Opin 24:1209–1216
55. Riccioni G, D'Orazio N, Salvatore C et al (2012) Carotenoids and vitamins C and E in the prevention of cardiovascular disease. Int J Vitam Nutr Res 82:15–26
56. Oliveira BF, Veloso CA, Nogueira-Machado JA, Martins Chaves M (2012) High doses of in vitro beta-carotene, alpha-tocopherol and ascorbic acid induce oxidative stress and secretion of IL-6 in peripheral blood mononuclear cells from healthy donors. Curr Aging Sci 5:148–156
57. Kjertakov M, Hristovski R, Racaj M (2013) The use of dietary supplement among soldiers from the macedonian special operations regiment. J Spec Oper Med 13:19–24
58. Taverne YJ, Bogers AJ, Duncker DJ, Merkus D (2013) Reactive oxygen species and the cardiovascular system. Oxid Med Cell Longev 2013:862423
59. Lauver DA, Kaissarian NM, Lucchesi BR (2013) Oral pretreatment with liposomal glutathione attenuates reperfusion injury in rabbit isolated hearts. J Cardiovasc Pharmacol 61:233–239
60. Radu F, Leggett RE, Schuler C, Levin RM (2011) The effect of antioxidants on the response of the rabbit urinary bladder to in vitro ischemia/reperfusion. Mol Cell Biochem 355:65–73
61. Llesuy S, Milei J, Picone V et al (1995) Effect of vitamins A and E on ischemia-reperfusion damage in rabbit heart. Mol Cell Biochem 145:45–51
62. Milei J, Boveris A, Llesuy S et al (1986) Amelioration of adriamycin-induced cardiotoxicity in rabbits by prenylamine and vitamins A and E. Am Heart J 111:95–102
63. Otero-Losada M, Grana DR, Müller A et al (2011) Lipid profile and plasma antioxidant status in sweet carbonated beverage-induced metabolic syndrome in rat. Int J Cardiol 146:106–109
64. Vázquez-Medina JP, Popovich I, Thorwald MA et al (2013) Angiotensin receptor-mediated oxidative stress is associated with impaired cardiac redox signaling and mitochondrial function in insulin-resistant rats. Am J Physiol Heart Circ Physiol 305:H599–H607
65. Bryan S, Baregzay B, Spicer D et al (2013) Redox-inflammatory synergy in the metabolic syndrome. Can J Physiol Pharmacol 91:22–30
66. Song Y, Cook NR, Albert CM et al (2009) Effects of vitamins C and E and beta-carotene on the risk of type 2 diabetes in women at high risk of cardiovascular disease: a randomized controlled trial. Am J Clin Nutr 90:429–437
67. Otero-Losada M, Vila S, Azzato F, Milei J (2013) Antioxidants supplementation in elderly cardiovascular patients. Oxid Med Cell Longev 2013:408260
68. Carrasquedo F, Glanc M, Fraga CG (1999) Tissue damage in acute myocardial infarction: selective protection by vitamin E. Free Rad Biol Med 26:1587–1590
69. Weber P, Bendich A, Machlin LJ (1997) Vitamin E and human health: rationale for determining recommended intake levels. Nutrition 13:450–460
70. Suwannalert P, Boonsiri P, Khampitak T et al (2007) The levels of lycopene, alpha-tocopherol and a marker of oxidative stress in healthy northeast Thai elderly. Asia Pac J Clin Nutr 16:27–30
71. Kelly GS (2002) The interaction of cigarette smoking and antioxidants, Part 2: alpha-tocopherol. Altern Med Rev 7:500–511
72. Ueda S, Yasunari K (2006) What we learnt from randomized clinical trials and cohort studies of antioxidant vitamin? Focus on vitamin E and cardiovascular disease. Curr Pharm Biotechnol 7:69–72
73. Steinhubl SR (2008) Why have antioxidants failed in clinical trials? Am J Cardiol 101:14–19
74. Sachidanandam K, Fagan SC, Ergul A (2005) Oxidative stress and cardiovascular disease: antioxidants and unresolved issues. Cardiovasc Drug Rev 23:115–132
75. Gerss J, Köpcke W (2009) The questionable association of vitamin E supplementation and mortality, inconsistent results of different meta-analytic approaches. Cell Mol Biol 55:1111–1120

76. Olmedilla B, Granado F, Southon S et al (2002) A European multicentre, placebo-controlled supplementation study with alpha-tocopherol, carotene-rich palm oil, lutein or lycopene: analysis of serum responses. Clin Sci (Lond) 102:447–456
77. Junqueira VB, Barros SB, Chan SS et al (2004) Aging and oxidative stress. Mol Aspects Med 25:5–16
78. Miwa K, Fujita M (2006) Sex difference in effects of smoking on serum vitamin E concentrations in a young population. J Cardiol 48:201–207
79. Franzini L, Ardigò D, Valtueña S et al (2012) Food selection based on high total antioxidant capacity improves endothelial function in a low cardiovascular risk population. Nutr Metab Cardiovasc Dis 22:50–57
80. Giacco R, Clemente G, Cipriano D et al (2010) Effects of the regular consumption of wholemeal wheat foods on cardiovascular risk factors in healthy people. Nutr Metab Cardiovasc Dis 20:186–194
81. Ohrvall M, Sundlof G, Vessby B (1996) Gamma, but not alpha, tocopherol levels in serum are reduced in coronary heart disease patients. J Inter Med 239:111–117
82. Singh PP, Chandra A, Mahdi F et al (2010) Reconvene and reconnect the antioxidant hypothesis in human health and disease. Indian J Clin Biochem 25:225–243
83. Marchioli R, Levantesi G, Macchia A et al (2006) Vitamin E increases the risk of heart failure after myocardial infarction: results from the GISSI Preventive Trial. J Cardiovasc Med 6:347–350
84. Dotan Y, Pinchuk I, Litchenberg D, Leshno M (2009) Decision analysis supports the paradigm that indiscriminate supplementation of vitamin E does more harm than good. Arterioscl Thromb Vasc Biol 29:1304–1309
85. Milei J, Forcada P, Fraga CG et al (2007) Relationship between oxidative stress, lipid peroxidation, and ultrastructural damage in patients with coronary artery disease undergoing cardioplegic arrest/reperfusion. Cardiovasc Res 73:710–719
86. Romaschin AD, Wilson GJ, Thomas U et al (1990) Subcellular distribution of peroxidized lipids in myocardial reperfusion injury. Am J Physiol 259:H116–H123
87. Ambrosio G, Zweier JL, Flaherty JT (1991) The relationship between oxygen radical generation and impairment of myocardial energy metabolism following post-ischemic reperfusion. J Mol Cell Cardiol 23:1359–1374

Chapter 14
Oxidative Stress Influence in the Development of Pulmonary Arterial Hypertension

Adriane Belló-Klein, Alex Sander Araujo, Paulo Cavalheiro Schenkel, and Bruna Gazzi de Lima Seolin

Abstract This review addresses pulmonary arterial hypertension (PAH), an incurable disease that determines high morbidity and mortality. Diagnosis is usually performed at advanced stages of the disease, because symptoms are unspecific. Treatment is expensive and do not promote reversion of the disease, only demonstrates some improvement in patients quality of life. Oxidative stress is one of the mechanisms involved in the pathogenesis of PAH, contributing to the development of pulmonary vascular remodeling and consequent increase of pulmonary pressure. This results in an enhanced right ventricle (RV) afterload, determining RV hypertrophy which progresses to RV failure. Literature has shown increased oxidative stress not only in pulmonary vessels, but also in lungs and RV of pulmonary hypertensive patients and experimental animals. The use of experimental models has contributed to the understanding of the pathophysiology of PAH, and to the development of new therapeutic strategies. Therapeutics focused on the modulation of oxidative stress has been considered very promising. Some of these therapeutic interventions constitute of substances that modulate gene expression promoting antioxidant adaptations.

Keywords Monocrotaline • Oxidative stress • Pulmonary arterial hypertension

1 Introduction

Pulmonary hypertension (PH) is a chronic disease of a varied etiology, fast evolution and bad prognosis, being related to high morbidity and mortality [1, 2]. Clinically, PH is defined by the verification of a mean pulmonary arterial pressure (mPAP) equal or above 25 mmHg at rest, obtained through cardiac catheterism [3].

A. Belló-Klein (✉) • A.S. Araujo • P.C. Schenkel • B.G. de Lima Seolin
Laboratory of Cardiovascular Physiology, Department of Physiology, Instituto de Ciências Básicas da Saúde, Universidade Federal do Rio Grande do Sul,
Rua Sarmento Leite 500, Porto Alegre, RS CEP 90050170, Brazil
e-mail: belklein@ufrgs.br

© Springer International Publishing Switzerland 2016
R.J. Gelpi et al. (eds.), *Biochemistry of Oxidative Stress*, Advances in Biochemistry in Health and Disease 16, DOI 10.1007/978-3-319-45865-6_14

The pulmonary endothelial dysfunction, characteristic of PAH, causes gradual reduction of lumen resulting in an increase in pulmonary vascular resistance (PVR). The consequence is an enhancement in right ventricle (RV) afterload, hypertrophy, and subsequently, dilation and failure [4, 5]. The indicative signs of PAH are generally attributed to impaired oxygen transport, abnormalities in gas exchange, and reduced cardiac output [6]. Besides being asymptomatic at the initial phase, the most common symptoms of PAH are dyspnea, fatigue, weakness, exercise intolerance, syncope, angina, increased abdominal size, and peripheral edema [7, 8]. Since these symptoms are very unspecific, diagnosis is determined at more advanced stages of this disease, when RV dysfunction already exists [9]. It is known that the most common cause of death in these patients is RV failure, and that RV function is the main predictor of morbidity and mortality in this population [10].

During the last twenty years, many studies were performed with the aim to promote improvement in PAH patients health. Besides the advances in therapeutics, PAH is a highly disabling condition and still without cure. By this way, understanding the adaptive changes produced by PH in pulmonary vasculature and its impact in different organs, especially RV, is of great importance to find more effective therapies [11]. Since oxidative stress is involved in the development of this disease, participating in endothelial and ventricular dysfunction, it emerges as one relevant target to be explored in PAH therapeutics.

2 Pulmonary Arterial Hypertension

2.1 *Classification*

In the 5· World Symposium on Pulmonary Hypertension held in 2013 in Nice, France, this disease was classified in five groups: Group 1 (pulmonary arterial hypertension), Group 2 (pulmonary hypertension caused by diseases of left heart), Group 3 (pulmonary hypertension caused by pulmonary disease and/or hypoxia), Group 4 (chronic thromboembolic pulmonary hypertension), and Group 5 (pulmonary hypertension by multifactorial or unclear mechanisms) (Table 14.1) [12]. This classification was constructed based on similarities in physiologic mechanisms, clinical conditions, hemodynamic characteristics and therapeutic approaches [13]. Group 1, pulmonary arterial hypertension (PAH), includes the most studied classification of this disease in terms of etiopathogenic and therapeutic aspects. Inside group 1, the most frequent disease associated to PAH, is schistosomiasis, which is prominent in certain world regions, such as Brazil and the Sub-Saharan Africa [2, 14].

2.2 *Epidemiology*

It is estimated that PAH occurs around 15–26 cases per million of people, with an incidence of 10.7 cases per million per year [1, 15], without any ethnic predominance [16]. In a recent review, Hoeper et al. [2] estimated that 50–70 million people

Table 14.1 Classification of pulmonary hypertension according to the 5th World Symposium of Pulmonary Hypertension (Nice, France; 2013)

Classification of pulmonary hipertension
1. Pulmonary arterial hypertension (PAH)
1.1 Idiopathic
1.2 Heritable
1.2.1 Bone morphogenic protein receptor type II (BMPR2)
1.2.2 ALK-1, endoglin (ENG), SMAD9, caveolin-1 (CAV1), KCNK3
1.2.3 Unknown
1.3 Drug and toxin induced
1.4 Associated with:
1.4.1 Connective tissue disease
1.4.2 Human immunodeficiency virus (HIV) infection
1.4.3 Portal hypertension
1.4.4 Congenital heart diseases
1.4.5 Schistosomiasis
1a. Pulmonary veno-occlusive disease and/or pulmonary capillary hemangiomatosis
1b. Persistent pulmonary hypertension of the newborn (PPHN)
2. Pulmonary hypertension due to left heart disease
2.1 Left ventricular systolic dysfunction
2.2 Left ventricular diastolic dysfunction
2.3 Valvular disease
2.4 Congenital/acquired left heart inflow/outflow tract obstruction and congenital cardiomyopathies
3. Pulmonary hypertension due to lung diseases and/or hypoxia
3.1 Chronic obstructive pulmonary disease
3.2 Interstitial lung disease
3.3 Other pulmonary diseases with mixed restrictive and obstructive pattern
3.4 Sleep-disordered breathing
3.5 Alveolar hypoventilation disorders
3.6 Chronic exposure to high altitude
3.7 Developmental lung diseases
4. Chronic thromboembolic pulmonary hypertension
5. Pulmonary hypertension with unclear multifactorial mechanisms
5.1 Hematologic disorders: chronic hemolytic anemia, myeloproliferative disorders, splenectomy
5.2 Systemic disorders: lymphangioleiomyomatosis, pulmonary histiocytosis and sarcoidosis
5.3 Metabolic disorders: Gaucher disease, glycogen storage disease and thyroid disorders
5.4 Others: chronic renal failure, fibrosing mediastinitis, segmental pulmonary hypertension and tumoral obstruction

Adapted from Simonneau et al. [12]

are affected by PAH worldwide and 100 thousand individuals are affected by PAH, not including the PAH associated to schistosomiasis and to HIV. These numbers are expected to rise continuously on the next decades, because of the increase in life expectancy.

Recent data suggest that PAH is now frequently diagnosed in old patients, with approximately 65 years old [17, 18]. Prevalence is highly variable, although data indicate a higher incidence in women [19].

In general, survival is about 10 years in 78 % of cases [20]. The main factors closely related to enhanced mortality are: RV failure, male gender, limitation to physical exercise [21], and black ethnic group [22].

3 Pulmonary Hypertension Models

In order to better understand the pathophysiology of PAH many experimental models were developed. The use of these pre-clinical models to study PAH allowed the development of all therapeutic alternatives that currently exist [4, 23]. These models were also useful in determining efficacy and safety of novel therapeutic strategies. The established models to reproduce PAH are: induction by monocrotaline, hypoxia, transgenic, knockout and inflammatory or surgical models [24]. No single model can mimic all the features of human PAH, however there are many acceptable models. In the last twenty years, two models have been central in the PAH research: exposition to hypoxia and monocrotaline (MCT) [25, 26]. MCT is the most used model to mimic PAH since its application is simple, providing easiness, quickness, and low cost in reproducing the disease. Because of this, the MCT-induced PAH model is perhaps the model that most contributed to the understanding of the pathophysiology of this disease and to the development of therapeutic treatments. Moreover, cardiovascular and pulmonary changes exhibited are very similar to those presented by PAH patients, as well as its functional and biochemical outcomes [24, 27].

3.1 The PAH Monocrotaline Model

This model was established in 1967, by feeding rats with seeds of *Crotalaria spectabilis* [28]. Plants of *Crotalaria* gender belong to the *Leguminosae* family, with more than 600 species, abundant in tropical and subtropical zones, including Africa, India, Mexico and Brazil [29]. These plants received this name because their dried pods, when scratched resemble the sound of the rattlesnake tail, whose gender is *Crotalus* [30]. In Brazil, more than 40 species where cataloged, being opportunistic and crop invasive plants, capable to produce human and cattle intoxication [31, 32]. A single subcutaneous or intraperitoneal MCT injection (60–100 mg/kg) results in a reliable method to induce PAH [24, 33, 34]. Nevertheless, the level of toxicity varies with age, gender, species, and dose of MCT [35].

3.2 Monocrotaline Action Mechanism

MCT action mechanism is not fully elucidated. However, it is known that MCT hepatic metabolism is responsible for its toxicity, being the dehydroxylation performed by the P450 enzymatic complex (subtype CYP3A4). After dehydroxylation, the active and highly reactive metabolite, dehydromonocrotaline, promotes structural and functional changes in lungs and pulmonary vasculature [36, 37]. It is known that dehydromonocrotaline accumulates in erythrocytes, enhancing by this way, the transport of liver metabolites to the lungs, where they are released in the pulmonary endothelium (Fig. 14.1) [38]. Contraction of the vascular smooth muscle of pulmonary arteries is observed after 4 h of MCT administration. Between 8 and 16 h after MCT administration an increase in arterial wall thickening is observed

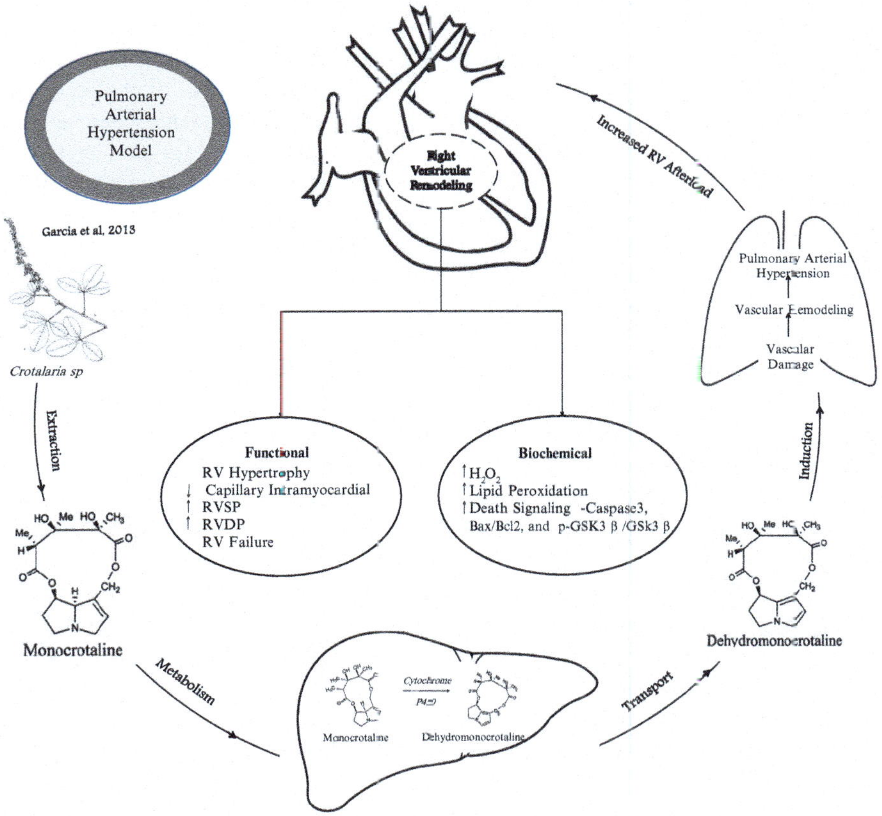

Fig. 14.1 Action mechanism of monocrotaline-induced pulmonary arterial hypertension, with the main functional and biochemical changes that occur in right ventricle (RV) remodeling subsequent to the increase in RV afterload promoted by elevated pulmonary pressure. *RVSP* right ventricle systolic pressure, *RVDP* right ventricle diastolic pressure

associated to the presence of an inflammatory infiltrate [39]. The inflammatory response triggers a diminished nitric oxide (NO) production [40] and increased endothelin-1 levels [41].

NO is a potent vasodilator released in the vascular lumen whose synthesis is catalized by nitric oxide synthases (NOS) enzymes [42]. Among NOS isoforms, there are the endothelial NOS (eNOS), present in endothelial cells in basal conditions, and the inducible NOS (iNOS), produced by macrophages in response to inflammatory cytokynes [43, 44]. Besides its vasodilating action, NO is an important inhibitor of platelet adhesion and aggregation in the vascular wall, inhibitor of smooth muscle cells proliferation and inflammatory response. By binding to the vascular smooth muscle receptors, NO causes vascular relaxation and an opposite action of endothelin [45, 46].

Endothelin-1 (ET-1) is a potent endogen vasoconstrictor, secreted mostly by endothelial cells as a biologically inactive peptide, that through the endothelin converting enzyme (ECE) is converted to active endothelin. Many factors stimulate ET-1 secretion, such as catecholamines [47], low shear stress [48], free radicals [49], TNF-α [50], and interleukins 1 and 6 (IL-1 e IL-6) [51, 52]. It is also known that ET-1 stimulates growth factors production [53] and presents mitogenic activity in smooth muscle cells [54]. ET-1 effects are exerted by stimulation of two G protein-coupled receptors: ET-A and ET-B [50]. ET-1, via ET-A binding, stimulates ROS production, especially superoxide anion, through its pro-inflammatory action [55]. When ET-1 acts via ET-B receptors in endothelial cells, it inhibits ECE and NO-mediated vasodilation [50, 56]. NO inhibits ET-1 effects and when NO bioavailability is decreased, ET-1 deleterious effects of vasoconstriction and vascular remodeling occur [57]. Moreover, in the situation of oxidative stress NO reacts with superoxide anion, in a diffusion limited reaction, without need of enzymatic catalysis. The product of this reaction is peroxynitrite, a powerful and highly unstable oxidant. Besides the deleterious effects caused to proteins and lipids through nitrosylation and peroxidation, peroxynitrite is capable of activating growth factors and cytokynes [43, 59]. The unbalance between NO and ET-1 promotes smooth muscle layer hypertrophy (tunica media), remodeling of basal membrane (tunica adventicia) and neointima formation, changes that culminate in an enhancement of pulmonary vascular resistance (PVR) [60, 61]. By this way, MCT leads to progressive appearance of severe and irreversible PAH [24]. Thus, the primary changes induced by MCT in pulmonary vascular bed result in an augmented PVR and, consequently, enhanced mean pulmonary arterial pressure (mPAP) [39]. With this significant increase in afterload, RV adapts in order to sustain cardiac output. It has been reported an acute activation of the sympathetic nervous system to restore or to preserve cardiac output, pressure levels, and to maintain cardiovascular homeostasis [62]. Consequently, an early increase in contractility index (dP/dt max) and in RV systolic pressure have been observed in MCT-induced PAH [63]. Simultaneously, there is a strong stimulus to a compensatory hypertrophy, aiming to reduce the wall stress imposed by enhanced PVR. This hypertrophy is characterized by an increased wall thickness, achieved by an enhanced protein synthesis and in parallel sarcomeres deposition [64]. Although beneficial in a short period, RV adaptations become ineffective over time and RV begins to present signs of failure. These signs include impairment in systolic function parameters, such as cardiac output decrease and elevated RV end-diastolic pressure [65]. Our group has observed similar oscillation in these parameters that evidence RV dysfunction in MCT-treated rats (Fig. 14.1) [66, 67]. Furthermore, we also observed that MCT-RV dysfunction is influenced by estrogen levels in rats. The RVEDP elevation was observed only in ovariectomized rats and it was attenuated after estrogen therapy [68]. This observation on gender differences contributes to explain the major susceptibility observed in women.

When exploring cardiac changes promoted by MCT at a longer term, our group found not only RV dysfunction but also a subsequent dysfunction in left ventricle (LV) [69]. One factor that could accelerate PAH progression, leading to a maladaptive cardiac remodeling, is the long term sympathetic hyperactivity [62]. In fact, norepinephrine action over cardiac cells may lead to increased reactive oxygen species (ROS) production [70], causing direct effects to myocardium, such as necrosis and inflammation [71]. By this way, exploring the role of ROS in the pathogenesis and physiopathology of PAH became a very relevant issue.

4 Oxidative Stress in PAH

Many reported studies support the involvement of oxidative stress in the pathogenesis of PAH [61]. The most relevant ROS involved in PAH seems to be superoxide radical. Superoxide is a charged molecule, resultant from the one-electron reduction of molecular oxygen. There are different sources of superoxide anion in PAH, such as NADPH oxidase activation, which may occur as a result of inflammatory process. Redout et al. [72] have demonstrated that the catalytic subunit gp91 phox of NADPH oxidase is increased in failing RV of rats with PAH induced by MCT. Same authors also identified an increase in the expression of complex II subunit B of mitochondrial respiratory chain, which seems to be important for ROS production in RV failure. Other sources of superoxide anion in PAH include uncoupled NOS, and xanthine oxidase [61]. Data from our group demonstrated that hydrogen peroxide levels, another ROS, increased in the RV of MCT-treated rats [66, 73].

The elevation of ROS levels leads to an enhanced oxidative damage to lipids, proteins and DNA. In fact, it was found increased levels of urinary isoprostanes, a robust marker of lipid peroxidation, in patients with pulmonary hypertension [74]. Our group detected increased systemic lipid peroxidation in rats with PAH induced by MCT [75]. As aforementioned, PAH affects pulmonary vasculature and this impacts RV function. Thus, it is relevant to explore oxidative stress markers not only in lungs, but also in RV. The lungs of patients with severe pulmonary hypertension show enhanced 8-hydroxyguanosine staining, a marker of oxidative damage to DNA [76]. In the rat model of PAH induced by MCT, it was also visualized an increased oxidative damage to lipids in lung homogenates [67]. It has been reported increased lipid peroxidation in RV of MCT-treated rats [73, 77].

As the first line of defense against ROS, it can be highlighted the action of superoxide dismutase (SOD), catalase (CAT), and glutathione peroxidase (GPx), antioxidant enzymes. The SOD activity diminished in the lungs of PAH patients indicates an impaired capacity to detoxify superoxide radical [76]. In fact, SOD enzyme activity may control superoxide anion levels, reducing its reactivity and capacity of causing oxidative damage. Confirming this idea, Kamezaki et al. [78] observed that extracellular SOD overexpression in lungs was able to ameliorate PAH induced by MCT. CAT activity was found to be increased in lung tissue of MCT-treated rats [67, 79]. However, CAT activity is variable in PH from different etiologies, being suggested that the decrease in its activity occurs at early stages of the disease [61]. Data regarding GPx activity in PAH are also inconsistent. Depending upon the duration and severity of oxidative stress, it shows a different response, varying from a compensatory augment until a decrease. Three weeks after MCT administration to rats, it was found a compensatory and adaptive elevation of lung GPx activity [79] .

In terms of RV antioxidant enzyme changes in MCT-treated rats, it was found that MCT (50 mg/kg i.p.) did not change antioxidant enzyme activities [77]. However, the dose of 60 mg/kg of MCT, observed at three different RV functional stages, promoted biphasic changes in antioxidant enzyme activities: increase during hypertrophic stage and decrease in RV failure stage [35]. Our group demonstrated that, three weeks after PAH induction by MCT (60 mg/kg i.p.), SOD activity was decreased in RV homogenates, while CAT was increased and GPx was unchanged [75]. In terms of non-enzymatic defenses, it was evidenced that MCT promoted a decrease in sulfhydryl groups in lungs of rats with PAH [80]. Vitamin E levels were shown to be decreased in RV of MCT-treated rats [77]. Data from our group showed decreased ascorbic acid concentration in the LV of MCT-treated rats, in the stage where functional impairment reached this ventricular chamber [69]. In female rats, reduced glutathione (GSH) content, which is the most abundant cellular non-enzymatic antioxidant, was negatively correlated with RVEDP [68]. This suggests that redox imbalance may have a role in RV dysfunction induced by MCT.

5 Therapeutic Interventions for PAH

There are clinical approaches to improve the health condition of patients that have developed PAH. Some drugs are available to treat this vascular injury. The first class is given by endothelin receptor antagonists that inhibit the signaling to endothelin-induced vasoconstriction, for example bosentan. The second class is the phosphodiesterase-5 inhibitors that reduce the degradation of cyclic guanosine monophosphate (cGMP) and increase vasodilating effects, such as sildenafil. The third group is provided by prostacyclin analogues which provoke a rapid relaxation process of pulmonary vessels (such as epoprostenol) [81]. Actually, the combined therapy seems to be the best choice in order to treat PH, especially in cases in which exists a relevant hemodynamic dysfunction [82].

Even though many pharmacologic options can be utilized to improve the clinical state of patients with PH, this disease does not have cure. Therefore, researching new therapeutic alternatives seems to be imperative. Our group demonstrated benefits with flavonoids treatment on MCT-induced PAH. One flavonoid, with antioxidant action, used by us was pterostilbene. This molecule is derived from blueberry (*Vaccinium* spp) and its protective effect is due to its ability in reducing ROS production. In our studies, it has been observed that pterostilbene administration attenuated experimental PAH. Likewise, it has been detected an amelioration in antioxidant status and a decreased oxidative stress in pterostilbene-treated animals (data not published). Other treatment rich in flavonoids used in our experiments was grape juice. Grape juice was administrated in a dose of 10 mL/kg/day, during 6 weeks, being 3 weeks prior and 3 weeks after MCT injection. MCT promoted an increase in lipid peroxidation, and a decreased protein expression of eNOS in lung homogenates [67]. These alterations were prevented by grape juice consumption, reducing lipid peroxidation and increasing eNOS expression. These biochemical changes were accompanied by an improved pulmonary arterioles remodeling, with less tunica media thickening. This could explain the reduction in lung congestion, and in RV end-diastolic pressure (suggesting decreased RV afterload) by grape juice treatment. Using the same protocol, the study of Mosele et al. [73] showed that grape juice administration decreased hydrogen peroxide concentrations and lipid peroxidation in RV of MCT-treated rats. Grape juice treatment also attenuated the elevation in PVR promoted by MCT. Although grape juice was not able to reduce RV hypertrophy induced by MCT, it modulated the expression of proteins involved in RV remodeling. In this sense, there was a decrease in the activation of cleaved caspase-3, suggesting a more favorable remodeling.

In this context, current studies have suggested that some exogenous molecules exert their antioxidant effects, acting as mild prooxidants that stimulate hormetic responses. Hormesis is defined as a phenomenon characterized by an adaptation to a low dose of the stressor agent, involving gene expression control of other antioxidant elements [83]. Phytochemicals, such as sulforaphane, brazilin, chalcone, zerumbone, resveratrol, and curcumin utilize this action mechanism. Sulforaphane is a phytochemical, extracted from broccoli, whose main feature is the induction of antioxidant enzymes, especially hemoxygenase 1, through NrF2 activation [84]. In a not published study from our group, sulforaphane attenuated hemodynamic changes induced by MCT in rats. This functional improvement was associated with a better oxidative status promoted by sulforaphane.

Other non-pharmacological strategies have been suggested in order to aid PAH treatment, such as exercise training. Our group verified that MCT-exercised rats decreased lipid peroxidation and augmented SOD activity in erytrhocytes, in the fourth week of disease. It was also verified that moderate exercise training improved survival in the group of animals that received MCT [75]. Using the same experimental protocol, Colombo et al. showed that exercise training produced an improvement in RV function, especially in terms of reducing the elevation of RV end-diastolic pressure (RVEDP) induced by MCT. Exercise training improved the structural remodeling of pulmonary artery, especially by reducing pulmonary artery thickness, without reducing lung congestion promoted by MCT administration. RV hypertrophy was also not reduced by aerobic exercise training, but an enhanced intramyocardial capillary volume and reduced interstitial fibrosis in RV was observed. Data from this study suggest that improvement of RV remodeling may be accomplished by signaling proteins influenced by exercise, such as GSK3-β [66]. Another study from our group corroborated these data, since it was found that exercise training was able to increase Akt phosphorylation and to decrease Bax/Bcl2 ratio and caspase-3 activation [85]. The cardioprotection offered by exercise training may be a consequence of a more favorable signaling to cardiac remodeling, accomplished by decreased ROS concentrations. This adaptive response could also constitute a hormetic effect.

6 Conclusion

In this review, the involvement of oxidative stress in the pathogenesis of PAH became clear. It was reported, through different markers, not only in experimental studies but also in patients, that oxidative stress is one of the mechanisms responsible for pulmonary endothelial dysfunction and consequent RV function impairment. However, many other mechanisms are involved in the PAH pathogenesis and were not the focus of this review. The successful utilization of some antioxidant strategies to mitigate this disease, such us the use of grape juice, reinforce the idea that oxidative stress participates in the mechanisms of this pathogenic process. New therapeutic approaches focusing oxidative stress modulation represent promising alternatives to be used as an adjuvant treatment for PAH patients.

Acknowledgements This work was supported by Brazilian Research Agencies: Conselho Nacional de Desenvolvimento Científico e Tecnológico (CNPq), Coordenação de Aperfeiçoamento de Pessoal de Nível Superior (CAPES) and Fundação de Amparo à Pesquisa do Rio Grande do Sul (FAPERGS). Dr. Belló-Klein and Dr. Araujo are recipients of research fellowships from CNPq.

References

1. Jansa P, Jarkovsky J, Al-Hiti H et al (2014) Epidemiology and long-term survival of pulmonary arterial hypertension in the Czech Republic: a retrospective analysis of a nationwide registry. BMC Pulm Med 14:45
2. Hoeper MM, Humbert M, Souza R et al (2016) A global view of pulmonary hypertension. Lancet 4:S2213–S2600
3. Hoeper MM, Liu X, Panaglia V et al (2013) Definitions and diagnosis of pulmonary hypertension. J Am Coll Cardiol 62:D42–D50
4. Schermuly RT, Kreisslmeier KP, Ghofrani HA et al (2004) Chronic sildenafil treatment inhibits monocrotaline-induced pulmonary hypertension in rats. Am J Resp Crit Care Med 169:39–45

5. Chin KM, Rubin LJ (2008) Pulmonary arterial hypertension. J Am Coll Cardiol 51:1527–1538
6. Groepenhoff H, Vonk-Noordegraaf A, Veerdonk MCV (2013) Prognostic relevance of changes in exercise test variables in pulmonary arterial hypertension. PLoS One 8:e72013
7. Matura LA, McDonough A, Carrol DL (2012) Cluster analysis of symptoms in pulmonary arterial hypertension: a pilot study. Eur J Cardiovasc Nurs 11:51–61
8. Rich S, Dantzker DR, Ayres SM (1987) Primary pulmonary hypertension. A national prospective study. Ann Intern Med 107:216–223
9. Malenfant S, Margaillan G, Loehr JE (2013) The emergence of new therapeutic targets in pulmonary hypertension: from now the near future. Expert Rev Respir Med 7:43–55
10. D'Alonzo GE, Barst RJ, Ayres SM et al (1991) Survival in patients with primary pulmonary hypertension: results from a national prospective registry. Ann Intern Med 115:343–349
11. Gurtu V, Michelakis ED (2015) Emerging therapies and future directions in pulmonary arterial hypertension. Can J Cardiol 31:489–501
12. Simonneau G, Gatzoulis MA, Adatia I et al (2013) Updated clinical classification of pulmonary hypertension. J Am Coll Cardiol 62:D34–D41
13. Simonneau G, Robbins IM, Beghetti M et al (2009) Updated clinical classification of pulmonary hypertension. J Am Coll Cardiol 54:S43–S54
14. Barreto AC, Franchi SM, Pereira AC (2005) Pulmonary arterial hypertension. Pathophysiology, genetic aspects and response to the chronic use of sildenafil. Arq Bras Cardiol 85:147–154
15. Peacock AJ, Murphy NF, McMurray JJ et al (2007) An epidemiological study of pulmonary arterial hypertension. Eur Resp J 30:104–109
16. Le Pavec J, Humbert M (2009) Idiopathic, familial and anorexigen-associated pulmonary arterial hypertension. In: Humbert M, Lynch JP (eds) Pulmonary Hypertension, vol 236. CRC Press, Boca Raton, pp 151–159
17. Hoeper MM, Huscher D, Ghofrani HA (2013) Elderly patients diagnosed with idiopathic pulmonary arterial hypertension: results from the COMPERA registry. Int J Cardiol 168:871–880
18. Hoeper MM, Huscher D, Pittrow D (2016) Incidence and prevalence of pulmonary arterial hypertension in Germany. Int J Cardiol 203:612–613
19. Hoeper MM, Gibbs JSR (2014) The changing landscape of pulmonary arterial hypertension and implications for patient care. Eur Resp Rev 23:450–457
20. Ogawa A, Ejiri K, Matsubar H (2014) Long-term patient survival with idiopathic/heritable pulmonary arterial hypertension treated at a single center in Japan. Life Sci 118:414–419
21. Humbert M, Sitbon O, Chaouat A et al (2010) Survival in patients with idiopathic, familial, and anorexigen – associated pulmonary arterial hypertension in the modern management era. Circulation 122:156–163
22. Mehari A, Valle O, Gillum R (2014) Trends in pulmonary hypertension mortality and morbidity. Pulm Med 2014:1–5
23. Hill NS, Warburton RR, Pietras L et al (1997) Nonspecific endothelin-receptor antagonist blunts monocrotaline-induced pulmonary hypertension in rats. J Appl Physiol 83:1209–1215
24. Ryan JJ, Marsboom G, Archer SL (2013) Rodent models of group 1 pulmonary hypertension. In: Humbert M, Evgenov OV, Stasch JP (eds) Pharmacotherapy of pulmonary hypertension, vol 218. Springer, Berlin, pp 107–140
25. Stenmark KR, Meyrick B, Galie N (2009) Animal models of pulmonary arterial hypertension: the hope for etiological discovery and pharmacological cure. Am J Physiol Lung Cell Mol Physiol 297:L1013–L1032
26. Colvin KL, Yeager ME (2014) Animal models of pulmonary hypertension: matching disease mechanisms to etiology of the human disease. J Pulm Resp Med 4:198–296
27. Brown L, Miller J, Dagger A et al (1998) Cardiac and vascular responses after monocrotaline-induced hypertrophy in rats. J Cardiovasc Pharmacol 31:108–115
28. Kay JM, Harris P, Heath D (1967) Pulmonary hypertension produced in rats by ingestion of Crotalaria spectabilis seeds. Thorax 22:176–179

29. Cheeke PR (1988) Toxicity and metabolism of pyrrolizidine alkaloids. J Anim Sci 66:2343–2350
30. Williams MS, Molyneux RJ (1987) Occurrence, concentration and toxicity of pyrrolizidine alkaloids in Crotalaria seeds. Weed Sci 35:476–481
31. Atal CK, Sawhney RS (1973) The pyrrolizidine alkaloids from Indian Crotalarias. Indian J Pharm 35:1–12
32. Palomino G, Vásquez R (1991) Cytogenetic studies in mexican populations of species of *Crotalaria* (Leguminosae - Papilionoideae). Cytologia 56:343–351
33. Sehgal PB, Mukhopadhyay S (2007) Dysfunctional intracellular trafficking in the pathobiology of pulmonary arterial hypertension. Am J Respir Cell Mol Biol 37:31–37
34. Urboniene D, Haber I, Fang YH et al (2010) Validation of high-resolution echocardiography and magnetic resonance imaging vs. high-fidelity catheterization in experimental pulmonary hypertension. Am J Physiol Lung Cell Mol Physiol 299:L401–L412
35. Farahmand F, Hill MF, Singal PK (2004) Antioxidant and oxidative stress changes in experimental cor pulmonale. Mol Cell Biochem 260:21–29
36. Schultze AE, Roth RA (1998) Chronic pulmonary hypertension the monocrotaline model and involvement of the hemostatic system. J Toxicol Environ Health B Crit Rev 1:271–346
37. Wilson DW, Segall HL, Pan LC (1992) Mechanisms and pathology of monocrotaline pulmonary toxicity. Crit Rev Toxicol 22:307–325
38. Pan LC, Lamé MW, Morin D et al (1991) Red blood cells augment transport of reactive metabolites of monocrotaline from liver to lung in isolated and tandem liver and lung preparations. Toxicol Appl Pharmacol 110:336–346
39. Wilson DW, Segall HJ, Pan LC et al (1989) Progressive inflammatory and structural changes in the pulmonary vasculature of monocrotaline-treated rats. Microvasc Res 38:57–80
40. Mathew R, Zeballos GA, Tun H et al (1995) Role of nitric oxide and endothelin-1 in monocrotaline-induced pulmonary hypertension in rats. Cardiovasc Res 30:739–746
41. Frasch HF, Marshall C, Marshall BE (1999) Endothelin-1 is elevated in monocrotaline pulmonary hypertension. Am J Physiol 276:L304–L310
42. Lundberg JO, Weitzberg E, Gladwin MT (2008) The nitrate-nitrite-nitric oxide pathway in physiology and therapeutics. Nat Rev Drug Discov 7:156–167
43. Pacher P, Beckman JS, Liaudet L (2007) Nitric oxide and peroxynitrite in health and disease. Physiol Rev 87:315–424
44. Andrew PJ, Mayer B (1999) Enzymatic function of nitric oxide synthases. Cardiovasc Res 43:521–531
45. Salvemini D, Wang Z, Wyatt PS et al (1996) Nitric oxide: a key mediator in the early and late phase of carrageenan-induced rat paw inflammation. Br J Pharmacol 118:829–838
46. Dias-Junior CA, Cau BA, Tanus-Santos JE (2008) Role of nitric oxide in the control of the pulmonary circulation: physiological, pathophysiological, and therapeutic implications. J Braz Pneumol 34:412–419
47. Yanagisawa M, Kurihara H, Kimura S et al (1988) A novel potente vasoconstrictor peptide produced by vascular endothelial cells. Nature 332:411–415
48. Shao D, Park JE, Wort SJ (2011) The role of endothelin-1 in the pathogenesis of pulmonary arterial hypertension. Pharmacol Res 63:504–511
49. Motte S, McEntee K, Naeije R (2006) Endothelin receptor antagonists. Pharmacol Ther 110:386–414
50. Hynynen MM, Khalil RA (2006) The vascular endothelin system in hypertension: recent patents and discoveries. Recent Pat Cardiovasc Drug Discov 1:95–108
51. Maemura K, Kurihara H, Morita T et al (1992) Production of endothelin-1 in vascular endothelial cells is regulated by factors associated with vascular injury. Gerontology 38:29–35
52. Yamashita J, Ogawa M, Nomura K et al (1993) Interleukin 6 stimulates the production of immunoreactive endothelin 1 in human breast cancer cells. Cancer Res 53:464–467
53. Pedram A, Razandi M, Hu RM et al (1997) Vasoactive peptides modulate vascular endothelial cell growth factor production and endothelial cell proliferation and invasion. J Biol Chem 272:17097–17103

54. Zamora MA, Dempsey CE, Walchak SJ (1993) BQ123, um antagonista do receptor ETA, inibe a proliferação mediada por endotelina-1-de células do músculo liso da artéria pulmonar humana. Am J Respir Cell Mol Biol 9:429–433
55. Kowalczyk A, Kleniewska P, Kolodzieiczyk M et al (2015) The role of endothelin-1 and endothelin receptor antagonists in inflammatory response and sepsis. Arch Immunol Ther Exp (Warsow) 63:41–52
56. Galié N, Manes A, Branzi A (2004) The endothelin system in pulmonary arterial hypertension. Cardiovasc Res 61:227–237
57. Bourque SL, Davidge ST, Adam MA (2011) The interaction between endothelin-1 and nitric oxide in the vasculature: new perspectives. Am J Physiol Regul Integr Comp Physiol 300:R1288–R1295
58. Garcia M, Kawakita K, Miotto STS (2013) O gênero Crotalaria L. (Leguminosae, Faboideae, Crotalarieae) na planície de Inundação do Alto Rio Paraná, Brasil. R Bras Bioci 11:209–226
59. Oishi P, Grobe A, Benavidez E et al (2006) Inhaled nitric oxide induced NOS inhibition and rebound pulmonary hypertension: a role for superoxide and peroxynitrite in the intact lamb. Am J Physiol Lung Cell Mol Physiol 290:L359–L366
60. Humbert M, Morrell NW, Archer SL et al (2004) Cellular and molecular pathobiology of pulmonary hypertension. J Am Coll Cardiol 43:S13–S24
61. Aggarwal S, Gross CM, Sharma S et al (2013) Reactive oxygen species in pulmonary vascular remodeling. Compr Physiol 3:1011–1034
62. Triposkiadis F, Karayannis G, Giamouzis G et al (2009) The sympathetic nervous system in heart failure physiology, pathophysiology, and clinical implications. J Am Coll Cardiol 54:1747–1762
63. Austin ED, Lahm T, West J et al (2013) Gender, sex hormones and pulmonary hypertension. Pulm Circ 3:294–314
64. Bogaard HJ, Abe K, Vonk Noordegraaf A et al (2009) The right ventricle under pressure: cellular and molecular mechanisms of right-heart failure in pulmonary hypertension. Chest 135:794–804
65. Hessel MH, Steendiik P, Den Abel B et al (2006) Characterization of right ventricular function after monocrotaline-induced pulmonary hypertension in the intact rat. Am J Physiol Heart Circ Physiol 291:H2424–H2430
66. Colombo R, Siqueira R, Becker CU et al (2013) Effects of exercise on monocrotaline-induced changes in right heart function and pulmonary artery remodeling in rats. Can J Physiol Pharmacol 91:38–44
67. Lüdke AR, Mosele F, Caron-Lienert R et al (2010) Modulation of monocrotaline-induced cor pulmonale by grape juice. J Cardiovasc Pharmacol 55:89–95
68. Siqueira R, Campos C, Colombo R et al (2011) Influence of estrogen on pulmonary arterial hypertension: role of oxidative stress. Cell Biochem Funct 29:543–548
69. Leichsenring-Silva F, Tavares AM, Mosele F et al (2011) Association of the time course of pulmonary arterial hypertension with changes in oxidative stress in the left ventricle. Clin Exp Pharmacol Physiol 38:804–810
70. Amin JK, Xiao L, Pimental DR et al (2001) Reactive oxygen species mediate α-adrenergic receptor-stimulated hypertrophy in adult rat ventricular myocytes. J Mol Cell Cardiol 33:131–139
71. Behonick GS, Novak MJ, Nealley EW et al (2001) Toxicology update: the cardiotoxicity of the oxidative stress metabolites of catecholamines (aminochromes). J Appl Toxicol 21 S15–S22
72. Redout EM, Wagner MJ, Zuidwiik MJ (2007) Right-ventricular failure is associated with increased mitochondrial complex II activity and production of reactive oxygen species. Cardiovasc Res 75:770–781
73. Mosele F, Tavares AM, Colombo R (2012) Effects of purple grape juice in the redox-sensitive modulation of right ventricular remodeling in a pulmonary arterial hypertension model. J Cardiovasc Pharmacol 60:15–22
74. Cracowski JL, Cracowski C, Bessard G et al (2001) Increased lipid peroxidation in patients with pulmonary hypertension. Am J Respir Crit Care Med 164:1038–1042

75. Souza-Rabbo M, Silva LF, Auzani JA et al (2008) Effects of a chronic exercise training protocol on oxidative stress and right ventricular hypertrophy in monocrotaline-treated rats. Clin Exp Pharmacol and Physiol 35:944–948
76. Bowers R, Cool C, Murphy RC et al (2004) Oxidative stress in severe pulmonary hypertension. Am J Respir Crit Care Med 169:764–769
77. Pichardo J, Palace V, Farahmand F (1999) Myocardial oxidative stress changes during compensated right heart failure in rats. Mol Cel Biochem 196:51–57
78. Kamezaki F, Tasaki H, Yamashita K et al (2008) Gene transfer of extracellular superoxide dismutase ameliorates pulmonary hypertension in rats. Am J Respir Crit Care Med 177:219–226
79. Jin HF, Du SX, Zhao X et al (2008) Effects of endogenous sulfur dioxide on monocrotaline-induced pulmonary hypertension in rats. Acta Pharmacol Sin 29:1157–1166
80. Mathew R, Yuan N, Rosenfeld L et al (2002) Effects of monocrotaline on endothelial nitric oxide synthase expression and sulfhydryl levels in rat lungs. Heart Dis 4:152–158
81. Humbert M, Hossein-Ardeschir G (2015) The molecular targets of approved treatments for pulmonary arterial hypertension. Thorax 71:73–83
82. Sitbon O, Sattler C, Bertoletti L (2016) Initial dual oral combination therapy in pulmonary arterial hypertension. Eur Respir J 47:1727–1736
83. Calabrese V, Cornelius AT, Dinkova-Kostova I (2012) Cellular stress responses, hormetic phytochemicals and vitagenes in aging and longevity. Biochim Biophys Acta 1822:753–783
84. Surh YJ, Kundu JK, Na HK (2008) Nrf2 as a master redox switch in tuning the cellular signaling involved in the induction of cytoprotective genes by some chemopreventive phytochemicals. Planta Med 74:1526–1539
85. Colombo R, Siqueira R, Conzatti A et al (2015) Aerobic exercise promotes a decrease in right ventricle apoptotic proteins in experimental cor pulmonale. J Cardiovasc Pharmacol 66:246–253

Chapter 15
Redox Regulation of Vascular Remodeling

Leonardo Y. Tanaka, Denise de Castro Fernandes, and Francisco R.M. Laurindo

Abstract Vascular remodeling, the structural reorganization of whole-vessel circumference, is the main determinant of lumen caliber due to sustained blood flow changes and in vasculoproliferative diseases. Here we discuss the importance of redox processes as central determinants of vascular remodeling. The strong dependence of shear stress-associated remodeling on endothelial redox signaling indicates primary roles of NO biovailability as remodeling determinant. In addition, superoxide can limit NO bioavailability and counteract expansive remodeling, while hydrogen peroxide favors it. Nox NADPH oxidases appear essential oxidant sources in this context and reportedly mediate flow-induced remodeling. Also, vascular injury and repair is strongly modulated by redox processes, with oxidant generation as a major early response to injury. The neointima, in addition, is under sustained oxidative stress promoted at least partially by Noxes. Also, we provide evidence that superoxide dismutase under-activity supports constrictive remodeling in the injury/repair model. Furthermore, we showed that the peri/epicellular pool of protein disulfide isomerase (PDI) importantly orchestrates cytoskeletal/extracellular matrix organization to support an anti-constrictive remodeling effect during vessel repair. Overall, such redox modulation extrapolates known roles of redox signaling in individual cellular events towards a predictable synchronized responsiveness of whole vessel remodeling to redox-based interventions, with potentially significant therapeutic implications.

Keywords Vascular remodeling • Redox signaling • Nitric oxide • Superoxide dismutase • Protein disulfide isomerase • Shear stress • Hydrogen peroxide • Atherosclerosis

---

L.Y. Tanaka • D. de Castro Fernandes • F.R.M. Laurindo (✉)
Heart Institute (Incor), Vascular Biology Laboratory, School of Medicine, University of São Paulo, CEP05403-000 São Paulo, Brazil
e-mail: francisco.laurindo@incor.usp.br

© Springer International Publishing Switzerland 2016

R.J. Gelpi et al. (eds.), *Biochemistry of Oxidative Stress*, Advances in Biochemistry in Health and Disease 16, DOI 10.1007/978-3-319-45865-6_15

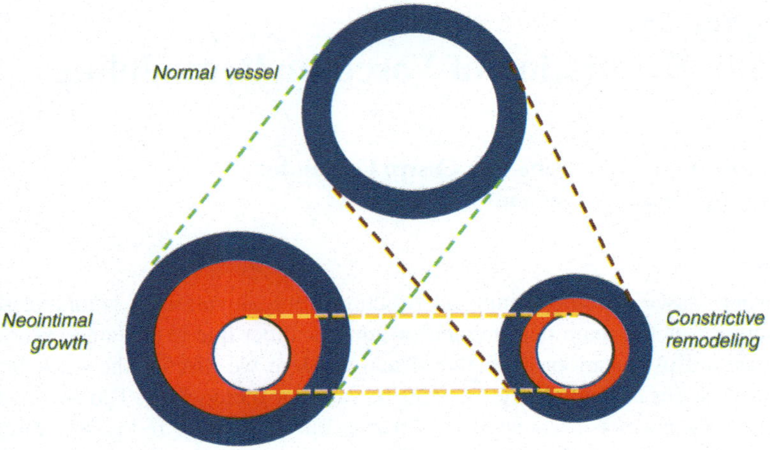

Fig. 15.1 Normal vessel (*top*) with lumen preservation might be affected in disease processes through two main structural alterations resulting in lumen narrowing (*yellow dotted line* at bottom arteries). (1) At left, by vascular wall thickening resulted from neointimal growth or plaque development (*red* layer). (2) At right, *red dotted line* depicts decreasing on total vessel area occurring in constrictive remodeling

1 Introduction

Vasculoproliferative processes underlie the vast majority of cardiovascular diseases, including atherosclerosis, hypertensive and diabetic vasculopathies and restenosis following interventional therapeutic procedures. Overall, such processes share common mechanisms with wound healing, a fundamental homeostatic response that recapitulates ontogenetic pathways [1]. In this context, the obstruction of vascular lumen and consequent blood flow limitation is a consequence of the combination of two mechanisms: neointima formation and vessel remodeling (Fig. 15.1). The neointima is the cell layer forming at the lumenal side of the internal elastic lamina, resulting from migration and proliferation of de-differentiated smooth muscle cells, neoendothelium, fibroblasts and possibly circulating or local stem cells, in addition to a substantial amount of collagen or non-collagen matrix [2]. The mechanisms for neointima formation and maturation have been extensively studied [3]. Vascular remodeling, however, remains a less understood process. In addition, the term "vascular remodeling" has been often used throughout the literature in a broader sense, comprising processes involving proliferation, migration and death of all distinct cell types, as well as extracellular matrix deposition and destruction [4]. In this chapter, we will focus on a *strictu sensu* definition of vascular remodeling [5], which consists of changes in whole vessel circumference, defined through the outward or inward displacement of all vessel layers, with the external elastic lamina circumference being the usual reference landmark (Fig. 15.1).

Remodeling can be expansive (also referred to as outward or positive) or constrictive (inward or negative). While the pathophysiology of remodeling involves an initial transient phase of vasomotricity changes, remodeling implies in the fixed resetting of a novel steady-state whole-vessel caliber involving matrix and cytoskeletal reorganization [6, 7]. Restenosis after balloon angioplasty has been a paradigm for constrictive vascular remodeling. This process is quite difficult to single out in small rodent models and thus most studies of vascular injury performed in mice and rats, can only address neointimal changes and are inappropriate for addressing *strictu sensu* remodeling. Since the development and routine use of intravascular stents, which counteract remodeling and successfully prevent restenosis despite enhancing neointima, there has been less interest in investigating the mechanisms underlying remodeling. However, such remodeling remains a significant disease process in several prevalent hard-to-treat conditions, including native atherosclerosis, diabetic or transplant vasculopathy. In initial atherosclerosis the extent of remodeling often correlates better than plaque mass with lumen caliber [8]. Vessel remodeling is often expansive in the early phases of plaque development (the so-called Glagov phenomenon) and this process is able to buffer lumen encroachment of plaques occupying up to about 40 % of whole lumen area [9]. However, in advanced fibrotic plaques, remodeling is usually constrictive, while plaques complicated by inflammation, which are prone to rupture and thrombosis, correlate with the emergence of expansive remodeling, which can even be a diagnostic sign [1]. In addition to large-vessel remodeling, cellular and mechano-biological mechanisms governing conductance vessel remodeling may be shared by those of small-vessel remodeling in hypertension [10]. Such importance in disease reflects the seminal roles of vascular remodeling in the physiological adaptations to shear stress, another paradigmatic example of vascular remodeling, and exercise [11]. Despite such relevance, there are substantial gaps in the understanding of vascular remodeling mechanisms. In this chapter, we discuss mechanisms of vascular remodeling, with strong emphasis on the role of redox processes.

2 Pathophysiology of *strictu sensu* Vascular Remodeling: Role of Redox Processes

A detailed discussion of the mechanisms underlying *strictu sensu* vascular remodeling is beyond the scope of this chapter and has been addressed in a previous review [5]. Here, we will focus on the role of redox processes. It is important to note, however, that *strictu sensu* remodeling often occurs on a background of other processes associated to *latu sensu* remodeling, as discussed above, which are related, for example, to atherosclerosis or vascular repair following angioplasty-induced injury. Such *latu sensu* remodeling processes include: a) endothelial cell proliferation, migration, senescence and cell death; b) vascular smooth muscle phenotypic changes, proliferation, migration, matrix secretion, senescence, apoptosis; c) adventitial fibroblast differentiation and activation, as well as inflammatory cell infiltration; and d) stem cell (circulating and/or local)/pericyte differentiation. The main structural readout of these processes is the buildup of a neointimal layer and new layers of elastic laminae, medial thickening and strong accumulation of extracellular matrix. The early immature neointima is composed of smooth muscle cells bearing a synthetic migratory and secretory phenotype, which, over time, differentiate towards a more contractile phenotype [12]. As a result, the mature neointima, as recapitulated in the model of vascular injury/repair [12], is composed of variably differentiated smooth muscle cells and stem cells, together with a large proportion of collagen extracellular matrix [2]. A substantial number of studies addressed individual redox-associated mechanisms underlying *latu sensu* remodeling processes [12]. As a general outcome, however, one might conclude that although redox processes individually modulate each of these mechanisms over a time course scale, it has been difficult, with some exceptions, to single out defined redox processes that orchestrate the complex collective *latu sensu* vascular response in a predictable way. This may be due to the complexity of the vascular response, concerning temporo-spatial discrimination of multiple localized events, as well as their distinct redox sensitivities.

The identification of endothelium [13] and nitric oxide [14] dependency of shear-induced remodeling raised the possible role of redox processes as *strictu sensu* remodeling mediators, as further suggested in many subsequent studies [11]. An important outcome of such studies has been the notion that such remodeling is directly linked to endothelial function so that preserved endothelium-associated relaxation (particularly via NO bioavailability) associates with expansive remodeling or antagonism of constrictive remodeling [6]. Redox processes related to shear stress will be discussed in the next section. In native atherosclerosis, as well as in vascular repair post-balloon injury, it is well known that constrictive remodeling is a more important determinant of restenosis and final lumen caliber than neointimal size [9, 15]. As discussed previously, such remodeling involves structural reorganization of the vessel and is not only a vasomotor-like process. Such structural correlates of *strictu sensu* remodeling share some common features with those of *latu sensu* remodeling. In particular, collagen content of the media is a clear structural correlate of constrictive remodeling [6, 16], as well as cytoskeletal reorganization [16]. Similarly, enhanced vascular wall apoptosis rates also seem to correlate with constrictive remodeling [17]. Conversely, expansive remodeling is often associated with inflammatory mediators and matrix degradation [1]. Studies from our and other laboratories suggest that redox processes play a role in the signaling program of vascular repair reaction. In particular, superoxide and other reactive oxygen species (ROS) production is prominent immediately after injury and underlies nuclear factor-kappaB (NFkB) activation, as well as several other signaling targets [18, 19]. Such early oxidant generation is attributed at least in part to nicotinamide adenine dinucleotide phosphate (NADPH) oxidases [18]. At later stages of vascular repair, oxidant production has also been documented to a lower degree in neointima and adventitia. Nox NADPH-oxidases are overexpressed during vascular repair, with Nox1 peaking earlier, during the proliferative phase, Nox4 increasing later during the late repair and Nox2 increasing throughout the whole response [20]. While antioxidant compounds have exhibited variable, but generally inconsistent effect on neointima formation [19], genetic deletion of specific Noxes [21] has been able to yield a more clear effect, with mice knockout for Nox1 exhibiting reduced neointima formation after arterial injury in association with reduced cofilin activation [22]. Analogous effects were observed in arterial injury models with Nox2 peptide inhibitors [23] or the knockout of Noxa-1, a regulatory subunit supportive of Nox1 activation [24]. The role of Nox4 appears to be related to support of cell migration through an yet unclear mechanism, probably related to cell adhesion [21]. While global Nox4 knockout, as well as Nox2 knockout mice, have not been tested with regard to neointimal formation, the VSMC-specific Nox4 knockout shows reduced neointimal size after injury [25]. In contrast with injury-induced vascular response, in native atherosclerosis the Noxes appear to play inconsistent or negative roles [26].

Close to nothing is known about the redox pathophysiology of stent-associated restenosis, which is attributable mainly to neointimal proliferation rather than *strictu sensu* remodeling [27]. Recently, a systems approach, associated with genome-wide screening, identified glutathione peroxidase (Gpx1) as major hub gene associated with in-stent stenosis. Loss of Gpx1 activity led to reductive stress associated with increased glutathione levels and phosphorylation of the tyrosine kinase ROS1, in close association with enhanced stenosis [28]. Overall, *strictu sensu* remodeling has been shown to exhibit a more consistent response to redox-related interventions, suggesting that redox processes more predictably orchestrate vascular remodeling than neointimal buildup. In analogy with shear stress-associated remodeling, *strictu sensu* remodeling and associated collagen deposition after balloon injury in atherosclerotic rabbits were also found to correlate with endothelial function [6], a highly redox-modulated variable. Accordingly, protective effects, particularly in the antagonism of constrictive remodeling, have been demonstrated with the antioxidant probucol in experimental models [29]. In particular, the randomized clinical trial Multivitamins and Probucol Study [30] showed that the antioxidant probucol, although not multivitamins, significantly prevented restenosis post-balloon angioplasty, essentially by preventing constrictive remodeling rather than neointimal proliferation [31]. Such redox-responsiveness is in line with the importance of redox signaling in cytoskeletal dynamics [32] and extracellular matrix organization [33], both being key processes in vessel remodeling. That is not to say that the chemical biology of vessel remodeling is simple: for example, the type of ROS may likely influence the outcome of remodeling. While superoxide production promotes decreased NO bioavailability and tends to promote constrictive remodeling [15], enhanced production of hydrogen peroxide associates with overall vascular expansion [16, 34].

3 Redox Modulation of Shear Stress-Induced Remodeling

In few other physiological systems the collective organ function is so closely dictated by subcellular functions as in the vascular system, and in few other systems, as well, redox signaling processes are so prominent cellular regulators. The paradigmatic example of this redox dependence is the biomechanical response of vascular cells and, more specifically, the endothelial response to shear stress, which is highly modulated by redox processes. These responses strongly affect cellular/tissue responses to physiological stimuli, as well as disease processes. In particular, arterial levels of laminar shear stress are widely regarded to be atheroprotective, while disturbed patterns of shear such as low levels of laminar shear or turbulence-associated oscillatory shear are proatherogenic [35]. These patterns are likely to explain the focal distribution of atherosclerotic plaques along the arterial tree [36].

There is an extensive literature about shear-related responses in endothelial cells [37]. Redox-related responses are involved at essentially every level of mechanobiological responses, acting as integrative elements of mechanosensing and mechanoeffector signaling. Shear stress is one of most potent and relevant physiological inducers of endothelial nitric oxide synthase (eNOS) expression and activation and eNOS-derived NO production dominates the redox landscape of shear responses. Laminar shear-associated eNOS activation is induced via serine 1179 phosphorylation promoted by Akt [38], which is able to yield NO flux even in the absence of high calcium levels, by increasing electron transfer from NADPH to L-arginine and by decreasing calmodulin sensitivity to calcium [38, 39]. In addition, shear stress promotes enhanced physical interaction between eNOS and the Hsp90 chaperone, which further supports NO production through conformational changes as well as serine 116 phosphorylation, which relieves eNOS from its inhibitory interaction with caveolin [39]. Enhanced stability of eNOS mRNA is also important [39], composing an overall picture of increased eNOS expression and NO availability.

In parallel with the strong NO response, one of the earliest responses to shear stress is superoxide production, as described originally by our group through in vitro and in vivo assays [40]. In these studies, enhanced formation of spin radical adducts with the spin trap DMPO (in vitro) or increased formation of ascorbyl radical in plasma were detected via EPR spectroscopy after sudden increases in shear stress induced by perfusion changes (in vitro) or vasodilator boluses (in vivo). Importantly, integrity of the endothelium was essential for the formation of these adducts [40]. The functional role of such early superoxide peak is unknown, but the intriguing possibility that it may help activate eNOS [41] deserves additional testing. Further studies by other groups largely confirmed these findings in distinct models, indicating that laminar shear promotes an early superoxide peak that later decreases as NO production rises in a sustained pattern. Conversely, oscillatory shear associates with sustained superoxide production and lower levels of bioavailable NO [42]. Nox NADPH oxidases appear to be important sources of shear-derived superoxide in these studies, as assessed through the requirement of the subunit p47phox for oxidant generation [43], thus pointing to roles of either Nox1 or Nox2 complexes, which rely on such cytosolic subunit. In fact, superoxide production during oscillatory shear in cultured cells associates with enhanced sustained expression of the subunits p22phox, p47phox and Nox2, but also of Nox4 [44], which has been suggested to have some atheroprotective roles [45]. Also, in carotid arteries subjected to abnormal high flow in vivo, increased Nox-derived superoxide is detectable [46]. In addition to Nox regulation, there is usually enhanced expression of antioxidant enzymes after a few hours of mechanical stimulation, including superoxide dismutases (Cu,Zn-SOD and Mn-SOD), glutathione peroxidase (GPx), peroxiredoxin 1 (Prx1) and thioredoxin 1. In addition, sustained laminar shear enhances the activity of glucose-6-phosphate dehydrogenase (G6PD), which promotes enhanced formation of NADPH, contributing to sustain glutathione levels. All these changes associate and/or result from activation of the transcription factor Nrf2, also induced by laminar shear, and overall there is a shift towards a more reduced cellular environment, which further contributes to sustain NO bioavailability [44].

In close parallel with redox responses mediating flow adaptation, arterial remodeling is also closely sensitive to redox processes. The essential requirement for the endothelium regarding flow-dependent remodeling was described in the model of partial carotid ligation, in which reduced flow through the left carotid promotes constrictive remodeling and a slightly compensatory enhancement in flow through the contralateral carotid promotes expansive remodeling, although the latter process is model-dependent [5]. In addition, such responses are significantly affected by knockout of each NOS isoform in mice: while both eNOS [46] and nNOS [47] deletions essentially abrogate vascular remodeling, deletion of each of the 3 isoforms exacerbate vessel wall thickness [48]. Importantly, shear-induced remodeling is also attenuated in mice knockout for the NADPH oxidase subunit p47phox, indicating that oxidative processes, even though associated with decreased NO bioavailability, are also required for remodeling [46, 49], possibly through effects in matrix remodeling. An improved model was described by Nam et al [49], in which residual flow decreases through the left carotid were adjusted to induce a pattern of oscillatory shear stress in such artery, leading to enhanced vessel wall thickness and constrictive remodeling. If these experiments are performed on a background of apolipoprotein E deficiency, there is development of a significant atheroma plaque with features of instability [49]. This model has also allowed to establish genomic studies of shear-responsive genes, as well as epigenetic modulation such as microRNAs and DNA-methylation. From these omics approaches, some specific mechanisms have been addressed in in vitro models of disturbed flow and novel therapeutic targets could be developed [50]. Importantly, redox-dependent signaling during in vivo remodeling closely follows patterns of in vitro laminar vs. disturbed blood flow. There is a loss of outward remodeling at the non-ligated contralateral carotid caused by deletion of NOS isoforms [46–48]. Intriguingly, there is enhanced positive remodeling, as well as wall thickness, associated with genetic overexpression of p22phox [51], maybe due to enhanced activation of multiple Nox isoforms (e.g., Nox1, Nox2 and Nox4). In addition, the inflammatory phenotype in the ligated left carotid artery is recapitulated through in vitro model of low and oscillatory shear, both showing important modulation by redox mechanisms, such as thioredoxin 1-dependent activation of NFkB and upregulation of the adhesion mediators ICAM-1 and VCAM-1 [52]. Moreover, the induction of endothelial dysfunction and atherosclerosis in such low blood flow model is inhibited by genetic deletion of p47phox, indicating Nox-dependent pathways [49].

As the model of flow-dependent remodeling constitutes a suitable one to understand mechanisms of *strictu sensu* remodeling, due to absence of superimposed

Table 15.1 Some redox-associated effects of shear-mediated vascular remodeling

Effector	Redox-associated mechanism	References
eNOS	Nitric oxide production	[46]
nNOS	Nitric oxide production	[47]
iNOS	Nitric oxide production	[48]
p22phox	ROS production	[51]
p47phox	Superoxide production	[46]
P2X4	Calcium-dependent activation of NO production	[53]
PECAM-1	Upstream signaling for nitric oxide production	[54]
MMP-9	Collagen-ECM organization and VSMC migration	[55]
NfkB	MMPs induction	[56]
VEGFR3	Activation of NfkB and MMP-9 induction	[57]
tTG	Inward remodeling effect inhibitable by NO	[58]
GPX1	Balanced redox homeostasis preventing ROS1 activation	[28]

atheromas or injuries, they offer a perspective to interrogate effectors of vessel remodeling. Some relevant redox-associated effectors of vascular remodeling are shown in Table 15.1. Overall, these considerations suggest that vascular remodeling may be dominated by dependence on endothelial function and associated redox events even for other more complex models involving disease processes.

4 Superoxide Dismutase and Vascular Remodeling

The above considerations suggest that redox processes may mediate vascular remodeling. While such studies concentrated mainly in sources of reactive intermediates, much less information is available with respect to the possible effect of antioxidant enzyme dysregulation. Our group provided evidence that superoxide dismutase under-activity contributes to constrictive remodeling in the model of balloon-induced injury and vascular repair-associated remodeling [15]. In this study, normolipidemic rabbits were submitted to iliac artery balloon overdistension with a coronary angioplasty-type balloon and were followed for 14 days after injury. At this stage, there is a residual proliferation in an already substantial neointima, the size of which is known to be about 55 % of the mature neointima at day 28 after injury. In addition, the period between day 7 and 14 after injury marks the most accelerated loss of lumen caliber, with about 50 % total caliber loss occurring within this time interval. Significant decreases in vascular SOD activity were observed at 7 and 14 days after injury (by 45 % and 34 %, respectively, versus control). Both extracellular SOD (ec-SOD) and Cu,Zn-SOD activities were reduced, as assessed with concanavalin-A column separation. However, the protein expressions were either unchanged (Cu,Zn-SOD) or increased (ec-SOD). The expression of inducible NOS (iNOS) is known to be increased during vascular repair [59], but it is unclear whether its activity leads to protective or deleterious signaling. We assessed the interplay between nitrogen species and oxidative stress. Immunoreactivity to nitrotyrosine was significantly increased in medial and neointimal layers at day 14 after injury, accompanied by evidence of enhanced protein expression of iNOS and neuronal NOS (nNOS), as well as iNOS mRNA. Importantly, intravenous administration of ec-SOD from days 7 to 14 after injury, in amounts adjusted to be sufficient to normalize SOD activity, minimized caliber loss by 59 % and reduced collagen accumulation in medial layer, while the neointima was unaltered. This picture is consistent with the proposal that sustained low vascular SOD activity has a key role in constrictive remodeling after injury. The mechanisms underlying this effect are not known in detail. However, our data suggest that ec-SOD replenishment rescued iNOS-derived NO bioactivity. This was suggested by ec-SOD induced large increase in nitrate (NO_3^-) levels in injured artery homogenates, accompanied by significant reduction in nitrotyrosine immunoreactivity. Such increase in NO_3^- levels was about 70 % inhibited by the specific iNOS antagonist 1400w, while the expression of nNOS expression was almost abolished, indicating that the later was unlikely to have contributed to NO_3^- production. This indicates that SOD function may critically determine whether iNOS induction is beneficial or deleterious in vivo. Another study reported that transfection of adenovirus carrying ec-SOD after balloon injury in hypercholesterolemic rabbits induced decrease in neointimal formation [60]. A remarkable feature of our study was that the effects of SOD replenishment were evident at a relatively late time frame (from 7 to 14 days post-injury), indicating that redox imbalances are not directly related to the injury process per se, but rather constitute an ongoing feature of the vascular repair as a determinant of constrictive remodeling [15].

5 Novel Roles of Protein Disulfide Isomerase in the Regulation of Vascular Remodeling

While the above results are consistent with a general role of oxidant generators and antioxidants in vascular remodeling, the mechanisms that organize and orchestrate redox-dependent pathways are unclear. Recently, work from our group provided evidence that the endoplasmic reticulum (ER) chaperone protein disulfide isomerase (PDIA1 or PDI), known to play important roles in redox homeostasis and signaling [61], may exert such an effect [16]. PDI, a thioredoxin superfamily protein, is the prototype of a >20-member family of multifunctional redox chaperones having, among many functions [62, 63],the main role of disulfide bond formation/isomerization in nascent ER proteins. Previous work from our group showed, through loss- and gain-of-function experiments, that PDI associates with and functionally regulates Nox NADPH oxidases. PDI is essentially required in vascular smooth muscle cells (VSMC) for Nox1 activation and expression in response to incubation with angiotensin II [64] or platelet-derived growth factor [32]. Analogous requirements for PDI were observed for macrophage and neutrophil Nox2 complexes [65,

66]. Moreover, PDI is necessary for platelet-derived growth factor-stimulated, Nox1-dependent,VSMC migration, while PDI silencing with siRNA associates with marked actin cytoskeleton disorganization and impaired activation of Rho-GTPases Rac1 and RhoA [32]. VSCM migration supports neointimal thickening [22] and atheroma development [67]. In this context, PDI might exert a proatherogenic role via Nox1. In contrast, however, PDI physically associates to other Noxes, in particular Nox4, which has been implicated in anti-atherogenic responses, mediating maintenance of the endothelial function [68] and differentiation of vascular smooth muscle cells [69].

PDI family proteins closely interplay with endoplasmic reticulum (ER) stress, a condition of impaired ER-dependent proteostasis which triggers a signaling network known as the "unfolded protein response" (UPR). The adaptive arm of the UPR relates to upregulation of ER chaperones and antioxidant defenses, while if ER stress is sustained or severe, proapoptotic signaling is triggered through the transcription factor CHOP, as well as JNK [70]. ER stress plays signaling roles during different stages of atherosclerosis development. Increased expression of ER chaperones and CHOP was demonstrated in atherosclerotic plaques [71]. Meanwhile, differential upregulation of the apoptotic arm or the incapacity to counteract the UPR was associated with accentuated markers of plaque vulnerability [71], the scenario of disease complications and progression to acute coronary events. Importantly, advanced human atheromas were associated with redox-mediated inactivation of PDI, most likely driven by low-density lipoprotein oxidation [72]. In our model of vascular repair after injury, inactivation of PDI (with siRNA in organ culture) was demonstrated to upregulate ER stress and apoptosis rates, and associated with lower expression of differentiation markers, indicating that PDI protects against stress and may support VSMC differentiation during vascular repair [16]. Overall, although PDI expression is usually considered not highly responsive to UPR induction, these evidences point out towards an adaptive role of PDI during disease conditions recapitulated by the vascular injury/repair model.

Bearing in mind that the canonical location of PDI is at the ER lumen, it is still unclear whether another intracellular PDI pool outside the ER participates in Nox regulation. In parallel, there is substantial evidence for an extracellular PDI pool, which comprises a pool of PDI attached to the cell surface through yet unclear mechanisms (epicellular), and another often indistinguishable pool of secreted PDI (pericellular). We coined the term "peri/epicellular" (pec) PDI to designate such extracellular PDI pools. PecPDI mediates many cell-surface and extracellular thiol redox-dependent processes [73]. The most important of such processes, as related to the vascular system, is intravascular thrombosis and platelet activation. PDI is a key mediator of injury-related thrombosis and both platelet and endothelial PDI are involved in these processes through mechanisms associated with signaling by distinct integrins [74]. Platelet/megakariocyte-specific PDI knockouts exhibit reduced thrombosis with negligible impairment of normal platelet adhesion and bleeding time [75]. Accordingly, inhibitors of PDI or other enzymes from the PDI family constitute innovative antithrombotic strategies that are undergoing extensive prospection and even clinical testing [73]. In addition to thrombosis, pecPDI has other functions including virus internalization, galectin binding [76],and metalloproteinase regulation [77]. Contrary to the ER pool, in which PDI exerts a typical thiol isomerase activity, pecPDI exerts predominant thiol reductase activity [74]. There is also evidence for a role of pecPDI as mediator of trans-nitrosation reactions that support nitric oxide bioavailability [78].

Considering the above-described characteristics of pecPDI and the redox sensitivity of vessel remodeling, we hypothesized that pecPDI pool regulates vascular remodeling. As a proof-of-concept test of our hypothesis, we assessed PDI immunoreactivity in autopsy atheroma specimens from patients dying from acute coronary events. Constrictive or expansive remodeling were defined in relation to segments with preserved appearance from the same arteries, assumed as reference controls [79]. There was decreased PDI immunoreactivity in plaques exhibiting constrictive remodeling and, in parallel, enhanced PDI expression in plaques exhibiting expansive remodeling. Such correlation between overall PDI expression and vessel remodeling in humans led us to further investigate PDI and specifically pecPDI modulation of vascular remodeling in the model of vascular repair post-injury. In normolipidemic rabbits submitted to balloon overdistension iliac artery injury, protein expression of PDI was markedly enhanced at 14 days post-injury (25-fold vs. baseline), while pecPDI pool exhibited a parallel increase. Neutralization of pecPDI with 2 distinct antibodies delivered in pluronic gel at the perivascular injury site led to significant decreases in arteriographic vessel caliber (by 25 %); in vivo experiments with optical coherence tomography confirmed such lumen loss, which was associated with whole vessel caliber reduction rather than neointimal growth (Fig. 15.2), as in fact confirmed by histomorphometry. Decrease in whole vessel circumference without increased neointima is consistent with constrictive remodeling induced by pecPDI neutralization. The occurrence of such constrictive remodeling was confirmed through marked changes in collagen structural organization, as well as actin cytoskeleton disorganization. We were able to identify integrin beta1 as a redox-modulated target of pecPDI, indicating a possible transmembrane mechanism through which pecPDI-induced redox changes can promote reorganization of intracellular cytoskeleton, as well as extracellular matrix. PecPDI neutralization decreased hydrogen peroxide detection in vessel segments, possibly through Nox-dependent mechanisms [16, 80]. Paradoxically, pecPDI is implicated in surface thiol reduction, the most likely mechanism by which it activates integrins [74]. Importantly, we also showed that pecPDI neutralization impairs actin stress fiber remodeling in cultured VSMC submitted to cyclic stretch, as well as endothelial cells undergoing shear stress. Thus, pecPDI emerges as a novel redox-dependent mediator of vascular lumen preservation during vasculoproliferative processes, mainly through anti-constrictive remodeling mechanisms. In addition, pecPDI may be a novel mediator of mechanoadaptation in vascular cells [16].

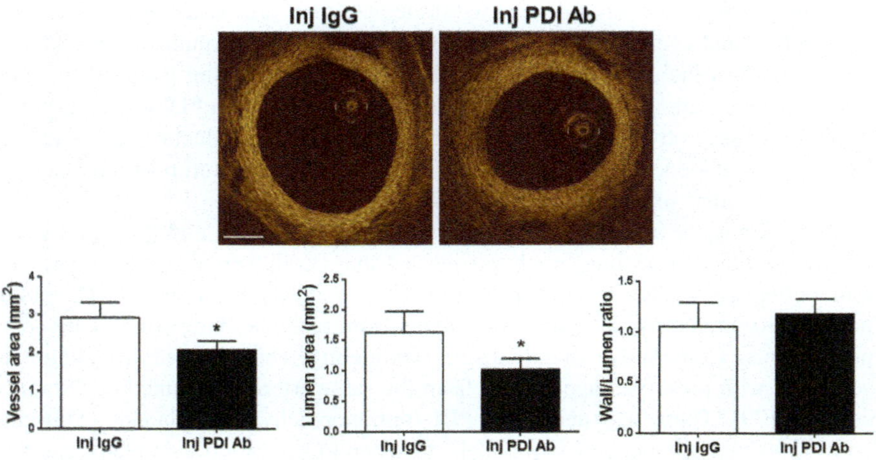

Fig. 15.2 Contribution of pecPDI to caliber maintenance in experimental model of vascular injury by balloon overdistension. In vivo perivascular application of neutralizing mouse-monoclonal anti-PDI (PDI Ab, clone RL90, 1 μg/ml) or control of mouse-immunoglobulin (IgG, 1 μg/ml) in pluronic gel (25 %) at 12th-day after balloon injury (Inj) and analyzed after 48 h. *Top*: Intraluminal optical coherence tomography measurement of control IgG and PDI Ab-treated arteries. *Bottom*: Graphs depicting quantification of vessel area, lumen area and wall/lumen ratio; $p < 0.05$ vs. Inj IgG, n = 7 (Modified from Ref. [16])

6 Conclusions

Strictu sensu remodeling is an important event in physiology and in the pathophysiology of vascular diseases. Despite the widespread use of stents in vascular therapeutic interventions, which antagonize constrictive remodeling, other disease conditions not amenable to such interventions still occur. In addition, expansive remodeling is a hallmark of atheroma inflammation and lesion instabilization. Thus, understanding mechanisms of vascular remodeling is relevant. In this context, the importance of redox processes as a fundamental process underlying remodeling (Fig. 15.3) extrapolates the known roles of redox signaling in individual cellular events to define a predictable responsiveness of whole vessel remodeling to redox-based interventions.

Whether such responsiveness is the result of syncronized collective cellular redox effects or whether it reflects mainly an endothelial effect is not clear at this point. Nevertheless, the strong dependence of shear stress-associated remodeling on endothelial function and redox signaling is an indication for a prime role of the endothelium in all forms of *strictu sensu* remodeling. Accordingly, NO bioavailability appears to be a central determinant of remodeling. Importantly, however, other oxidants may also contribute to such responses. Superoxide can limit NO bioavailability and counteract expansive remodeling, while hydrogen peroxide favors it. Nox NADPH oxidases appear essential sources of reactive intermediates

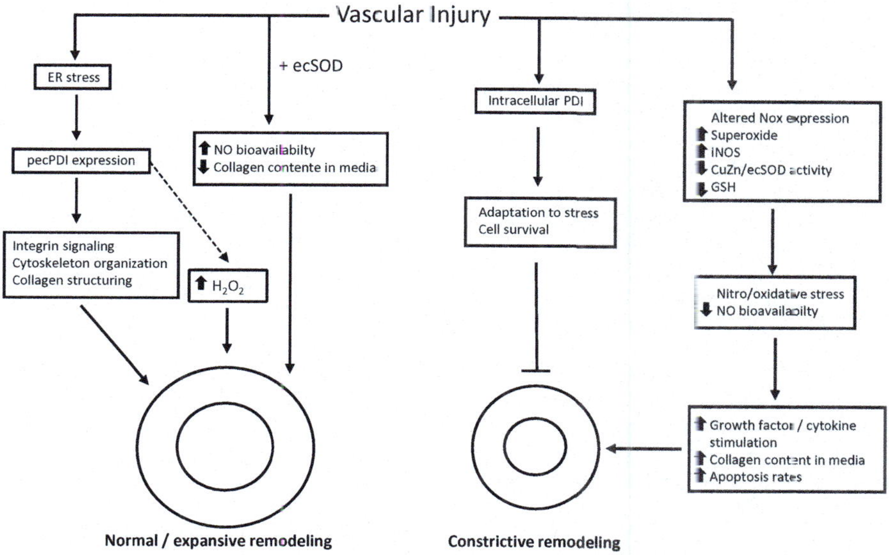

Fig. 15.3 Scheme depicting some major redox-dependent mechanisms involved with expansive or constrictive remodeling after vascular injury. *ER* endoplasmic reticulum, *pecPDI* peri/epicellular protein disulfide isomerase, *ecSOD* extracellular superoxide dismutase, H_2O_2 hydrogen peroxide, *iNOS* inducible nitric oxide synthase, *CuZn-SOD* cupper zinc superoxide dismutase, *GSH* reduced glutathione

in this context and reportedly mediate flow-induced remodeling. We provided evidence also that superoxide dismutase under-activity supports constrictive remodeling in the injury/repair model. Recently, using the same model, we showed that the peri/epicellular pool of protein disulfide isomerase (PDI) is an important determinant of remodeling, coordinating cytoskeletal/extracellular matrix organization to support an anti-constrictive remodeling effect. Collectively, these observations indicate that further research on redox mechanisms underlying *strictu sensu* remodeling is likely to yield fruitful knowledge and meaningful therapeutic advances.

References

1. Pasterkamp G, de Kleijn DP, Borst C (2000) Arterial remodeling in atherosclerosis, restenosis and after alteration of blood flow: Potential mechanisms and clinical implications. Cardiovasc Res 45:843–852
2. Newby AC (2006) Matrix metalloproteinases regulate migration, proliferation, and death of vascular smooth muscle cells by degrading matrix and non-matrix substrates. Cardiovasc Res 69:614–624
3. Laurindo FR, de Souza HP, de Pedro A, Janiszewski M (2002) Redox aspects of vascular response to injury. Meth Enzymol 352:432–454

4. Gibbons GH, Dzau VJ (1994) The emerging concept of vascular remodeling. N Engl J Med 330:1431–1438
5. Korshunov VA, Schwartz SM, Berk BC (2007) Vascular remodeling: hemodynamic and biochemical mechanisms underlying Glagov's phenomenon. Arterioscler Thromb Vasc Biol 27:1722–1728
6. Lafont A, Durand E, Samuel JL et al (1999) Endothelial dysfunction and collagen accumulation: two independent factors for restenosis and constrictive remodeling after experimental angioplasty. Circulation 100:1109–1115
7. Schiffers PM, Henrion D, Boulanger CM et al (2000) Altered flow-induced arterial remodeling in vimentin-deficient mice. Arterioscler Thromb Vasc Biol 20:611–616
8. Schoenhagen P, Tuzcu EM, Apperson-Hansen C et al (2006) Determinants of arterial wall remodeling during lipid-lowering therapy: serial intravascular ultrasound observations from the reversal of atherosclerosis with aggressive lipid lowering therapy (reversal) trial. Circulation 113:2826–2834
9. Glagov S, Weisenberg E, Zarins CK et al (1987) Compensatory enlargement of human atherosclerotic coronary arteries. N Engl J Med 316:1371–1375
10. Mulvany MJ (2012) Small artery remodelling in hypertension. Basic Clin Pharmacol Toxicol 110:49–55
11. Lehoux S (2006) Redox signalling in vascular responses to shear and stretch. Cardiovasc Res 71:269–279
12. Karimi Galougahi K, Ashley EA, Ali ZA (2016) Redox regulation of vascular remodeling. Cell Mol Life Sci 73:349–363
13. Langille BL, O'Donnell F (1986) Reductions in arterial diameter produced by chronic decreases in blood flow are endothelium-dependent. Science 231:405–407
14. Rudic RD, Shesely EG, Maeda N et al (1998) Direct evidence for the importance of endothelium-derived nitric oxide in vascular remodeling. J Clin Invest 101:731–736
15. Leite PF, Danilovic A, Moriel P et al (2003) Sustained decrease in superoxide dismutase activity underlies constrictive remodeling after balloon injury in rabbits. Arterioscler Thromb Vasc Biol 23:2197–2202
16. Tanaka LY, Araújo HA, Hironaka GK et al (2016) Peri/epicellular protein disulfide isomerase sustains vascular lumen caliber through an anticonstrictive remodeling effect. Hypertension 67:613–622
17. Walsh K, Smith RC, Kim HS (2000) Vascular cell apoptosis in remodeling, restenosis, and plaque rupture. Circ Res 87:184–188
18. Souza HP, Souza LC, Anastacio VM et al (2000) Vascular oxidant stress early after balloon injury: evidence for increased NAD(P)H oxidoreductase activity. Free Radic Biol Med 28:1232–1242
19. Azevedo LC, Pedro MA, Souza LC et al (2000) Oxidative stress as a signaling mechanism of the vascular response to injury: the redox hypothesis of restenosis. Cardiovasc Res 47:436–445
20. Szöcs K, Lassègue B, Sorescu D et al (2002) Upregulation of Nox-based NAD(P)H oxidases in restenosis after carotid injury. Arterioscler Thromb Vasc Biol 22:21–27
21. Lassègue B, San Martín A, Griendling KK (2012) Biochemistry, physiology, and pathophysiology of NADPH oxidases in the cardiovascular system. Circ Res 110:1364–1390
22. Lee MY, San Martin A, Mehta PK et al (2009) Mechanisms of vascular smooth muscle NADPH oxidase 1 (Nox1) contribution to injury-induced neointimal formation. Arterioscler Thromb Vasc Biol 29:480–487
23. Csányi G, Cifuentes-Pagano E, Al Ghouleh I (2011) Nox2 B-loop peptide, Nox2ds, specifically inhibits the NADPH oxidase Nox2. Free Radic Biol Med 51:1116–1125
24. Niu XL, Madamanchi NR, Vendrov AE et al (2010) Nox activator 1: a potential target for modulation of vascular reactive oxygen species in atherosclerotic arteries. Circulation 121:549–559

25. Tong X, Khandelwal AR, Qin Z et al (2015) Role of smooth muscle Nox4-based NADPH oxidase in neointimal hyperplasia. J Mol Cell Cardiol 89:185–194
26. Brandes RP, Weissmann N, Schröder K (2014) Nox family NADPH oxidases: molecular mechanisms of activation. Free Radic Biol Med 76:208–226
27. Watt J, Wadsworth R, Kennedy S, Oldroyd KG (2008) Pro-healing drug-eluting stents: a role for antioxidants? Clin Sci (Lond) 114:265–273
28. Ali ZA, de Jesus PV, Yuan K et al (2014) Oxido-reductive regulation of vascular remodeling by receptor tyrosine kinase ROS1. J Clin Invest 124:5159–5174
29. Schneider JE, Berk BC, Gravanis MB et al (1993) Probucol decreases neointimal formation in a swine model of coronary artery balloon injury. A possible role for antioxidants in restenosis. Circulation 88:628–637
30. Tardif JC, Côté G, Lespérance J et al (1997) Probucol and multivitamins in the prevention of restenosis after coronary angioplasty. Multivitamins and Probucol Study Group. N Engl J Med 337:365–372
31. Rodés J, Côté G, Lespérance J et al (1998) Prevention of restenosis after angioplasty in small coronary arteries with probucol. Circulation 97:429–436
32. Pescatore LA, Bonatto D, Forti FL et al (2012) Protein disulfide isomerase is required for platelet-derived growth factor-induced vascular smooth muscle cell migration, nox1 nadph oxidase expression, and rhogtpase activation. J Biol Chem 287:29290–29300
33. Eble JA, de Rezende FF (2014) Redox-relevant aspects of the extracellular matrix and its cellular contacts via integrins. Antioxid Redox Signal 20:1977–1993
34. Parastatidis I, Weiss D, Joseph G, Taylor WR (2013) Overexpression of catalase in vascular smooth muscle cells prevents the formation of abdominal aortic aneurysms. Arterioscler Thromb Vasc Biol 33:2389–2396
35. Pedrigi RM, Poulsen CB, Mehta VV et al (2015) Inducing persistent flow disturbances accelerates atherogenesis and promotes thin cap fibroatheroma development in D374Y-PCSK9 hypercholesterolemic minipigs. Circulation 132:1003–1012
36. Dai G, Kaazempur-Mofrad MR, Natarajan S et al (2004) Distinct endothelial phenotypes evoked by arterial waveforms derived from atherosclerosis-susceptible and -resistant regions of human vasculature. Proc Natl Acad Sci U S A 101:14871–14876
37. Kwak BR, Bäck M, Bochaton-Piallat ML et al (2014) Biomechanical factors in atherosclerosis: mechanisms and clinical implications. Eur Heart J 35:3013–3020
38. Dimmeler S, Fleming I, Fisslthaler B et al (1999) Activation of nitric oxide synthase in endothelial cells by Akt-dependent phosphorylation. Nature 399:601–605
39. Qian J, Fulton D (2013) Post-translational regulation of endothelial nitric oxide synthase in vascular endothelium. Front Physiol 4:347
40. Laurindo FR, de Pedro M, Barbeiro HV et al (1994) Vascular free radical release. Ex vivo and in vivo evidence for a flow-dependent endothelial mechanism. Circ Res 74:700–709
41. Zhang Q, Malik P, Pandey D et al (2008) Paradoxical activation of endothelial nitric oxide synthase by NADPH oxidase. Arterioscler Thromb Vasc Biol 28:1627–1633
42. Lu X, Kassab GS (2004) Nitric oxide is significantly reduced in ex vivo porcine arteries during reverse flow because of increased superoxide production. J Physiol 561:575–582
43. Hwang J, Saha A, Boo YC et al (2003) Oscillatory shear stress stimulates endothelial production of O_2^- from p47phox-dependent NAD(P)H oxidases, leading to monocyte adhesion. J Biol Chem 278:47291–47298
44. Nigro P, Abe J, Berk BC (2011) Flow shear stress and atherosclerosis: a matter of site specificity. Antioxid Redox Signal 15:1405–1414
45. Gray SP, Di Marco E, Kennedy K et al (2016) Reactive oxygen species can provide atheroprotection via NOX4-dependent inhibition of inflammation and vascular remodeling. Arterioscler Thromb Vasc Biol 36:295–307
46. Castier Y, Brandes RP, Leseche G et al (2005) p47phox-dependent NADPH oxidase regulates flow-induced vascular remodeling. Circ Res 97:533–540

47. Morishita T, Tsutsui M, Shimokawa H et al (2002) Vasculoprotective roles of neuronal nitric oxide synthase. FASEB J 16:1994–1996
48. Yogo K, Shimokawa H, Funakoshi H et al (2014) Different vasculoprotective roles of NO synthase isoforms in vascular lesion formation in mice. Arterioscler Thromb Vasc Biol 20:E96–E100
49. Nam D, Ni CW, Rezvan A, Suo J et al (2009) Partial carotid ligation is a model of acutely induced disturbed flow, leading to rapid endothelial dysfunction and atherosclerosis. Am J Physiol Heart Circ Physiol 297:H1535–H1543
50. Simmons RD, Kumar S, Jo H (2016) The role of endothelial mechanosensitive genes in atherosclerosis and omics approaches. Arch Biochem Biophys 591:111–131
51. Khatri JJ, Johnson C, Magid R et al (2004) Vascular oxidant stress enhances progression and angiogenesis of experimental atheroma. Circulation 109:520–525
52. Go YM, Son DJ, Park H et al (2009) Disturbed flow enhances inflammatory signaling and atherogenesis by increasing thioredoxin-1 level in endothelial cell nuclei. PLoS One 9:e108346
53. Yamamoto K, Sokabe T, Matsumoto T et al (2005) Impaired flow-dependent control of vascular tone and remodeling in P2X4-deficient mice. Nat Med 12:133–137
54. Chen Z, Tzima E (2009) PECAM-1 is necessary for flow-induced vascular remodeling. Arterioscler Thromb Vasc Biol 29:1067–1073
55. Galis ZS, Johnson C, Godin D et al (2002) Targeted disruption of the matrix metalloproteinase-9 gene impairs smooth muscle cell migration and geometrical arterial remodeling. Circ Res 91:852–859
56. Castier Y, Ramkhelawon B, Riou S et al (2009) Role of NF-kappaB in flow-induced vascular remodeling. Antioxid Redox Signal 11:1641–1649
57. Baeyens N, Nicoli S, Coon BG et al (2015) Vascular remodeling is governed by a VEGFR3-dependent fluid shear stress set point. Elife 4
58. Bakker EN, Buus CL, Spaan JA et al (2005) Small artery remodeling depends on tissue-type transglutaminase. Circ Res 96:119–126
59. González-Fernández F, López-Farré A, Rodríguez-Feo JA et al (1998) Expression of inducible nitric oxide synthase after endothelial denudation of the rat carotid artery: role of platelets. Circ Res 83:1080–1087
60. Laukkanen MO, Kivelä A, Rissanen T et al (2002) Adenovirus-mediated extracellular superoxide dismutase gene therapy reduces neointima formation in balloon-denuded rabbit aorta. Circulation 106:1999–2003
61. Laurindo FR, Pescatore LA, de Castro Fernandes C (2012) Protein disulfide isomerase in redox cell signaling and homeostasis. Free Radic Biol Med 52:1954–1969
62. Wilkinson B, Gilbert HF (2004) Protein disulfide isomerase. Biochim Biophys Acta 1699:35–44
63. Hatahet F, Ruddock LW (2009) Protein disulfide isomerase: A critical evaluation of its function in disulfide bond formation. Antioxid Redox Signal 11:2807–2850
64. Fernandes DC, Manoel AH, Wosniak J, Laurindo FR (2009) Protein disulfide isomerase overexpression in vascular smooth muscle cells induces spontaneous preemptive NADPH oxidase activation and Nox1 mRNA expression: effects of nitrosothiol exposure. Arch Biochem Biophys 484:197–204
65. Santos CX, Stolf BS, Takemoto PV (2009) Protein disulfide isomerase (PDI) associates with NADPH oxidase and is required for phagocytosis of leishmania chagasi promastigotes by macrophages. J Leukoc Biol 86:989–998
66. de A Paes AM, Veríssimo-Filho S, Guimarães LL et al (2011) Protein disulfide isomerase redox-dependent association with p47(phox): evidence for an organizer role in leukocyte NADPH oxidase activation. J Leukoc Biol 90:799–810
67. Schwartz SM (1997) Perspectives series: cell adhesion in vascular biology. Smooth muscle migration in atherosclerosis and restenosis. J Clin Invest 99:2814–2816
68. Schürmann C, Rezende F, Kruse C et al (2015) The NADPH oxidase Nox4 has anti-atherosclerotic functions. Eur Heart J 36:3447–3456

69. Clempus RE, Sorescu D, Dikalova AE et al (2007) Nox4 is required for maintenance of the differentiated vascular smooth muscle cell phenotype. Arterioscler Thromb Vasc Biol 27:42–48
70. Ron D, Walter P (2007) Signal integration in the endoplasmic reticulum unfolded protein response. Nat Rev Mol Cell Biol 8:519–529
71. Myoishi M, Hao H, Minamino T et al (2007) Increased endoplasmic reticulum stress in atherosclerotic plaques associated with acute coronary syndrome. Circulation 116:1226–1233
72. Muller C, Bandemer J, Vindis C et al (2013) Protein disulfide isomerase modification and inhibition contribute to ER stress and apoptosis induced by oxidized low density lipoproteins. Antioxid Redox Signal 18:731–742
73. Flaumenhaft R, Furie B, Zwicker JI (2015) Therapeutic implications of protein disulfide isomerase inhibition in thrombotic disease. Arterioscler Thromb Vasc Biol 35:16–23
74. Essex DW (2009) Redox control of platelet function. Antioxid Redox Signal 11:1191–1225
75. Kim K, Hahm E, Li J et al (2013) Platelet protein disulfide isomerase is required for thrombus formation but not for hemostasis in mice. Blood 122:1052–1061
76. Bi S, Hong PW, Lee B, Baum LG (2011) Galectin-9 binding to cell surface protein disulfide isomerase regulates the redox environment to enhance t-cell migration and HIV entry. Proc Natl Acad Sci U S A 108:10650–10655
77. Willems SH, Tape CJ, Stanley PL et al (2010) Thiol isomerases negatively regulate the cellular shedding activity of adam17. Biochem J 428:439–450
78. Zai A, Rudd MA, Scribner AW, Loscalzo J (1999) Cell-surface protein disulfide isomerase catalyzes transnitrosation and regulates intracellular transfer of nitric oxide. J Clin Invest 103:393–399
79. Taylor AJ, Burke AP, Farb A et al (1999) Arterial remodeling in the left coronary system: the role of high-density lipoprotein cholesterol. J Am Coll Cardiol 34:760–767
80. Furlan-Freguia C, Marchese P, Gruber A et al (2011) P2x7 receptor signaling contributes to tissue factor-dependent thrombosis in mice. J Clin Invest 121:2932–2944

Chapter 16
Mitochondrial Complex I Inactivation After Ischemia-Reperfusion in the Stunned Heart

Laura B. Valdez, Silvina S. Bombicino, Darío E. Iglesias, Ivana A. Rukavina-Mikusic, and Verónica D'Annunzio

Abstract Mitochondrial complex I (NADH-ubiquinone oxidoreductase) catalyzes the transfer of two electrons from NADH *via* flavin mononucleotide (FMN) and a series of iron-sulfur centers (Fe-S) to ubiquinone (UQ) in a reaction associated with proton translocation across the inner membrane, contributing to the proton-motive force. Complex I produces superoxide anion (O_2^-) through the autoxidation reaction of the flavin-semiquinone (FMNH•) with molecular oxygen. Superoxide reacts with nitric oxide (NO) to yield peroxynitrite ($ONOO^-$), a strong oxidant and nitrating compound. When the steady-state concentration of $ONOO^-$ is enhanced, tyrosine nitration, protein oxidation and damage to Fe-S centers take place, leading to a sustained complex I inhibition. Dysfunction of complex I was found in a number of clinical conditions such as Parkinson's disease, ischemia-reperfusion, endotoxic shock, and aging. We have shown that the ventricular dysfunction observed in myocardial stunning is associated with a mitochondrial dysfunction that includes partial inactivation of complex I and mitochondrial nitric oxide synthase (mtNOS) activities, oxidative and nitrosative damages and increased H_2O_2 and $ONOO^-$ production rates. Moreover, adenosine proved to be effective in attenuating ventricular dysfunction and also in protecting from mitochondrial dysfunction and complex I syndrome.

Keywords Stunned heart • Complex I • Mitochondrial nitric oxide synthase (mtNOS) • Hydrogen peroxide • Nitric oxide • Superoxide anion • Peroxynitrite

---

L.B. Valdez (✉) • S.S. Bombicino • D.E. Iglesias • I.A. Rukavina-Mikusic
Institute of Biochemistry and Molecular Medicine (IBIMOL; UBA-CONICET), Physical Chemistry Division, School of Pharmacy and Biochemistry, University of Buenos Aires, Buenos Aires, Argentina
e-mail: lbvaldez@ffyb.uba.ar

V. D'Annunzio
Institute of Biochemistry and Molecular Medicine (IBIMOL; UBA-CONICET), Pathology Division, School of Medicine, University of Buenos Aires, Buenos Aires, Argentina
e-mail: vdannunzio@gmail.com

© Springer International Publishing Switzerland 2016

R.J. Gelpi et al. (eds.), *Biochemistry of Oxidative Stress*, Advances in Biochemistry in Health and Disease 16, DOI 10.1007/978-3-319-45865-6_16

1 Introduction

Cardiac ischemic syndromes include a group of entities classified taking into account the duration of ischemia and its pathophysiology. These entities range from transient ischemic episodes without functional alterations [1, 2] to prolonged ischemia following myocardial infarction, with cell death and necrosis, eventually resulting in ventricular remodeling [3]. When ischemia is sustained for less than 20 min, it leads to a reversible post-ischemic ventricular dysfunction called stunned myocardium [1, 4]. Myocardial stunning is defined as a contractile and diastolic dysfunction that persists during reperfusion after a short ischemic episode. In the stunned heart the changes are reversible and the contractile reserve is preserved without myocardial necrosis [4, 5]. Although myocardial stunning is a fully reversible phenomenon, contractile dysfunction persists for hours or even days.

2 Mitochondrial Complex I

Production of energy in aerobic cells is provided by the action of the mitochondrial respiratory chain and the ATP synthase. Reducing equivalents from pyridine nucleotides (NADH) generated in several catabolic pathways are routed into the energy-converting respiratory chain via complex I. The mitochondrial complex I (NADH:ubiquinone-oxidoreductase) catalyses the reversible reaction:

$$\mathrm{NADH} + \mathrm{UQ} + \mathrm{H}^{+} + 4\,\mathrm{H}^{+}\mathit{in} \leftrightarrows \mathrm{NAD}^{+} + \mathrm{UQH}_2 + 4\,\mathrm{H}^{+}\mathit{out}$$

Ubiquinone (UQ) and ubiquinol (UQH_2), anchored in the membrane by the isoprenoid side chain are reduced and oxidized, respectively, and $[H^+]_{in}$ and $[H^+]_{out}$ are protons pumped from the matrix to the intermembrane space. This reaction provides the NAD^+ required for the steady-state operation of the Krebs cycle, and serves as the major entry point for feeding the respiratory chain with the reducing equivalents needed for the generation of the H^+ electrochemical gradient and subsequent ATP synthesis. Complex I itself is one of the H^+ pumps that generate mitochondrial $\Delta\mu_{H+}$. The mammalian enzyme is the most complex among all known oxido-reductases. It is composed of 45 different polypeptides assembled in a single unit as the result of the coordinated operation of nuclear and mitochondrial genomes [6–9]. Complex I contains one flavin mononucleotide (FMN) in the active site for NADH oxidation, a chain of 7 iron-sulfur clusters (Fe-S) that link the flavin with the quinone binding site, and an unusually positioned Fe-S cluster on the opposite side of the flavin, separated from the chain of clusters. The last cluster in the chain (cluster N2) donates electrons to the bound quinone (UQ) [8, 9]. All redox centers (FMN and Fe-S clusters) are located within core subunits in the hydrophilic domain of the enzyme, whereas H^+ translocation is carried out by several Na^+/H^+antiporter-like subunits. Dysfunction of complex I was reported in a number of clinical conditions such as Parkinson's disease, ischemia-reperfusion, neuromuscular disorders, endotoxic shock, Leber's optic neuropathy and aging.

3 Complex I and Reactive Oxygen and Nitrogen Species

Complex I produces superoxide anion (O_2^-)through the autoxidation reaction of the flavin-semiquinone (FMNH•) with molecular oxygen [10, 11]. It is understood that the ubisemiquinone (UQH•) autoxidation contribution, in complex I, is negligible. The O_2^- production rate by complex I is increased by inhibition of electron transfer with rotenone [12] or by complex I dysfunction [13–19].

In the mitochondrial matrix, O_2^- is metabolized by the dismutation reaction catalyzed by Mn-superoxide dismutase (Mn-SOD) that produces O_2 and hydrogen peroxide (H_2O_2). The latter species is involved in the regulation of redox sensitive signaling [20, 21]. Isolated respiring mitochondria produce H_2O_2 at rates that depend on the redox state of the components of the respiratory chain and, consequently, on the mitochondrial metabolic state and the presence of inhibitors [12, 22]. In addition, O_2^- reacts with nitric oxide (NO), this latter species produced through the mitochondrial nitric oxide synthase (mtNOS), the isoenzyme of the NOS family located in the mitochondrial inner membrane [23–25]. Nitric oxide is produced at a rate of about 1.0–1.5 nmol NO. min^{-1}. mg $protein^{-1}$ and kept at a steady-state level of about 10^{-9}–10^{-8} M in the mitocondrial matrix [22, 26]. The diffusion controlled reaction ($k = 1.9 \times 10^{10}$ M^{-1} s^{-1}) between NO and O_2^- yields peroxynitrite ($ONOO^-$), a strong oxidant and nitrating compound. Peroxynitrite is normally reduced by mitochondrial reductants, as NADH, ubiquinol (UQH_2) and GSH and kept at an intramitochondrial steady-state level of about 5–10 nM [27]. When the steady-state concentration of $ONOO^-$ is increased to about 20–50 nM, tyrosine nitration, protein oxidation and damage to Fe-S centers take place. The process leads to complex I inhibition and to increased generation of O_2^- by complex I.

4 Ventricular Dysfunction in the Stunned Heart

In myocardial stunning the infarct area is not significant and the functional injury is completely reversible. We have reported a mitochondrial dysfunction in rabbit heart in a mild condition of ischemia/reperfusion (I/R) that simulates the condition of human stunned myocardium [18]. The impairment in mitochondria and in heart function develops gradually along the steps of ischemia (15 min I / 0 min R), early reperfusion (15 min I / 5 min R), and late reperfusion (15 min I / 30 min R). We confirmed that the contractile state, determined through left ventricular developed pressure (LVDP) of rabbit hearts exposed to I/R, recovered only 63 % of the initial pre-ischemic values (107 ± 7 mm Hg) at the end of reperfusion. When left-ventricle respiration rates were assessed in 1-mm^3 tissue cubes, heart O_2 uptake (1.57 ± 0.10 μmol O_2. min^{-1}. g $heart^{-1}$) decreased successively in the steps of ischemia (15/0), early reperfusion (15/5), and late reperfusion (15/30), reaching 0.85 ± 0.08 μmol O_2. min^{-1}. g $heart^{-1}$, that is an overall decrease of 46 % in the last step. Moreover, and as shown in Fig. 16.1, a linear correlation was obtained between LVDP and heart O_2

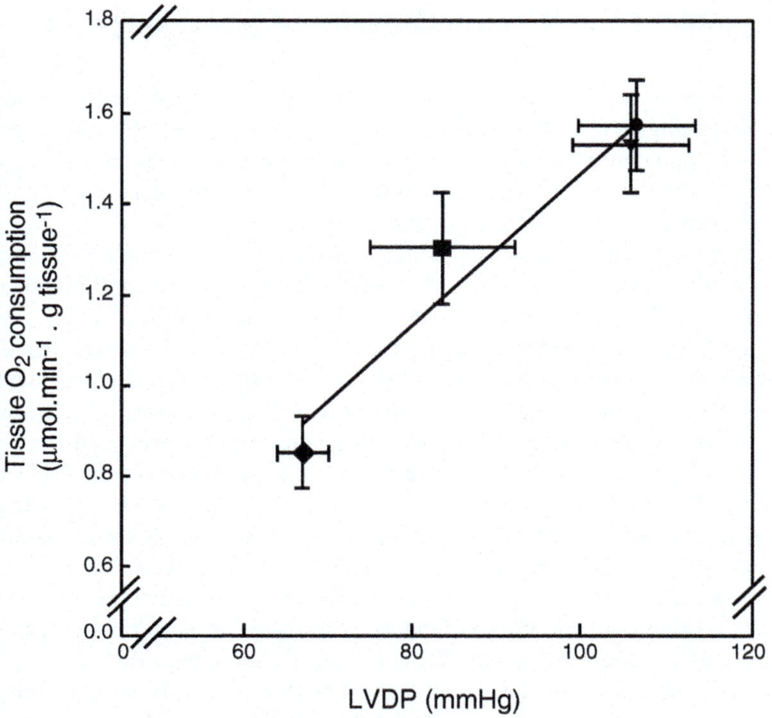

Fig. 16.1 Linear correlation between left-ventricle developed pressure (LVDP) and heart O_2 uptake ($r^2 = 0.947$) in rabbit heart exposed to ischemia/reperfusion: 0/0 (●); 15/0 (▼); 15/5 (■); 15/30 (♦)

consumption ($r^2 = 0.947$) supporting the idea that heart strictly depends on O_2 availability to achieve the contraction force needed to carry out the mechanical work.

Regarding diastolic function, we previously observed a clear reduction in the relaxation rate during the maximum transitory systolic overshoot. This alteration continued up to 90 s of reperfusion, even when the contractile state has already begun to decrease and was tending to reach its stabilization periods. After 90 s of reperfusion, the relaxation rate starts to normalize until it reaches the pre-ischemic values after 5 min of reperfusion. We also detected that myocardial stiffness (represented by left ventricular end diastolic pressure, LVEDP) was already increased after 60 s of reperfusion and that it continued to increase up to 30 min of reperfusion. Interestingly, this increase in the stiffness is negatively correlated to the maximum contractility achieved, *i.e.*, in those animals where the transitory systolic overshoot is greater, the increase in LVEDP is lower. Additionally, the treatment with adenosine in the perfusion medium (0.03 μg. kg^{-1}. min^{-1}, in Krebs-Henseleit buffer) before ischemia and during reperfusion, not only improved the post-ischemic recovery after reperfusion reaching a value of 97 ± 6 mmHg in LVDP, but also avoided the decrease in tissue O_2 uptake observed after 30 min reperfusion.

5 Mitochondrial Complex I Dysfunction in the Stunned Heart

Mitochondrial state 3 respiration, the active respiration yielding ATP, was decreased by 32 % when malate-glutamate were used as complex I substrates with the corresponding decrease in respiratory control (6.2 to 4.4). This effect was not observed when succinate was used as complex II substrate, indicating that mild I/R selectively affects complex I enzymatic activity. Moreover, complex I activity was gradually reduced up to a maximal 28 % inhibition after 30 min reperfusion, a time point where complex II and complex IV activities did not modify. In addition, the reduction observed in complex I activity was associated with a decline in mtNOS activity. Accordingly, linear correlations between state 3 O_2 consumption and mitochondrial complex I-III activity ($r^2 = 0.939$) and between state 3 O_2 consumption and mtNOS activity ($r^2 = 0.932$) were observed (Fig. 16.2) indicating that the pattern observed for the decline in mitochondrial O_2 uptake using malate-glutamate is also observed for the reduction of complex I and mtNOS activities. This data is in agreement with the reported functional interaction between complex I proteins and mtNOS [18, 19, 28–30]. Regarding this point, Poderoso's group [28] showed that not only complex IV but also complex I proteins immunoprecipitate with mtNOS, indicating molecular interactions among mtNOS and complexes I and IV. Furthermore, the functional association between complex I and mtNOS agrees with the observations by Parihar

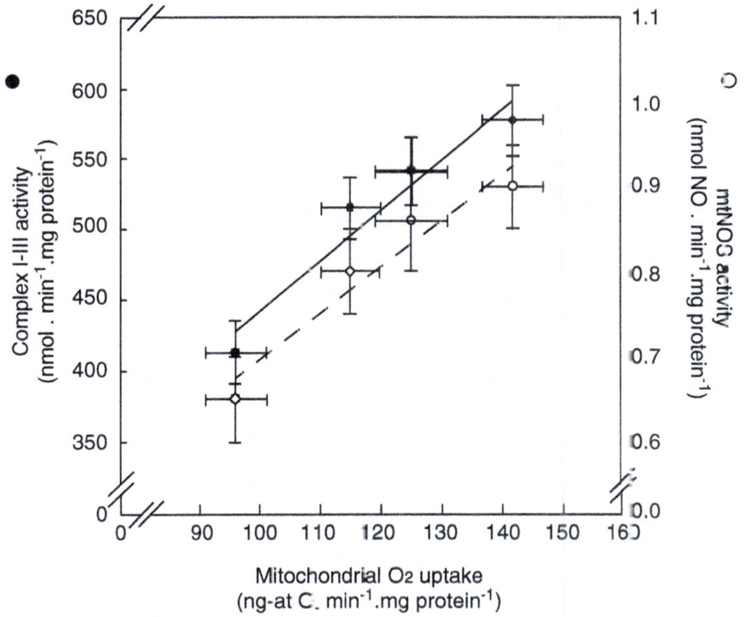

Fig. 16.2 Linear correlation between malate-glutamate-supported state 3 O_2 consumption with complex I ($r^2 = 0.939$) and mtNOS ($r^2 = 0.932$) activities

et al. [29] and Boveris's group [13, 31], who proposed that mtNOS is structurally adjacent to complex I. The impairment of complex I activity was also evidenced by the successive increases in H_2O_2 production rates when malate-glutamate were used as substrates. The rate of H_2O_2 release in state 4 was 78 % higher in the last step of the process (15/30) compared with control mitochondria (0/0) (0.18 ± 0.02 nmol H_2O_2. min^{-1}. mg $protein^{-1}$). The activity (53 ± 5 U. mg $protein^{-1}$) and the active concentration (7.4 ± 0.7 μM) of Mn-SOD were not modified in the stunned heart; then, the observed increase in H_2O_2 generation by complex I implies an enhancement in O_2^- production. This, in turn, explains an increased level of $ONOO^-$even with a decreased NO generation by mtNOS. This point of view was supported by the estimation of O_2^- and NO steady-states values and the $ONOO^-$production rates.

6 Steady-States of O_2^- and NO and Production Rate of $ONOO^-$

Taking into account the experimental values of NO and H_2O_2 production rates at each time of ischemia/reperfusion and the Mn-SOD concentration in the mitochondrial matrix (7.4 ± 0.7 μM), the steady-state concentrations of O_2^- and NO and the $ONOO^-$ production rate were calculated (Table 16.1). In the mitochondrial matrix, O_2^- is consumed by two diffusion controlled reactions: the disproportionation reaction catalyzed by Mn-SOD to produce H_2O_2 and O_2 and the reaction with NO to produce $ONOO^-$. Mitochondrial NO is produced by mtNOS and released into the mitochondrial matrix where it reacts with O_2^-, ubiquinol (UQH_2) and cytochrome oxidase (cyt aa_3-Fe^{3+}). The following equations describe the intramitochondrial NO and O_2^- reactions:

$$O_2^- + NO \rightarrow ONOO^- \qquad k_1 = 1.9 \times 10^{10}\ M^{-1}s^{-1}$$

$$2O_2^- + 2H^+ \rightarrow H_2O_2 + O_2 \qquad k_2 = 2.3 \times 10^9\ M^{-1}s^{-1}$$

The previous dismutation reaction can be spontaneous or catalyzed. The rate constant given above corresponds to the second order and diffusion limited SOD-catalyzed reaction.

Table 16.1 O_2^- and NO steady-states concentrations and $ONOO^-$ production rates in myocardial stunning

I/R (min)	$[O_2^-]_{ss}$ (10^{-11} M)	$[NO]_{ss}$ (10^{-9} M)	$d[ONOO^-]/dt$ ($nM.s^{-1}$)
0/0	4.8	9.1	8.4
15/0	5.0	8.7	8.3
15/5	6.4	8.1	9.8
15/30	8.1	6.6	10.1

$$NO + UQH_2 \rightarrow UQH^{\bullet} + H^+ + NO^- \qquad k_3 = 1.5 \times 10^4\ M^{-1}s^{-1}$$

$$NO + cytaa_3 - Fe^{2+} \rightarrow cyt\ aa_3 - Fe^{3+} + NO^- \qquad k_4 = 4.0 \times 10^7\ M^{-1}s^{-1}$$

In the steady-state and by definition, the production rate is equal to the consumption rate of a chemical species. The NO and O_2^- steady-state concentrations ($[NO]_{ss}$ and $[O_2^-]_{ss}$) were calculated from Eqs. (16.1) and (16.2), respectively, by mathematical iteration, and using the rate constants: $k_1 = 1.9 \times 10^{10}\ M^{-1}\ s^{-1}$ [32]; $k_2 = 2.3 \times 10^9\ M^{-1}\ s^{-1}$ [33]; $k_3 = 1.5 \times 10^4\ M^{-1}\ s^{-1}$ [34]; and $k_4 = 4 \times 10^7\ M^{-1}\ s^{-1}$ [35]

$$[O_2^-]_{ss} = 2d[H_2O_2]/dt/(k_1[NO] + k_2[Mn-SOD]) \tag{16.1}$$

$$[NO]_{ss} = d[NO]/dt/(k_1[O_2^-] + k_3[UQH_2] + k_4[cytaa_3 - Fe^{2+}]) \tag{16.2}$$

The O_2^- production rate was calculated from the experimental H_2O_2 production rates and taking into account the 2:1 stoichiometry of the disproportionation reaction of O_2^- to H_2O_2. Nitric oxide diffusion to and from cytosol was not included in Eq. (16.2). Ubiquinol and cytochrome aa_3 contents were taken as 277 μM and 5.6 μM, respectively, considering a mitochondrial matrix volume of 7 0 μl.mg $protein^{-1}$ [36, 37]. In addition, $ONOO^-$ production rates were calculated considering the steady-state concentrations of NO and O_2^- in the mitochondrial matrix, calculated from Eqs. (16.1) and (16.2) and the second order rate constant $k_1 = 1.9 \times 10^{10}\ M^{-1}\ s^{-1}$ (Eq. 16.3)

$$d[ONOO^-]/dt = k_1[O_2^-][NO] \tag{16.3}$$

As Table 16.1 shows, the $[O_2^-]_{ss}$ calculated at 15 min of ischemia is similar to the control one (0/0), but it was increased by 32 % and 67 % after 5 and 30 min of reperfusion, respectively. Since no changes in Mn-SOD concentration were observed at any time of ischemia and reperfusion, the H_2O_2 production rate values are the result of changes in the O_2^- production rates. In this way, an increase in O_2^- production rate leads to a significant enhancement in $[O_2^-]_{ss}$. Moreover, the increase in H_2O_2 production rates and in $[O_2^-]_{ss}$ agree with the enhancement in phospholipid (42 %) and protein oxidation (17 %) products, that were observed after 15 min of ischemia and 30 min reperfusion. Accordingly, a linear regression ($r^2 = 0.940$) was obtained between heart mitochondrial H_2O_2 production sustained by complex I substrates and lipid peroxidation measured as TBARS (Fig. 16.3).

In addition, we observed a decrease in $[NO]_{ss}$ at each time of I/R (Table 16.1) in agreement with the reduction in the NO production rate. Despite the decrease in $[NO]_{ss}$, $ONOO^-$ production was increased by 20 % after 15 min of ischemia and 30 min reperfusion, due to the more significant enhancement in $[O_2^-]_{ss}$. This result agrees with the increase (70 %) in tyrosine nitration observed when hearts were exposed to 15 min of ischemia and 30 min reperfusion [18]. In physiological situations, only about 15–20 % of the O_2^- generated in mitochondria is catabolized through its reaction with NO. But this pathway consumes about 80 % of mitochon-

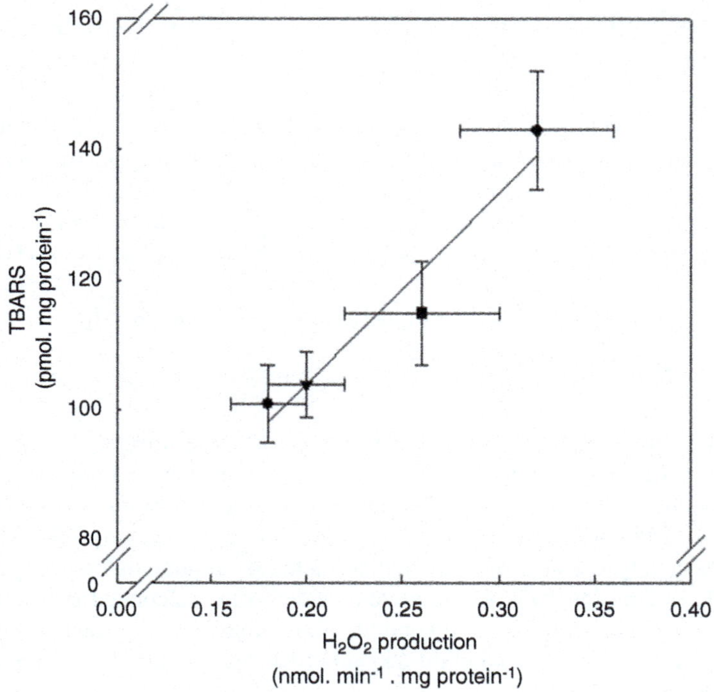

Fig. 16.3 Linear correlation between mitochondrial H_2O_2 production rates using malate-glutamate with lipid oxidation products, measured as TBARS, ($r^2 = 0.940$) of rabbit heart exposed to ischemia/reperfusion: 0/0 (●); 15/0 (▼); 15/5 (■); 15/30 (♦)

drial NO. However, after I/R and in some pathological situations in which enhancements in $[O_2^-]_{ss}$ without modifications in Mn-SOD expression were observed, as it is the case in the stunned heart, this metabolic pathway could be exacerbated, increasing the decomposition of O_2^- and leading to an enhancement in $ONOO^-$ generation. To note, reperfusion with adenosine protected from the impairment of complex I and mtNOS activities produced by I/R and from the enhancement of mitochondrial phospholipid peroxidation and tyrosine nitration. The improvement in the contractile response by adenosine treatment was associated with a decrease in nitrotyrosine formation that occurs simultaneously with an increase in Akt and troponin I phosphorylation [38]. Concerning this point, it is known that Ca^{2+} overload is responsible, at least in part, for the myocardial stunning [39]. Adenosine may attenuate myocardial stunning by inhibiting Ca^{2+} influx through stimulation of A_1 receptors [5, 40].

7 Complex I Syndrome

The mitochondrial dysfunction observed in rabbit myocardial stunning is in agreement with the concept of complex I syndrome as it was described by Boveris, Carreras and Poderoso [41]. The complex I syndrome includes decreases in tissue O_2 uptake, malate-glutamate-supported mitochondrial respiration and complex I and mtNOS activities, together with increases in O_2^- and H_2O_2 production rates and phospholipid and protein oxidation and protein nitration products. Complex I syndrome implies a significant decline in cardiomyocyte capacity to generate enough energy for keeping the ionic composition for myocardial contraction. This reversible condition establishes a period in which energy production becomes a limiting step for cardiomyocyte homeostasis and function. The molecular mechanism responsible for complex I syndrome is probably accounted by a series of processes and reactions that lead to complex I inactivation. The lack of oxygen during ischemia leads to a high reduction state of the components of the mitochondrial respiratory chain (Fig. 16.4). Upon reoxygenation, a burst of O_2^- and H_2O_2 production occurs owing to the increased autoxidation rate of the most important intramitochondrial sources of O_2^- (UQH• and FMNH•). The enhancement in O_2^- and H_2O_2 production rates and steady-states leads to enhancement of the free-radical mediated reactions of lipid peroxidation and protein oxidation that drive, through free-radical mediated reactions, to the oxidation of complex I polypeptide chains. In addition, the formation of $ONOO^-$ at the vicinity of NADH-dehydrogenase active center provides a pathway leading to complex I nitration and inactivation. It have been shown that complex I inhibition by nitrosation, protects mitochondria during hypoxia and reoxygenation and protects cardiomyocytes during ischemia/reperfusion [42, 43]. There is evidence that the S-nitrosation of complex I proteins is associated with cyto-protection from the damage caused by I/R [44–46]. It is understood that the reactions that inactivate complex I change some of the weak native non-covalent intermolecular bonding forces and synergistically promote covalent cross-linking with protein inactivation. It has been known that complex I is highly sensitive to steroids and detergents [36], a fact that is now interpreted as the impairment of the non-covalent bonding that holds complex I proteins together. In addition, complex I is a target for reversible S-glutathionylation, a redox sensitive covalent modification that involves formation of a disulfide bridge between glutathione and a protein cysteine thiol [47–50]. Several subunits in the N-module of complex I have been identified as key sites for regulation by S-glutathionylation, leading to a decrease in complex I activity. Specifically Ndusf1 (~75 kDa) and Ndufv1 (~51 kDa) have been identified as major S-glutathionylation targets [48, 51, 52]. It is worth mentioning that the accessory subunit Ndufa11, which plays a role in complex I assembly and stability, has also been shown to be modified by S-glutathionylation in isolated mouse hearts subjected to I/R [53].

Consequently, a transient reversible inhibition of complex I reduces the reactivation of mitochondria during the first minutes in the early period of reperfusion in the ischemic heart. Simultaneously, there is increased ROS production, oxidative damage and tissue necrosis.

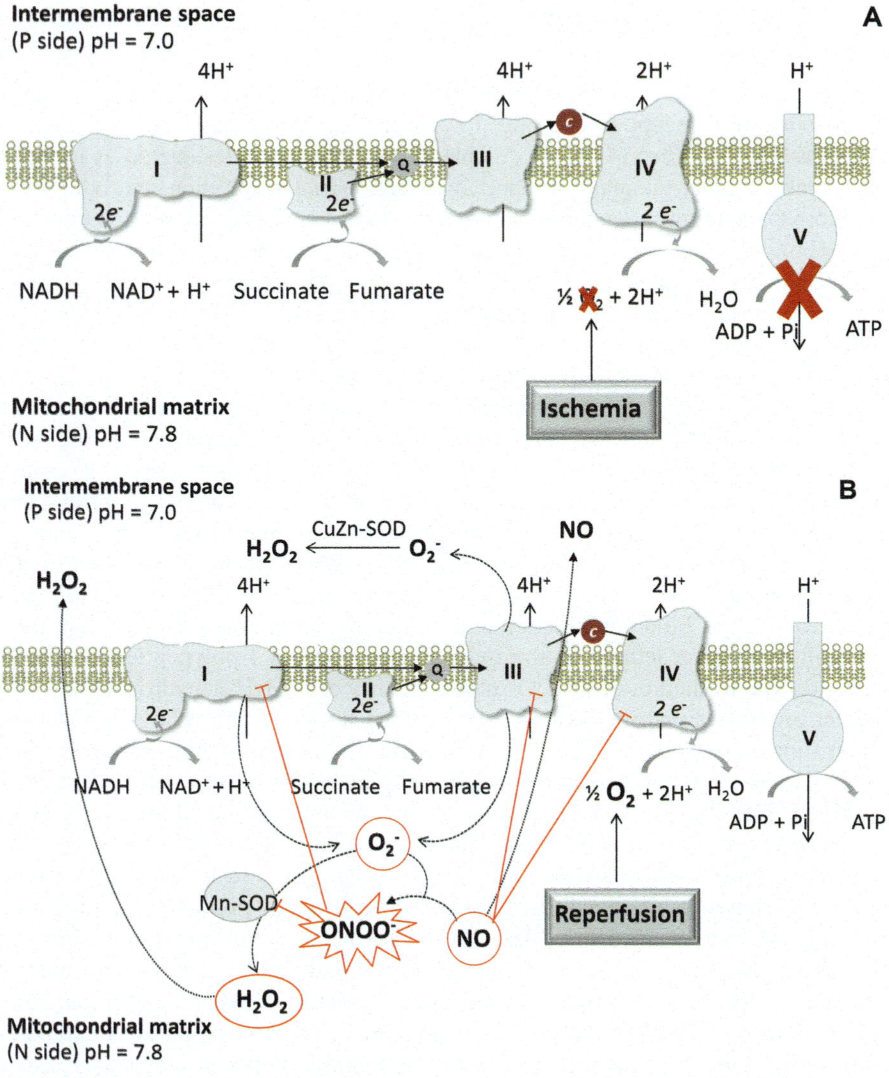

Fig. 16.4 Scheme ilustrating the mitochondrial metabolism of O_2^-, NO and $ONOO^-$ in ischemia (**a**) and reperfusion (**b**)

8 Conclusions

The ventricular dysfunction observed in myocardial stunning is associated with a mitochondrial dysfunction that shows a series of changes described as complex I syndrome with a partial inactivation of complex I and mtNOS activities, a oxidative and/or nitrosative damage and increased O_2^-, H_2O_2 and $ONOO^-$ production rates. All the mentioned processes are reversible and return to pre-ischemic values in hours. Moreover, compounds that proved to be effective in attenuating ventricular dysfunction with reduction of myocardium stunning, such as adenosine, also protected from the complex I syndrome observed upon reperfusion.

Acknowledgment This work was supported by grants from the University of Buenos Aires (UBACYT 20020110100140, 20020130100721 and 20020130100557), Agencia Nacional de Promoción Científica y Tecnológica (PICT 2013-0373, 2014-0964), and Consejo Nacional de Investigaciones Científicas y Técnicas (PIP 11220110100444).

References

1. Bolli R, Zhu W, Thornby J et al (1988) Time course and determinants of recovery of function after reversible ischemia in conscious dogs. Am J Physiol 254:H102–H114
2. Bolli R (1990) Mechanism of myocardial "stunning". Circulation 82:723–738
3. Olivetti G, Carpasso J, Meggs L et al (1991) Cellular basis of chronic ventricular remodeling after myocardial infarction in rats. Circulation 68:856–869
4. Heyndrickx GG, Baig H, Nellens P et al (1978) Depression of regional blood flow and wall thickening after brief coronary occlusions. Am J Physiol 234:H653–H659
5. Donato M, Gelpi RJ (2003) Adenosine and cardioprotection during reperfusion-an overview. Mol Cell Biochem 251:153–159
6. Vinogradov AD, Grivennikova VG (2001) The mitochondrial complex I: progress in understanding the catalytic properties. IUBMB Life 52:129–134
7. Hunte C, Zickermann V, Brandt U (2010) Functional modules and structural basis of conformational coupling in mitochondrial complex I. Science 329:448–451
8. Baradaran R, Berrisford JM, Minhas GS, Sazanov LA (2013) Crystal structure of the entire respiratory complex I. Nature 494:443–448
9. Hirst J (2013) Mitochondrial complex I. Annu Rev Biochem 82:551–575
10. Turrens JF, Boveris A (1980) Generation of superoxide anion by the NADH dehydrogenase of bovine heart mitochondria. Biochem J 191:421–427
11. Boveris A, Cadenas E (2000) Mitochondrial production of hydrogen peroxide regulation by nitric oxide and the role of ubisemiquinone. IUBMB Life 50:245–250
12. Boveris A, Chance B (1973) The mitochondrial generation of hydrogen peroxide. Biochem J 134:617–630
13. Navarro A, Boveris A (2009) Brain mitochondrial dysfunction and oxidative damage in Parkinson's disease. J Bioenerg Biomembr 41:517–521
14. Navarro A, Boveris A, Bández MJ et al (2009) Human brain cortex: mitochondrial oxidative damage and adaptative response in Parkinson disease and in dementia Lewy bodies. Free Rad Biol Med 46:1574–1580
15. Hensley K, Kotake Y, Sang H et al (2000) Dietary choline restriction causes complex I dysfunction and increased H_2O_2 generation in liver mitochondria. Carcinogenesis 21:983–989
16. Navarro A, Bandez MJ, Gomez C et al (2010) Effects of rotenone and pyridaben on complex I electron transfer and on mitochondrial nitric oxide synthase functional activity. J Bioenerg Biomembr 42:405–412
17. Navarro A, Bandez MJ, Lopez-Cepero JM et al (2011) High doses of vitamin E improve mitochondrial dysfunction in rat hyppocampus and frontal cortex upon aging. Am J Physiol Regul Integr Comp Physiol 300:R827–R834
18. Valdez LB, Zaobornyj T, Bombicino S et al (2011) Complex I syndrome in myocardial stunning and the effect of adenosine. Free Radic Biol Med 51:1203–1212
19. Valdez LB, Zaobornyj T, Bombicino SS et al (2011) Regulation of heart mitochondrial nitric oxide synthase (mtNOS) by oxygen. In: Cadenas S (ed) Mitochondrial pathophysiology. Transworld Research Network, Kerala, pp 29–42

20. Yin F, Boveris A, Cadenas E (2014) Mitochondrial energy metabolism and redox signaling in brain aging and neurodegeneration. Antioxid Redox Signal 20:353–371
21. Yin F, Cadenas E (2015) Mitochondria: the cellular hub of the dynamic coordinated network. Antioxid Redox Signal 22:961–964
22. Boveris A, Valdez LB, Zaobornyj T, Bustamante J (2006) Mitochondrial metabolic states regulate nitric oxide and hydrogen peroxide diffusion to the cytosol. Biochim Biophys Acta 1757:535–542
23. Tatoyan A, Giulivi C (1998) Purification and characterization of a nitric oxide synthase from rat liver mitochondria. J Biol Chem 273:11044–11048
24. Giulivi C, Poderoso JJ, Boveris A (1998) Production of nitric oxide by mitochondria. J Biol Chem 273:11038–11043
25. Elfering SL, Sarkela TM, Giulivi C (2002) Biochemistry of mitochondrial nitric oxide synthase. J Biol Chem 277:38079–38086
26. Valdez LB, Zaobornyj T, Boveris A (2006) Mitochondrial metabolic states and membrane potential modulate mtNOS activity. Biochim Biophys Acta 1757:166–172
27. Valdez LB, Alvarez S, Lores-Arnaiz S et al (2000) Reactions of peroxynitrite in the mitochondrial matrix. Free Radic Biol Med 29:349–356
28. Franco MC, Arciuch VG, Peralta JG et al (2006) Hypothyroid phenotype is contributed by mitochondrial complex I inactivation due to translocated neuronal nitric oxide synthase. J Biol Chem 281:4779–4786
29. Parihar MS, Nazarewicz RR, Kincaid E et al (2008) Association of mitochondrial nitric oxide synthase activity with respiratorty chain complex I. Biochem Biophys Res Commun 366:23–28
30. Parihar MS, Parihar A, Villamena FA et al (2008) Inactivation of mitochondrial respiratory chain complex I leads mitochondrial nitric oxide synthase to become pro-oxidative. Biochem Biophys Res Commun 367:761–767
31. Valdez LB, Boveris A (2007) Mitochondrial nitric oxide synthase, a voltage-dependent enzyme, is responsible for nitric oxide diffusion to cytosol. Front Biosci 12:1210–1219
32. Kissner R, Nauser T, Bugnon P et al (1997) Formation and properties of peroxynitrite as studied by laser flash photolysis, high-pressure stopped-flow technique and pulse radiolysis. Chem Res Toxicol 10:1285–1292
33. McCord JM, Fridovich I (1969) Superoxide dismutase. An enzymatic function for erythrocuprein (hemocuprein). J Biol Chem 244:6049–6055
34. Poderoso JJ, Carreras MC, Schöpfer F et al (1999) The reaction of nitric oxide with ubiquinol: kinetic properties and biological significance. Free Radic Biol Med 26:925–935
35. Antunes F, Boveris A, Cadenas E (2004) On the mechanism and biology of cytochrome oxidase inhibition by nitric oxide. Proc Natl Acad Sci U S A 101:16774–16779
36. Boveris A, Stoppani AO (1970) Inhibition of electron and energy transfer in mitochondria by 19-nor-ethynyltestosterone acetate. Arch Biochem Biophys 141:641–655
37. Costa LE, Boveris A, Koch OR, Taquini AC (1988) Liver and heart mitochondria in rats submitted to chronic hypobaric hypoxia. Am J Physiol 255:C123–C129
38. Venardos KM, Zatta AJ, Marshall T et al (2009) Reduced L-arginine transport contributes to the pathogenesis of myocardial ischemia-reperfusion injury. J Cell Biochem 108:156–168
39. Marban E, Kitakaze M, Koretsune Y et al (1990) Quantification of [Ca^{2+}] in perfused hearts: critical evaluation of 5F-BAPTA and the nuclear magnetic resonance method as applied for the study of ischemia and reperfusion. Circ Res 66:1255–1267
40. Ogawa T, Miura T, Kazuaki S, Limura OJ (1996) Activation of adenosine receptors before ischemia enhances tolerance against myocardial stunning in the rabbit heart. J Am Coll Cardiol 27:225–233
41. Boveris A, Carreras MC, Poderoso JJ (2010) The regulation of cell energetics and mitochondrial signaling by nitric oxide. In: Ignarro L (ed) Nitric oxide. Elsevier Academic Press, London, pp 441–482
42. Chen Q, Moghaddas S, Hoppel CL, Lesnefsky EJ (2006) Reversible blockade of electron transport during ischemia protects mitochondria and decreases myocardial injury following reperfusion. J Pharmacol Exp Ther 319:1405–1412

43. Shiva S, Sack MN, Greer JJ et al (2007) Nitrite augments tolerance to ischemia/reperfusion injury via the modulation of mitochondrial electron transfer. J Exp Med 204:2089–2102
44. Duranski MR, Greer JJM, Dejam A et al (2005) Cytoprotective effects of nitrite during in vivo ischemia–reperfusion of heart and liver. J Clin Invest 115:232–1240
45. Prime TA, Blaikie FH, Evans C et al (2009) A mitochondria-targeted S-nitrosothiol modulates respiration, nitrosates thiols, and protects against ischemia-reperfusion injury. Proc Natl Acad Sci U S A 106:10764–10769
46. Chouchani ET, Methner C, Nadtochiy SM et al (2013) Cardioprotection by S-nitrosation of a cysteine switch on mitochondrial complex I. Nature Med 19:753–759
47. Hurd TR, Costa NJ, Dahm CC et al (2005) Glutathionylation of mitochondrial proteins. Antioxid Redox Signal 7:999–1010
48. Mailloux RJ, Willmore WG (2014) S-glutathionylation reactions in mitochondrial function and disease. Front Cell Dev Biol http://dx.doi.org/10.3389/fcell.2014.00068
49. Chang AH, Sancheti H, Garcia J et al (2014) Respiratory substrates regulate S-nitrosylation of mitochondrial proteins through a thiol-dependent pathway. Chem Res Toxicol 27:794–804
50. Drose S, Brandt U, Wittig I (2014) Mitochondrial respiratory chain complexes as sources and targets of thiol-based redox-regulation. Biochim Biophys Acta 1844:1344–1354
51. Hurd TR, Requejo R, Filipovska A et al (2008) Complex I within oxidatively stressed bovine heart mitochondria is glutathionylated on Cys-531 and Cys-704 of the 75-kDa subunit: potential role of CYS residues in decreasing oxidative damage. J Biol Chem 283:24801–24815
52. Chen J, Chen CL, Rawale S et al (2010) Peptide-based antibodies against glutathione-binding domain suppress superoxide production mediated by mitochondrial complex I. J Biol Chem 285:3168–3180
53. Kumar V, Kleffmann T, Hampton MB et al (2013) Redox proteomics of thiol proteins in mouse heart during ischemia/reperfusion using ICAT reagents and mass spectrometry. Free Radic Biol Med 58:109–117

Chapter 17
Reactive Oxygen Species Are Involved in Myocardial Remote Ischemic Preconditioning

Martín Donato, Diamela T. Paez, Pablo Evelson, and Ricardo Jorge Gelpi

Abstract Ischemic heart disease is the leading cause of death worldwide. There has been a continued search for better therapeutic strategies that would reduce myocardial ischemia/reperfusion injury. Remote ischemic preconditioning (rIPC) was first introduced in 1993 by Przyklenk et al who reported that brief regional occlusion-reperfusion episodes in one vascular bed of the heart render protection to remote myocardial tissue. Subsequently, different studies have showed that rIPC applied to the kidney, liver, mesentery, and skeletal muscle, have all exhibited cardioprotective effects. The main purpose of this chapter is to summarize the advances in understanding the molecular mechanisms of rIPC, including those related to oxidative stress. Detailed understanding of the pathways involved in cardioprotection induced by rIPC is expected to lead to the development of new drugs to reduce the consequences of prolonged ischemia.

Keywords Myocardial infarction • Remote preconditioning • Cardioprotection

---

M. Donato • R.J. Gelpi (✉)
Institute of Cardiovascular Pathophysiology (INFICA), Department of Pathology, Faculty of Medicine, University of Buenos Aires, Buenos Aires, Argentina

Institute of Biochemistry and Molecular Medicine (IBIMOL, UBA-CONICET), Faculty of Medicine, University of Buenos Aires, Buenos Aires, Argentina
e-mail: rgelpi@fmed.uba.ar

D.T. Paez
Institute of Cardiovascular Pathophysiology (INFICA), Department of Pathology, Faculty of Medicine, University of Buenos Aires, Buenos Aires, Argentina

P. Evelson
Institute of Biochemistry and Molecular Medicine (IBIMOL, UBA-CONICET), Faculty of Pharmacy and Biochemistry, University of Buenos Aires, Buenos Aires, Argentina

© Springer International Publishing Switzerland 2016
R.J. Gelpi et al. (eds.), *Biochemistry of Oxidative Stress*, Advances in Biochemistry in Health and Disease 16, DOI 10.1007/978-3-319-45865-6_17

1 Introduction

Ischemic heart disease is the leading cause of death and disability worldwide. Therefore, novel therapeutic strategies are required to protect the heart against ischemia/reperfusion injury, preserve myocardial function, prevent heart failure, and improve clinical outcomes in patients with ischemic heart disease.

In 1993, Przyklenk et al. [1] demonstrated that brief cycles of ischemia/reperfusion of the circumflex coronary artery protected remote virgin myocardium in the left anterior descending coronary artery territory. This observation extended the concept of classical ischemic preconditioning (IPC) described by Murry et al. [2] in 1986, to protect the heart at a distance or 'remote ischemic preconditioning' (rIPC). The rIPC phenomenon has been described in different organs and tissues such as kidney and brain, emerging as a strategy of inter-organ protection against the effects of acute ischemia/reperfusion injury.

Thus, in the last 20 years, the concept of rIPC has evolved from being an experimental observation, whose underlying mechanisms continue to elude investigators, to a clinical application which offers the therapeutic potential to benefit patients with ischemic heart disease. Despite intensive investigation the mechanisms underlying rIPC remain unclear. The current hypothesis divides the mechanistic pathway of rIPC into three components (Fig. 17.1):

1. Remote organ or tissue: rIPC stimulates the release of autacoids within the remote organ or tissue that activates a local afferent neural pathway [3].
2. The connecting pathway: Different authors have described that the cardioprotective signal transference from the remote organ to the heart is the result of the action of humoral factors [4], neural pathways [5], or a neurohumoral interaction [6]. These hypotheses are not mutually exclusive and are probably part of the same mechanisms.
3. Target organ: The protective factor activate an intracellular signaling pathways in the target organ which mediates the rIPC protective effects.

2 Neural Pathway of rIPC

The involvement of a neurogenic pathway in remote cardioprotection has been demonstrated by different authors [7, 8]. Pretreatment with the ganglion blocker hexamethonium abolished remote cardioprotection in rats through 15 min of mesenteric arterial occlusion [7]. Furthermore, rIPC activates a neural afferent pathway (Femoral and sciatic nerves and spinal cord) and the cardioprotective signal reaches the heart through the vagus nerve (efferent pathway) [9].

Thus, experimental and clinical studies have demonstrated that rIPC protection is dependent on an intact neural pathway to the remote organ or tissue with local resection of the neural pathway abolishing rIPC. The current paradigm has proposed that in response to the rIPC stimulus, autacoids such as adenosine [10],

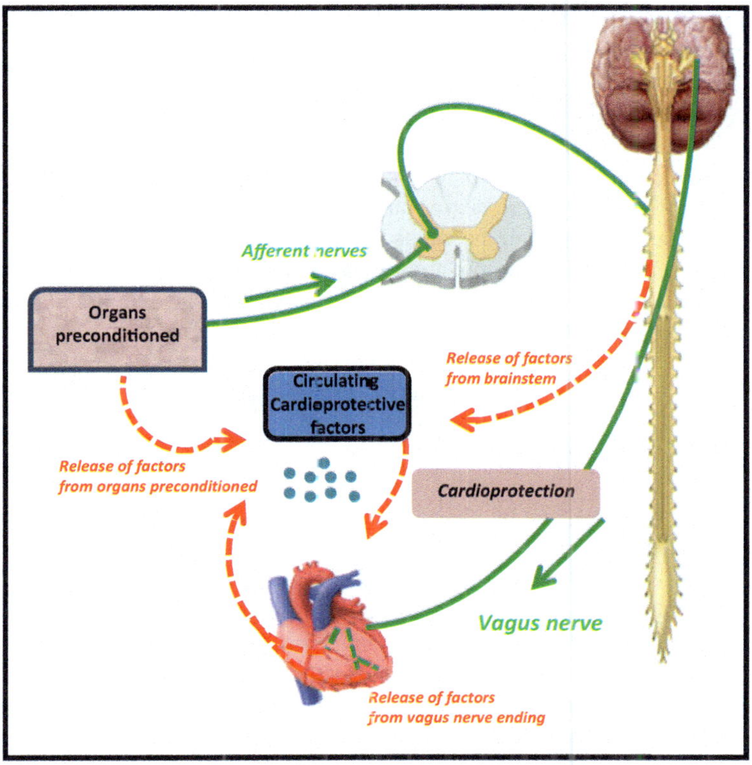

Fig. 17.1 This figure shows the link between the neural pathway (*green solid lines*) and humoral pathway (*broken red lines*) in the mechanism of rIPC. Cycles of brief ischemia/reperfusion induced the local release of factors, which then activate local sensory afferent neurons. A study shown the participation of the neurons in the dorsal motor vagal nucleus (DMVN) in rIPC mechanism, this provides parasympathetic innervation of the left ventricle. The potential sites of cardioprotective factor(s) release include: (1) from the conditioned limb itself, (2) from the central nervous system, (3) from pre-/post-ganglionic parasympathetic nerve endings within the heart (*broken green lines*); and (4) from a non-conditioned remote organ/tissue receiving parasympathetic innervation

bradykinin [11] are produced in the remote organ or tissue resulting in the nitric oxide dependent stimulation of local afferent sensory nerves. In this sense, the administration of adenosine into the femoral artery resulted in the production of a cardioprotective plasma dialysate, in patients undergoing coronary angiography, confirming the findings in experimental animal studies that adenosine acted as a "trigger" for limb rIPC. On the other hand, it has suggested that the sensory arm of the neural pathway leading from the remote organ or tissue may be recruited by the activation of Transient receptor potential vanilloid (TRPV) [12]; and different experimental studies have demonstrated that the activation of these fibers by topical capsaicin or nociceptive stimuli can mimic the rIPC cardioprotection. However, the neural components of the pathway downstream of this sensory afferent neural via in the remote organ or tissue remain unclear. Jones et al. [13] described that cardioprotection elicited by peripheral nociception was blocked by spinal transection at T7 but not C7, suggesting that direct stimulation of cardiac nerves may be responsible for conveying the cardioprotective signal to the heart. Basalay et al demonstrated that rIPC activates a neural pathway, and the signal reaches the heart through the vagus nerves [14]. As we mentioned, and in accordance with the pioneer findings of the Gourine's group, we showed that rIPC activates a neural afferent pathway (Femoral and sciatic nerves and spinal cord) and the cardioprotective signal reaches the heart through the vagus nerve (efferent pathway) and acetylcholine activates the classic ischemic preconditioning (IPC) phenomenon when acting on the muscarinic receptors [9]. However, the remotely activated signal transductions participating in the rIPC intracellular mechanism remain unclear.

3 Humoral Pathway of rIPC

The earliest experimental evidence for a blood-borne cardioprotective factor released by rIPC was provided in 1999 by Dickson et al. [4], who demonstrated that the cardioprotection elicited by IPC could be transferred via blood transfusion to a non-preconditioned animal. The existence of a circulating cardioprotective factor was first demonstrated in a model of porcine transplantation [15], where hind-limb preconditioning in an acceptor animal provided a significant cardioprotection to the subsequently transplanted and denervated donor heart. A similar type of study was performed by Kristiansen et al. [16] who demonstrated that hearts excised from a rat that had been received a rIPC protocol experienced a smaller infarct size, when it was subjected to a prolonged episode of ischemia and reperfusion on a Langendorff system. Subsequent studies confirmed the presence of a circulating factor and postulated a number of candidate, that including calcitonin gene-related peptide [17], opioids [18], endogenous cannabinoids [19], and hypoxia inducible factor-1α (HIF-1α) [20].

Although the actual identity of the factor remains unclear, some authors suggested that the factor may be a peptide <30 kDa [21]. In an isolated rabbit heart model [22], plasma from remotely preconditioned animals was cardioprotective when perfused into an isolated naïve heart. The authors concluded that the factor molecular weight is <15 kDa. Alternatively, the endogenous mediator may activate an afferent neural pathway within the remote organ to confer cardioprotection, as is the case with adenosine, bradykinin and CGRP.

Novel candidates for the blood-borne cardioprotective factors of rIPC were proposed, each with varying degrees of experimental evidence: including (1) stromal derived factor-1α or SDF-1α [23]; (2) exosomes [24]; nitrite [25]; (3) microRNA-144 [26]; (4) HIF-1α [20] Apolipoprotein α-I [27]. However, these studies have failed to demonstrate that the cardioprotective factor was actually responsible for the beneficial effect.

4 How Do the Neural and Humoral Pathways Interact to Mediate rIPC?

The neural and humoral pathways of rIPC could interact to mediate the protective effect, however the underlying mechanism of this relationship has not been describe (see Fig. 17.1 for a hypothetical scheme). Studies from Redington's and Botker's groups have suggested the link between these two pathways in the setting of rIPC. They use plasma dialysate harvested from animals or humans treated with rIPC to demonstrated a reduction in infarct size in a naïve animal hearts.

Redington et al. [28] have shown that the cardioprotective plasma dialysate can be produced in animals and human in response to neural stimulation that include direct nerve stimulation, transcutaneous electrical nerve stimulation, electro-acupuncture and topical capsaicin Botker et al. [29] demonstrated that diabetic patients with a peripheral sensory neuropathy do not produce the cardioprotective plasma dialysate in response to rIPC protocol, when compared to diabetic patients without sensory neuropathy. Therefore, all this evidence suggests that the cardio-protective factor is produced downstream of the neural pathway. However, the question is: Where along the neural pathway is the cardioprotective factor released into the blood?; and which cell is responsible for its release?.

5 Myocardial Mechanisms of Remote Ischemic Preconditioning

Once the cardioprotective signal has been conveyed from the remote preconditioning organ to the heart, intracellular signal transduction mechanisms are recruited within cardiomyocytes. In this sense, some authors [6, 30] suggested that rIPC activates a signalling mechanism similar to that described for ischemic preconditioning (IPC), while others show that the cardioprotection conferred by rIPC follows a different pattern [31]. Recently, Heusch G carefully reviewed the signal transduction pathways involved in the different ischemic conditioning phenomena and noted that there are still some unsolved problems when studying the myocardial protection. Particularly, the absence of a temporal description of the cardioprotective signals involved [32].

A few years ago, Downey et al. proposed a classification of the signals participating in IPC that follows a logical/causal sequence of events and meets the temporal sequence of the preconditioning protocol [33, 34]. They defined a trigger as a factor released during the preconditioning ischemia periods that activates the cardioprotection phenomenon, and defined mediators as factors that transmit the cardioprotective signal during the prolonged myocardial ischemia to one or more end effectors, which are responsible for attenuating the irreversible injury during the lethal ischemic insult and/or during the subsequent reperfusion period. From this point of view, several studies evaluated the intracellular signalling pathway involved

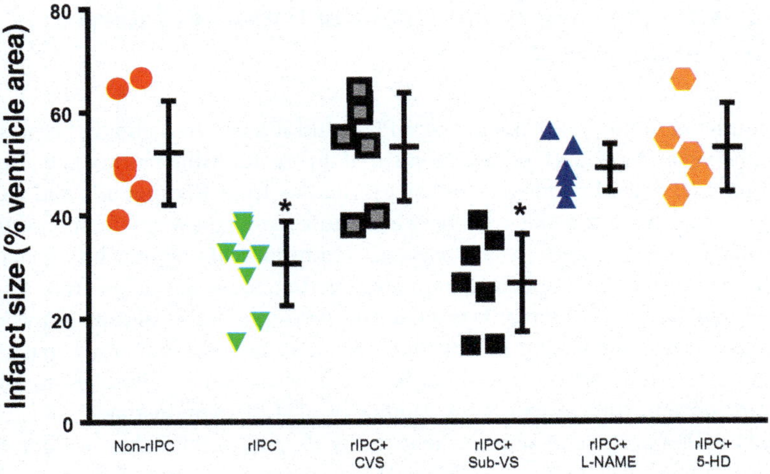

Fig. 17.2 Infarct size, expressed as a percentage of the left ventricular area. The rIPC significantly reduced the infarct size and this effect was abolished by the different treatments (CVS, L-NAME, 5-HD). *rIPC* remote ischemic preconditioning, *CVS* cervical vagal section, *Sub-VS* sub-diaphragmatic vagal section.*$p < 0.05$ vs. each other group

in the rIPC cardioprotection [35–37]. However, these authors have not considered the timing sequence of signalling activation.

In a recently study [38], we evaluated the signaling pathway that is activated at heart level, before myocardial ischemia. We hypothesized that pre-ischemic activation of muscarinic receptors induces Akt phosphorylation and through this pathway, the phosphorylation of the endothelial nitric oxide synthase (eNOS). As a consequence of a higher production of nitric oxide (NO), mitochondrial K^+_{ATP} channels (mK^+_{ATP}) open and mitochondrial production of hydrogen peroxide (H_2O_2) increases.

In order to demonstrate this hypothesis, we performed experiments in isolated rat hearts. As we expected, rIPC (Fig. 17.2) significantly reduce infarct size. The cervical vagal section (CVS) completely abolished the beneficial effect of rIPC, however the Sub-diaphragmatic vagal section (Sub-VS) did not modify the rIPC effect, thus demonstrating that denervation of other organs, different from the heart; do not contribute to the loss of rIPC.

Since activation of muscarinic receptors can increase NO synthesis, we studied a possible involvement of NO in the observed infarct size reduction afforded by the rIPC. In this regard, administration of L-NAME, during the rIPC protocol, completely abolished the protective effect of rIPC, pointing out a central role of NO in the rIPC cardioprotection. Given that NO could induce the mK^+_{ATP} channels to open, we administered 5-HD before the rIPC protocol. The mK^+_{ATP} channels blocker completely abolished the effect of rIPC, thus providing evidence of an involvement of mK^+_{ATP} channels in the rIPC.

Activation of the muscarinic receptors induces phosphorylation of Akt enzyme [39]. To address this issue in a rIPC group phosphrylation of Akt in cardiac tissue

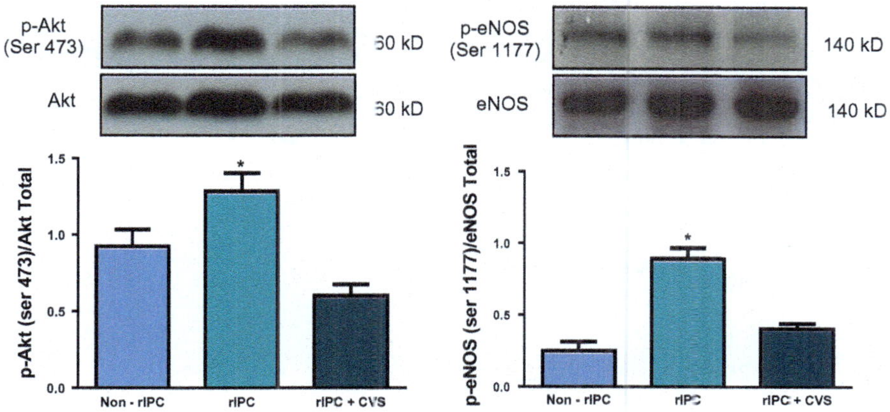

Fig. 17.3 In *Panel A*, cardiac expression of phosphorylated Akt (Ser 473) can be observed in the Non-rIPC, rIPC and the rIPC + CVS groups, immediately after hindlimb rIPC. The rIPC induced a significant increase of the Akt phosphorylation, which was abolished by CVS. In *Panel B* of the same figure, expression of the phosphorylated eNOS (Ser 1177) can be observed. The rIPC induced a significant increase of the phosphorylation of this enzyme, which was abolished by CVS. $*p < 0.05$. *rIPC* remote ischemic preconditioning, *CVS* cervical vagal section

was determined at the end of the hindlimb preconditioning protocol (Fig. 17.3, Panel A). Importantly, rIPC induced a significant increase of the cardiac Akt phosphorylation, which was abolished by the CVS, in hearts that were not yet subjected to ischemia/reperfusion.

Since eNOS (Ser-1177) can be phosphorylated by Akt, we studied phosphor-eNOS expression. In these experiments, rIPC induced a significant increase in the phosphorylation of this enzyme, which was abolished by CVS, before the myocardial ischemia (Fig. 17.3, Panel B). Taken together, these results clearly indicate involvement of the Akt-eNOS pathway in the heart as triggers of the rIPC mechanism before the ischemia/reperfusion cardiac insult.

Reactive oxygen species (ROS) have been shown to be toxic but also function as signalling molecules. It has been suggested that mitochondrial ROS production might play a relevant role in IPC [40], but this issue was not addressed in rIPC. Figure 17.4 (Panel A) shows a representative trace during an initial stabilization period of the reaction mixture and after the addition of isolated mitochondria from the following groups: Non-rIPC and rIPC. As it can be seen, an increased H_2O_2 release in the rIPC group is evident. Panel B shows that CVS, L-NAME and the mK^+_{ATP} channels blockade with 5-HD attenuated the H_2O_2 release.

Clearly, rIPC induces activation of Akt enzyme and eNOS phosphorylation, mK^+_{ATP} channels opening, and mitochondrial H_2O_2 production in the heart before the index myocardial ischemia. Therefore, they could be considered as rIPC triggers. In addition, the protective effect of rIPC was abolished by CVS but not by SVS, reinforcing the hypothesis of a parasympathetic vagal pathway.

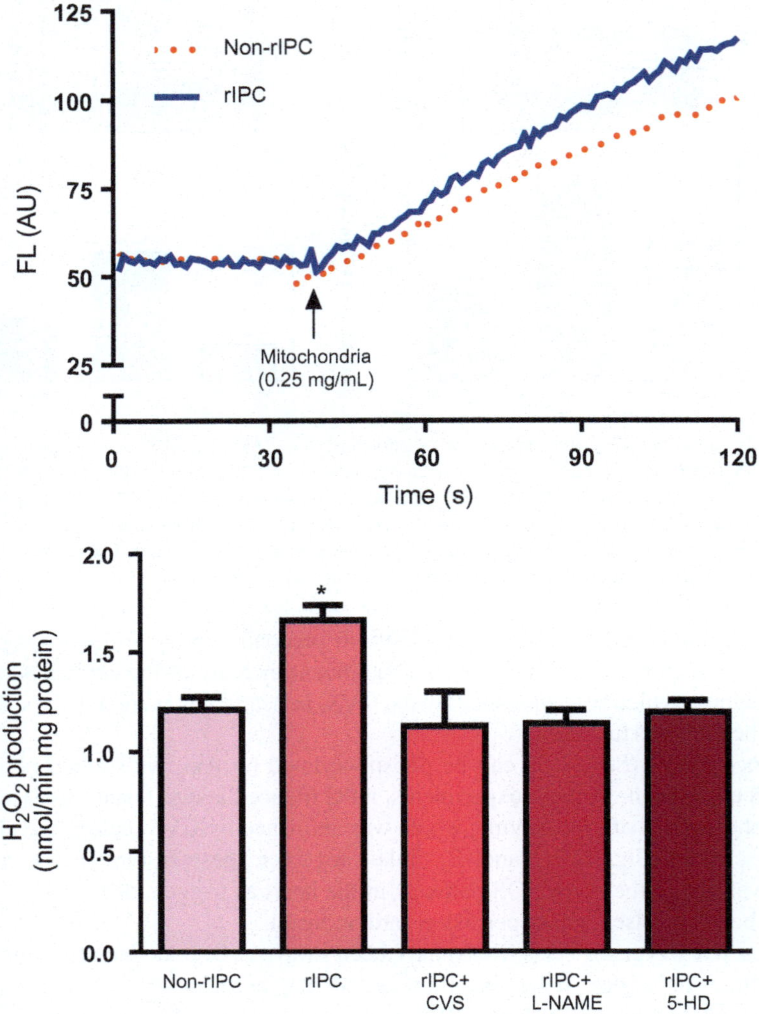

Fig. 17.4 *Panel A* shows a representative trace during an initial stabilization period and after the addition of isolated mitochondria from Non-rIPC or rIPC rat hearts. *Panel B* shows the mitochondrial H_2O_2 production rate mean in the different study groups. CVS, L-NAME and 5-HD abolished the mitochondrial H_2O_2 production rate. *rIPC* remote ischemic preconditioning, *CVS* cervical vagal section. *FL* Fluorescence intensity. *$p < 0.05$

Different authors have evaluated the possible intracellular mechanisms involved in rIPC pathway. However, most of these works have studied mechanisms at time points that are different from the ones used in this study: during early reperfusion [41], in late reperfusion [42], and also using different animal species [9, 43, 44], making difficult to establish a comparison. In agreement with our findings, Li SJ et al showed that rIPC induces the activation of the PI3K/Akt and inhibition of

GSK3β before myocardial ischemia [35]. However, the authors proposed that rIPC would induce release of a factor to the bloodstream, which activates the PI3K/Akt/GSK3β pathway at cardiac level, suggesting a myocardial mechanism similar to IPC. The results of our study support the existence of a vagal efferent pathway, and they extend the current knowledge by showing that rIPC could activate the Akt pathway. Furthermore, the denervation of other organs, different from the heart; do not contribute to the loss of rIPC, since that SVS did not abolish the rIPC cardioprotection.

Akt phosphorylation would lead to activation of the eNOS enzyme and, subsequently, to mK^{+}_{ATP} channels opening and mitochondrial changes. The role of NO in the rIPC mechanism is difficult to evaluate because it could participate in the tissue or organ where the preconditioning stimulus is originated and/or at cardiac level as part of the cardioprotection signalling. In this last case, it is well known that the infusion of a drug able to increase the NO bioavailability puts the heart into a preconditioned state [45]. Our results demonstrate that the rIPC activation involves eNOS, since by administering L-NAME we abolished the protective effect, but more important they demonstrate that there is a cardiac eNOS phosphorylation immediately after the rIPC protocol, before myocardial ischemia. In addition, an increase in the NO production is capable of acting directly on the mitochondria inhibiting the respiratory chain and favoring ROS production [46], or activates the cGMP/PKG pathway, which was involved in the IPC cardioprotection mechanism.

Several authors showed that IPC induces mK^{+}_{ATP} opening [47] however this fact has not been evaluated for rIPC. Our results shown that the administration of 5-HD abolished the rIPC protective effect, thus demonstrating an important role of these channels in rIPC. The mK^{+}_{ATP} opening produces a higher K^{+} influx to the mitochondria, decreasing the mitochondrial membrane potential and leading to an increase of H_2O_2 mitochondrial production [40].

6 Role of Reactive Oxygen Species in rIPC Protection

ROS are thought to mediate the oxygen toxicity because of their greater chemical reactivity with regard to oxygen. They also operate as intracellular signalling molecules, a function that has been widely documented but is still controversial. In the setting of acute myocardial ischemia/reperfusion injury, oxidative stress plays a dual role. Its detrimental role is as a mediator of lethal reperfusion injury, however, its beneficial signalling role is believed to mediate the cardioprotective effects elicited by both IPC and postconditioning [48].

As mention, three components of rIPC can be distinguished: the signal generation, the transfer of the signal to the target organ, and its response to the transferred signal resulting in cardioprotection.

Weinbrenner et al. suggested a possible beneficial signalling role for ROS in the setting of rIPC. They showed that the administration of a free radical scavenger abolished the protection elicited by rIPC [49]. These results are in agreement with our findings in which rIPC induces a higher mitochondrial H_2O_2 production before the myocardial ischemia, and this effect is attenuates by the inhibition of the NO production with L-NAME and with the blockade of mK^{+}_{ATP} channels with 5-HD.

In this regard, H_2O_2 could act as a second messenger of the rIPC protective signal. Besides, ROS production by the respiratory chain has been shown to activate mK^{+}_{ATP} [50]. Furthermore, ROS generated by the mitochondria could activate other sensitive redox enzymes, among them, the PKCε, which is one of the most important kinases participating in the IPC mechanism [51].

It has been described in humans that a short episode of forearm ischemia/reperfusion performed with a blood pressure cuff is capable of modulating the composition of nitric oxide (NO•)/nitrite levels in the blood [52, 53]. Nitrite is not only the oxidation product of NO• but also a key reservoir for NO• in blood and cellular compartments [54, 55]. The half-life of nitrite in plasma is unknown but calculated for humans to be approximately 35 min [56].

In an elegant study, Rassaf et al. [57] evaluated in healthy volunteers whether the ischemic phase or the reactive hyperemia with the resulting shear stress during rIPC are responsible for the modulation of plasma nitrite levels. They determined that endothelial eNOS is responsible for nitrite generation during reactive hyperemia, which is then transported to the myocardium. In additional experiments, the authors assessed the response of the target organ to the transferred signal. Using myoglobin (Mb) knockout mouse, they demonstrated that the nitrite generated during reactive hyperemia is converted to bioactive NO• with subsequent modification of mitochondria by S-nitrosation, which then ultimately confer the cardioprotective effects [58]. Transfer experiments of plasma from healthy volunteers subjected to rIPC of the arm identified plasma nitrite as a cardioprotective agent in isolated Langendorff mouse heart preparations exposed to ischemia/reperfusion.

Finally, a higher H_2O_2 mitochondrial production before ischemia would protect the heart against an exacerbated production of ROS during reperfusion [39], being another important underlying mechanism of the rIPC protective effect. A scheme combining our results and current knowledge regarding the intracellular mechanisms activated by rIPC in the heart is depicted in Fig. 17.5.

On the other hand, there is experimental evidence describing the protective properties of NO, despite the limitations of endocrine movement to a remote site. Endogenous NO seems to play a pivotal role in mechanism of rIPC in reducing liver damage, and this is abrogated by treatments with the NO scavenger carboxy-2-phenyl-4,4,5,5-tetramethylimidazoline-1-oxyl-3-oxide (cPTIO) and inhibited in the eNOS knockout mouse [59]. Tokuno et al. [55] have involved iNOS activation as a trigger for delayed rIPC of the heart using cerebral ischemia as preconditioning stimulus. The cardioprotective effect was seen 24 h later and was absent in iNOS knockout mice. Further studies demonstrated that NO is necessary for the development of ischemia-induced delayed protection against myocardial infarction [60]. Although it is clear that NOS and NO seem to participate in the process of rIPC, the mechanism for NO transport to a distant site and the nature of the endocrine rIPC mediator have remained unknown.

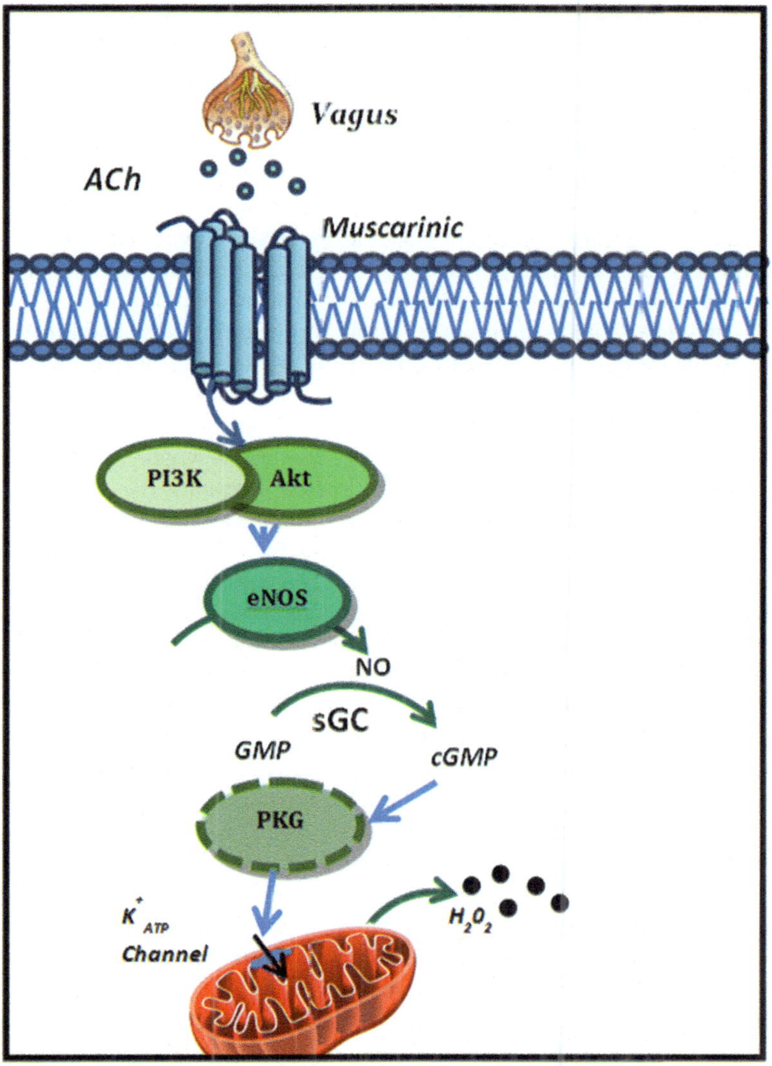

Fig. 17.5 Schematic illustration of the intracellular pathways activated by remote ischemic preconditioning before myocardial ischemia. Acetylcholine, released from cardiac vagal nerve endings, activates muscarinic receptors located in the cardiomyocyte plasma membrane, inducing the phosphorylation of Akt and eNOS enzymes. Subsequently, activation of soluble guanylate cyclase and protein kinase G could lead to mK^{+}_{ATP} channels opening and increasing of H_2O_2 mitochondrial production. This way, H_2O_2 could act as a second messenger of the rIPC protective signal. *Ach* acetylcholine, *eNOS* endothelial nitric oxide synthase, *NO* nitric oxide, *sGC* soluble guanylate cyclase, *PKG* protein kinase G, mK^{+}_{ATP} mitochondrial K^{+}_{ATP} channels

Rassaf et al. [57] evaluated the mechanism of rIPC and explore the possible identity of the circulating endocrine mediator. They first find in humans that the levels of plasma nitrite increase after brachial artery occlusion and release (reactive hyperemia). This is caused by eNOS activation with NO formation and oxidation to the more stable nitrite. The authors performed studies in mice and showed that nitrite levels increase. Inhibition of NO with cPTIO or in eNOS knockout mice prevents the rise in nitrite and rIPC effects on myocardial infarction. This association was confirmed by infusions of nitrite to match levels observed with rIPC. Finally, they infuse human plasma with and without rIPC into the isolated heart model of ischemia/reperfusion and showed that the increase nitrite account for effects.

Although these studies suggest that nitrite forms during rIPC and travels in the plasma to the heart, how is it then converted in the heart back into NO?. During ischemia, nitrite is reduced to NO and N_2O_3 by different nitrite reductase enzyme systems [61, 62]. Mitochondrial NO and S-nitrosothiols formed from nitrite dynamically and reversibly inhibit complex I during reperfusion, which limits ROS formation from complexes I and III. This ultimately prevents the opening of the mitochondrial permeability transition pore and the release of cytochrome c. It has been shown that the site of nitrosation is on cysteine 39 of the ND3 (NADH dehydrogenase, subunit 3) subunit of complex I. Several enzymes are required to convert nitrite into NO during organ ischemia. In the heart, deoxygenated myoglobin acts as a functional nitrite reductase. Nitrite-dependent NO formation is significantly decreased in myoglobin-deficient hearts and nitrite administration reduces myocardial infarction with abrogated effects in the myoglobin knockout mice. Rassaf et al. [57] showed that the effect of rIPC is inhibited in the myoglobin knockout mouse, providing additional evidence that the mediator factor of this effect is nitrite, which is produced in the extremity and travels in blood to the heart, where it is reduced by myoglobin to produce NO.

In a different experimental model of spinal cord ischemic injury, Dong et al. [63] showed that the beneficial effect of limb rIPC was attenuated by administration of a free-radical scavenger before rIPC. Additional, rIPC induced an increase in the activity of catalase and superoxide dismutase (SOD) in the serum. The increase in catalase and SOD activities was accompanied by a transient increase of serum malondialdehyde levels. The free radicals scavenger treatment abolished the increase in catalase and SOD activity induced by rIPC. This indicates that the increase in antioxidant enzyme activities is closely related to the ROS generated by rIPC.

7 Conclusions

Strong experimental evidence supports protection by rIPC from myocardial ischemia/reperfusion injury and other organs. However the mechanisms for local release of the protective signal at the remote site and the contributions of neuronal and humoral pathways are not yet clear, not only in signal release, but also in signal transfer to the target organ and protective signal transduction within the target organ. Thus, in this chapter we summarized the more recent advances in the molecular mechanisms of rIPC, particularly those related to oxidative stress.

A better understanding of the complex signaling involved in transduction of the rIPC signal from the remote organ and tissue to the protected target may allow the imminent discovery of novel pharmacological agents to directly activate the protective signaling pathways.

References

1. Przyklenk K, Bauer B, Ovize M, Kloner RA et al (1993) Regional ischemic 'preconditioning' protects remote virgin myocardium from subsequent sustained coronary occlusion. Circulation 87:893–899
2. Murry CE, Jennings RB, Reimer KA (1986) Preconditioning with ischemia: a delay of lethal cell injury in ischemic myocardium. Circulation 74:1124–1136
3. Wever KE, Warlé MC, Wagener FA, Van der Hoorn JW et al (2011) Remote ischaemic preconditioning by brief hind limb ischaemia protects against renal ischaemia-reperfusion injury: the role of adenosine. Nephrol Dial Transplant 26:3108–3117
4. Dickson EW, Lorbar M, Porcaro WA, Fenton RA et al (1999) Rabbit heart can be "preconditioned" via transfer of coronary effluent. Am J Physiol 277:H2451–H2457
5. Mastitskaya S, Marina N, Gourine A, Gilbey MP et al (2012) Cardioprotection evoked by remote ischaemic preconditioning is critically dependent on the activity of vagal pre-ganglionic neurones. Cardiovasc Res 95:487–494
6. Hausenloy DJ, Yellon DM (2008) Remote ischaemic preconditioning: underlying mechanisms and clinical application. Cardiovasc Res 79:377–386
7. Kingma JG Jr, Simard D, Voisine P, Rouleau JR (2011) Role of the autonomic nervous system in cardioprotection by remote preconditioning in isoflurane-anaesthetized dogs. Cardiovasc Res 89:384–391
8. Steensrud T, Li J, Dai X, Manlhiot C, Kharbanda RK et al (2010) Pretreatment with the nitric oxide donor SNAP or nerve transection blocks humoral preconditioning by remote limb ischemia or intra-arterial adenosine. Am J Physiol Heart Circ Physiol 299:H1598–H1603
9. Donato M, Buchholz B, Rodríguez M, Pérez V et al (2013) Role of the parasympathetic nervous system in cardioprotection by remote hindlimb ischaemic preconditioning. ExpPhysiol 98:425–434
10. Takaoka A, Nakae I, Mitsunami K, Yabe T et al (1999) Renal ischemia/reperfusion remotely improves myocardial energy metabolism during myocardial ischemia via adenosine receptors in rabbits: effects of "remote preconditioning". J Am Coll Cardiol 33:556–564
11. Schoemaker RG, van Heijningen CL (2000) Bradykinin mediates cardiac preconditioning at a distance. Am J Physiol Heart Circ Physiol 278:H1571–H1576
12. Pickard J, Bøtker H, Crimi G, Davidson B et al (2015) Remote ischemic conditioning: from experimental observation to clinical application: report from the 8th Biennial Hatter Cardiovascular Institute Workshop. Basic Res Cardiol 110:453–467
13. Jones WK, Fan GC, Liao S, Zhang JM et al (2009) Peripheral nociception associated with surgical incision elicits remote nonischemic cardioprotection via neurogenic activation of protein kinase C signaling. Circulation 120(11 Suppl):S1–S9
14. Basalay M, Barsukevich V, Mastitskaya S, Mrochek A et al (2012) Remote ischaemic pre- and delayed postconditioning similar degree of cardioprotection but distinct mechanisms. Exp Physiol 97:908–917

15. Skyschally A, Gent S, Amanakis G, Schulte C et al (2015) Across-species transfer of protection by remote ischemic preconditioning with species-specific myocardial signal transduction by reperfusion injury salvage kinase and survival activating factor enhancement pathways. Circ Res 117:279–288
16. Kristiansen SB, Henning O, Kharbanda RK, Nielsen-Kudsk JE et al (2005) Remote preconditioning reduces ischemic injury in the explanted heart by a KATP channel-dependent mechanism. Am J Physiol Heart Circ Physiol 288:H1252–H1256
17. Wolfrum S, Nienstedt J, Heidbreder M, Schneider K et al (2005) Calcitonin gene related peptide mediates cardioprotection by remote preconditioning. Regul Pept 127:217–224
18. Patel HH, Moore J, Hsu AK, Gross GJ (2002) Cardioprotection at a distance: mesenteric artery occlusion protects the myocardium via an opioid sensitive mechanism. J Mol Cell Cardiol 34:1317–1323
19. Hajrasouliha AR, Tavakoli S, Ghasemi M, Jabehdar-Maralani P et al (2008) Endogenous cannabinoids contribute to remote ischemic preconditioning via cannabinoid CB2 receptors in the rat heart. Eur J Pharmacol 579:246–252
20. Kant R, Diwan V, Jaggi AS, Singh N et al (2008) Remote renal preconditioning-induced cardioprotection: a key role of hypoxia inducible factor-prolyl 4-hydroxylases. Mol Cell Biochem 312:25–31
21. Lang SC, Elsasser A, Scheler C, Vetter S et al (2006) Myocardial preconditioning and remote renal preconditioning–identifying a protective factor using proteomic methods? Basic Res Cardiol 101:149–158
22. Przyklenk K, Whittaker P (2011) Remote ischemic preconditioning: current knowledge, unresolved questions, and future priorities. J Cardiovasc Pharmacol Ther 16:255–259
23. Davidson SM, Selvaraj P, He D, Boi-Doku C, Yellon RL et al (2013) Remote ischaemic preconditioning involves signalling through the SDF-1alpha/CXCR4 signalling axis. Basic Res Cardiol 108:377
24. Giricz Z, Varga ZV, Baranyai T, Sipos P et al (2014) Cardioprotection by remote ischemic preconditioning of the rat heart is mediated by extracellular vesicles. J Mol Cell Cardiol 68:75–78
25. Rassaf T, Ferdinandy P, Schulz R (2014) Nitrite in organ protection. Br J Pharmacol 171:1–11
26. Li J, Rohailla S, Gelber N, Rutka J et al (2014) MicroRNA-144 is a circulating effector of remote ischemic preconditioning. Basic Res Cardiol 109:423
27. Hibert P, Prunier-Mirebeau D, Beseme O, Chwastyniak M et al (2013) Apolipoprotein a-I is a potential mediator of remote ischemic preconditioning. PLoS One 8:e77211
28. Redington KL, Disenhouse T, Strantzas SC, Gladstone R et al (2012) Remote cardioprotection by direct peripheral nerve stimulation and topical capsaicin is mediated by circulating humoral factors. Basic Res Cardiol 107:241
29. Jensen RV, Stottrup NB, Kristiansen SB, Botker HE (2012) Release of a humoral circulating cardioprotective factor by remote ischemic preconditioning is dependent on preserved neural pathways in diabetic patients. Basic Res Cardiol 107:285
30. Lim SY, Hausenloy DJ (2012) Remote ischemic conditioning: from bench to bedside. Front Physiol 3:27
31. Heinen NM, Pütz VE, Görgens JI, Huhn R et al (2011) Cardioprotection by remote ischemic preconditioning exhibits a signaling pattern different from local ischemic preconditioning. Shock 36:45–53
32. Heusch G (2015) Molecular basis of cardioprotection: signal transduction in ischemic pre-, post-, and remote conditioning. Circ Res 116:674–699
33. Yellon DM, Downey JM (2003) Preconditioning the myocardium: from cellular physiology to clinical cardiology. Physiol Rev 83:1113–1151
34. Downey JM, Krieg T, Cohen MV (2008) Mapping preconditioning's signaling pathways: an engineering approach. Ann NY Acad Sci 1123:187–196
35. Li SJ, Wu YN, Kang Y, Yin YQ et al (2010) Noninvasive limb ischemic preconditioning protects against myocardial I/R injury in rats. J Surg Res 164:162–168

36. Breivik L, Helgeland E, Aarnes EK, Mrdalj J et al (2011) Remote postconditioning by humoral factors in effluent from ischemic preconditioned rat hearts is mediated via PI3K/Akt-dependent cell-survival signaling at reperfusion. Basic Res Cardiol 106:135–145
37. Wolfrum S, Schneider K, Heidbreder M, Nienstedt J et al (2002) Remote preconditioning protects the heart by activating myocardial PKC epsilon-isoform. Cardiovasc Res 55:583–589
38. Donato M, Goyeneche MA, Garces M, Marchini T et al (2016) Myocardial triggers involved in activation of remote ischaemic preconditioning. Exp Physiol 101:708–716
39. Krieg T, Landsberger M, Alexeyev MF, Felix SB et al (2003) Activation of Akt is essential for acetylcholine to trigger generation of oxygen free radicals. Cardiovasc Res 58:196–202
40. Kalogeris T, Bao Y, Korthuis RJ1 (2014) Mitochondrial reactive oxygen species: a double edged sword in ischemia/reperfusion vs preconditioning. Redox Biol 2:702–714
41. Hausenloy DJ, Iliodromitis EK, Andreadou I, Papalois A et al (2012) Investigating the signal transduction pathways underlying remote ischemic conditioning in the porcine heart. Cardiovasc Drugs Ther 26:87–93
42. Dow J, Bhandari A, Simkhovich BZ. Hale SL et al (2012) The effect of acute versus delayed remote ischemic preconditioning on reperfusion induced ventricular arrhythmias. J Cardiovasc Electrophysiol 23:1374–1383
43. Brandenburger T, Huhn R, Galas A, Pannen BH et al (2014) Remote ischemic preconditioning preserves Connexin 43 phosphorylation in the rat heart in vivo. J Transl Med 12:228
44. Konstantinov IE, Li J, Cheung MM, Shimizu M et al (2005) Remote ischemic preconditioning of the recipient reduces myocardial ischemia-reperfusion injury of the denervated donor heart via a Katp channel-dependent mechanism. Transplantation 79:1691–1695
45. Sun J, Aponte AM, Kohr MJ, Tong G et al (2013) Essential role of nitric oxide in acute ischemic preconditioning: S-nitros(yl)ation versus sGC/cGMP/PKG signaling? Free Radic Biol Med 54:105–112
46. Antunes F, Boveris A, Cadenas E (2004) On the mechanism and biology of cytochrome oxidase inhibition by nitric oxide. Proc Natl Acad Sci U S A 101:16774–16779
47. Yao Z, Gross GJ (1993) Role of nitric oxide, muscarinic receptors, and the ATP-sensitive K+ channel in mediating the effects of acetylcholine to mimic preconditioning in dogs. Circ Res 73(6):1193–1201
48. D'Autréaux B, Toledano M (2007) ROS as signalling molecules: mechanisms that generate specificity in ROS homeostasis. Nat Rev Mol Cell Biol 8:813–824
49. Weinbrenner C, Schulze F, Sárváry L, Strasser RH (2004) Remote preconditioning by infrarenal aortic occlusion is operative via delta1-opioid receptors and free radicals in vivo in the rat heart. Cardiovasc Res 61:591–599
50. Tullio F, Angotti C, Perrelli MG, Penna C et al (2013) Redox balance and cardioprotection. Basic Res Cardiol 108:392
51. Fornazari M, de Paula JG, Castilho RF, Kowaltowski AJ (2008) Redox properties of the adenoside triphosphate-sensitive K+ channel in brain mitochondria. J Neurosci Res 86:1548–1556
52. Neye N, Enigk F, Shiva S, Habazettl H et al (2012) Inhalation of NO during myocardial ischemia reduces infarct size and improves cardiac function. Intensive Care Med 38:1381–1391
53. Birnbaum Y, Hale SL, Kloner RA (1997) Ischemic preconditioning at a distance: reduction of myocardial infarct size by partial reduction of blood supply combined with rapid stimulation of the gastrocnemius muscle in the rabbit Circulation 96:1641–1646
54. Gho BC, Schoemaker RG, Van den Doel MA, Duncker DJ et al (1996) Myocardial protection by brief ischemia in noncardiac tissue. Circulation 94:2193–2200
55. Tokuno S, Hinokiyama K, Tokuno K, Löwbeer C et al (2002) Spontaneous ischemic events in the brain and heart adapt the hearts of severely atherosclerotic mice to ischemia. Arterioscler Thromb Vasc Biol 22:995–1001
56. Walsh SR, Tang T, Sadat U, Dutka DP et al (2007) Cardioprotection by remote ischaemic preconditioning. Br J Anaesth 99:611–616

57. Rassaf T, Totzeck M, Hendgen-Cotta UB, Shiva S, Heusch G et al (2014) Circulating nitrite contributes to cardioprotection by remote ischemic preconditioning. Circ Res 114:1601–1610
58. Chouchani ET, Methner C, Nadtochiy SM et al (2013) Cardioprotection by S-nitrosation of a cysteine switch on mitochondrial complex I. Nat Med 19:753–759
59. Abu-Amara M, Yang SY, Quaglia A, Rowley P et al (2011) Nitric oxide is an essential mediator of the protective effects of remote ischaemic preconditioning in a mouse model of liver ischaemia/reperfusion injury. Clin Sci (Lond) 121:257–266
60. Bolli R (2000) The late phase of preconditioning. Circ Res 87:972–983
61. Sparacino-Watkins CE, Tejero J, Sun B, Gauthier MC et al (2014) Nitrite reductase and nitric-oxide synthase activity of the mitochondrial molybdopterin enzymes mARC1 and mARC2. J Biol Chem 289:10345–10358
62. Tejero J, Gladwin MT (2014) The globin superfamily: functions in nitric oxide formation and decay. Biol Chem 395:631–639
63. Dong HL, Zhang Y, Su BX, Zhu ZH et al (2010) Limb remote ischemic preconditioning protects the spinal cord from ischemia-reperfusion injury: a newly identified nonneuronal but reactive oxygen species-dependent pathway. Anesthesiology 112:881–891

Chapter 18
Inhaled Particulate Matter Leads to Myocardial Dysfunction

Pablo Evelson, Timoteo Marchini, Mariana Garces, Lourdes Cáceres, Natalia Magnani, and Silvia Alvarez

Abstract Epidemiological studies have shown that the exposure to environmental particulate matter (PM) is associated with increased cardiopulmonary mortality rates. Daily changes in PM concentration have also been positively correlated with increased hospitalizations due to lower respiratory diseases, ischemic cardiovascular events, arrhythmias, and heart failure. Human and animal models have also shown a pulmonary and systemic inflammatory response and oxidative stress associated with the exposure to environmental particles which could, in turn, alter heart oxygen metabolism and cardiovascular function. Given that mitochondria play an essential role in cellular O_2 and energetic metabolism, it has been suggested that mitochondrial dysfunction is a key feature in the development of cardiac alterations during the exposure to PM. This chapter is focused in the discussion of the different mechanisms triggered by PM exposure that may lead to myocardial dysfunction, emphasizing the role of the systemic proinflammatory mediators released after PM inhalation.

Keywords Air pollution • Particulate matter • Cardiac dysfunction • Inflammation • Oxidative stress • Mitochondria

1 Introduction

The World Health Organization reported that in 2012, 7 million deaths were recorded, one in eight of total global deaths, as a result of air pollution exposure [1]. This finding more than doubles previous estimates and confirms that air pollution is now the world's largest single environmental health risk. This mortality has been pointed out by several epidemiological studies, which have shown a positive correlation between decreased air quality and adverse health effects [2]. In the

P. Evelson (✉) • T. Marchini • M. Garces • L. Cáceres • N. Magnani • S. Alvarez
Universidad de Buenos Aires. Consejo Nacional de Investigaciones Científicas y Técnicas, Instituto de Bioquímica y Medicina Molecular (IBIMOL), Facultad de Farmacia y Bioquímica, Buenos Aires, Argentina
e-mail: pevelson@ffyb.uba.ar

© Springer International Publishing Switzerland 2016

R.J. Gelpi et al. (eds.), *Biochemistry of Oxidative Stress*, Advances in Biochemistry in Health and Disease 16, DOI 10.1007/978-3-319-45865-6_18

previously mentioned report, the WHO estimates that 80 % of outdoor air pollution-related premature deaths were due to cardiovacular diseases, such as ischaemic heart disease and strokes, while 14 % of deaths were due to chronic obstructive pulmonary disease or acute lower respiratory infections; and 6 % of deaths were due to lung cancer. Increased morbidity and mortality rates have been found to be associated not only with chronic air pollution exposures, but also with short-term daily exposures as well [3]. Air pollution is a complex heterogeous mix whose complexity is increased due to the variation in its components, between places and over time [4]. The most important pollutants in ambient air which are of concern regarding health effects, include sulfur dioxide (SO_2), nitrogen oxides (NOx), carbon monoxide (CO), volatile organic compounds (VOCs), and particulate matter (PM) [5]. Air pollution derives from a variety of sources, of which the combustion of fossil-fuel products utilized for road transport and electricity generation is the most important one.

Air pollutants can be classified by their source, chemical composition, size, and mode of release into indoor or outdoor environments. When taking into account their source of origin, pollutants can be divided into two main groups: the ones directly emitted into the atmosphere, which are known as primary pollutants. Secondary pollutants are those formed in the air as a result of chemical reactions with other pollutants and atmospheric gases. Examples of primary pollutants are CO and SO_2, while ozone is an example of a secondary pollutant. Others can be considered as both primary and secondary pollutants, such as nitrogen dioxide (NO_2) and some PM. They are both emitted directly into the atmosphere, and formed from other pollutants. For example, some of the NO_2 is emitted directly from power stations and vehicle exhaust, while most is formed by the oxidation of nitric oxide (NO) in the air. Fine PM is emitted directly from a number of natural and anthropogenic sources as well as being formed within the atmosphere. Secondary PM is formed mainly from the oxidation of SO_2 and NO_2. The reactions of hydrocarbons in the air to produce organic particles observed in some areas are an important source of secondary PM. Thus, it is accepted that air pollution exists as a complex mixture and that effects attributed to different pollutants such as ozone (O_3), NOx, or PM, may be influenced by the underlying toxicity of the full mixture of all air pollutants [6].

Suspended PM is composed of solid and liquid particles that come not only from anthropogenic sources as vehicle exhaust and fossil combustion, but from natural sources, such as forest fires, volcanic emissions and sea spray as well [7]. Of the different air pollutants, it is accepted that PM is the major concern from a health perspective [1]. It is now well established that PM cause respiratory diseases and increase lung cancer, as well as death due to cardiovascular diseases [8]. PM consists of a mixture of particle components, including traffic and combustion-derived carbon-centered particles, secondary particles (nitrates and sulphates), wind-blown dust of geological origin and potentially containing endotoxins, and biological particles (spores, pollen) with their associated allergens [9]. Largely because of the complex nature of PM, it has been measured and regulated based primarily on mass within defined size ranges. Based on the size, PM is classified into three main groups: coarse, also known as "thoracic" particles (with an aerodynamic diameter between 2.5 and 10 μm, PM_{10}), fine (aerodynamic diameter between 0.1 and 2.5 μm, $PM_{2.5}$) and ultrafine (aerodynamic diameter less than 0.1 μm). PM_{10} deposit predominantly in the lower airways (nose and throat) and are cleared by exhalation, mucociliary transport and swallowing. Fine and ultrafine particles can penetrate into the lower airways and alveoli and are more closely related with the adverse health effects of PM than coarser particles [10]. The fine fraction or "respirable" particles, smaller than 2.5 μm ($PM_{2.5}$), can penetrate into the gas-exchange region of the lung and the ultrafine fraction (sometimes called nanoparticles) which contribute little to particle mass but which are most abundant in terms of numbers and offer a very large surface area, with increasing degrees of lung penetration.

2 PM Adverse Health Effects

The exposure to PM, at levels experienced by populations throughout the world, contributes to pulmonary and cardiac disease through multiple mechanistic pathways that are complex and interdependent [11, 12]. Experimental evidence suggests a series of events that are triggered by pollution-induced pulmonary inflammatory reactions and oxidative stress [13–15] with an associated risk of vascular dysfunction, altered cardiac autonomic function, and ischemic cardiovascular and obstructive pulmonary diseases [16, 17].

3 Pulmonary Events After PM Exposure

Pulmonary inflammation after PM exposure is well documented in animals as well as humans [18–21]. A critical component of the inflammatory response to particles in the lung, is the release of cytokines from activated macrophages and lung epithelial cells, resulting in neutrophil recruitment. This response may be caused by the deposition of PM into the alveolar space in the lung, inducing the release of cytokines from alveolar macrophages. Cultured human alveolar macrophages produce TNF-α and proinflammatory cytokines such as granulocyte-macrophage colony-stimulating factor, IL-6, and IL-1 after phagocytosing PM [22]. The same observations were registered in different animal models [21, 23]. The release of proinflammatory mediators from PM-exposed macrophages appears to be important in stimulating cytokine release from lung epithelial cells, thus amplifying the inflammatory response [24].

The activation of inflammatory cells leads to the generation of reactive oxygen and nitrogen species. Inflammatory lung diseases are characterized by activation of epithelial cells and resident macrophages and the recruitment and activation of neutrophils, eosinophils, monocytes, and lymphocytes. Immune cells such as neutrophils, eosinophils, and macrophages contain a membrane bound enzyme,

NADPH oxidase, which is induced during the inflammatory response. NADPH oxidase produces superoxide anion ($O_2^{\bullet-}$) using O_2. These reactive oxygen species (ROS) are removed by superoxide dismutase, which is present at the surface of the epithelial cells of the lung [15]. The product of this reaction is hydrogen peroxide (H_2O_2), which can easily diffuse through plasma membranes. H_2O_2 can therefore initiate intracellular signaling pathways, or contribute to the generation of other ROS. It is understood that oxidative stress caused by the activation of the inflammatory system, comprised by alveolar macrophages and neutrophils, plays an important role in the deleterious effects of PM in multicellular organisms [25]. ROS are generated during phagocytosis of the particles leading to an enhancement of oxidative stress [26].

Another proposed mechanism that may lead to oxidative damage is the direct generation of ROS at the surface of the particles. This is supported by the concept that the particle surface offers a unique physicochemical interface to catalyze reactions resulting in oxidant production. The interaction of PM with membrane components was recognized by the presence of free radicals and oxidants on the particle surface [27]. Finally, PM can also contain a large number of soluble metals that have the ability of redox cycling. The involvement of transition metals, such as Fe, Va, Cr, Mn, Co, Ni and Cu, which are able to catalyze Fenton-type reactions and generate hydroxyl radicals, has been proposed [28]. Indeed, the production of oxidative damage, as measured by TBARS levels, showed significant correlations with the overall metal content and with the content of individual metals such as Fe and Va. This effect is prevented by the presence of antioxidants such as dimethylthiourea or chelating agents such as deferoxamine [18, 29]. These and other works have led to the suggestion that the dose of bioavailable transition metal, rather than particulate mass, may be the principal determinant of the acute inflammatory response [30].

4 Cardiac Effects Caused by Exposure to PM

It has been postulated that pulmonary PM exposure may trigger a cardiovascular response through three different mechanisms (Fig. 18.1) [31–33]. The first hypothesis proposes that PM deposited in the lung acts through a neural mechanism to alter central nervous system function. In the lung, nociceptive neurons are stimulated by ROFA [34]. Cardiac autonomic function is also altered by PM exposure, suggesting that central input to the heart is altered [35]. Acute electrocardiographic changes after PM exposure suggest an activated neural mechanism [36]. The second proposed mechanism is that ROFA deposited in the lung may enter in the systemic circulation and directly interact with different organs. It has been reported that after lung exposure, PM deposits has been found in several non-pulmonary tissues, including the blood, ventricular microvascular walls, liver, spleen, heart, and brain [37, 38]. PM deposition was found in the cardiac arteriolar wall of dogs where polymorphonuclear leukocyte (PMNL) margination and microthrombi were also

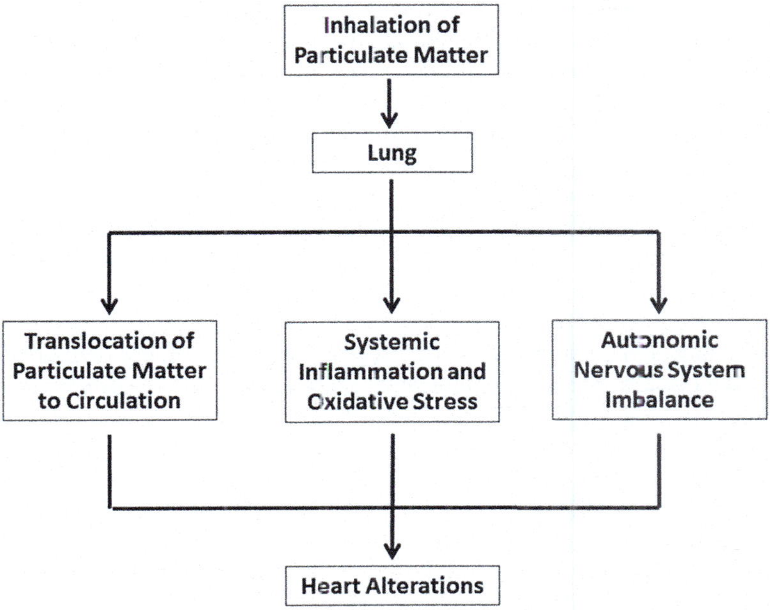

Fig. 18.1 Proposed mechanisms of heart adverse effects induced by PM exposure

observed [37]. When PM is exposed to cells or tissues *in vitro*, cytokine and tumor necrosis factor-alpha (TNF-α) production, cytotoxicity via endotoxins, oxidative stress, and smooth muscle relaxation was observed [39–42]. The third postulated mechanism is that PM deposited in the lung initiates a local inflammatory response that develops into a systemic inflammatory response, characterized by alterations in circulating factors and cells associated with inflammation. Circulating interleukin IL-1 and IL-6 are increased in humans exposed to PM [42]. IL-1, TNF-α, and the immune-related transcription factor nuclear factor kB are elevated in the brain of mice exposed to PM [43]. Furthermore, in blood samples from healthy humans exposed to PM, increases in the levels of immature PMNL, neutrophils, and platelets were observed [44, 45]. Although the evidence for these three proposed hypothesis is significant, the basic mechanisms responsible for the biological effects in the heart remain unclear.

5 Oxygen Metabolism, Mitochondrial Function and Cardiac Function: Role of Systemic Inflammation

Numerous studies in humans and animal models have shown a pulmonary and systemic inflammatory response and oxidative stress associated with PM exposure which can alter heart O_2 metabolism and cardiovascular function [46]. Given that mitochondria play an essential role in cellular O_2 and energetic metabolism, several authors suggested that mitochondrial dysfunction is a key feature in the development of cardiac alterations during the exposure to air pollution PM [47, 48]. Cardiac contraction and relaxation have a continuous energy requirement, consuming more energy than any other organ. Because of a mismatch in ATP supply and demand, decreased levels of high-energy phosphates have been reported in the failing human heart [49] hampering with the transference of chemical energy to contractile work [50]. Most of this energy is produced in mitochondria by oxidative phosphorylation, a process that involves electron-transfer reactions through the mitochondrial respiratory chain complexes at the inner mitochondrial membrane. In this context, any alteration triggered by PM inhalation in these multienzymatic complexes, in the electrochemical H+ gradient that they generate across the inner membrane or in F_0–F_1 ATP synthase activity, could lead to a deficient ATP production resulting in a bioenergetic dysfunction and organ failure [51].

Taking into account that inhaled PM could alter heart oxidative metabolism, the need of an adequate energy supply to sustain proper contractile work, and the crucial role of mitochondria in both O_2 and energetic metabolism, we evaluated cardiac O_2 metabolism and contractile function, focused on mitochondrial function, in a mice model of acute exposure to Residual Oil Fly Ash (ROFA). ROFA is the inorganic residue that remains after the incomplete oxidation of such carbonaceous materials and contributes to PM in urban air [52]. Diverse PM surrogates have been assayed in different animal models in order to study the biological effects of PM exposure. Among them, ROFA has been particularly useful given that it is especially rich in soluble transition metals (namely iron, nickel and vanadium), and because of its low concentration of organic compounds [53]. Therefore, ROFA is the most frequently used combustion-derived particle in order to evaluate the contribution of transitions metals in the biological effects of PM inhalation [54]. Moreover, ROFA particles often present an aerodynamic diameter smaller than 2.5 μm ($PM_{2.5}$), a size that have been shown to be more closely associated with PM adverse health effects than coarser particles ($PM_{10-2.5}$) [55]. Swiss mice were intranasally instilled with either ROFA (1.0 mg/kg body weight) or saline solution. The amount of particles administered in this study was 25 μg per mouse. The selected dose falls within the range of concentrations consistently used in several animal studies [56, 57] and it can be considered as a medium to low dose when compared with pollution levels attained in large cities worldwide [58]. Three hours after ROFA exposure, tissue O_2 consumption was significantly decreased, as well as mitochondrial rest (state 4) and active (state 3) respiration. These findings were associated with decreased complex II activity, mitochondrial depolarization and deficient ATP production. Even though basal contractility was not modified, isolated perfused hearts failed to properly respond to isoproterenol in ROFA-exposed mice. As it is

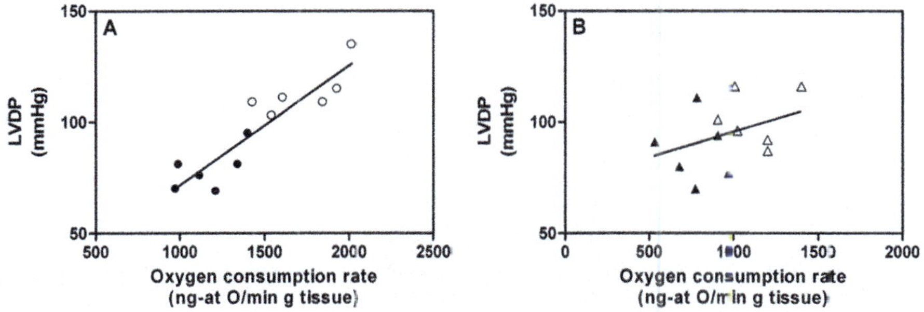

Fig. 18.2 Linear correlation between heart O_2 consumption and Left Ventricular Developed Pressure (LVDP) measurements. (**a**) Control animals in basal conditions (•) and after a β-adrenergic stimulus (○) ($r^2 = 0.827$, $p < 0.0001$). (**b**) ROFA-exposed mice in basal conditions (▲) and after a β-adrenergic stimulus (△) ($r^2 = 0.140$) (Taken with permission from Ref. [59])

shown in Fig. 18.2, tissue O_2 consumption rates positively correlated with cardiac contractile state in controls, but not in treated mice [59].

These results indicate that the myocardium fails to properly sustain contractile work when work output is increased in mice exposed to PM. Interestingly, pretreatment with infliximab, a chimeric monoclonal antibody that blocks TNF-α biological activity [60], recovered the positive correlation between cardiac contractile state and O_2 consumption [61]. These findings support the notion that systemic inflammation is a key pathway in the alterations in cardiac function observed after PM exposure.

6 Conclusions

The mechanisms of PM health effects in the cardiovascular system are still poorly understood. A better understanding of the mechanisms underlying PM induced health problems would allow a more targeted approach to face the toxic effects of PM, and could possibly provide different ways to decrease individual sensitivity to PM. Epidemiological data and studies performed in different animal models suggest a variety of possible mechanisms, including a cardiac effect induced, at least in part, by proinflammatory mediators that affects oxygen metabolism and mitochondrial function. The whole series of events are summarized in Fig. 18.3. The current hypothesis is that the lung inflammation induced by inhalation of ambient particles implies a systemic inflammatory response that leads to cardiac dysfunction. These changes may contribute to the increased morbidity and mortality associated to polluted areas.

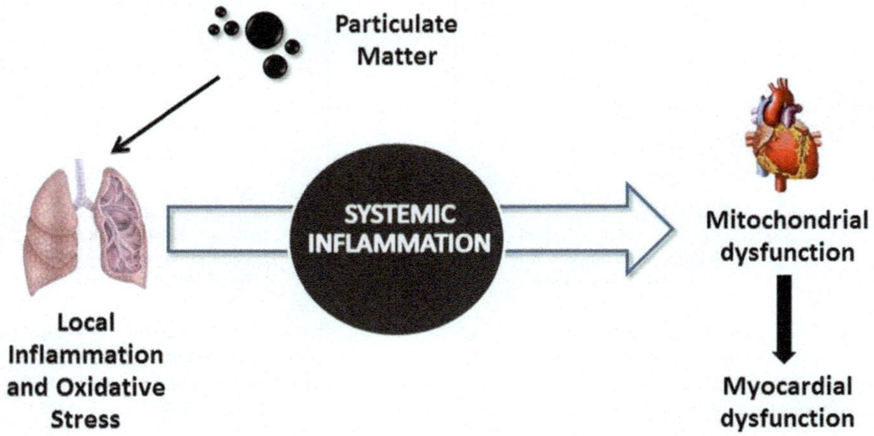

Fig. 18.3 Proposed mechanism of PM effects on cardiac tissue

References

1. World Health Organization (WHO) (2014) WHO air quality guidelines for particulate matter, ozone, nitrogen dioxide and sulfur dioxide. Global update 2014
2. Nel A (2005) Atmosphere. Air pollution-related illness: effects of particles. Science 308:804–806
3. Brook RD (2008) Cardiovascular effects of air pollution. Clin Sci (Lond) 115:175–187
4. D'Amato G, Cecchi L (2008) Effects of climate change on environmental factors in respiratory allergic diseases. Clin Exp Allergy 38:1264–1274
5. Ghio AJ, Huang YC (2004) Exposure to concentrated ambient particles (CAPs): a review. Inhal Toxicol 16:53–59
6. Kelly FJ, Fussell JC (2015) Air pollution and public health: emerging hazards and improved understanding of risk. Environ Geochem Health 37:631–649
7. Kelly FJ, Fussell JC (2012) Size, source and chemical composition as determinants of toxicity attributable to ambient particular matter. Atm Environ 60:504–526
8. Pope CA, Muhlestein JB, May HT et al (2006) Ischemic heart disease events triggered by short-term exposure to fine particulate air pollution. Circulation 114:2443–2448
9. Brook RD, Rajagopalan S, Pope CA et al (2010) Particulate matter air pollution and cardiovascular disease. An update of the Scientific Statement from the American Heart Association. Circulation 121:2331–2378
10. Schins RJ, Lightbody P, Borm T et al (2004) Inflammatory effects of coarse and fine particulate matter in relation to chemical and biological constituents. Toxicol Appl Pharmacol 184:1–11
11. Donaldson K, Stone V, Seaton A et al (2001) Ambient particle inhalation and the cardiovascular system: potential mechanisms. Environ Health Perspect 109:523–527
12. Zareba W, Nomura A, Couderc JP (2001) Cardiovascular effects of air pollution: what to measure in ECG? Environ Health Perspect 109:533–538
13. Pope CA III (2000) Epidemiology of fine particulate air pollution and human health: biologic mechanisms and who's at risk? Environ Health Perspect 108:713–723
14. Souza MB (1998) Respiratory changes due to long-term exposure to urban levels of air pollution. Chest 113:1312–1318

15. Magnani N, Marchini T, Tasat D et al (2011) Time course of lung oxidative metabolism after exposure to ambient particles. Biochem Byophys Res Comm 412:667–672
16. Dockery DW (1994) Epidemiologic evidence of cardiovascular effects of particulate air pollution. Environ Health Perspect 109:483–486
17. Schwartz J (2001) Air pollution and blood markers of cardiovascular risk. Environ Health Perspect 109:405–409
18. Dreher KL, Jaskot RH, Lehmann JR et al (1997) Soluble transition metals mediate residual oil fly ash-induced acute lung injury. J Toxicol Environ Health 50:285–305
19. Nurkiewicz TR, Porter DW, Barger M et al (2004) Particulate matter exposure impairs systemic microvascular endothelium-dependent dilation. Environ Health Perspect 112:1299–1306
20. Schaumann F, Borm PJ, Herbrich et al (2004) Metal-rich ambient particles (particulate matter 2.5) cause airway inflammation in healthy subjects. Am J Respir Crit Care Med 170:898–903
21. Marchini T, Magnani ND, Paz ML et al (2014) Time course of systemic oxidative stress and inflammatory response induced by an acute exposure to Residual Oil Fly Ash. Toxicol Appl Pharmacol 274:274–282
22. Oury T, Chang L, Marklund S et al (1994) Immunocytochemical localization of extracellular superoxide dismutase in human lung. Lab Invest 70:889–898
23. Jimenez L, Drost E, Gilmour P et al (2000) PM(10)-exposed macrophages stimulate a proinflammatory response in lung epithelial cells via TNF-alpha. Am J Physiol 282:L237–L248
24. Tao F, Gonzalez-Flecha B, Kobzik L (2003) Reactive oxygen species in pulmonary inflammation by ambient particulates. Free Radic Biol Med 35:327–340
25. Gurgueira SA, Lawrence J, Coull B et al (2002) Rapid increases in the steady-state concentration of reactive oxygen species in the lungs and heart after particulate air pollution inhalation. Environ Health Perspect 110:749–755
26. Behndig AF, Mudway IS, Brown JL et al (2006) Airway antioxidant and inflammatory responses to diesel exhaust exposure in healthy humans. Eur Respyr J 27 359–365
27. Fubini B, Fenoglio I, Ceschino R et al (2004) Relationship between the state of the surface of four commercial quartz flours and their biological activity in vitro and in vivo. Int J Hyg Environ Health 207:89–104
28. Chen LC, Lippmann M (2009) Effects of metals within ambient air particulate matter (PM) on human health. Inhal Toxicol 21:1–31
29. Sorensen M, Schins R, Hertel O et al (2005) Transition metals in personal samples of PM2.5 and oxidative stress in human volunteers. Cancer Epidemiol Biomarkers Prev 14:1340–1343
30. Samet J, DeMarini D, Malling H (2004) Biomedicine. Do airborne particles induce heritable mutations? Science 304:971–972
31. Brook RD, Franklin B, Cascio W et al (2004) Air pollution and cardiovascular disease: a statement for healthcare professionals from the Expert Panel on Population and Prevention Science of the American Heart Association. Circulation 109:2655–2671
32. Nemmar A, Hoet PH, Vanquickenborne B (2002) Passage of inhaled particles into the blood circulation in humans. Circulation 105:411–414
33. Oberdörster G, Sharp Z, Atudorei V et al (2002) Extrapulmonary translocation of ultrafine carbon particles following whole-body inhalation exposure of rats. J Toxicol Environ Health 65:1531–1543
34. Veronesi B, Oortgiesen M, Roy J et al (2000) Vanilloid (capsaicin) receptors influence inflammatory sensitivity in response to particulate matter. Toxicol Appl Pharmacol 169:66–76
35. Chen LC, Hwang JS (2005) Effects of subchronic exposures to concentrated ambient particles (CAPs) in mice. IV. Characterization of acute and chronic effects of ambient air fine particulate matter exposures on heart-rate variability. Inhal Toxicol 17:209–216
36. Wichers LB, Nolan JP, Winsett DW et al (2004) Effects of instilled combustion-derived particles in spontaneously hypertensive rats. Part I: Cardiovascular responses. Inhal Toxicol 16:391–405

37. Calderon-Garciduenas L, Gambling TM, Acuna H et al (2001) Canines as sentinel species for assessing chronic exposures to air pollutants: cardiac pathology. Toxicol Sci 61:356–367
38. Kreyling WG, Semmler M, Erbe F et al (2002) Translocation of ultrafine insoluble iridium particles from lung epithelium to extrapulmonary organs is size dependent but very low. J Toxicol Environ Health 65:1513–1530
39. Bagate K, Meiring JJ, Cassee FR et al (2004) The effect of particulate matter on resistance and conductance vessels in the rat. Inhal Toxicol 16:431–436
40. Dye JA, Adler KB, Richards JH et al (1997) Epithelial injury induced by exposure to residual oil fly-ash particles: role of reactive oxygen species? Am J Respir Cell Mol Biol 17:625–633
41. Li N, Sioutas C, Cho A et al (2003) Ultrafine particulate pollutants induce oxidative stress and mitochondrial damage. Environ Health Perspect 111:455–460
42. van Eeden SF, Tan WC, Suwa T et al (2001) Cytokines are involved in the systemic inflammatory response induced by exposure to particulate matter air pollutants (PM10). Am J Respir Crit Care Med 164:826–830
43. Campbell A, Oldham M, Becaria A et al (2005) Particulate matter in polluted air may increase biomarkers of inflammation in mouse brain. Neurotoxicology 26:133–140
44. Salvi S, Blomberg A, Rudell B et al (1999) Acute inflammatory responses in the airways and peripheral blood after short-term exposure to diesel exhaust in healthy human volunteers. Am J Respir Crit Care Med 159:702–709
45. Tan WC, Qiu D, Liam BL et al (2000) The human bone marrow response to acute air pollution caused by forest fires. Am J Respir Crit Care Med 161:1213–1217
46. Kim K, Kabir E, Kabir S (2015) A review on the human health impact of airborne particulate matter. Environ Int 74:136–143
47. Knight-Lozano CA, Young CG, Burow DL et al (2002) Cigarette smoke exposure and hypercholesterolemia increase mitochondrial damage in cardiovascular tissues. Circulation 105:849–854
48. Alvarez S, Evelson P (2007) Nitric oxide and oxygen metabolism in inflammatory conditions: sepsis and exposition to polluted ambients. Front Biosci 12:964–974
49. Beer M, Seyfarth T, Sandstede J (2002) Absolute concentrations of high-energy phosphate metabolites in normal, hypertrophied, and failing human myocardium measured noninvasively with 31P-SLOOP magnetic resonance spectroscopy. J Am Coll Cardiol 40:1267–1274
50. Neubauer S (2007) The failing heart: an engine out of fuel. N Engl J Med 356:1140–1151
51. Nicholls DG, Ferguson SJ (2002) Bioenergetics, 3rd edn. Academic, London
52. Ghio AJ, Silbajoris R, Carson JL et al (2002) Biologic effects of oil fly ash. Environ Health Perspect 110:89–94
53. Schroeder WH, Dobson M, Kane D et al (1987) Toxic trace elements associated with airborne particulate matter: a review. JAPCA 37:1267–1285
54. Shuster-Meiseles T, Shafer MM, Heo J et al (2016) ROS-generating/ARE-activating capacity of metals in roadway particulate matter deposited in urban environments. Environ Res 146:252–262
55. Upadhyay D, Panduri V, Ghio A et al (2003) Particulate matter induces alveolar epithelial cell DNA damage and apoptosis: role of free radicals and mitochondria. Am J Respir Cell Mol Biol 29:180–187
56. Magnani N, Marchini T, Vanasco V et al (2013) Reactive oxygen species produced by NADPH oxidase and mitochondrial dysfunction in lung after an acute exposure to Residual Oil Fly Ashes. Toxicol Appl Pharmacol 270:31–38
57. Di Pietro A, Visalli G, Baluce B et al (2011) Multigenerational mitochondrial alterations in pneumocytes exposed to oil fly ash metals. Int J Hyg Environ Health 214:138–144
58. Yosgizaki Akinaga LM, Lichtenfels AJ, Carvalho-Oliveira R et al (2009) Effects of chronic exposure to air pollution from Sao Paulo city on the coronary of swiss mice from birth to adulthood. Toxicol Pathol 37:306–314

59. Marchini T, Magnani N, D'Annunzio V et al (2013) Impaired cardiac mitochondrial function and contractile reserve following an acute exposure to environmental particulate matter. Biochim Biophys Acta 1830:2545–2552
60. Knight DM, Trinh H, Le J et al (1993) Construction and initial characterization of a mouse-human chimeric anti-TNF antibody. Mol Immunol 30:1443–1453
61. Marchini T, D'Annunzio V, Paz ML et al (2015) Selective TNF-α targeting with infliximab attenuates impaired oxygen metabolism and contractile function induced by an acute exposure to air particulate matter. Am J Physiol Heart Circ Physiol 309:H1621–H1628

Part III
Neuronal Function and Neurodegeneration

Chapter 19
Neurovascular Coupling Mediated by Neuronal Derived-Nitric Oxide: Mechanisms in Health and Dysfunction with Impact on Aging and Alzheimer's Disease

Cátia F. Lourenço, Ana Ledo, Rui M. Barbosa, and João Laranjinha

Abstract Neurovascular coupling is an intricate mechanism whereby local blood flow is strictly adjusted in time and space to neuronal activity. Intimately associated to glutamatergic neurotransmission, it is a complex mechanism that relies on the concerted communication of neurons and vascular cells, with assumed enrolment of other cell types. Nitric oxide (•NO) is uniquely suited to integrate the activity of all components of the neurovascular unit. Due to its hydrophobicity and reduced size, •NO can diffuse in the brain tissue and integrate the activity of multiple cells irrespective of their physical connection to the producing cell.

In this chapter we review evidence supporting the involvement of •NO derived directly from neuronal activity as a result of glutamatergic neurotransmission in neurovascular coupling. Furthermore, we discuss dysfunction of •NO-mediated neurovascular coupling as a fundamental process in the aging brain and in Alzheimer's disease, emphasizing the putative role of oxidative stress.

Keywords Nitric oxide • Neurovascular coupling • Alzheimer's Disease • Aging • Oxidative stress

1 Introduction

The brain accounts for 20 % of resting O_2 consumption in the adult human, despite only representing 2 % of total body mass [1, 2]. To sustain functional and structural integrity, the brain requires blood supply to be attuned to the physiological demands imposed by neural activation with both high temporal and regional precision [3].

C.F. Lourenço • A. Ledo • R.M. Barbosa • J. Laranjinha (✉)
Center for Neuroscience and Cell Biology, Faculty of Pharmacy, University of Coimbra, Coimbra, Portugal
e-mail: laranjin@ci.uc.pt

© Springer International Publishing Switzerland 2016

R.J. Gelpi et al. (eds.), *Biochemistry of Oxidative Stress*, Advances in Biochemistry in Health and Disease 16, DOI 10.1007/978-3-319-45865-6_19

The existence of a mechanism that integrates variations of neuronal activity with local blood supply was firstly suggested by Roy and Sherrington in 1890. By recording intracranial pressure simultaneously with systemic blood pressure they concluded that "chemical products of cerebral metabolism contained in the lymph which bathes the walls of the arterioles of the brain can cause variations of the caliber of the cerebral vessel" and proposed that the brain possessed an "intrinsic mechanism by which its vascular supply can be varied locally in correspondence with local variations in functional activity" [4]. This conjecture outlines the concept of neurovascular coupling and supports a premise that is still valid nowadays: besides extrinsic control, cerebral blood flow (CBF) is uniquely regulated by intrinsic brain factors. Neurovascular coupling is the basis for functional brain imaging techniques such as blood oxygen level-dependent (BOLD) functional magnetic resonance imaging (fMRI) and yet, in spite of over a century's worth of research, the underlying mechanisms remain to be fully elucidated [3, 5]. Nitric oxide (•NO), a free radical intercellular diffusible messenger, was implicated in the regulation of CBF soon after its identification as the endothelial-derived relaxing factor (EDRF) and as a neuromodulator [6, 7]. However, three decades later, the role played by •NO in neurovascular coupling is still ambiguous.

In this chapter we review the current knowledge concerning the contribution of •NO to the mechanisms underlying neurovascular coupling, exploiting the indirect and direct evidence that supports the involvement of •NO in this process. We further discuss the impairment of •NO-mediated neurovascular coupling in aging and Alzheimer's disease, giving particular emphasis to the putative role of oxidative stress.

2 Neurovascular Coupling: Basic Tenets

The concept of neurovascular coupling describes a concerted communication between neurons and vascular cells that likely involves the cooperation of other cell types, including astrocytes and pericytes, that together form the neurovascular unit [5]. It was originally hypothesized that active neurons control the vascular supply of glucose and O_2 through the generation of a metabolic signal, establishing a negative-feedback mechanism. A drop in O_2 and glucose concentration or a rise in CO_2 concentration would induce a local increase in CBF to suppress that unbalance. However, data obtained from manipulation of glucose and O_2 concentrations, as well as the lack of acidification of extracellular medium accompanying the neuronal activity, rebutted this hypothesis [8, 9].

Nowadays it has been recognized that neuronal activity *per se* elicits the increase of CBF in a feedforward mechanism. Indeed, although other neurotransmitters may be involved [10], there is a large consensus that glutamate synaptically released during neuronal activation triggers neurovascular coupling [3, 5, 11]. Glutamate is the main excitatory neurotransmitter in the brain and can activate both G-protein coupled (metabotropic) receptors (mGluR) and ligand-gated cationic (ionotropic) channels (iGluR) [12]. The first evidence of vasodilatory effect of glutamate in intact cerebral circulation was obtained from the observation of an increased diameter of pial arterioles following glutamate application to the exposed surface of the piglet cerebral cortex [13]. This fundamental observation has been consistently demonstrated in other animal models (reviewed in [14]). Furthermore, diverse experimental approaches support the hypothesis that the vasodilatory effects of glutamate are not restricted to the cerebral cortex. Glutamate receptor activation by electrical or pharmacological stimulation promotes vasodilation in a variety of brain regions, including the cerebellum [15, 16], hippocampus [17, 18] and striatum [19]. While the involvement of glutamate is well established, the mechanisms by which it promotes vasodilation are more questionable. It has been suggested that brain endothelial cells express functional glutamate receptors and that glutamate can mediate vasodilation by acting directly at the microvasculature level, but this hypothesis has scarce experimental support [20]. The endothelium may, instead, be implicated in retrograde propagation of vasodilation [21].

More robust evidence claim glutamate vasodilatory action relying on activation of glutamate receptors either in neurons or astrocytes, with subsequent activation of Ca^{2+}-dependent intracellular signaling pathways and production of vasoactive molecules that target smooth muscle cells in arterioles and/or pericytes in capillaries, ultimately promoting vasodilation and increase of CBF. Among the vasoactive substances proposed to be involved in the vasodilation associated to neurovascular coupling are •NO, K^+, arachidonic acid metabolites and adenosine [22].

Before discussing the pathways that contribute to the regulation of neurovascular coupling, in particular those involving •NO, one must note that it is conceivable that several molecules and/or pathways may cooperate to translate the need for metabolic substrates imposed by neuronal activity into changes of CBF. Moreover, as discussed later, these mechanisms are expectedly distinct amongst different brain regions, reflecting diverse neuroanatomic and functional properties.

3 Nitric Oxide in the Neurovascular Coupling

3.1 Nitric Oxide and Glutamatergic Transmission

In the brain, •NO signalling is intimately associated with glutamatergic transmission. At glutamatergic synapses, •NO synthesis involves the stimulation of ionotropic glutamate receptors, particularly N-methyl-D-aspartate (NMDA)-subtype that, upon activation, allow the influx of Ca^{2+}. Binding of Ca^{2+} to calmodulin activates the neuronal isoform of nitric oxide synthase (nNOS), which catalyzes the conversion of L-arginine to L-citrulline and •NO, provided that the substrates (L-arginine, O_2) and several cofactors (NADPH, FMN, FAD, tetrahydrobiopterin, heme) are available [23, 24]. Neuronal NOS, in particular the α-splice variant, possesses an N-terminal PDZ motif [25], which allows the enzyme to bind to other PDZ-containing proteins, such as the synaptic density scaffold protein PSD-95 [26]. The functional impact of this association is highly relevant, as PSD-95 binds simultaneously to the NR2 subunit of the NMDA receptor [27, 28], thus forming a supramolecular complex that places the Ca^{2+}-dependent nNOS under the direct effect of Ca^{2+} influx through the activated NMDA receptor channel [29] and regulates •NO production via specific adaptor proteins [30, 31]. Associated to glutamatergic transmission, •NO has been implicated in a wide range of physiological processes, including neurovascular coupling, learning and memory formation, neural differentiation and development, nociception and drug addiction [32].

3.2 *Additional Sources of Nitric Oxide in the Brain*

In the brain, •NO can also be produced by other members of the NOS family, namely the endothelial NOS (eNOS) and the inducible NOS (iNOS). In general terms, eNOS, like nNOS, is a Ca^{2+}-calmodulin-dependent enzyme and is constitutively expressed mainly in the endothelial cells. The iNOS is expressed upon demand in multiple cell types, including glial cells, in response to inflammatory stimuli and, due to its high affinity for calmodulin, can produce •NO under basal Ca^{2+} concentration [33].

More recently, a novel pathway for non-enzymatic •NO production has been proposed in connection with the endogenous univalent reduction of nitrite. In fact, •NO can be generated from nitrite under acidic and reducing conditions such as those occurring during hypoxia (reviewed in [34]). In this context, evidence suggests that •NO generated from nitrite can be utilized to support normal cerebrovascular physiology [35].

3.3 *Determinant Features of Nitric Oxide with Relevance for Neurovascular Coupling*

What makes •NO a unique candidate to mediate neurovascular coupling? The immediate answer is that, in addition to being released upon neuronal activation, •NO is a potent vasodilator. The elevation of •NO from environmental pollutant to a biological signaling molecule occurred from its identification as the endothelial derived relaxing factor (EDRF) and, since then, it has been implicated in the regulation of vascular tone in a large number of tissues [36]. The main mechanism underlying the vasodilatory effect of •NO involves the activation of soluble guanylate cyclase (sGC) in smooth muscle cells and the production of the second messenger cGMP. The subsequent activation of cGMP-dependent protein kinase promotes the decrease intracellular $[Ca^{2+}]$, resulting in dephosphorylation of myosin light chain and relaxation of the smooth muscle cells [37]. Also, •NO promotes vasodilation through cGMP-independent mechanisms such as activation of calcium-dependent potassium channels [38], stimulation of the sarco/endoplasmic reticulum calcium

(Ca^{2+}) ATPase (SERCA) [39] and by affecting the synthesis of other vasoactive molecules [40], as discussed below in more detail.

The distinguishable features of ˙NO as neuromodulator derive from its low molecular weight and hydrophobicity, which make ˙NO not only highly diffusible but also membrane permeable [41]. Indeed, *in vivo* studies in the brain support that cellular membranes may act as low-resistance pathways that facilitate ˙NO diffusion in nervous tissue, consequently increasing its diffusional field [42]. In acute hippocampal slices, by measuring endogenously produced ˙NO via the activation of NMDA-type glutamate receptors, we found that the activation of multiple nNOS-containing neurons within a 50 μm radius resulted in a diffusional spread of ˙NO close to 400 μm [43]. Under *in vivo* conditions, where operant blood flow promotes the scavenging of ˙NO by hemoglobin, the diffusional field of ˙NO is expectedly lower as suggested by the comparatively lower half-life found in the intact brain of anesthetized rats [44]. Still, evidence supports that ˙NO produced in a finite volume by multiple sources can diffuse and integrate the activity of multiple cells (neurons, asctrocytes, endothelial cells, etc) without the requirement of a physical connection to the ˙NO producing cell [45]. In the hippocampus, the mean distance between arterioles and NADPH-diaphorase-stained nerve fibers along the longitudinal axis of the *CA1* pyramidal layer ranges from 70 to 150 μm [17], which is consistent with the hypothesis that ˙NO produced by neurons can diffuse to the smooth muscle cells of arterioles and promote vasodilation. Furthermore, in the somatosensory cortex NOS-positive interneurons have been detected with their neurites in intimate contact with local blood vessels. According to this study, in layers I-III, 28 % of the interneurons located within a 50 μm radius of blood vessel express NOS [46].

The distance that ˙NO can diffuse has paramount importance for its role as a volume transmitter in the brain [47], a feature that may also have implications for neurovascular coupling. Given that this is a complex process involving different cell types and putatively several signaling cascades, the volume signaling of ˙NO, along with the multiplicity of molecular targets with which it can interact, may contribute to modulate and integrate different pathways contributing to neurovascular coupling. For instance, it was suggested that ˙NO can regulate gap junctional intercellular communication and thus contribute to upstream vasodilation, which is an important component of neurovascular coupling [39].

These arguments position ˙NO as a unique candidate to mediate the tight coupling between neuronal activity and local blood perfusion. Before exploring concrete experimental evidence that supports the role of ˙NO in neurovascular coupling, we should emphasize a further peculiarity of ˙NO-mediated neurovascular coupling related to the intrinsic regulation of this pathway. The increase of CBF linked to neuronal activation entails an increase of the circulating red blood cells that, ultimately, lead to enhanced ˙NO scavenging [44]. Thus, one can envisage that the signaling conveyed by ˙NO is translated to an increase in CBF, which in turn inactivates ˙NO, thus contributing to shape the underlying signaling pathway in a self-regulated manner.

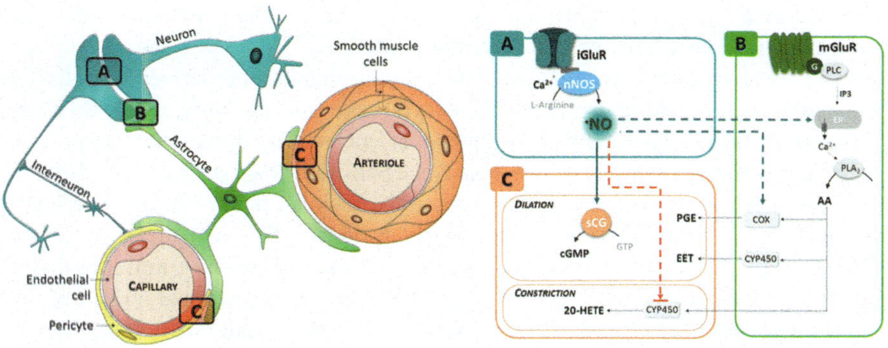

Fig. 19.1 Constituents of the neurovascular unit and schematic representation of the signaling pathways by which •NO may impact over neurovascular coupling. In neurons (**a**), synaptically released glutamate activates ionotropic glutamate receptors (iGluR), allowing the influx of Ca^{2+} and activating neuronal nitric oxide synthase (nNOS). The •NO released may directly promotes vasorelaxation and increase in the cerebral blood flow by activating sGC in smooth muscle cells of arterioles and/or in the pericytes surrounding capillaries (**c**, *solid blue line*). Indirectly, •NO may promotes vasodilation by modulating the signaling pathways involving arachidonic acid metabolites (*dashed lines*). In astrocytes (**b**), activation of metabotropic glutamate receptors (mGluR), by raising $[Ca^{2+}]_i$ activates phospholipase A_2 (PLA_2) which promotes the release of AA from membranes. •NO can either stimulate the activity of cyclooxygenase (COX), favoring the synthesis of prostaglandins (PGE) (*blue dashed line*), and inhibit the activity of cytochrome P450 hydroxylase, limiting the production of the vasoconstrictor 20-hydroxyeicosatetraenoic acid (20-HETE) (*red dashed line*)

3.4 Evidence for the Involvement of Nitric Oxide in Neurovascular Coupling

In the following sections we will explore the evidence supporting the role of •NO in neurovascular coupling, either directly as a mediator, or indirectly by modulation of other pathways. The conceptual framework of these lines of evidence is schematized in Fig. 19.1.

3.4.1 Indirect Evidence Provided by Pharmacological Approaches

The first experimental data implicating •NO in neurovascular coupling dates from the early 90s, when a number of pharmacological studies showed that NOS inhibitors could effectively reduce the cerebrovascular response coupled to neuronal activation. In anesthetized rats and mice, the competitive antagonism of NOS was shown to attenuate the increase of CBF in the somatosensory cortex in response to vibrissae stimulation [48], in the motor cortex in response to peripheral nerve stimulation [49] and in the cerebellar cortex in response to afferent electric stimulation [50, 51]. The same effect was observed in the cerebrovascular responses elicited by NMDA and glutamate [52–54]. These observations were later corroborated by the systemic administration of a relatively selective inhibitor for neuronal NOS. Under different experimental paradigms of neuronal activation, 7-nitroindazole (7-NI) successfully perturbed the hemodynamic responses associated with neurovascular coupling, suggesting the involvement of the neuronal isoform of NOS [15, 18, 55–57]. In parallel to the reports supporting the involvement of •NO in neurovascular coupling, other evidence arose in the opposite direction. Some studies reported the absence of effect of NOS inhibitors on the increase in cerebral blood flow following somatosensory stimulation, thus suggesting the lack of •NO participation [58–60]. Of note, the latter studies were performed in awake restrained animals, a condition associated to altered cerebrovascular regulation [61]. A further level of complexity was added by authors arguing that NOS inhibition may hamper neuronal activity, thereby affecting the coupled cerebrovascular response. Although evidence supports that •NO may act as an autocrine messenger affecting the neuron's intrinsic excitability [62], many studies have demonstrated that the effect of NOS inhibition on evoked potentials is absent or negligible [63–65].

In sum, while these pharmacological approaches provided relevant insights into the involvement of •NO in neurovascular coupling, they also added controversy. The interpretation of data using such approaches may be imprecise considering that the efficacy of pharmacological agents in complex biological is not complete. Supporting this notion, 7-NI does not maximally inhibit nNOS activity nor does it affect all brain regions to the same extent [66].

3.4.2 Indirect Evidence Based on NOS Knockout Mice

A complementary strategy to investigate the role of •NO in neurovascular coupling has relied on the use of genetically modified animals lacking selected isoforms of NOS. Curiously, the data collected from these animal models is unconvincing. In nNOS knockout mice, the increase of CBF in the cerebellar cortex, evoked either by electric stimulation or by glutamate superfusion, was significantly attenuated when compared to wild-type mice [67]. In turn, in the somatosensory cortex the hemodynamic response induced by whisker stimulation was found to be similar to that observed in wild-type mice and unresponsive to NOS inhibitors. It has been speculated that in mice lacking expression of nNOS neurovascular coupling is subserved by parallel or redundant pathways [68].

In this context, it should be emphasized that in the mouse brain there were identified three different NH_2-terminal nNOS protein variants, generated by the alternative 5′-end splicing of nNOS mRNA: nNOSα, nNOSβ and nNOSγ. The above-mentioned nNOS knockout mice were generated by targeted disruption of exon 2 (coding for PDZ domain) of the NOS1 gene [69], and thus only lack nNOSα. While nNOSα accounts for over 95 % of nNOS catalytic activity [70], nNOS knockout mice maintain a considerable amount of nNOS immunoreactivity in many brain regions. Coherently, by measuring •NO in the hippocampus with chemically modified microelectrodes, we have detected •NO transients in response to glutamatergic activation in these nNOS knockout mice (Fig. 19.2). This could relate to the

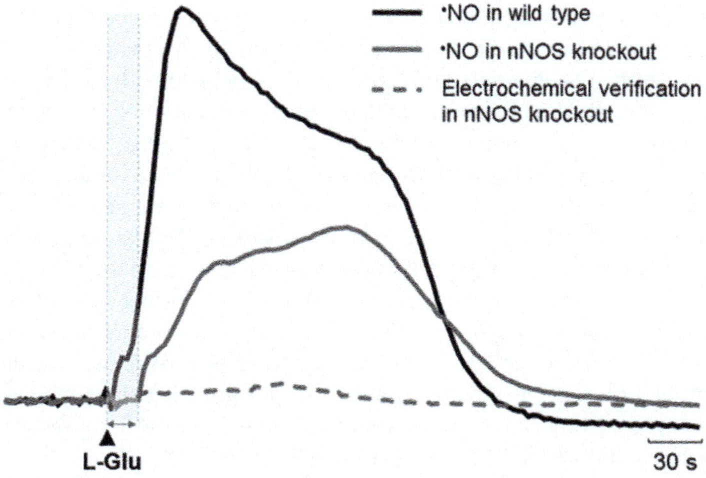

Fig. 19.2 Nitric oxide concentration gradients in the hippocampus of wild type mice (*black line*) and nNOS knockout mice (*gray line*) elicited by glutamatergic stimulation. •NO was measured directly using •NO selective microelectrodes as previously described [89]. Glutamate was pressure ejected at the time indicated by the arrow (20 mM, 25 nL). *Dashed line* represents the electrochemical verification of recorded signal in the nNOS knockout mice: at a low oxidation potential (+0.4 V vs Ag/AgCl), insufficient to oxidize •NO but able to oxidize other electroactive molecules, only a residual current is measured, validating •NO as the contributor for to the recorded signal. Note the diminished amplitude of the •NO signal and delayed onset delay in the nNOS knockout mice, as compared to the wild type mice, suggesting diverse pathways for •NO synthesis

cytosolic located nNOSβ variant, shown to be overexpressed in some brain regions of these nNOS knockout mice [70]. Concordantly, the mRNA of nNOSβ was still detectable in both the CA1-3 and dentate gyrus subregions of the hippocampus of nNOS knockout mice [71]. As depicted in Fig. 19.2, the temporal and concentration profile of the •NO transients in the hippocampus of nNOS knockout mouse differed from those detected in the wild-type, supporting alternative pathways for •NO synthesis in the former. Thus, data obtained from knockout models must be interpreted with caution due to potential mechanisms that might replace or compensate for the lost nNOS activity. This may result either from pre-existing redundant pathways or altered expression in the setting of the gene deletion [72]. Supporting this hypothesis, evidence suggests that eNOS compensates for a lack of nNOS in penile erection in the nNOS knockout mice [73] and that nNOS may compensate for eNOS in the pial vessel response to acetylcholine of eNOS knockout mice [74].

3.4.3 Direct Evidence: Temporal and Spatial Coupling Between Nitric Oxide Dynamics and Cerebral Blood Flow Changes

In spite of intensive research over the last decades, using both pharmacological approaches and genetically modified animals, the role of •NO in the neurovascular coupling has to date not been entirely clarified. The reasons largely rely on the experimental limitations imposed by the effusive nature of •NO (making hard to follow its concentration dynamics) and by neurovascular coupling itself: a spatial, temporal, and amplitude association has to be established between vascular changes and neuronal activation-dependent •NO dynamics in order to support the role of neuronal •NO as a direct mediator of neurovascular coupling. A step forward in the understanding of the contribution •NO in neurovascular coupling was made by the *in vivo* simultaneous measurements of •NO dynamics and cerebral blood flow changes during neuronal activation. Buerk and collaborators used •NO-selective microelectrodes to measure •NO produced in the rat somatosensory cortex upon forepaw electrical stimulation while simultaneously recording changes in the cerebral blood flow [75]. They observed transient increases of •NO and CBF upon stimulation, with increase in •NO preceding that of CBF. However, the interdependency of both events was not addressed nor was the cellular source of •NO investigated.

Inspired by this work, we have investigated the role of •NO in neurovascular coupling in the hippocampus, attempting to establish the sequence of events between neuronal activation and blood flow responses and, thus, identify the source of •NO. By simultaneously measuring •NO and CBF in response to glutamatergic activation coupled to pharmacological modulation of the process, we collected robust data supporting a direct role of neuronal-derived •NO in neurovascular coupling [18]. The observation that both MK-801, a NMDA receptor antagonist, and 7-NI individually elicited a dramatic inhibition of •NO production and that this effect was translated into limited CBF changes, in the absence of a significant effect of L-NIO (eNOS inhibitor), supported the interdependency of both events and identified the neurogenic origin of •NO. In turn, the inhibition of the CBF response by ODQ, a selective inhibitor of sGC, identified the pathway for neurovascular coupling process in the hippocampus: •NO acts directly via the canonic cGMP-dependent vasodilation pathway (Fig. 19.3). Our data strongly support the notion of •NO as a direct mediator of neuron-to-vessel signaling in the hippocampus [18]. However, we cannot completely rule out the hypothesis that the signaling pathways involved in the CBF response to NMDA receptor activation is more complex than simple •NO diffusion from neurons to vascular smooth muscle [19].

3.4.4 Modulatory Role of Nitric Oxide in Neurovascular Coupling

Several studies have suggested that •NO, rather than acting directly as a mediator, exerts its role by modulating other operant pathways in neurovascular coupling. Until recently, a major paradigm supported astrocytes as intermediaries bridging neuronal activity and hemodynamic changes [5], an idea raised by the observation of Ca^{2+} transients in these cells associated to the increase in cerebral blood flow [76]. Astrocytes have a privileged anatomical location to relay information from neurons directly to blood vessels (see Fig. 19.1) and form an extensive network interconnected by gap junctions that is ideally suited to propagate information upstream along the vascular tree [77]. The involvement of astrocytes is intimately connected to the activation of mGluR, namely mGluR1 and mGluR5, and increase

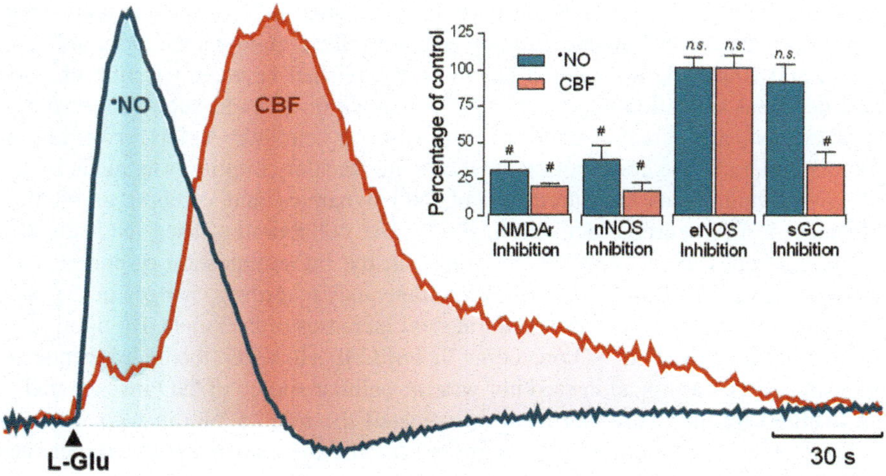

Fig. 19.3 Temporal correlation between ·NO concentration gradients and cerebral blood flow changes in response to glutamatergic activation. ·NO and cerebral blood flow were simultaneously measured in the hippocampus of urethane-anesthetized Wistar rats as previously described [18]. Glutamate was locally delivered at the time indicated by the *arrow*. Insert: Pharmacological modulation of the ·NO signals and cerebral blood flow changes using inhibitors of NMDA receptors (MK-801, 1 mg/kg i.p), nNOS (7-NI, 50 mg/kg i.p), eNOS (L-NIO, 40 mg/Kg i.p) and sGC (ODQ, 25 pmol locally) (Adapted from Ref. [18])

in intracellular [Ca^{2+}], which upon activation of phospholipase A2 leads to the release of arachidonic acid from plasma membrane phospholipids. Depending on the downstream pathway, arachidonic acid can be metabolized to prostaglandins (PGE) (via cyclooxygenase – COX) [76] and/or epoxyeicosatrienoic acids (EET) (via cytochrome P450 epoxygenase) in astrocytes [78, 79]. Arachidonic acid can also be released to smooth muscle cells where it is converted, via cytochrome P450 hydroxylase, to a potent vasoconstrictor, 20-hydroxyeicosatetraenoic acid (20-HETE) [80]. Nitric oxide has been implicated as modulator of the above-mentioned pathways by regulating the heme-containing enzymes involved in the metabolism of arachidonic acid. Indeed, ·NO can bind the heme of cytochrome P450, thereby inactivating the enzyme [81]. In this context, there is evidence supporting that ·NO-mediated vasodilation is partially dependent on the inhibition of 20-HETE synthesis [40, 82]. Conversely, ·NO can regulate COX activity favoring the synthesis of PGE [83] and thus vasodilation. Furthermore, it is suggested that ·NO may elicit intercellular Ca^{2+} increase via ryanodine receptor activation and influence the Ca^{2+} wave initiation and propagation [84].

Despite all the evidence, the role of astrocytes in the neurovascular through the mGluR- pathway has recently been contested based on two critical pieces of evidence: i) measureable changes of intracellular [Ca^{2+}] are detected later after the onset of vasodilation [85, 86]; and ii) mGluR5 receptor are absent in adult brain [87]. While these questions remain unsolved, it is evident that astrocytes are central players in neurovascular coupling. For instance, astrocytes may impact ·NO-mediated vasodilation by making arginine available to neurons. Arginine, a substrate for •NO synthesis, is released by astrocytes upon glutamatergic activation [88].

3.5 Regional Specificities of the Role of Nitric Oxide in Neurovascular Coupling

Considering that nNOS is scattered through the brain and that its regulation might be modulated as a function of local metabolic/redox status [89], it is possible that neuronal •NO is not necessarily equally dominant in controlling CBF as a function of neuronal activity in distinct areas of the brain. In fact, while in the hippocampus [17, 18] and cerebellum [16, 56] •NO seems to play a critical role by directly mediating vasodilation coupled to neuronal activity, this may not be the case for the cerebral cortex. In this brain region, •NO appears to lack a direct mediator role, and rather acts as a modulator of other pathways (see section 3.3.4) This is partially supported by the observation that the attenuation of hemodynamic response promoted by NOS inhibition can be restored by •NO donors in the somatosensory cortex [90], but not in the cerebellar cortex [91]. Thus, although •NO may be required, neurovascular coupling in the cerebral cortex may not linearly depend on its dynamic fluctuations. The quantitative relationship between •NO signals and CBF changes observed upon glutamatergic stimulation in the hippocampus and cerebral cortex further strengthens the concept of a distinct role for •NO in neurovascular coupling across brain regions [18]. This is predictable considering that, as mentioned previously, although nNOS is present in virtually every area of the CNS, there are significant variations expression and activity levels amongst the regions, with highest expression levels being found in the cerebellum [92]. Also, a pseudo-quantitative analysis of nNOS expression revealed a higher density in the hippocampus compared to the cerebral cortex, which was translated into higher •NO fluxes upon glutamatergic activation in the former region [89]. In the cerebral cortex NOS expressing neurons correspond to a very small proportion (0.5–2 %) of cortical neurons, 80 % of which are GABAergic. [93]. In the hippocampus, NOS expressing neurons are more abundant, comprising 2–60 % of the neurons depending on the layer and sub-region [94]. Furthermore, while many are also GABAergic, it should be noted that nNOS expressing interneurons are considered to be the most abundant interneuron subpopulation in the hippocampus, in opposition to what is observed in the cerebral cortex [95].

3.6 Obstacles in Studying the Role of Nitric Oxide in the Neurovascular Coupling

It is easily recognizable from the evidence scrutinized so far that the complete understanding of the role played by such an elusive free radical as •NO in neurovascular coupling has still to be achieved, mainly due to methodological limitations. As discussed above, the experimental approach used can constrain the results obtained, and pharmacological approaches or use of transgenic lines *per se* cannot provide reliable answers. Among others, the type of preparation (slices *versus* whole brain), the age of the animals, the use and type of anesthetic, the type/strength of stimuli, the afferent pathway targeted may have a huge contribution to the discrepancies reported. Underlying these problems, the quantitative measuring of •NO dynamics still remains a technological challenge. The unorthodox properties of •NO add further complexity to this matter. While the use of optogenetic approaches has been suggested as a promising strategy to study neurovascular coupling [96], the direct and reliable measurement of •NO is crucial to unraveling its role in the underlying mechanisms of neurovascular coupling.

4 Dysfunctional Neurovascular Coupling in Brain Aging and Alzheimer's Disease

Failure in the regulation of the tight coupling between neuronal activity and blood supply is associated with neuronal dysfunction. Multiple lines of evidence, ranging from clinical imaging to epidemiological and pharmacotherapeutical studies suggest that the disturbance of neurovascular coupling occurs during brain aging and age-related neuropathologies and is closely associated with cognitive decline [3, 97]. This is the case of Alzheimer's disease (AD), a common age-related neurodegenerative disorder characterized by progressive cognitive decline associated to neuronal loss in brain areas linked to memory processing. Although AD is classically considered an inherently neuronal disorder, convincing data support the occurrence of concomitant cerebrovascular dysfunction [98]. Indeed, several lines of evidence support cerebral hypoperfusion as a preclinical condition in AD [3]. Also, numerous clinical studies recognize a negative correlation between global cerebral blood flow and the age of healthy subjects [99–102]. In addition to resting blood flow, the adjustment of CBF in response to neuronal activation is described to be impaired in several experimental models [103–105] and AD patients [106–108], as well as during non-pathological aging [102, 109, 110]. Yet, the impact of these alterations is not fully resolved and questions related to the cause or consequence dilemma in neuronal dysfunction still persist. Using rodent models, we have observed that •NO-dependent neurovascular coupling is impaired during non-pathological aging and in AD and the dysfunction appears to be primarily cerebrovascular rather than neuronal in its origin. By simultaneously measuring •NO and CBF in the hippocampus of triple transgenic mouse model of AD (3xTg-AD mice) and in Fischer 344 rats (aging model) we observed that the increase in CBF elicited by glutamatergic activation was decreased both as a function of age and AD genotype, in spite of a barely unchanged •NO signaling. In 3xTg-AD mice the impairment is associated with a shift in the CBF response relative to the •NO signal, reflected by both a larger delay to the onset of CBF rise and a decreased amplitude

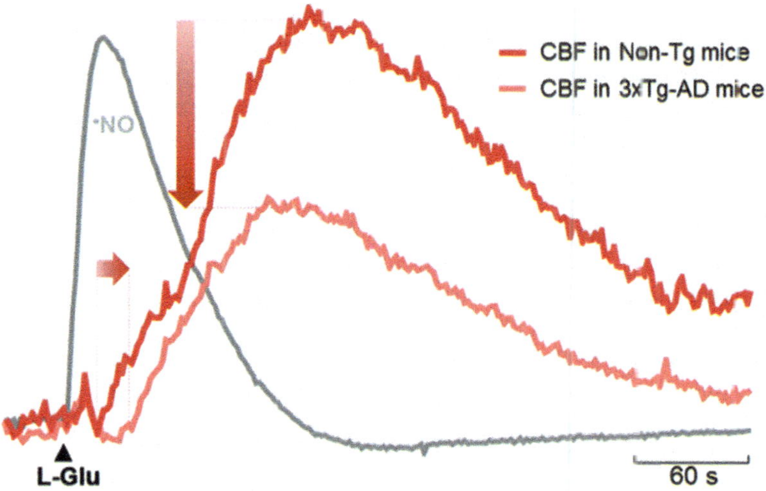

Fig. 19.4 Impairment of •NO-dependent neurovascular coupling in AD elucidated by the simultaneous measurements of •NO concentration dynamics and CBF changes in the hippocampus of 12 months-old Non-Tg mice and 3xTg-AD mice in response to L-glutamate. L-glutamate (0.5 nmol, 1 s) was locally applied at time indicated by the *black arrow* (Adapted from Ref. [111])

of CBF change (Fig. 19.4) [111]. Identical observations were obtained during aging in F344 rats. Whereas an age-dependent decline in •NO concentration dynamics upon glutamatergic activation is apparent in several brain regions [112], this did not occur in the *dentate gyrus* of the hippocampus. In turn, in this sub-region, the coupled cerebral blood flow changes were significantly attenuated during age progression (unpublished data). These observations suggest that the impairment of cerebral blood flow response compromises •NO-mediated neurovascular coupling in AD and aging, thus supporting the notion of cerebrovascular dysfunction as a fundamental process in both conditions.

4.1 Role of Oxidative Stress

The causes for cerebrovascular dysfunction and impairment of neurovascular coupling in AD and non-pathological brain aging can be diverse. However, convergent evidence based on 3-nitrotyrosine immunoreactivity and hydroethidine microfluorography in cerebral microvessels highlight changes in the redox balance of the cellular milieu towards oxidative stress as a pivotal and early factor in neurovascular deregulation [3, 113]. It is suggested that β-amyloid peptide (Aβ), the classical culprit in AD, induces the production of reactive oxygen and nitrogen species (RNOS) in the vasculature prior to brain parenchyma [114]. Amongst the multiple potential sources of RNOS in the brain, a significant contribution has been attributed to the nicotinamide adenine dinucleotide phosphate (NADPH) oxidase family, whose activation is reported to occur in AD and non-pathologic aging [115] and in response to cerebral hypoperfusion [116]. The superoxide anion ($O_2^{-\bullet}$) generated by NADPH oxidase is suggested to negatively impact neurovascular regulation primarily by decreasing $^{\bullet}$NO bioavailability, thus hampering its binding to sGC and consequent vasodilation. Indeed, the impairment of $^{\bullet}$NO-mediated neurovascular coupling in the hippocampus of 3xTg-AD mice and aged Fischer 344 rats can be mimicked in healthy young rats by intracerebroventricular injection of 2,3-dimethoxy-1,4-naphthoquinone (DMNQ), a redox-cycle agent that triggers intracellular $O_2^{-\bullet}$ production [117]. We observed that DMNQ promotes a decrease of CBF response while barely affecting glutamate-induced $^{\bullet}$NO dynamics (unpublished data). Superoxide dismutase (SOD) mimetics have shown potential to abrogate the attenuation of cerebral blood flow elicited by whisker stimulation observed in aged [118] and AD mice [113, 114] further corroborating the role of $O_2^{-\bullet}$ in the neurovascular coupling impairment.

Another important aspect of the interruption the $^{\bullet}$NO signaling by $O_2^{-\bullet}$ relates with diffusion limited reaction of $^{\bullet}$NO and $O_2^{-\bullet}$ itself and the formation of peroxynitrite ($ONOO^-$), a potent oxidant and nitrating agent that can alter the biological activity of lipids, proteins and nucleic acids. One of the most relevant cytotoxic pathways for $ONOO^-$ action involves the nitration of proteins, mainly in tyrosine residues, yielding 3-nitrotyrosine [119]. A growing body of evidence shows enhanced 3-nitrotyrosine immunoreactivity in brains from AD patients [120] with functional consequences in terms of nitration of several relevant proteins [121]. For instance, the mitochondrial isoform of SOD (MnSOD), a central intracellular $O_2^{-\bullet}$ scavenger, is reported to be nitrated in AD [122], a modification shown to lead to enzyme inactivation [119] and thus to enhanced nitroxidative stress. Peroxynitrite can further impair neurovascular coupling both by altering other mechanisms for vasodilation and by promoting structural alterations in the blood vessels [115].

5 Conclusions

In the present chapter we reviewed the concept of neurovascular coupling and discussed direct and indirect evidence supporting the role of $^{\bullet}$NO as a key regulator of the complex mechanism which guarantees the fine-tuning between neuronal activity, energetic demand and vascular response. Intimately associated to glutamatergic neurotransmission, both indirect (inhibition of NOS and NOS knockout mouse) and direct evidence stemming from direct *in vivo* recording of $^{\bullet}$NO concentration dynamics simultaneously with CBF support the notion that neuronal-derived $^{\bullet}$NO is a mediator of neurovascular coupling. Despite compelling evidence supporting a direct role in modulation of vascular tone via sGC activation in smooth muscle cells, the possible convergence of other $^{\bullet}$NO-dependent modulatory pathways encompassing the activity of astrocytes should and cannot be excluded.

Finally, aberrant RNOS formation is likely involved as a central issue in the dysfunction of neurovascular coupling observed in the aging brain or pathological conditions such as Alzheimer's disease.

References

1. Magistretti PJ (2006) Neuron-glia metabolic coupling and plasticity. J Exp Biol 209:2304–2311
2. Magistretti PJ, Pellerin L, Rothman DL, Shulman RG (1999) Energy on demand. Science 283:496–497
3. Iadecola C (2004) Neurovascular regulation in the normal brain and in Alzheimer's disease. Nat Rev Neurosci 5:347–360
4. Roy CS, Sherrington CS (1890) On the regulation of the blood-supply of the brain. J Physiol 11(85–158):117
5. Attwell D, Buchan AM, Charpak S, Lauritzen M et al (2010) Glial and neuronal control of brain blood flow. Nature 468:232–243
6. Iadecola C (1993) Regulation of the cerebral microcirculation during neural activity: is nitric oxide the missing link? Trends Neurosci 16:206–214
7. Gally JA, Montague PR, Reeke GN Jr, Edelman GM (1990) The NO hypothesis: possible effects of a short-lived, rapidly diffusible signal in the development and function of the nervous system. Proc Natl Acad Sci U S A 87:3547–3551
8. Powers WJ, Hirsch IB, Cryer PE (1996) Effect of stepped hypoglycemia on regional cerebral blood flow response to physiological brain activation. Am J Physiol 270:H554–H559
9. Lindauer U, Leithner C, Kaasch H, Rohrer B et al (2010) Neurovascular coupling in rat brain operates independent of hemoglobin deoxygenation. J Cereb Blood Flow Metab 30:757–768
10. Duchemin S, Boily M, Sadekova N, Girouard H (2012) The complex contribution of NOS interneurons in the physiology of cerebrovascular regulation. Front Neural Circuits 6:51
11. Bonvento G, Sibson N, Pellerin L (2002) Does glutamate image your thoughts? Trends Neurosci 25:359–364
12. Kew JN, Kemp JA (2005) Ionotropic and metabotropic glutamate receptor structure and pharmacology. Psychopharmacology (Berl) 179:4–29
13. Busija DW, Leffler CW (1989) Dilator effects of amino acid neurotransmitters on piglet pial arterioles. Am J Physiol 257:H1200–H1203
14. Busija DW, Bari F, Domoki F, Louis T (2007) Mechanisms involved in the cerebrovascular dilator effects of N-methyl-d-aspartate in cerebral cortex. Brain Res Rev 56:89–100
15. Yang G, Iadecola C (1996) Glutamate microinjections in cerebellar cortex reproduce cerebrovascular effects of parallel fiber stimulation. Am J Physiol 271:R1568–R1575
16. Rancillac A, Rossier J, Guille M, Tong XK et al (2006) Glutamatergic Control of Microvascular Tone by Distinct GABA Neurons in the Cerebellum. J Neurosci 26:6997–7006
17. Lovick TA, Brown LA, Key BJ (1999) Neurovascular relationships in hippocampal slices: physiological and anatomical studies of mechanisms underlying flow-metabolism coupling in intraparenchymal microvessels. Neuroscience 92:47–60
18. Lourenço CF, Santos RM, Barbosa RM, Cadenas E et al (2014) Neurovascular coupling in hippocampus is mediated via diffusion by neuronal-derived nitric oxide. Free Radic Biol Med 73:421–429
19. Bhardwaj A, Northington FJ, Carhuapoma JR, Falck JR et al (2000) P-450 epoxygenase and NO synthase inhibitors reduce cerebral blood flow response to N-methyl-D-aspartate. Am J Physiol Heart Circ Physiol 279:H1616–H1624

20. LeMaistre JL, Sanders SA, Stobart MJ, Lu L et al (2012) Coactivation of NMDA receptors by glutamate and D-serine induces dilation of isolated middle cerebral arteries. J Cereb Blood Flow Metab 32:537–547
21. Chen BR, Kozberg MG, Bouchard MB, Shaik MA et al (2014) A critical role for the vascular endothelium in functional neurovascular coupling in the brain. J Am Heart Assoc 3:e000787
22. Girouard H (1985) Iadecola C (2006) Neurovascular coupling in the normal brain and in hypertension, stroke, and Alzheimer disease. J Appl Physiol 100:328–335
23. Garthwaite J, Boulton CL (1995) Nitric oxide signaling in the central nervous system. Annu Rev Physiol 57:683–706
24. Bredt DS, Glatt CE, Hwang PM, Fotuhi M et al (1991) Nitric oxide synthase protein and mRNA are discretely localized in neuronal populations of the mammalian CNS together with NADPH diaphorase. Neuron 7:615–624
25. Tochio H, Zhang Q, Mandal P, Li M et al (1999) Solution structure of the extended neuronal nitric oxide synthase PDZ domain complexed with an associated peptide. Nat Struct Biol 6:417–421
26. Brenman JE, Chao DS, Gee SH, McGee AW et al (1996) Interaction of nitric oxide synthase with the postsynaptic density protein PSD-95 and alpha1-syntrophin mediated by PDZ domains. Cell 84:757–767
27. Kornau HC, Schenker LT, Kennedy MB, Seeburg PH (1995) Domain interaction between NMDA receptor subunits and the postsynaptic density protein PSD-95. Science 269:1737–1740
28. Niethammer M, Kim E, Sheng M (1996) Interaction between the C terminus of NMDA receptor subunits and multiple members of the PSD-95 family of membrane-associated guanylate kinases. J Neurosci 16:2157–2163
29. Christopherson KS, Hillier BJ, Lim WA, Bredt DS (1999) PSD-95 assembles a ternary complex with the N-methyl-D-aspartic acid receptor and a bivalent neuronal NO synthase PDZ domain. J Biol Chem 274:27467–27473
30. Jaffrey SR, Snowman AM, Eliasson MJ, Cohen NA et al (1998) CAPON: a protein associated with neuronal nitric oxide synthase that regulates its interactions with PSD95. Neuron 20:115–124
31. Nedvetsky PI, Sessa WC, Schmidt HH (2002) There's NO binding like NOS binding: protein-protein interactions in NO/cGMP signaling. Proc Natl Acad Sci U S A 99:16510–16512
32. Dawson VL, Dawson TM (1998) Nitric oxide in neurodegeneration. Prog Brain Res 118:215–229
33. Daff S (2010) NO synthase: structures and mechanisms. Nitric Oxide 23:1–11
34. Pereira C, Ferreira NR, Rocha BS, Barbosa RM et al (2013) The redox interplay between nitrite and nitric oxide: From the gut to the brain. Redox Biol 1:276–284
35. Piknova B, Kocharyan A, Schechter AN, Silva AC (2011) The role of nitrite in neurovascular coupling. Brain Res 1407:62–68
36. Moncada S, Higgs EA (2006) The discovery of nitric oxide and its role in vascular biology. Br J Pharmacol 147(Suppl 1):S193–S201
37. Murad F (1986) Cyclic guanosine monophosphate as a mediator of vasodilation. J Clin Invest 78:1–5
38. Cohen RA, Adachi T (2006) Nitric-oxide-induced vasodilatation: regulation by physiologic s-glutathiolation and pathologic oxidation of the sarcoplasmic endoplasmic reticulum calcium ATPase. Trends Cardiovasc Med 16:109–114
39. Adachi T, Weisbrod RM, Pimentel DR, Ying J et al (2004) S-Glutathiolation by peroxynitrite activates SERCA during arterial relaxation by nitric oxide. Nat Med 10:1200–1207
40. Mulligan SJ, MacVicar BA (2004) Calcium transients in astrocyte endfeet cause cerebrovascular constrictions. Nature 431:195–199
41. Lancaster JR Jr (1997) A tutorial on the diffusibility and reactivity of free nitric oxide. Nitric Oxide 1:18–30

42. Santos RM, Lourenco CF, Gerhardt GA, Cadenas E et al (2011) Evidence for a pathway that facilitates nitric oxide diffusion in the brain. Neurochem Int 59:90–96
43. Ledo A, Barbosa RM, Gerhardt GA, Cadenas E et al (2005) Concentration dynamics of nitric oxide in rat hippocampal subregions evoked by stimulation of the NMDA glutamate receptor. Proc Natl Acad Sci U S A 102:17483–17488
44. Santos RM, Lourenco CF, Pomerleau F, Huettl P et al (2011) Brain nitric oxide inactivation is governed by the vasculature. Antioxid Redox Signal 14:1011–1021
45. Laranjinha J, Santos RM, Lourenço CF, Ledo A et al (2012) Nitric oxide signaling in the brain: translation of dynamics into respiration control and neurovascular coupling. Ann N Y Acad Sci 1259:10–18
46. Cauli B, Tong XK, Rancillac A, Serluca N et al (2004) Cortical GABA interneurons in neurovascular coupling: relays for subcortical vasoactive pathways. J Neurosci 24:8940–8949
47. Garthwaite J (2016) From synaptically localized to volume transmission by nitric oxide. J Physiol 594:9–18
48. Dirnagl U, Lindauer U, Villringer A (1993) Role of nitric oxide in the coupling of cerebral blood flow to neuronal activation in rats. Neurosci Lett 149:43–46
49. Northington FJ, Matherne GP, Berne RM (1992) Competitive inhibition of nitric oxide synthase prevents the cortical hyperemia associated with peripheral nerve stimulation. Proc Natl Acad Sci U S A 89:6649–6652
50. Akgoren N, Fabricius M, Lauritzen M (1994) Importance of nitric oxide for local increases of blood flow in rat cerebellar cortex during electrical stimulation. Proc Natl Acad Sci U S A 91:5903–5907
51. Iadecola C, Li J, Ebner TJ, Xu X (1995) Nitric oxide contributes to functional hyperemia in cerebellar cortex. Am J Physiol 268:R1153–R1162
52. Meng W, Tobin JR, Busija DW (1995) Glutamate-induced cerebral vasodilation is mediated by nitric oxide through N-methyl-D-aspartate receptors. Stroke 26:857–862; discussion 863
53. Faraci FM, Breese KR (1993) Nitric oxide mediates vasodilatation in response to activation of N-methyl-D-aspartate receptors in brain. Circ Res 72:476–480
54. Pelligrino DA, Gay RL 3rd, Baughman VL, Wang Q (1996) NO synthase inhibition modulates NMDA-induced changes in cerebral blood flow and EEG activity. Am J Physiol 271:H990–H995
55. Bonvento G, Cholet N, Seylaz J (2000) Sustained attenuation of the cerebrovascular response to a 10 min whisker stimulation following neuronal nitric oxide synthase inhibition. Neurosci Res 37:163–166
56. Yang ST, Chang HH (1998) Nitric oxide of neuronal origin mediates NMDA-induced cerebral hyperemia in rats. Neuroreport 9:415–418
57. Faraci FM, Brian JE Jr (1995) 7-Nitroindazole inhibits brain nitric oxide synthase and cerebral vasodilatation in response to N-methyl-D-aspartate. Stroke 26:2172–2175; discussion 2176
58. Adachi K, Takahashi S, Melzer P, Campos KL et al (1994) Increases in local cerebral blood flow associated with somatosensory activation are not mediated by NO. Am J Physiol 267:H2155–H2162
59. Greenberg JH, Sohn NW, Hand PJ (1999) Nitric oxide and the cerebral-blood-flow response to somatosensory activation following deafferentation. Exp Brain Res 129:541–550
60. Wang Q, Kjaer T, Jorgensen MB, Paulson OB et al (1993) Nitric oxide does not act as a mediator coupling cerebral blood flow to neural activity following somatosensory stimuli in rats. Neurol Res 15:33–36
61. Lacombe P, Seylaz J (1984) Significance of the cerebrovascular effects of immobilization stress in the rabbit. J Cereb Blood Flow Metab 4:397–406
62. Artinian L, Zhong L, Yang H, Rehder V (2012) Nitric oxide as intracellular modulator: internal production of NO increases neuronal excitability via modulation of several ionic conductances. Eur J Neurosci 36:3333–3343

63. Hoffmeyer HW, Enager P, Thomsen KJ, Lauritzen MJ (2007) Nonlinear neurovascular coupling in rat sensory cortex by activation of transcallosal fibers. J Cereb Blood Flow Metab 27:575–587
64. Brozickova C, Otahal J (2013) Effect of an inhibitor of neuronal nitric oxide synthase 7-nitroindazole on cerebral hemodynamic response and brain excitability in urethane-anesthetized rats. Physiol Res 62(Suppl 1):S57–S66
65. Burke M, Buhrle C (2006) BOLD response during uncoupling of neuronal activity and CBF. Neuroimage 32:1–8
66. Kalisch BE, Connop BP, Jhamandas K, Beninger RJ et al (1996) Differential action of 7-nitro indazole on rat brain nitric oxide synthase. Neurosci Lett 219:75–78
67. Yang G, Zhang Y, Ross ME, Iadecola C (2003) Attenuation of activity-induced increases in cerebellar blood flow in mice lacking neuronal nitric oxide synthase. Am J Physiol Heart Circ Physiol 285:H298–H304
68. Ma J, Ayata C, Huang PL, Fishman MC et al (1996) Regional cerebral blood flow response to vibrissal stimulation in mice lacking type I NOS gene expression. Am J Physiol 270:H1085–H1090
69. Huang PL, Dawson TM, Bredt DS, Snyder SH et al (1993) Targeted disruption of the neuronal nitric oxide synthase gene. Cell 75:1273–1286
70. Eliasson MJ, Blackshaw S, Schell MJ, Snyder SH (1997) Neuronal nitric oxide synthase alternatively spliced forms: prominent functional localizations in the brain. Proc Natl Acad Sci U S A 94:3396–3401
71. Putzke J, Seidel B, Huang PL, Wolf G (2000) Differential expression of alternatively spliced isoforms of neuronal nitric oxide synthase (nNOS) and N-methyl-D-aspartate receptors (NMDAR) in knockout mice deficient in nNOS alpha (nNOS alpha(Delta/Delta) mice). Brain Res Mol Brain Res 85:13–23
72. Huang PL, Lo EH (1998) Genetic analysis of NOS isoforms using nNOS and eNOS knockout animals. Prog Brain Res 118:13–25
73. Burnett AL, Nelson RJ, Calvin DC, Liu JX et al (1996) Nitric oxide-dependent penile erection in mice lacking neuronal nitric oxide synthase. Mol Med 2:288–296
74. Meng W, Ma J, Ayata C, Hara H et al (1996) ACh dilates pial arterioles in endothelial and n e u r o n a l NOS knockout mice by NO-dependent mechanisms. Am J Physiol 271:H1145–H1150
75. Buerk DG, Ances BM, Greenberg JH, Detre JA (2003) Temporal dynamics of brain tissue nitric oxide during functional forepaw stimulation in rats. Neuroimage 18:1–9
76. Zonta M, Angulo MC, Gobbo S, Rosengarten B et al (2003) Neuron-to-astrocyte signaling is central to the dynamic control of brain microcirculation. Nat Neurosci 6:43–50
77. Giaume C, Liu X (2012) From a glial syncytium to a more restricted and specific glial networking. J Physiol Paris 106:34–39
78. Harder DR, Alkayed NJ, Lange AR, Gebremedhin D et al (1998) Functional hyperemia in the brain: hypothesis for astrocyte-derived vasodilator metabolites. Stroke 29:229–234
79. Iliff JJ, Wang R, Zeldin DC, Alkayed NJ (2009) Epoxyeicosanoids as mediators of neurogenic vasodilation in cerebral vessels. Am J Physiol Heart Circ Physiol 296:H1352–H1363
80. Metea MR, Newman EA (2006) Glial cells dilate and constrict blood vessels: a mechanism of neurovascular coupling. J Neurosci 26:2862–2870
81. Roman RJ (2002) P-450 metabolites of arachidonic acid in the control of cardiovascular function. Physiol Rev 82:131–185
82. Sun CW, Falck JR, Okamoto H, Harder DR et al (2000) Role of cGMP versus 20-HETE in the vasodilator response to nitric oxide in rat cerebral arteries. Am J Physiol Heart Circ Physiol 279:H339–H350
83. Fujimoto Y, Uno E, Sakuma S (2004) Effects of reactive oxygen and nitrogen species on cyclooxygenase-1 and -2 activities. Prostaglandins Leukot Essent Fatty Acids 71:335–340
84. Willmott NJ, Wong K, Strong AJ (2000) A fundamental role for the nitric oxide-G-kinase signaling pathway in mediating intercellular Ca(2+) waves in glia. J Neurosci 20:1767–1779

85. Nizar K, Uhlirova H, Tian P, Saisan PA et al (2013) In vivo stimulus-induced vasodilation occurs without IP3 receptor activation and may precede astrocytic calcium increase. J Neurosci 33:8411–8422
86. Bonder DE, McCarthy KD (2014) Astrocytic Gq-GPCR-linked IP3R-dependent Ca2+ signaling does not mediate neurovascular coupling in mouse visual cortex in vivo. J Neurosci 34:13139–13150
87. Sun W, McConnell E, Pare JF, Xu Q et al (2013) Glutamate-dependent neuroglial calcium signaling differs between young and adult brain. Science 339:197–200
88. Gensert JM, Ratan RR (2006) The metabolic coupling of arginine metabolism to nitric oxide generation by astrocytes. Antioxid Redox Signal 8:919–928
89. Lourenço CF, Ferreira NR, Santos RM, Lukacova N et al (2014) The pattern of glutamate-induced nitric oxide dynamics in vivo and its correlation with nNOS expression in rat hippocampus, cerebral cortex and striatum. Brain Res 1554:1–11
90. Lindauer U, Megow D, Matsuda H, Dirnagl U (1999) Nitric oxide: a modulator, but not a mediator, of neurovascular coupling in rat somatosensory cortex. Am J Physiol 277:H799–H811
91. Yang G, Iadecola C (1997) Obligatory role of NO in glutamate-dependent hyperemia evoked from cerebellar parallel fibers. Am J Physiol 272:R1155–R1161
92. Bredt DS, Hwang PM, Snyder SH (1990) Localization of nitric oxide synthase indicating a neural role for nitric oxide. Nature 347:768–770
93. Valtschanoff JG, Weinberg RJ, Kharazia VN, Schmidt HH et al (1993) Neurons in rat cerebral cortex that synthesize nitric oxide: NADPH diaphorase histochemistry, NOS immunocytochemistry, and colocalization with GABA. Neurosci Lett 157:157–161
94. Valtschanoff JG, Weinberg RJ, Kharazia VN, Nakane M et al (1993) Neurons in rat hippocampus that synthesize nitric oxide. J Comp Neurol 331:111–121
95. Tricoire L, Vitalis T (2012) Neuronal nitric oxide synthase expressing neurons: a journey from birth to neuronal circuits. Front Neural Circuits 6:82
96. Bregestovski P, Zilberter Y (2014) Optogenetics to help exploring the cerebral blood flow regulation. Front Pharmacol 5:107
97. Zlokovic BV (2011) Neurovascular pathways to neurodegeneration in Alzheimer's disease and other disorders. Nat Rev Neurosci 12:723–738
98. de la Torre JC (2004) Is Alzheimer's disease a neurodegenerative or a vascular disorder? Data, dogma, and dialectics. Lancet Neurol 3:184–190
99. Fisher JP, Hartwich D, Seifert T, Olesen ND et al (2013) Cerebral perfusion, oxygenation and metabolism during exercise in young and elderly individuals. J Physiol 591:1859–1870
100. Schultz SK, O'Leary DS, Boles Ponto LL, Watkins GL et al (1999) Age-related changes in regional cerebral blood flow among young to mid-life adults. Neuroreport 10:2493–2496
101. Krejza J, Mariak Z, Walecki J, Szydlik P et al (1999) Transcranial color Doppler sonography of basal cerebral arteries in 182 healthy subjects: age and sex variability and normal reference values for blood flow parameters. AJR Am J Roentgenol 172:213–218
102. Fabiani M, Gordon BA, Maclin EL, Pearson MA et al (2014) Neurovascular coupling in normal aging: a combined optical, ERP and fMRI study. Neuroimage 85(Pt 1):592–607
103. Niwa K, Carlson GA, Iadecola C (2000) Exogenous A beta1-40 reproduces cerebrovascular alterations resulting from amyloid precursor protein overexpression in mice. J Cereb Blood Flow Metab 20:1659–1668
104. Shin HK, Jones PB, Garcia-Alloza M, Borrelli L et al (2007) Age-dependent cerebrovascular dysfunction in a transgenic mouse model of cerebral amyloid angiopathy. Brain 130:2310–2319
105. Mueggler T, Baumann D, Rausch M, Staufenbiel M et al (2003) Age-dependent impairment of somatosensory response in the amyloid precursor protein 23 transgenic mouse model of Alzheimer's disease. J Neurosci 23:8231–8236
106. Ruitenberg A, den Heijer T, Bakker SL, van Swieten JC et al (2005) Cerebral hypoperfusion and clinical onset of dementia: the Rotterdam Study. Ann Neurol 57:789–794

107. Alsop DC, Detre JA, Grossman M (2000) Assessment of cerebral blood flow in Alzheimer's disease by spin-labeled magnetic resonance imaging. Ann Neurol 47:93–100
108. Xu G, Antuono PG, Jones J, Xu Y et al (2007) Perfusion fMRI detects deficits in regional CBF during memory-encoding tasks in MCI subjects. Neurology 69:1650–1656
109. Jessen SB, Mathiesen C, Lind BL, Lauritzen M (2015) Interneuron deficit associates attenuated network synchronization to mismatch of energy supply and demand in aging mouse brains. Cereb Cortex (*in press*)
110. Balbi M, Ghosh M, Longden TA, Jativa Vega M et al (2015) Dysfunction of mouse cerebral arteries during early aging. J Cereb Blood Flow Metab 35:1445–1453
111. Lourenço CF, Ledo A, Dias C, Barbosa RM et al (2015) Neurovascular and neurometabolic derailment in aging and Alzheimer's disease. Front Aging Neurosci 7:103
112. Ledo A, Lourenço CF, Caetano M, Barbosa RM et al (2015) Age-associated changes of nitric oxide concentration dynamics in the central nervous system of Fisher 344 rats. Cell Mol Neurobiol 35:33–44
113. Hamel E, Nicolakakis N, Aboulkassim T, Ongali B et al (2008) Oxidative stress and cerebrovascular dysfunction in mouse models of Alzheimer's disease. Exp Physiol 93:116–120
114. Park L, Anrather J, Forster C, Kazama K et al (2004) Abeta-induced vascular oxidative stress and attenuation of functional hyperemia in mouse somatosensory cortex. J Cereb Blood Flow Metab 24:334–342
115. Chrissobolis S, Faraci FM (2008) The role of oxidative stress and NADPH oxidase in cerebrovascular disease. Trends Mol Med 14:495–502
116. Choi DH, Lee KH, Kim JH, Seo JH et al (2014) NADPH Oxidase 1, a Novel Molecular Source of ROS in Hippocampal Neuronal Death in Vascular Dementia. Antioxid Redox Signal 21:533–550
117. Brunmark A, Cadenas E (1989) Redox and addition chemistry of quinoid compounds and its biological implications. Free Radic Biol Med 7:435–477
118. Park L, Anrather J, Girouard H, Zhou P et al (2007) Nox2-derived reactive oxygen species mediate neurovascular dysregulation in the aging mouse brain. J Cereb Blood Flow Metab 27:1908–1918
119. Radi R (2004) Nitric oxide, oxidants, and protein tyrosine nitration. Proc Natl Acad Sci U S A 101:4003–4008
120. Sultana R, Poon HF, Cai J, Pierce WM et al (2006) Identification of nitrated proteins in Alzheimer's disease brain using a redox proteomics approach. Neurobiol Dis 22:76–87
121. Smith MA, Richey Harris PL, Sayre LM, Beckman JS et al (1997) Widespread peroxynitrite-mediated damage in Alzheimer's disease. J Neurosci 17:2653–2657
122. Aoyama K, Matsubara K, Fujikawa Y, Nagahiro Y et al (2000) Nitration of manganese superoxide dismutase in cerebrospinal fluids is a marker for peroxynitrite-mediated oxidative stress in neurodegenerative diseases. Ann Neurol 47:524–527

Chapter 20
Oxidative Stress and Neurodegeneration

Juana M. Pasquini, Laura A. Pasquini, and Hector R. Quintá

Abstract Multiple sclerosis, a highly disseminated chronic inflammatory demyelinating disease, entails progressive neuroaxonal degeneration and is one of the most common causes of progressive disability affecting young people. The mechanisms involved in oxidative stress-mediated neurodegeneration in MS patients include free radical production from different sources: (a) mitochondria forced to produce high levels of energy for axonal transport upon myelin sheath loss, (b) immune cells activated upon demyelination and neurodegeneration, and (c) myelin deficiencies in producing ATP synthesis outside mitochondria. In addition, oxidative stress is amplified by iron released into the extracellular space from myelin breakdown and degenerated macrophages and microglia. The normal neuronal polarization and development in each region of the central nervous system depend on the normal function of actin cytoskeleton dynamics. This dynamics is primarily affected when there is a deregulation in the intraneuronal production of reactive oxygen species. These reactive oxygen species promote oxidation of filamentous actin (cytoskeleton depolymerization) and, therefore, axonal collapse. In summary, preventing oxidative stress is crucial to maintain the normal function of the central nervous system.

Keywords Demyelination • Neurodegeneration • Microglia • Oxidative stress • Oligodendrocytes • Mical • Hydrogen peroxide • Semaphorin 3A

1 Introduction. Oxidative Stress in Demyelinating Diseases

Central nervous system (CNS) demyelinating diseases exhibit a wide range of clinical presentations, with different pathological findings, temporal evolution, severity and response to treatment. Among them, multiple sclerosis (MS), a highly disseminated, chronic, inflammatory demyelinating disease, entails progressive neuroaxonal degeneration and is one of the most common causes of progressive disability

J.M. Pasquini (✉) • L.A. Pasquini • H.R. Quintá
Departamento de Química Biológica and Instituto de Química y Fisicoquímica Biológica (IQUIFIB, UBA-CONICET), Facultad de Farmacia y Bioquímica, Universidad de Buenos Aires, Buenos Aires, Argentina
e-mail: jpasquin@qb.ffyb.uba.ar

© Springer International Publishing Switzerland 2016

R.J. Gelpi et al. (eds.), *Biochemistry of Oxidative Stress*, Advances in Biochemistry in Health and Disease 16, DOI 10.1007/978-3-319-45865-6_20

affecting young people. Despite variable disease courses, most patients initially present a relapsing-remitting pattern (relapsing-remitting MS; RRMS). After 10–15 years, this pattern becomes progressive in up to 50 % of untreated patients, with clinical symptoms slowly but constantly causing deterioration over a period of many years (secondary progressive MS; SPMS). In about 15 % MS patients, however, disease progression is relentless as from onset (primary progressive MS; PPMS). The transition to a progressive disease course is thought to occur when a threshold of neuronal and axonal loss is reached and the compensatory capacity of the CNS is surpassed. Abundant evidence supports the notion that axonal degeneration and neuronal loss, rather than inflammation, are the main pathophysiological substrates of permanent clinical deficits [1, 2]. Although the clinical sequelae of axonal loss are more apparent in the later stages of MS, axonal injury occurs early in acute MS lesions [3]. Despite the fact that demyelination with loss of trophic support from oligodendrocytes clearly contributes to axonal degeneration [4, 5], both axonal and neuronal injury may occur without demyelination [6–11].

A central role has been widely assigned to oxidative stress in the pathogenesis of MS [12] on the basis of biochemical analyses of cerebrospinal fluid/blood samples, tissue homogenates [13–16] and animal models [17, 18]. Among them, oxidized DNA, lipids and protein adducts were frequently detected in active MS lesions [18, 19] and were spatially and quantitatively correlated with apoptotic oligodendrocytes and neurodegeneration [20]. So far, the mechanisms involved in oxidative stress-mediated neurodegeneration in MS patients include free radical production from different sources: (a) mitochondria forced to produce high levels of energy for axonal transport upon myelin sheath loss, (b) immune cells activated upon demyelination and neurodegeneration, and (c) myelin deficiencies in producing ATP synthesis outside mitochondria. In addition, oxidative stress is amplified by iron released into the extracellular space from myelin breakdown and degenerated macrophages and microglia (Fig. 20.1). These sources will be discussed in detail in the following sections.

1.1 Mitochondrial Oxidative Stress in Demyelinated Axons

Mitochondria, the most efficient producers of energy in the form of ATP, meet energy demands from axons, placed at relatively great distances from the neuronal cell body, and produce excessive free radicals when functionally compromised [21]. Axonal transport is highly energy-demanding and thus extremely sensitive to fluctuations in energy supply. Recent work has shown that oligodendrocytes supply lactate to myelinated axons as a metabolic substrate for mitochondria to generate ATP, a deficient process upon demyelination [5, 22]. Myelination makes axons metabolically efficient, while demyelination generates the disruption of ion channel compartmentalization and the loss of both saltatory conduction and microenvironmental protection from extra-axoplasmic changes. Electrogenic machinery redistribution and the loss of metabolic substrates from oligodendrocytes are thought to

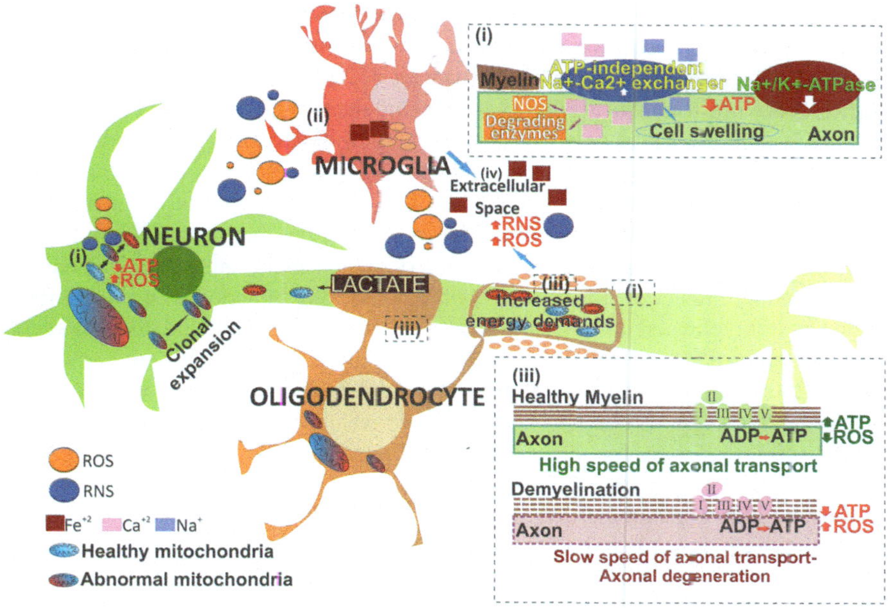

Fig. 20.1 Mechanisms involved in oxidative stress-mediated neurodegeneration in demyelinating diseases: (*i*) mitochondria are forced to produce high levels of energy for axonal transport, (*ii*) the immune cells are activated upon demyelination and neurodegeneration, (*iii*) demyelination produces deficiencies in oxidative phosphorylation in the mitochondria and outside of these organelles, and (*iv*) oxidative stress is amplified by the iron released into the extracellular space from myelin breakdown and from degenerated macrophages and microglia

boost energy needs in demyelinated axons, which might be responsible for the increase in mitochondrial density observed in hypomyelination, dysmyelination and demyelination [23–26]. This increase in the number of mitochondria may also entail a higher risk of reactive oxygen species (ROS) production leading to mitochondrial dysfunction [21], which is considered to occur in axonal pathology and degeneration after demyelination or dysmyelination [27–29].

Axonal mitochondria consist of two populations: large stationary and small motile mitochondria. Mitochondrial size, complex IV activity and the speed of motile mitochondria are increased in demyelinated axons. Recent work identifies this response termed "axonal mitochondrial response to demyelination" as a compensatory phenomenon [30]. Mitochondria, generated in the cell body, are transported to the axon (anterograde movement controlled by kinesins) to replace worn out mitochondria. Damaged axonal mitochondria are transported back to the cell body (retrograde movement controlled by dyneins) for degradation. The increase of axonal mitochondria volume following acute demyelination requires the presence of syntaphilin, an axonal molecule that immobilizes stationary mitochondria to microtubules, and protects against axonal degeneration [31].

Inflammatory products and free radicals injure neuronal mitochondria during the pre-progressive stage of MS, forming mitochondrial DNA (mtDNA) deletions in white and grey matter. In MS patients, the mutant to wild-type mtDNA ratio (heteroplasmy) in metabolically active postmitotic neurons expands with disease progression and age by clonal mtDNA expansion [32]. These abnormal mitochondria suffer aberrant placement to the axon causing energy failure in the demyelinated axon. Myelin sheaths are only transiently and inefficiently restored during MS and, although beneficial, remyelination only partially reverses the axonal metabolic changes that accompany demyelination [26]. The mitochondrial content within axons decreases in acute and chronic demyelination but fails to reach myelination values, which reflects a partial amelioration of axonal energy demands. The reasons for the scarce impact of this regeneration process may include: (a) a shorter intermode in remyelinated axons entails more nodes of Ranvier and thus increases sodium influx during impulse conduction, with a concomitant higher energy demand; (b) remyelinated sheaths are comparatively thinner; and (c) oligodendroglial metabolic support to axons is not fully recovered. Seventy-five percent of total neural energy is used for ionic equilibrium during information processing [33], which renders ion channel functions particularly vulnerable to energy shortages. Axonal sodium channel distribution is affected by demyelination and the resulting excessive sodium needs to be extruded by Na+/K + -ATPase, an energy-dependent process itself [34]. This increase in axonal sodium may alternatively be reversed by the ATP-independent Na + -Ca2+ exchanger, which in turn triggers a catastrophic increase in axonal calcium, leading to autolytic protease activation [35, 36]. Therefore, both energy shortage and demyelination associated to chronic CNS inflammation generate dysfunctional ion channels, which subsequently evoke downstream mechanisms mostly converging on Ca2+ overload. This overload appears to be chiefly responsible for neurotoxicity, which spearheads a vicious circle by activating degrading enzymes, compromising mitochondrial function and impairing axonal transport, all leading to even higher Ca2+ levels [37].

1.2 Oxidative Stress Induced by Activated Macrophages and Microglia

Activated macrophages and microglia have been centrally implicated in ongoing neurodegeneration and active inflammation and breakdown of the blood-brain barrier have been detected in the brains of MS patients. ROS and reactive nitrogen species (RNS) such as nitric oxide (NO and $ONOO^-$) are induced by macrophages and microglia and released into the extracellular space [38] in MS and experimental models, while their oxidation products correlate with inflammation and axonal damage [20, 39, 40]. NO diffuses across membranes and competes with O_2 for mitochondrial cytochrome c oxidase binding, thus decreasing respiratory chain function [41, 42]. ROS and RNS induce covalent alterations and consequent mutations in mtDNA, which is more vulnerable than nuclear DNA [43, 44]. These mtDNA mutations inhibit the efficiency of oxidative phosphorylation and further increase ROS production, thus initiating a vicious circle [45].

1.3 Oxidative Stress Generated from Myelin

Oxidative phosphorylation has been reported to occur not only inside but also outside the mitochondrion [46–51]. In particular, myelin has the capacity to produce aerobic metabolism and thus generate ATP, which might be transferred to the axon to meet its energetic needs. Also, isolated myelin can produce high levels of oxidation products such as MDA, 4-HNE and H_2O_2 through complexes I and III and contains active antioxidant enzymatic defenses, e.g. superoxide dismutase, catalase and glutathione peroxidase. Uncoupling compounds or Complex I inhibitors increase the rate of ROS generation and produce oxidative stress. Endogenous non-enzymatic antioxidants cannot keep the redox balance in pathological conditions like demyelinating diseases, that ultimately lead to axonal damage. These events may even constitute a never-ending vicious circle in which oxidative stress damage in the lipid-rich sheath in turn promotes more ROS production by the uncoupled electronic transport chain [52].

1.4 Iron and Oxidative Stress Amplification

Iron is mostly stored by ferritin [53] in the myelin sheaths [54, 55] of the human brain and is physiologically accumulated until reaching a plateau at 40–50 years [56]. In this context, myelin sheath breakdown [57, 58] and subsequent iron release into the extracellular space have been suggested to amplify the first wave of oxidative stress in MS lesions [59]. Specifically, by shifting between the ferrous and ferric states, iron induces the production of highly reactive hydroxyl radicals ($HO^{\bullet}$). Iron is phagocyted by macrophages and microglia, which later degenerate [59, 60] and release iron into the extracellular space, initiating an additional wave of oxidative stress [61, 62] with a probably significant contributon to oxidative neuronal damage [63]. Supporting data show that iron accumulation correlates with early axonal injury and that it is particularly prominent in active acute MS lesions [64]. In contrast, chronic MS exhibits a significant decrease in iron levels. It is worth pointing out that iron is key component in several cellular processes and CNS homeostasis, which should constitute a warning sign against the use of iron-chelating therapies in MS patients.

2 Cuprizone-Induced Demyelination and Oligodendroglial Susceptibility to Oxidative Stress

Cuprizone (CPZ), a copper chelator, added to the diet induces demyelination of several brain regions [65, 66]. In particular, 5–6-weeks of CPZ administration leads to almost complete demyelination of the corpus callosum in an acute demyelination; while a subsequent normal chow diet triggers a spontaneous endogenous remyelination process. In contrast, 12-13 weeks or even longer CPZ feeding, in a chronic demyelination, results in an impaired spontaneous remyelination process [67, 68]. In chronic demyelination, axons exhibit a shorter diameter [69], axonal damage, detected by SMI-32+ staining [70], and marked changes in locomotion, working memory and anxiety [71]. The CPZ model thus mimics some of the aspects of axonal degeneration in chronic and progressive MS [72]. Remarkably, *in vitro* experiments in rat primary glial cell cultures treated with CPZ demonstrated that microglia, astrocytes and oligodendrocyte precursors cells are unaffected by CPZ, while mature oligodendrocytes show evidence of toxicity as a consequence of the CPZ treatment [73, 74]. Experimentally, a 3-week diet of 0.2 % CPZ induces megamitochondria formation in oligodendrocytes [75–77], but not in neurons, astrocytes, cardiac, kidney, Kupffer or fat storing cells [78–81]. The factors responsible for this enlargement of mitochondria may include fission inhibition [80, 82–85] and exposure to high levels of ROS and RNS. In turn, this free radical formation during CPZ treatment may be due to: (a) oxidative phosphorylation uncoupling [86, 87], (b) reduction of the mitochondrial cristae or reaction surface [86], (c) decreases in cytochrome content [88], and (d) inhibition of the activities of complex II and complex IV [74, 81, 88–93]. Overall, CPZ treatment produces a high level of oligodendroglial oxidative stress. Moreover, oligodendrocytes themselves are particularly susceptible to oxidative stress. The factors responsible for this susceptibility include [94]: (a) scavenging of O_2^- and conversion to H_2O_2 is low in oligodendrocytes as compared with other neurons. In oligodendrocytes Mn-SOD is scarce [95, 96] and Cu,Zn-SOD decreases its activity upon treatment with CPZ [97–101]; (b) glutathione peroxidase activity needs GSH as an electron donor, both are intrinsically low in oligodendrocytes [97, 101] and are further reduced upon CPZ treatment [76]; (c) oligodendrocytes exhibit high levels of intracellular ferritin-bound Fe^{3+}. However, excessive cytosolic O_2^- produces a reduction and a shift from Fe^{3+} to Fe^{2+}, which initiates Haber-Weiss chemistry and initiates lipid peroxidation in the mitochondrial matrix or cytosol [96, 99, 100]. Supporting the idea, lipid peroxidation has been detected as a malondialdehyde increase in CPZ-treated oligodendrocytes [101]; (d) oligodendroglial cell redox potential and glycerol production, both regulated by $NADH_2$, are significantly higher upon CPZ treatment [76]; (e) oxidation and lipid synthesis in peroxisomes increase the levels of H_2O_2, which is normally eliminated by glutathione peroxidase and catalase, this latter being lower in oligodendrocytes upon CPZ treatment [101]. Furthermore, the number of peroxisomes increases during oligodendroglial differentiation for myelin lipid synthesis, leading to higher H_2O_2 production and oxidative stress [99]; (f) metallothioneins (MT) possess important anti-oxidant functions. Upon CPZ treatment, astrocytes increase MT1 and MT2 but oligodendrocytes display low MT levels [76, 95].

3 Oxidative Stress-Mediated Neurodegeneration and Oligodendroglial Vulnerability in Hypoxic-Ischemic White Matter Damage

White matter injury (WMI) is the main form of brain injury in premature birth [102] and periventricular leukomalacia (PVL), found in almost half of affected children [103], and is related to immature oligodendrocyte susceptibility to oxidative stress [104]. Hypoxia-ischemia (H/I) occurs in the immature brain and leads to PVL, in which oligodendroglial precursor cell (OPC) death in the non-myelinated cerebral white matter is considered a key factor resulting in hypomyelination [105, 106]. In turn, hypomyelination causes axon conduction failure, which leads to deficits in motor, sensory and/or cognitive functions depending on the location of the affected axons. It is well known that different molecules participate in OPC differentiation and maturation, which might induce positive signals for axonal recovery. In particular, apotransferrin (aTf) has been proven to act at several critical stages during OPC development and to be necessary for normal myelination [107, 108]. Neonatal rat models of white matter damage produced by H/I have shown that aTf provides neuroprotection to OPC and promote corpus callosum remyelination [109, 110]. In addition, aTf treatment has been proven to increase the number of OPC marker-positive cells and reduce cell death in the neonatal subventricular zone subjected to H/I [111]. All in all, aTf seems to promote OPC maturation and myelin recovery after H/I by decreasing iron-mediated toxicity and inducing new OPC from the subventicular zone (SVZ). In H/I animal models, in vitro real-time caspase-3 assays, which sense intracellular caspase-3 activity and stain the cell nucleus, have revealed that oligodendrocytes (OLG) apoptotic cell death and morphological changes taking place simultaneously. In turn, spinning disc confocal inverted microscope observations of cultured OPC incubated in a top chamber with 5 % CO_2 and 1 % O_2 at 37 °C rendered a significant increase in caspase-3-positive cells after 6 h culture in a low O_2 atmosphere. Parallel experiments performed with hypoxic OPC grown in the presence/absence of aTf showed a significant reduction in caspase labeling when aTf was added to the culture medium. Additionally, and to establish the impact of low O_2 at distinct stages of OLG maturation, OPCs treated with aTf and subjected to hypoxic culture conditions after switching the cells to a mitogen-free medium exhibited a general decline in early immunocytochemical markers (e.g. NG2) and an increase in intermediate (e.g. 04) and mature (e.g. GC and MBP) markers, consistent with OPC differentiation [111]. Interestingly, in the more mature differentiated OLG (OL) population subjected to hypoxic culture conditions, no significant differences were found among the experimental groups. This in vitro data suggest that aTf is effective in protecting immature OL during a hypoxic event and reveal that mature OL are more resistant to apoptotic death under hypoxic culture conditions. In addition, and under the same experimental conditions the following changes were found: a significant decrease in the number of OPC expressing immature markers, such as Sox2 and NG2, and an increase in cells expressing OL mature markers such as CC1 and MBP after 12 h of aTf treatment. These effects were more evident under low O_2 concentration, which suggests that aTf further promotes OPC differentiation under hypoxic culture conditions. Moreover, since mature OL are more resistant to low O_2 concentrations, the results indicate that aTf might be protecting OPCs against the hypoxic insult by promoting their differentiation [112].

Different studies have established that OPCs are selectively targeted by oxidative stress and have concluded that free radicals injure the OLG lineage in vitro as well as in vivo [113–120]. On the other hand, it has been clearly demonstrated that glutathione depletion produces a downstream increase in reactive oxygen species (ROS), which are the cause of OPC death [115] and different in vitro studies have found that caspase-mediated death of mixed populations of OPCs and pre-OLs takes place after oxidative stress in vitro [119, 121, 122]. The mechanisms of cell death induced by glutathione depletion are of great interest, as they may show how oxidative damage kills cells and thus lead to ways of preventing cell death. Depletion of the intracellular cystine results in a reduction in glutathione synthesis, and disruption of glutathione-dependent antioxidant mechanisms. As previously mentioned, OPCs are highly sensitive to death induced by glutathione depletion, being also dependent on 12-lipooxygenase activity. It should be pointed out that arachidonic acid is toxic to OPCs and that this toxicity can be blocked by 12-lipooxygenase inhibitors [123].

3.1 Neurodegeneration in Synucleinopathies

In the neurodegenerative diseases collectively called synucleinopathies [124], aggregation of α-synuclein (α-syn) is also present in oligodendrocytes. α-Syn is taken from the extracellular environment and may contribute to abnormal accumulation, hence promoting aggregation of endogenous α-syn. α-Syn can be degraded either by the ubiquitin-proteasome system or by autophagy [125, 126]. Oxidative stress and mitochondrial impairment have been implicated in this aggregation process and in inclusion body diseases, such as Parkinson's disease [127]. Different studies have demonstrated that α-syn aggregation is enhanced after oxidative stress [127, 128], which seems to promote α-syn uptake and oligomer formation [129].

4 Axonal Degeneration by Oxidative Signaling

4.1 General Overview

Several studies have been carried out to understand the critical processes of axonal degeneration. However, recent evidence seems to demonstrate that intra-neuronal production of reactive oxygen species (ROS), in particular hydrogen peroxide (H_2O_2), is one of the primary events triggering axonal degeneration. The changes in either H_2O_2 production or utilization could be due to various factors, ranging from cellular metabolic alterations to different kinds of traumatic injury, such as spinal cord or brain injuries. The increased and non-physiological concentration of H_2O_2 promotes deleterious effects on the neuronal shape, starting with structural and functional alterations which finally lead to neuronal death. The main mechanism by which H_2O_2 induces neuronal degeneration is linked to the impossibility of neurons to promote axonal outgrowth. The molecular signaling from H_2O_2 production to the inhibition of axonal outgrowth involves different proteins which regulate cytoskeleton dynamics, among which Mical is key in promoting post-injury H_2O_2 production. The high levels of this reactive oxygen species promote filamentous actin depolymerization and axonal degeneration.

4.2 H_2O_2: Physiological and Non-physiological Implications

Actin cytoskeleton dynamics is critical to develop correct neuronal polarization [130] and development, which in turn have been recently postulated to depend on suitable H_2O_2 concentrations [131]. Regarding axonal structure, the physiological concentrations of intra-neuronal H_2O_2 do not affect the cytoskeleton dynamics and the growth cone, thus present a normal filopodial development, in terms of number and length [132]. However, in non-physiological conditions, cytoskeleton dynamics in the growth cone presents alterations in both filopodial length and filopodial number, inflicted by an increase of H_2O_2 production which is triggered by the interaction between Semaphorin 3A (Sema3A) and Neuropilin-1 (NRP1)/PlexinA4 receptor promoting the activation of Mical pathway (Fig. 20.2) [133].

4.3 Molecular Signaling of Axonal Degeneration via H_2O_2 Production

Axonal damage is the result of a wide range of pathologies, from traumatic axotomy to metabolic or immune response failure, as in multiple sclerosis. Alterations in the axonal structure are the source of synaptic disconnection and thus the initiation of a neurodegenerative process [134]. Particularly in traumatic injuries, the initial

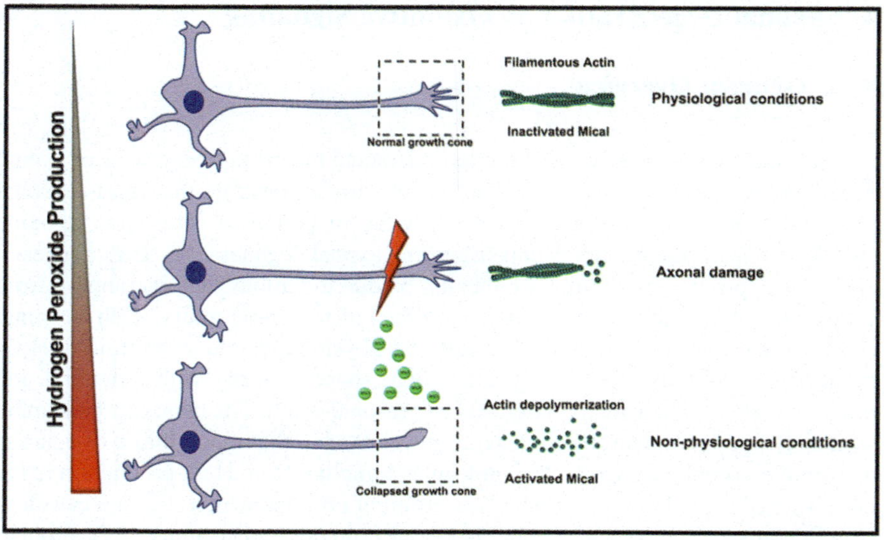

Fig. 20.2 Hydrogen peroxide production: physiological and non-physiological conditions

damage is followed by axonal degeneration, commonly named "die back", which is promoted by a collapse in cytoskeleton dynamics [135]. This collapse is triggered by a specific interaction between a transmembrane receptor complex named Neuropilin-1 (NRP1)/PlexinA4 and Semaphorin 3A (Sema3A), which is secreted at the lesion site by meningeal fibroblasts after injury [136, 137]. Secreted Sema3A binds to the NRP1/PlexinA4 receptor complex on the neuronal surface (axons and soma) [138, 139]. The whole complex is then internalized and PlexinA4 cytoplasmic domain activates Mical protein, which in turn triggers an increase in H_2O_2 production and concomitant filamentous actin oxidation at its methionine residues (M44 and M47), followed by axonal die back [140–143].

5 Conclusions

Reactive oxygen species, in particular H_2O_2, have different CNS and intra-neuronal physiological functions directly linked and derived from their physiological concentrations. However, increased levels of ROS and H_2O_2 to non-physiological levels immediately place a dyshomeostatic situation. Changes in the homeostasis of these reactive species are produced by endogenous cellular metabolic changes and also associated to an extracellular signaling triggered by traumatic injury in the nervous system (Fig. 20.2). As a consequence, the future of traumatic neurology should go out in search of antioxidant therapies to prevent this kind of harmful response.

References

1. Bjartmar C, Kinkel PR, Kidd G et al (2001) Axonal loss in normal-appearing white matter in a patient with acute MS. Neurology 57:1248–1252
2. Tallantyre EC, Bo L, Al-Rawashdeh O et al (2009) Greater loss of axons in primary progressive multiple sclerosis plaques compared to secondary progressive disease. Brain 132:1190–1199
3. Kuhlmann T, Lingfeld G, Bitsch A et al (2002) Acute axonal damage in multiple sclerosis is most extensive in early disease stages and decreases over time. Brain 125:2202–2212
4. Funfschilling U et al (2012) Glycolytic oligodendrocytes maintain myelin and long-term axonal integrity. Nature 485:517–521
5. Lee Y, Morrison BM, Li Y et al (2012) Oligodendroglia metabolically support axons and contribute to neurodegeneration. Nature 487:443–448
6. Frischer JM, Bramow S, Dal-Bianco A et al (2009) The relation between inflammation and neurodegeneration in multiple sclerosis brains. Brain 132:1175–1189
7. Kornek B, Storch MK, Weissert R et al (2000) Multiple sclerosis and chronic autoimmune encephalomyelitis: a comparative quantitative study of axonal injury in active, inactive, and remyelinated lesions. Am J Pathol 157:267–276
8. Kutzelnigg A, Lucchinetti CF, Stadelmann C et al (2005) Cortical demyelination and diffuse white matter injury in multiple sclerosis. Brain 128:2705–2712
9. DeLuca GC, Ebers GC, Esiri MM (2004) Axonal loss in multiple sclerosis: a pathological survey of the corticospinal and sensory tracts. Brain 127:1009–1018
10. Bitsch A, Schuchardt J, Bunkowski S et al (2000) Acute axonal injury in multiple sclerosis. Correlation with demyelination and inflammation. Brain 123:1174–1183
11. DeLuca GC, Williams K, Evangelou N et al (2006) The contribution of demyelination to axonal loss in multiple sclerosis. Brain 129:1507–1516
12. Smith KJ, Lassmann H (2002) The role of nitric oxide in multiple sclerosis. Lancet Neurol 1:232–241
13. Koch MW, Ramsaransing GSM, Arutjunyan AV et al (2006) Oxidative stress in serum and peripheral blood leukocytes in patients with different disease courses of multiple sclerosis. J Neurol 253:483–487
14. Vladimirova O, O'Connor J, Cahill A et al (1998) Oxidative damage to DNA in plaques of MS brains. Mult Scler 4:413–418
15. Hammann KP, Hopf HC (1986) Monocytes constitute the only peripheral blood cell population showing an increased burst activity in multiple sclerosis patients. Int Arch Allergy Applied Immunol 81:230–234
16. Greco A, Minghetti L, Sette G et al (1999) Cerebrospinal fluid isoprostane shows oxidative stress in patients with multiple sclerosis. Neurology 53:1876–1879
17. Nikic I, Merkler D, Sorbara C et al (2011) A reversible form of axon damage in experimental autoimmune encephalomyelitis and multiple sclerosis. Nat Med 17:495–499
18. van Horssen J, Witte ME, Schreibelt G, de Vries HE (2011) Radical changes in multiple sclerosis pathogenesis. Biochim Biophys Acta 1812:141–150
19. van Horssen J, Schreibelt G, Drexhage J et al (2008) Severe oxidative damage in multiple sclerosis lesions coincides with enhanced antioxidant enzyme expression. Free Radic Biol Med 45:1729–1737
20. Haider L, Fischer MT, Frischer JM et al (2011) Oxidative damage in multiple sclerosis lesions. Brain 134:1914–1924
21. Campbell GR, Mahad DJ (2011) Mitochondria as crucial players in demyelinated axons: lessons from neuropathology and experimental demyelination. Autoimmune Dis 2011:262847. doi:10.4061/2011/262847
22. Fünfschilling U, Supplie LM, Mahad D et al (2012) Glycolytic oligodendrocytes maintain myelin and longterm axonal integrity. Nature 485:517–521

23. Andrews H, White K, Thomson C et al (2006) Increased axonal mitochondrial activity as an adaptation to myelin deficiency in the shiverer mouse. J Neurosci Res 83:1533–1539
24. Hogan V, White K, Edgar J et al (2009) Increase in mitochondrial density within axons and supporting cells in response to demyelination in the Plp1 mouse model. J Neurosci Res 87:452–459
25. Sathornsumetee S, McGavern DB, Ure DR, Rodriguez M (2000) Quantitative ultrastructural analysis of a single spinal cord demyelinated lesion predicts total lesion load, axonal loss, and neurological dysfunction in a murine model of multiple sclerosis. Am J Pathol 157:1365–1376
26. Zambonin JL, Zhao C, Ohno N et al (2011) Increased mitochondrial content in remyelinated axons: implications for multiple sclerosis. Brain 134:1901–1913
27. Dutta R, McDonough J, Yin X et al (2006) Mitochondrial dysfunction as a cause of axonal degeneration in multiple sclerosis patients. Ann Neurol 59:478–489
28. Kalman B (2006) Role of mitochondria in multiple sclerosis. Curr Neurol Neurosci Rep 6:244–252
29. Smith KJ, McDonald WI (1999) The pathophysiology of multiple sclerosis: the mechanisms underlying the production of symptoms and the natural history of the disease. Philos Trans R Soc Lond B Biol Sci 354:1649–1673
30. Mahad DJ, Ziabreva I, Campbell G et al (2009) Mitochondrial changes within axons in multiple sclerosis. Brain 132:1161–1174
31. Ohno N, Kidd GJ, Mahad D et al (2011) Myelination and axonal electrical activity modulate the distribution and motility of mitochondria at CNS nodes of Ranvier. J Neurosci 31:7249–7258
32. Larsson NG (2010) Somatic mitochondrial DNA mutations in mammalian aging. Ann Rev Biochem 79:683–706
33. Howarth C, Gleeson P, Attwell D (2012) Updated energy budgets for neural computation in the neocortex and cerebellum. J Cereb Blood Flow Metab 32:1222–1232
34. Smith KJ (2007) Sodium channels and multiple sclerosis: roles in symptom production, damage and therapy. Brain Pathol 17:230–242
35. Tsutsui S, Stys PK (2013) Metabolic injury to axons and myelin. Exp Neurol 246:26–34
36. Waxman SG (2006) Ions, energy and axonal injury: towards a molecular neurology of multiple sclerosis. Trends Mol Med 12:192–195
37. Graham R, Campbell J, Worrall T, Mahad DJ (2014) The central role of mitochondria in axonal degeneration in multiple sclerosis. Multiple Sclerosis J 20:1806–1813
38. Babior BM (2004) NADPH oxidase. Curr Opin Immunol 16:42–47
39. Zeis T, Probst A, Steck AJ et al (2009) Molecular changes in white matter adjacent to an active demyelinating lesion in early multiple sclerosis. Brain Pathol 19:459–466
40. Nikić I, Merkler D, Sorbara C et al (2011) A reversible form of axon damage in experimental autoimmune encephalomyelitis and multiple sclerosis. Nat Med 17:495–499
41. Jekabsone A, Neher JJ, Borutaite V, Brown GC (2007) Nitric oxide from neuronal nitric oxide synthase sensitises neurons to hypoxia-induced death via competitive inhibition of cytochrome oxidase. J Neurochem 103:346–356
42. Haider L (2015) Inflammation, iron, energy failure and oxidative stress in the pathogenesis of multiple sclerosis. Oxid Med Cell Longev 2015:725370. doi:10.1155/2015/725370
43. DiMauro S, Schon EA, Carelli V, Hirano M (2013) The clinical maze of mitochondrial neurology. Nat Rev Neurol 9:429–444
44. Yakes FM, VanHouten B (1997) Mitochondrial DNA damage is more extensive and persists longer than nuclear DNA damage in human cells following oxidative stress. Proc Natl Acad Sci U S A 94:514–519
45. Witte ME, Geurts JJG, de Vries HE et al (2010) Mitochondrial dysfunction: a potential link between neuroinflammation and neurodegeneration? Mitochondrion 10:411–418
46. Panfoli I, Calzia D, Bianchini P et al (2009) Evidence for aerobic metabolism in retinal rod outer segment disks. Int J Biochem Cell Biol 41:2555–2565

47. Ravera S, Panfoli I, Calzia D et al (2009) Evidence for aerobic ATP synthesis in isolated myelin vesicles. Int J Biochem Cell Biol 41:1581–1591
48. Ravera S, Panfoli I, Aluigi MG et al (2011) Characterization of myelin sheath F_0-F_1-ATP synthase and its regulation by IF. Cell Biochem Biophys 59:63–70
49. Ravera S, Bartolucci M, Calzia D et al (2013) Tricarboxylic acid cycle-sustained oxidative phosphorylation in isolated myelin vesicles. Biochimie 95:1991–1998
50. Ravera S, Nobbio L, Visigalli D et al (2013) Oxydative phosphorylation in sciatic nerve myelin and its impairment in a model of dysmyelinating peripheral neuropathy. J Neurochem 126:82–92
51. Bartolucci M, Ravera S, Garbarino G et al (2015) functional expression of electron transport chain and FoF1-ATP synthase in optic nerve myelin sheath. Neurochem Res 40(11):2230–2241
52. Ravera S, Bartolucci M, Cuccarolo P et al (2015) Oxidative stress in myelin sheath: the other face of the extramitochondrial oxidative phosphorylation ability. Free Radic Res 49:1156–1164
53. Halliwell B (2001) Role of free radicals in the neurodegenerative diseases: therapeutic implications for antioxidant treatment. Drugs Aging 18:685–716
54. Hulet SW, Hess EJ, Debinski W et al (1999) Characterization and distribution of ferritin binding sites in the adult mouse brain. J Neurochem 72:868–874
55. Hulet SW, Powers S, Connor JR (1999) Distribution of transferrin and ferritin binding in normal and multiple sclerotic human brains. J Neurol Sci 165:48–55
56. Hallgren B, Sourander P (1958) The effect of age on the non haemin iron in the human brain. J Neurochem 3:41–51
57. Barnett MH, Prineas JW (2004) Relapsing and remitting multiple sclerosis: pathology of the newly forming lesion. Ann Neurol 55:458–468
58. Lassmann H (2011) The architecture of inflammatory demyelinating lesions: implications for studies on pathogenesis. Neuropathol Appl Neurobiol 37:698–710
59. Hametner S, Wimmer I, Haider L et al (2013) Iron and neurodegeneration in the multiple sclerosis brain. Ann Neurol 74:848–861
60. Lopes KO, Sparks DL, Streit WJ (2008) Microglial dystrophy in the aged and Alzheimer's disease brain is associated with ferritin immunoreactivity. Glia 56:1048–1060
61. Craelius W, Migdal MW, Luessenhop CP et al (1982) Iron deposits surrounding multiple sclerosis plaques. Arch Pathol Lab Med 106:397–399
62. Bagnato F, Hametner S, Yao B et al (2011) Tracking iron in multiple sclerosis: a combined imaging and histopathological study at 7 Tesla. Brain 134:3599–3612
63. Friese MA, Schattling B, Fugger L (2014) Mechanisms of neurodegeneration and axonal dysfunction in multiple sclerosis. Nat Rev Neurol 10:225–238
64. Lassmann H, Van Horssen J, Mahad D (2012) Progressive multiple sclerosis: pathology and pathogenesis. Nat Rev Neurol 8:647–656
65. Acs P, Kipp M, Norkute A et al (2009) 17β-estradiol and progesterone prevent cuprizone provoked demyelination of corpus callosum in male mice. Glia 57:807–814
66. Clarner T, Parabucki A, Beyer C, Kipp M (2011) Corticosteroids impair remyelination in the corpus callosum of cuprizone-treated mice. J Neuroendocrinol 23:601–611
67. Kipp M, Clarner T, Dang J et al (2009) The cuprizone animal model: new insights into an old story. Acta Neuropathol 118:723–736
68. Harsan LA, Steibel J, Zaremba A et al (2008) Recovery from chronic demyelination by thyroid hormone therapy: myelinogenesis induction and assessment by diffusion tensor magnetic resonance imaging. J Neurosci 28:14189–14201
69. Mason JL, Langaman C, Morell P et al (2001) Episodic demyelination and subsequent remyelination within the murine central nervous system: changes in axonal caliber. Neuropathol Appl Neurobiol 27:50–58
70. Lindner M, Fokuhl J, Linsmeier F et al (2009) Chronic toxic demyelination in the central nervous system leads to axonal damage despite remyelination. Neurosci Lett 453:120–125

71. Zhang H, Zhang Y, Xu H et al (2013) Locomotor activity and anxiety status, but not spatial working memory, are affected in mice after brief exposure to cuprizone. Neurosci Bull 29:633–641
72. Manrique-Hoyos N, Jürgens T, Grønborg M et al (2012) Late motor decline after accomplished remyelination: impact for progressive multiple sclerosis. Ann Neurol 71:227–244
73. Pasquini LA, Calatayud CA, Bertone U et al (2007) The neurotoxic effect of cuprizone on oligodendrocytes depends on thepresence of pro-inflammatory cytokines secreted by microglia. Neurochem Res 32:279–292
74. Benardais K, Kotsiari A, Skuljec J et al (2013) Cuprizone [bis(cyclohexyliden-hydrazide)] is selectively toxic for mature oligodendrocytes. Neurotoxic Res 24:244–250
75. Acs P, Komoly S (2012) Selective ultrastructural vulnerability in the cuprizone-induced experimental demyelination. Ideggyogy Sz 65:266–270
76. Biancotti JC, Kumar S, de Vellis J (2008) Activation of inflammatory response by a combination of growth factors in cuprizone-induced demyelinated brain leads to myelin repair. Neurochem Res 33:2615–2628
77. Hiremath MM, Saito Y, Knapp GW et al (1998) Microglial/macrophage accumulation during cuprizone-induced demyelinationin in C57BL/6 mice. J Neuroimmunol 92:38–49
78. Asano M, Wakabayashi T, Ishikawa K, Kishimoto H (1978) Mechanism of the formation of megamitochondria by copper-chelating agents IV. Role of the fusion phenomenon in the cuprizone-induced megamitochondrial formation. Acta Pathol Jpn 28:205–213
79. Ludwin SK (1978) Central nervous system demyelination and remyelination in the mouse: an ultrastructural study of cuprizone toxicity. Lab Invest 39:597–612
80. Tandler B, Hoppel CL (1973) Division of giant mitochondria during recovery from cuprizone intoxication. J Cell Biol 56:266–272
81. Wakabayashi T, Asano M, Ishikawa K, Kishimoto H (1978) Mechanism of the formation of megamitochondria by copper-chelating agents V. Further studies on isolated megamitochondria. Acta Pathol Jpn 28:215–223
82. Flatmark T, Kryvi H, Tangeras A (1980) Induction of megamitochondria by cuprizone (biscyclohexanone oxaldihydrazone). Evidence for an inhibition of the mitochondrial division process. Eur J Cell Biol 23:141–148
83. Miwa S, Lawless C, von Zglinicki T (2008) Mitochondrial turnover in liver is fast in vivo and is accelerated by dietary restriction: application of a simple dynamic model. Aging Cell 7:920–923
84. Wakabayashi T, Asano M, Kurono C (1975) Mechanism of the formation of megamitochondria induced by copper-chelating agents II. Isolation and some properties of megamitochondria from the cuprizone-treated mouse liver. Acta Pathol Jpn 25:39–49
85. Wakabayashi T, Asano M, Ishikawa K, Kishimoto H (1977) Cuprizone induced megamitochondrial formation and membrane fusion. J Electron Microsc (Tokyo) 26:137–140
86. Tandler B, Hoppel CL (1975) The failure of supplemental dietary copper to preventcuprizone-induced alterations in mouse hepatocytes. Beitr Pathol 156:56–64
87. Wagner T, Rafael J (1977) Biochemical properties of liver megamitochondria induced by chloramphenicol or cuprizone. Exp Cell Res 107:1–13
88. Kimberlin RH, Collis SC, Walker CA (1976) Profiles of brain glycosidase activity in cuprizone-fed Syrian hamsters and in scrapie-affected mice, rats, Chinese and Syrian hamsters. J Comp Pathol 86:135–142
89. Millet V, Moiola CP, Pasquini JM et al (2009) Partial inhibition of proteasome activity enhances remyelination after cuprizone-induced demyelination. Exp Neurol 217:282–296
90. Nelecz MJ, Wroniszewska A, Famulski KS, Wojtczak L (1982) Changes in the mitochondrial surface potential during cuprizone-induced formation of megamitochondria. Eur J Cell Biol 27:289–295
91. Petronilli V, Zoratti M (1990) A characterization of cuprizone-induced giant mouse liver mitochondria. J Bioenerg Biomembr 22:663–677
92. Venturini G (1973) Enzymic activities and sodium, potassium and copper concentrations in mouse brain and liver after cuprizone treatment in vivo. J Neurochem 21:1147–1151

93. Praet J, Guglielmetti C, Berneman Z et al (2014) Cellular and molecular neuropathology of the cuprizone mouse model: clinical relevance for multiple sclerosis. Neurosci Biobehav Rev 47:485–505
94. Bernardo A, Greco A, Levi G, Minghetti L (2003) Differential lipid peroxidation, Mn-superoxide dismutase and bcl-2 expression contribute to the maturation-dependent vulnerability of oligodendrocytes to oxidative stress. J Neuropathol Exp Neurol 62:509–519
95. Witherick J, Wilkins A, Scolding N, Kemp K (2010) Mechanisms of oxidative damage in multiple sclerosis and a cell therapy approach to treatment. Autoimmune Dis. doi:10.4061/2011/164608
96. Ljutakova SG, Russanov EM (1985) Differences in the in vivo effects of cuprizone on superoxide dismutase activity in rat liver cytosol and mitochondrial intermembrane space. Acta Physiol Pharmacol Bulg 11:56–61
97. Russanov EM, Ljutakova SG (1980) Effect of cuprizone on copper exchange and superoxide dismutase activity in rat liver. Gen Pharmacol 11:535–538
98. Zhang Y, Xu H, Jiang W et al (2008) Quetiapine alleviates the cuprizone-induced white matter pathology in the brain of C57BL/6 mouse. Schizophr Res 106:182–191
99. Juurlink BH, Thorburne SK, Hertz L (1998) Peroxide-scavenging deficit underlies oligodendrocyte susceptibility to oxidative stress. Glia 22:371–378
100. Marrif H, Juurlink BH (2003) Differential vulnerability of oligodendrocytes and astrocytes to hypoxic-ischemic stress. Adv Mol Cell Biol 31:857–867
101. Xu H, Yang HJ, Li XM (2013) Differential effects of antipsychotics on the devel-opment of rat oligodendrocyte precursor cells exposed to cuprizone. Eur Arch Psychiatry Clin Neurosci 264:121–129
102. Volpe JJ (2009) The encephalopathy of prematurity--brain injury and impaired brain development inextricably intertwined. Semin Pediatr Neurol 16:167–178
103. Bax M, Tydeman C, Flodmark O (2006) Clinical and MRI correlates of cerebral palsy: the European Cerebral Palsy Study. JAMA 296:1602–1608
104. Ferriero DM, Miller SP (2010) Imaging of selective vulnerability in the developing nervous system. J Anat 217:429–435
105. Rezaie P, Dean A (2002) Periventricular leukomalacia, inflammation and white matter lesions within the developing nervous system. Neuropathol 22:106–132
106. Haynes RL, Folkerth RD, Keefe RJ et al (2003) Nitrosative and oxidative injury to premyelinating oligodendrocytes in periventricular leukomalacia. J Neuropathol Exp Neurol 62:441–450
107. Paez PM, Marta CB, Moreno MB et al (2002) Apotransferrin decreases migration and enhances differentiation of oligodendroglial progenitor cells in an in vitro system. Dev Neurosci 24:47–58
108. Paez PM, Garcia CI, Davio C et al (2004) Apotransferrin promotes the differentiation of two oligodendroglial cell lines. Glia 45:207–217
109. Vannucci RC, Vannucci SJ (1997) A model of perinatal hypoxic ischemic brain damage. Ann N Y Acad Sci 835:234–249
110. Vannucci RC, Vannucci SJ (2005) Perinatal hypoxic–ischemic brain damage: evolution of an animal model. Dev Neurosci 27:81–86
111. Vannucci SJ (1999) Rat model of perinatal hypoxic–ischemic brain damage. J Neurosci Res 55:158–163
112. Guardia-Clausi M, Pasquini LA, Soto EF, Pasquini JM (2010) Apotransferrin-induced recovery after hypoxic/ischaemic injury on myelination. ASN Neurol 19:e00048. doi:10.1042/AN20100020
113. Paez PM, Fulton DJ, Spreuer V et al (2009) Regulation of store-operated and voltage-operated Ca channels in the proliferation and death of oligodendrocyte precursor cells by golgi proteins. ASN Neurol 1:1–15
114. Guardia-Clausi M, Paez PM, Campagnoni AT et al (2012) Intranasal administration of Tf protects and repairs the neonatal white matter after a cerebral hypoxic-ischemic event. Glia 60:1540–1554

115. Back SA, Gan X, Li Y et al (1998) Maturation-dependent vulnerability of oligodendrocytes to oxidative stress-induced death by glutathione depletion. J Neurosci 16:6241–6253
116. Back SA, Han BH, Luo NL et al (2002) Selective vulnerability of late oligodendrocyte progenitors to hypoxia–ischemia. J Neurosci 22:455–463
117. Baud O, Haynes RF, Wang H et al (2004) Developmental up-regulation of MnSOD in rat oligodendrocytes confers protection against oxidative injury. Eur J Neurosci 20:29–40
118. Fern R, Möller T (2000) Rapid ischemic cell death in immature oligodendrocytes: a fatal glutamate release feedback loop. J Neurosci 20:34–42
119. Fragoso G, Martínez-Bermúdez AK, Liu HN et al (2004) Developmental differences in HO-induced oligodendrocyte cell death: roles of glutathione, mitogen-activated protein kinases and caspase 3. J Neurochem 90:392–404
120. Lin S, Rhodes PG, Lei M et al (2004) α-Phenyl-n-tert-butyl-nitrone attenuates hypoxic-ischemic white matter injury in the neonatal rat brain. Brain Res 1007:132–141
121. Mronga T, Stahnke T, Goldbaum O, Richter-Landsberg C (2004) Mitochondrial pathway is involved in hydrogen peroxide induced apoptotic cell death of oligodendrocytes. Glia 46:446–455
122. Sánchez-Gómez MV, Alberdi E, Ibarretxe G et al (2003) Caspase-dependent and caspase-independent oligodendrocyte death mediated by AMPA and kainate receptors. J Neurosci 23:9519–9528
123. Wang H, Li J, Follett PL et al (2004) Lipoxygenase plays a key role in cell death caused by glutathione depletion and arachidonic acid in rat oligodendrocytes. Eur J Neurosci 20:2049–2058
124. Jellinger KA, Lantos PL (2010) Papp-Lantos inclusions and the pathogenesis of multiple system atrophy: an update. Acta Neuropathol 119:657–667
125. Riedel M, Goldbaum O, Schwarz L et al (2010) 17-AAG induces cytoplasmic α-synuclein aggregate clearance by induction of autophagy. PLoS One 18:e8753
126. Vogiatzi T, Xilouri M, Vekrellis K, Stefanis L (2008) Wild type α-synuclein is degraded by chaperone-mediated autophagy and macroautophagy in neuronal cells. J Biol Chem 283:23542–23556
127. Jellinger KA (2010) Neuropathology in Parkinson's disease with mild cognitive impairment. Acta Neuropathol 120:829–830
128. Pukass K, Richter-Landsberg C (2014) Oxidative stress promotes uptake, accumulation, and oligomerization of extracellular α-synuclein in oligodendrocytes. J Mol Neurosci 52:339–352
129. Riedel M, Goldbaum O, Richter-Landsberg C (2009) α-Synuclein promotes the recruitment of tau to protein inclusions in oligodendroglial cells: effects of oxidative and proteolytic stress. J Mol Neurosci 39:226–234
130. Bradke F, Dotti CG (1999) The role of local actin instability in axon formation. Science 283:1931–1934
131. Wilson C, Nunez MT, Gonzalez-Billault C (2015) Contribution of NADPH-oxidase to the establishment of hippocampal neuronal polarity in culture. J Cell Sci 128:2989–2995
132. Wilson C, Gonzalez-Billault C (2015) Regulation of cytoskeletal dynamics by redox signaling and oxidative stress: implications for neuronal development and trafficking. Front Cell Neurosci 9:381. doi:10.3389/fncel.2015.00381
133. Morinaka A, Yamada M, Itofusa R et al (2011) Thioredoxin mediates oxidation-dependent phosphorylation of CRMP2 and growth cone collapse. Sci Signal 4(170):ra26. doi:10.1126/scisignal.2001127
134. Casas C, Isus L, Herrando-Grabulosa M et al (2015) Network-based proteomic approaches reveal the neurodegenerative, neuroprotective and pain-related mechanisms involved after retrograde axonal damage. Sci Rep 5:9185. doi:10.1038/srep09185
135. Quinta HR, Pasquini LA, Pasquini JM (2015) Three-dimensional reconstruction of corticospinal tract using one-photon confocal microscopy acquisition allows detection of axonal disruption in spinal cord injury. J Neurochem 133:113–124

136. Pasterkamp RJ, Anderson PN, Verhaagen J (2001) Peripheral nerve injury fails to induce growth of lesioned ascending dorsal column axons into spinal cord scar tissue expressing the axon repellent semaphorin 3. Eur J Neurosci 13:457–471
137. Pasterkamp RJ, Giger RJ, Ruitenberg MJ et al (1999) Expression of the gene encoding the chemorepellent semaphorin 3 is induced in the fibroblast component of neural scar tissue formed following injuries of adult but not neonatal CNS. Mol Cell Neurosci 13:143–166
138. Kolodkin AL, Matthes DJ, Goodman CS (1993) The semaphorin genes encode a family of transmembrane and secreted growth cone guidance molecules. Cell 75:1389–1399
139. Quinta HR, Pasquini JM, Rabinovich GA, Pasquini LA (2014) Glycan-dependent binding of galectin-1 to neuropilin-1 promotes axonal regeneration after spinal cord injury. Cell Death Differ 21:941–955
140. Quinta HR, Pasquini JM, Rabinovich GA, Pasquini LA (2014) Axonal regeneration in spinal cord injury: key role of galectin-1. Medicina (Buenos Aires) 74:321–325
141. De Winter F, Oudega M, Lankhorst AJ et al (2002) Injury-induced class 3 semaphorin expression in the rat spinal cord. Exp Neurol 175:61–75
142. Giridharan SS, Caplan S (2014) MICAL-family proteins: complex regulators of the actin cytoskeleton. Antioxid Redox Signal 20 2059–2073
143. Hung RJ, Terman JR (2011) Extracellular inhibitors, repellents, and semaphorin/plexin/MICAL-mediated actin filament disassembly. Cytoskeleton (Hoboken) 68:415–433

Chapter 21
Mitochondrial DNA Damage in Autism

Sarah Rose, Sarah Wong, and Cecilia Giulivi

Abstract Mitochondria are organelles involved in essential roles in intermediary metabolism, perinatal neurodevelopment, immunity, bioenergetics, neurotransmitter metabolism, among other critical pathways. As such, mitochondrial dysfunction (MD) has deleterious effects with the potential of contributing to neurological diseases or enhancing their morbidity (e.g., autism and schizophrenia). Therefore, accumulation of mitochondrial damage is interpreted as a key element of the development of aging as well as neurodegenerative diseases. With the rise in the prevalence of autism spectrum disorders (ASD), there has been an increased interest in the etiology and contributors of this disorder. MD caused by genetics alone or by gene and environment interactions, may play a role in the etiology of ASD and holds promise for developing future therapies and/or interventions to help manage its symptoms or delaying its onset. Here we explore findings from our research and others analyzing the role of mitochondrial DNA damage as a contributor to ASD morbidity.

Keywords Oxidative stress • CHARGE • *Pten* • Polybrominated diphenyl ethers (PBDEs) • Fetal imprinting • Immunity • Ketogenic diet

1 Introduction

Autism spectrum disorders (ASD) encompass a complex neurodevelopmental condition associated with stereotyped and repetitive behaviors and characterized by abnormal social interaction, verbal and non-verbal communication, and minimal

S. Rose • S. Wong
Department of Molecular Biosciences, School of Veterinary Medicine, University of California Davis, Davis, CA 95616, USA

C. Giulivi (✉)
Department of Molecular Biosciences, School of Veterinary Medicine, University of California Davis, Davis, CA 95616, USA

Medical Investigations of Neurodevelopmental Disorders (MIND) Institute, University of California Davis, Sacramento, CA 95817, USA
e-mail: cgiulivi@ucdavis.edu

© Springer International Publishing Switzerland 2016
R.J. Gelpi et al. (eds.), *Biochemistry of Oxidative Stress*, Advances in Biochemistry in Health and Disease 16, DOI 10.1007/978-3-319-45865-6_21

interest in the surrounding environment [1]. Limited scientific advances have been made regarding the causes of ASD. However, there is a general consensus that both genetic and environmental components contribute to this condition [2]. While many aspects of the symptoms and etiology of ASD are studied, the role of the mitochondria in the bioenergetics of ASD has only recently been recognized [1, 3]. Diagnosis of ASD has become an increasingly prevalent health and social issue, with an estimated ASD prevalence of 1 in 42 boys and 1 in 189 girls in 2010, significantly increased from the incidence reported in 2008 (1 in 54 boys and 1 in 252 girls [4, 5]). In a study on California children, it was determined that changes in diagnostic criteria, inclusion of milder cases, genetic causes or an earlier age at diagnosis, could not be the sole reason for this dramatic increase in ASD [5]. This strongly suggests that yet unknown environmental factors have the potential to rise autism in our populations [6–8]. Indeed, a study on identical twins reported that exposure to environmental factors plays a more critical role than genetic heritability in ASD [9].

2 Lower Oxidative Phosphorylation Capacity and Increased Oxidative Stress in ASD

Oxidative phosphorylation provides most, if not all, ATP to the brain; therefore, a dysfunction in mitochondrial pathways could negatively affect neurodevelopment, a process heavily based on mitochondria-derived ATP [1]. In this regard, we hypothesized that impaired mitochondrial dysfunction (MD) causes oxidative stress influencing processes highly reliant on energy, such as neurodevelopment, and further contribute to autism onset or severity [1]. We investigated the link between mitochondrial dysfunction (MD) and ASD by drawing study participants from the Childhood Autism Risks from Genetics and Environment (CHARGE) study at the University of California Davis. The CHARGE study is an ongoing population-based, case-control that was originally launched in 2003. It includes clinically confirmed autism cases and also age-matched, genetically unrelated, typically developing controls. The children were between 2 to 5 years old and their diagnoses were confirmed through two different clinical examinations, the Autism Diagnostic Observation Schedule-Generic (ADOS-G) and the Autism Diagnostic Inventory-Revised (ADI-R). The ADOS-G is a semistructured, standardized evaluation of verbal and nonverbal communication, and of the creative use of materials, conducted by observing researchers in individuals with symptoms suspected of autism spectrum disorders [10]. On the other hand, the ADI-R provides a diagnostic algorithm for the definitions of autism by [11] and [12].

MD and mitochondrial DNA (mtDNA) anomalies were evaluated in lymphocytes from 10 children with autism with high severity scores (8 and above) and 10 typically neurodeveloping (TD) controls from this study. While mitochondrion typically possesses 2 to 10 copies of mtDNA, this number can be altered according to the energy needs of the cell with changing physiological conditions, and is not always coupled with mitochondrial replication [13, 14]. For example, in response to

Table 21.1 Lymphocyte mitochondrial activities in autism spectrum disorders (ASD)[a]

	Oxidase			F_0F_1-ATP synthase	Pyruvate dehydrogenase complex
	NADH	Succinate	Cytochrome *c*		
ASD	4.4	5	9	111	1.0
TD	12	7	9	147	2.3
99 % CI	7–17	4.4–9.6	3.3–14.9	68–225	1.4–3.2

[a]Extracted from Giulivi et al. [1]. The values in the table are mean values; TD, typical neurodeveloping controls. CI, confidence index

increased oxidative stress, a cell may increase the copy number of mtDNA without the increase of the oxidative phosphorylation capacity [15–20]. However, increases in copy number have their own detrimental effects and have been shown to be associated with defective transcription, respiratory chain insufficiency, and age associated accumulation of mtDNA deletions [21]. In our study, the lymphocytic mitochondrial function was evaluated by comparing oxidative phosphorylation (OXPHOS) capacity, mtDNA deletions and copy number, and mitochondrial rate of hydrogen peroxide production in the study group. Plasma lactate and pyruvate in the subject group were also assessed given that mitochondrial myopathies are generally followed by increases in the lactate-to-pyruvate ratio.

Malfunctioning or abnormal lymphocytic mitochondrial activities in children with autism were observed in this exploratory study as determined by the following parameters: (i) mtDNA overreplication and/or deletions, (ii) reduction of Complex I activity alone or accompanied by other Complexes, (iii) heightened mitochondrial rate of hydrogen peroxide production, (iv) decreased pyruvate dehydrogenase activity and (v) high lactate-to-pyruvate ratios in 20 % of the subjects [1]. The results showed that the rates of hydrogen peroxide production in lymphocytic mitochondria from children with autism were higher than levels recorded from controls at both Complex I (0.15 vs. 0.07 nmol of $H_2O_2 \times [\text{min} \times \text{mg of protein}]^{-1}$, respectively; $p = 0.03$) and Complex III (0.34 vs 0.16 nmol of $H_2O_2 \times [\text{min} \times \text{mg of protein}]^{-1}$; $p = 0.02$). Thus, lymphocytic mitochondria from children with ASD show a lower OXPHOS capacity, and have a role in increasing cellular oxidative stress (Table 21.1).

Table 21.1 shows decreased levels of key mitochondrial activities that are consistent with increased oxidative stress in non-activated immune cells from ASD subjects. This opens the door to explore the role of reactive oxygen species (ROS) in mediating mitochondrial or cellular damage [22]. In support of this notion, our laboratory and others have reported increased biomarkers of oxidative stress [23] in lymphocytes [1, 22, 24], granulocytes [25], post-mortem brain samples [26–29], lymphoblasts [30] and plasma from affected subjects [31]. Increased oxidative stress occurs when the production of ROS exceeds the antioxidant capacity. This mitochondrial damage establishes a feed-forward cycle of further damage. The increased oxidative stress observed in ASD can result from increased mitochondrial ROS production [1], the presence of deleterious polymorphisms in antioxidant enzymes [23], lower expression of antioxidant enzymes and/or factors such as nuclear factor erythroid 2-related factor (Nrf2) [25–29], and increased neuroinflammation [32]. However, whether these biomarkers of oxidative stress reflect significant biological contributions to phenotype development, or if they arise as a consequence of ASD is still unknown. As such, the precise role of oxidative stress in the etiology of autism is still undefined [22].

We hypothesized that damage to mtDNA, as judged by the increased deletions, could be mediated by mitochondria-derived ROS. A study done in 2013 examined the link between damage to mtDNA and ROS levels in ASD. Using participants from the CHARGE study, the mtDNA deletions were measured in peripheral blood monocytic cells (PBMC) by quantitative polymerase chain reaction (qPCR). In the majority of patients with single and multiple deletions, cytochrome *b* (CYTB) and/or NADH dehydrogenase subunit 4 (ND4) gene are deleted, whereas NADH dehydrogenase subunit 1 (ND1) is rarely deleted. Therefore, by the gene ratios of CYTB to ND1 and ND4 to ND1, mtDNA deletions were assessed to ascertain if the ASD condition show a pattern of increased oxidative damage to mtDNA [33]. The measurements were also used to determine if the putative mtDNA damage was *de novo* (when compared to their parents living in the same household) or inherited (if outcomes followed paternal and/or maternal patterns). In addition, sequence variants in mtDNA segments from the PBMC were also evaluated in ASD and TD children and their mothers, since mtDNA is maternally inherited. Increased mtDNA damage in ASD children was evidenced by (i) a 2-fold higher frequency of mtDNA deletions, (ii) 2.4-fold higher number of GC → AT transitions (GC sites experience more oxidative damage), and (iii) 1.6-fold higher frequency of G,C,T → A transitions [22]. These results suggested a high incidence of mitochondrial DNA polymerase gamma incorporating mainly A at bypassed apurinic/apyrimidinic sites, probably originated from oxidative stress to DNA. Consistent with the maternal inheritance of mtDNA, the same sequence variants observed in children were also observed in their mothers, regardless of diagnosis, suggesting that children with ASD have an inherited template with increased oxidative damage. However, the frequency of mtDNA deletions in ASD children (mainly male) was similar to that of their fathers. When stratified into paternal decades of age, the ratio of ASD fathers with mtDNA deletions vs. TD fathers was 1 during decade 3, 1.4 at decade 4, and 1.5 at decade 5 ($p < 0.05$). Given that the mean age was not significantly different between fathers of ASD and TD children, the residual effect of age as a confounding factor is precluded. Thus, our findings suggest that fathers over the age of 30 and with deficits in mtDNA repair systems and/or antioxidant defenses confers an increased risk for the child to develop ASD [34].

While these previous studies showed the prevalence of deletions and mutations in mtDNA from PBMC, a significant increase in the mtDNA copy number was also observed in tissues more relevant to ASD such as brain. Indeed, mtDNA copy number was higher in the frontal cortex (1.6 fold; $p = 0.004$) and temporal cortex (1.14 fold; $p = 0.04$) in post-mortem samples of children with ASD when compared to TD children. Taken together, there is evidence for the existence of biomarkers of mtDNA damage in both peripheral blood cells and brain, giving a strong rationale for continuing the research using peripheral cells, a less invasive sampling for children than muscle biopsies or other tissues normally used for diagnosing mitochondrial diseases. Furthermore, it allows exploring the correlation and possible causation between MD and ASD, while searching for possible causes of MD in ASD.

3 Mitochondrial and Genomic Mutations in Autism

Since 1998 there have been numerous case reports of MD co-occurring with ASD. Several of them have identified the genetic mutation responsible for the MD as well as measuring peripheral biochemical markers of MD [35–39]. The mitochondrial genome is a small circular double stranded DNA which encodes 13 proteins required by the respiratory chain to produce energy (ND1-ND6 including ND4L, CYTB, COX1-3, ATP6/8) [40]. In mammals, mitochondria are transmitted maternally, meaning that the risk of inheriting mtDNA mutations increases with the mother's proportion of mutated mtDNA [41]. There is anywhere between 10 and 10,000 copies of mtDNA in every cell and if a mutation arises it may result in a condition called heteroplasmy. This is defined as a mixture of both wild type and mutant mtDNA co-existing in the cell [41]. When these heteroplasmic cells divide, the mutant and wild type mtDNA are randomly distributed to daughter cells during mitotic segregation, so the relative amounts of mutated mtDNA in a given tissue can change over time. Additionally, around 1,500 nuclear genes encoded by genomic DNA (gDNA) are required for mitochondrial function [40]. Therefore, pathogenic mutations in either mitochondrial or genomic DNA can result in MD [42]. In addition, mitochondrial mutations have a higher incidence than nuclear mutations most likely due to the lack of histones, different DNA repair capacity and the proximity of mtDNA to the site of ROS production in mitochondria [42]. Although the majority of mtDNA mutations are inherited through maternal lines, they can also occur sporadically [43]. A study using ~3000 subjects based on a ten-point mutations in mtDNA found that the pathogenic *de novo* mtDNA mutation rates are greater than 1 in 200 live births [41].

Among mutations in the nuclear genome, more than 100 have been linked to ASD including the phosphatase and tensin homolog on chromosome ten (*Pten*). This gene encodes for PTEN, a phosphatase with tumor suppressor activity that is mutated in many human cancers [44]. The loss of both alleles seems to be connected with increased tumorigenesis and *Pten* pathogenic mutations and have been linked to benign macrocephaly, hamartomas, seizures, mental retardation, Alzheimer's disease and autism [44–56]. *Pten* mutations have been shown to be more prevalent in cases of autism where patients also exhibit macrocephaly [52]. A study by Kwon et al. showed that conditional *Pten* null mice resulted in impaired social interaction with no inclination for social novelty and reduced learning, less than average nest-forming activity, as well as irregular anxiety levels [57]. These same mice showed activation of the Akt/mTOR/S6k pathway and inactivation of GSK3β suggesting that nonstandard activation of the PI3K/Akt pathway in specific neuronal populations could be a causative of the observed macrocephaly and behavioral abnormalities [58]. This was determined by sampling brain matter in three regions, the cortex, hippocampus and cerebellum from heterozygous *Pten*$^{+/loxP}$ (HET) and heterozygous *Nse*-cre$^{+/-}$; *Pten*$^{+/loxP}$ (HET-CRE) mice aged 4–29 weeks and contrasting mitochondrial activities. Although in earlier weeks no social behavior change was observed, by 20–29 weeks the mice displayed abnormal social behavior (social avoidance, inability to recognize familiar mouse, and repetitive self-grooming), macrocephaly, increased oxidative stress, 50 % decreased cytochrome *c* oxidase (CCO) activity, and increased mtDNA deletions in cerebellum and hippocampus. This MD was the result of a downregulation of p53-signaling pathway evaluated by low protein expression of p21 (65 % of controls) and the CCO chaperone SCO2 (47 % of controls), two p53-downstream targets. This mechanism was then affirmed in *Pten*-deficient striatal neurons and HCT116 cells with differing *p53* gene dosage [59]. By comparing mitochondrial complex activities, mitochondrial mass and social conduct of mice, this study supported the association between aberrant social behavior caused by a *Pten* allele deletion and defects in bioenergetics [59]. This suggested a unique pathogenic mechanism of the *Pten-p53* axis that impairs mitochondrial function which is later amplified by the accumulation of mtDNA deletions as seen in ASD, and suggests this as a potential contributor to the etiology of autism.

It has been recently reported that ASD segregates with increased incidence of copy number variants. Copy number variants (CNV) are segments of DNA that vary in size from 50 base pairs to several megabases. This occurs due to reasons such as, deletions, insertions, duplications, inversions or complex recombinations. Several studies have found an increased prevalence of CNVs in children with ASD [60–63], but more notably, several of these CNV were found to be *de novo*. *De novo* mutations could arise from numerous reasons such as environmental factors that have yet to be identified [9]. Other studies failed to find significant differences in CNV frequencies between controls and ASD cases [64–66]. Thus, the *de novo* CNV may characterize only a small subset of ASD cases. CNV may play a more significant role in the normal functions of the human body. For instance, highly heterogeneous *de novo* CNV analyzed showed to affect several loci and possibly accounting for 5–8 % of certain forms of ASD [67, 68]. Further analysis of these unusual CNV show the involvement of these loci in activities such as axon targeting, synapse development, and neuron motility [69]. CNV have been linked to many clinical features such as severe neurological symptoms, a range of autism spectrum from mild to extreme and also to behavioral disorders outside of autism. The great variety in expressing CNV causes difficulties in diagnosing a sole CNV as a single cause of ASD.

4 Environmental Factors and Gene/Environment Interactions in ASD

Environmental factors have been found to aggravate prior genetic predispositions towards increased MD. Over the past 25 years, the use of polybrominated diphenyl ethers (PBDE), a group of flame-retardants, has been recognized with significant levels of accumulation in the environment and in human samples. The global production of PBDE has reached 148 million pounds per year [70]. Studies in the U.S. populations have shown the existence of PBDE in human breast milk, adipose tissue and blood, with women in Northern California having 3 to 25 times higher levels of PBDEs than women in many other parts of the world [71–73]. PBDEs with less than five Br substitutions (i.e., 2,2′,4,4′-tetrabromodiphenyl ether or BDE-47 and 2,2′,4,5′-tetrabromodiphenyl ether or BDE-49) are present at high levels and cause concern about their effects on health. BDE-49 has been reported to be one of the more biologically active PBDE even at lower concentrations than others in the group such as BDE-47. However, its capability to trigger mitochondrial disease, especially in cases where there is already some degree of MD, is unknown. A recent study by our laboratory compared the effects of BDE-49 in brain mitochondria and neuronal progenitor striatal cells (NPC) [8]. BDE-49 was found to uncouple neuronal mitochondria at concentrations lower than 0.1 nM. Other tests showed that concentrations greater than 1 nM inhibited the electron transport through mixed inhibitions (IC_{50} = 6 nM) and complex IV by noncompetitive inhibition (IC_{50} = 40 nM) [8]. These effects were enhanced in *Pten*-deficient neurons. The activity of complex IV in *Pten*-inhibited cells was reduced to one half as compared with *Pten*-sufficient cells. The addition of BDE-49 at concentrations equal to its K_i resulted in an additional 20 % inhibition (for a total of 70 %; $p = 0.05$). Enzymatic activity decrease thresholds for cells in culture are set at 20 % (major criterion) and 30 % (minor criterion) of control activities for diagnosing mitochondrial respiratory chain diseases, thus making 20 % inhibition quantitatively and biologically relevant [74]. Environmental chemicals even in extremely small concentrations can have deleterious effects on mitochondria, especially in those mitochondria that already show some degree of dysfunction.

5 Role of Mitochondrial Dysfunction in Perinatal Development and Impact on ASD

Exposures to either viral mimetics [75] or environmental chemicals and pharmaceuticals may influence fetal development with the potential of resulting in disorders later in life [76–78]. The increasing environmental PBDE concentrations cannot only trigger MD but they are also likely to cause developmental neurotoxicity through maternal transfer. Reports have shown that chronic, low-level maternal and fetal exposures to specific PBDEs (like BDE-49) during pregnancy could affect signaling systems necessary for activity-dependent dendritic growth and proper development of excitatory and inhibitory networks in the fetus [79, 80]. This abnormal development and consequent MD linked to PBDEs [8] may lead to an imbalance of excitatory and inhibitory neurotransmission which has been implicated in the etiology of several syndromic and idiopathic developmental disorders including autism [81].

The role of prenatal exposure to a viral mimetic and autism development was supported by research completed by our laboratory in which pregnant mice were treated with poly(I:C), a viral mimetic, with the adult offspring evaluated for mitochondrial outcomes [75]. A 45 % decrease in mitochondrial ATP production was observed in the adult offspring from poly(I:C)-treated females chiefly credited to a lower complex I activity with ASD-like behavior [82]. No differences were observed when comparing the coupling of electron transport to ATP synthesis, or the O_2 uptake under uncoupled conditions. The study reported for the first time that maternal immune activation caused by poly(I:C) in early gestation leads to a long-lasting effects in the bioenergetics of adult offspring and can result in behavioral impairments in the offspring similar to ASD and schizophrenia [75]. Several studies have suggested that maternal exposure to various pathogens, including viruses, can cause fetal imprinting of mitochondria-mediated metabolic responses and significantly increase the fetus risk for developing ASD or schizophrenia [83–89].

It has been found that exposure to Toll-like receptor (TLR) ligands can lead to maternal hypertension, vascular dysfunction, and proteinuria in pregnant animals but not in nonpregnant animals [90–92], suggesting the occurrence of a differential immune response pathway during pregnancy. A condition that pregnant women may develop from dysfunctional placenta, called preeclampsia, shows greater expression of *TLR3*, *TLR2*, *TLR4*, and *TLR9* than women with normal pregnancies and symptoms such as trophoblastic inclusions [93, 94]. When placentas from mothers of children later diagnosed with ASD were tested against a control group of placentas of mothers of children diagnosed as TD, a significant increase in trophoblastic inclusions were reported. The data imply a role of TLR signaling in altered fetal programming that can have lasting effects into adult life [95].

6 Effect of Mitochondrial Dysfunction and of Immunity on ASD

Besides their essential role in a number of critically important pathways in development, mitochondria may also impact the immune response and vice versa [96–98]. For instance, human neutrophil mitochondria are involved in several vital cell functions such as chemotaxis, respiratory burst activity, preservation of cell shape, and apoptosis [99–102]. Considering that both maternal diet and immune activation have an impact on fetal metabolic and immune programming, and as mitochondria are inherited via oocyte from the mother, it is hypothesized that the prenatal exposure of mothers to an immunogenic response will result in fetal imprinting leading to postnatal deficits in the bioenergetics of immune cells [103–105]. Additional studies on immune responses in ASD were conducted during childhood to give a wider understanding beyond prenatal immune development. Children diagnosed with ASD tended to show an impaired immune system as well as deficits in bioenergetics when compared with TD children [25]. Granulocytes were sampled from

ASD children with severity scores of ≥7 enrolled in the CHARGE study at University of California Davis and from TD children that were age-, race-, and gender-matched with the ASD children. Certain mitochondrial functions such as oxidative phosphorylation capacity, markers of oxidative stress (ROS production, mtDNA deletions) and immune response to phorbol 12-myristate 13-acetate–induced oxidative burst were evaluated in these granulocytes. The OXPHOS capacity of granulocytes in children with ASD was shown to be 3-fold lower than in TD children. There was also evidence of increased oxidative stress in cells from ASD children, the rate mitochondrial ROS production was 1.6-fold higher and the mtDNA copy number per cell was 1.5-fold higher [25].

7 Potential Therapies for ASD Based on Mitochondrial Dysfunction

Mitochondria are influential organelles whose morphology, composition, and activity adapt to changes in response to pathological and physiological signals such as those introduced by ketogenic diets (KGD). Several reports document changes in mitochondrial number or function in a variety of biological systems, from *in vitro* to *in vivo*, when exposed to KGD or KGD-mimetics. ASD has been associated to metabolic dysfunction, autism is a common trait of epilepsy-associated diseases, such as the Landau–Kleffner, Dravet, and Rett syndromes [2]. The KGD diet is essentially a high-fat regime with acceptable protein amount for growth but inadequate levels of carbohydrates for metabolic needs, forcing the body to primarily use fat as a fuel source [106]. As well as causing hormone changes, the diet also initiates an insulin reduction in response to reduction of plasma glucose, thus limiting glucose utilization. Instead of acetyl-CoA convertion to CO_2 and H_2O during the standard Krebs cycle, the increased disproportion of fatty acid mobilization and the capacity of the Krebs cycle cause the surplus acetyl-CoA to be converted into ketone bodies (KB), that are used as fuel by peripheral tissues instead of glucose. KB can also cross the blood-brain barrier [107] to be used as the main fuel source for the brain during fasting periods [108]. The low-carbohydrate intake forces the body to sustain systemic glycemia by hepatic gluconeogenesis from non-carbohydrate precursors (e.g., lactate, glucogenic amino acids, and glycerol). By providing alternative sources of acetylCoA, KGD is the dietary mediation for inborn genetic disorders in pyruvate dehydrogenase (*PDH*) and glucose transporter 1 (*GLUT1*), that has proven effective also in other metabolic alterations, including phosphofructokinase deficiency and glycogenosis type V (McArdle disease) [2]. Ketogenic diet has been used for more than 80 years in epilepsy treatment [109, 110] particularly in children and adolescents [106, 111] and has shown lessened frequencies of seizures [112, 113] and improvements in developmental progress [114]. The use of KGD is being investigated for the management of Alzheimer's, Parkinson's diseases and ASD [115, 116]. ASD has been associated to metabolic dysfunction and autism is a common trait of epilepsy-associated diseases, and syndromes treated with KGD, making this diet a possibility for reducing the symptoms of ASD.

Beneficial effects of KGD in children showing ASD symptoms have been described in two independent studies [117, 118]. The first, evaluated the role of KGD on 30 ASD children [117]. The John Radcliffe diet (a modified medium-chain triglyceride diet with a caloric distribution of 30 % in medium-chain triglyceride oil, 30 % fresh cream, 11 % saturated fat, 19 % carbohydrates, and 10 % proteins) was administered for 6 months, with intervals of 4 weeks interrupted by two diet-free weeks. Out of the group of 30, 12 children did not comply or did not tolerate the diet. Of the rest of the subjects, two children with milder autistic behaviors, showed a marked improvement (as assessed by total Childhood Autism Rating Scale score, concentration and learning abilities, and social behavior and interactions). The other 16 children displayed mild to moderate improvements. Remarkably, the beneficial effects of KGD persisted even after termination of the trial. The second study [118] reports the administration of a gluten- and casein-free modified-KGD (1.5:1 lipid:non-lipid ratio; medium-chain and polyunsaturated fatty acids) for two full years to a 12-year-old child with ASD, seizures and substantial comorbidities associated with a family history of metabolic and immune disturbances. This diet was deemed such a success that anticonvulsant medication doses were reduced without worsening of seizures, with improvements in seizure activity, amended electroencephalogram, cognitive and social skills, language function, and total resolution of stereotypies. However, many medications were administered during the diet along with other complications such as a significant weight loss and transitioning to puberty, so it is difficult to assess the sole role of the diet without the clinical background. The improvement shown in these studies may be due to enhancing mitochondrial biogenesis as shown in murine models [119, 120]. In cell cultures, an extrapolated diet was shown to increase citrate synthase and complex I activity in SH-SY5Y cells [121]. However, since the lower limit for an accepted mitochondrial disorder is an OXPHOS outcome lower of at least 30 % of control values, the proliferation of mitochondrial mass would need to equal at least 30 % in order to rescue the impaired ATP production of ASD individuals. These studies show promising results of enhancement of mitochondrial function that seemed to prompt improvement in social behaviors and learning abilities.

8 Conclusions

While mitochondrial or nuclear DNA pathogenic mutations have been characterized in cases with ASD, they do not seem to be major contributors to ASD incidence. However, it is becoming clear that there is a subgroup of individuals with autism and MD, with or without a currently known genetic defect. Therefore, rare genetic variants or single nucleotide polymorphisms in combination with environmental exposures or dietary factors, may confound MD and the ensuing increased oxidative stress. While no cure currently exists for individuals with mitochondrial diseases, there is a successful study that exchanged a defective mitochondrial genome with a functional genome in oocysts, effectively replacing damaged mitochondria with fully operational mitochondria [122]. Intriguing research on therapies and studies on the etiology and symptoms of MD have been reviewed in hopes to create new ways to manage MD [123, 124] similar to those seen in ASD. Although additional research is needed, there are several proposed directions for new therapies such as increasing respiratory chain function through the proliferation of wild-type mitochondria, the supplemental use of coenzyme Q10, mtDNA targeting endonucleases and the limitation of exposure to aggravating environmental factors such as BDE-49 [125].

A proposed method to increase mitochondrial mass is through the upregulation of *PGC-1α*, increasing the total number of mitochondria inside cells, in the hope of returning mitochondrial bioenergetics to average levels. However, this therapy assumes that the mutation or defect responsible for the respiratory chain deficits leaves residual oxidative phosphorylation levels [125]. Moraes *et al.* used a muscle-specific knockout of the *COX10* assembly factor to show that *PGC-1α* upregulation protects the muscle from MD in mice [126]. The enhancement of mitochondrial biogenesis has been found effective in treating MD in several common human conditions [127–129] and may hold possibilities for the management of ASD. A second proposed therapy involves the administration of coenzyme Q10, usually along with vitamin C, vitamin K1, vitamin B complex, and L-carnitine. This particular therapy is popular due to the reports of safety, even at high doses, and for its roles as antioxidant and for providing respiratory chain components [130]. A third and intriguing approach includes the use of a restriction endonuclease that can selectively target mutated mtDNA. The DNA-binding selectivity is based on the binding properties of zinc fingers (ZF), consisting of histidine and cysteine residues that bind zinc. It is the Zn-ligated section of the polypeptide that binds to sequence specific DNA [131]. When this ZF is attached to an endonuclease it has the potential to cleave almost any nuclear DNA sequence [132]. The next logical step would be to formulate an mtDNA-specific ZF nuclease. Another interesting proposed direction for managing MD in society would be to limit the amount of exposure to chemicals, e.g. the toxic fire retardants, by careful evaluation of their cost-benefit use. The genetic and environmental components of MD have significant contributions to observed situations and each must be taken into account in future research in order to fully appreciate this disorder of mitochondrial malfunction.

Acknowledgements We thank the families that participated in this study as well as the technical and intellectual contributions over the years of Catherine Ross-Inta, Alicja Omanska-Klusek, Gyu Song, Yi-Fan Zhang, Daniel Sakaguchi, Casey Bronec, Erica Santos, Nadia Dueñas, Joel Ramirez, Prithvi Bomdica, Allegra De Sandri, and Drs. Eleonora Napoli, Siming Liu, Kazu Kato and Yasuko Fujisawa. We are grateful for the scientific collaborations with Drs. Irva Hertz-Picciotto, Sally Ozonoff, Randi Hagerman, and Flora Tassone that made these studies possible. The research in this article was supported by grants from the Simons Foundation (SFARI 271406) and National Institute of Environmental Health Sciences R01-ES011269, R01-ES015359, R01-ES020392, and P30-ES023513. The National Institute of Child and Human Development: U45-HD079125; EPA Science to Achieve Results (STAR): R-829388 and R833292; and the University of California Davis MIND Institute. The funders had no role in study design, data collection and analysis, decision to publish, or preparation of the manuscript.

References

1. Giulivi C, Zhang YF, Omanska-Klusek A et al (2010) Mitochondrial dysfunction in autism. JAMA 304:2389–2396
2. Napoli E, Duenas N, Giulivi C (2014) Potential therapeutic use of the ketogenic diet in autism spectrum disorders. Front Pediatr 2:69
3. Oliveira G, Diogo L, Grazina M et al (2005) Mitochondrial dysfunction in autism spectrum disorders: a population-based study. Dev Med Child Neurol 47:185–189
4. Baio J (2012) Prevalence of autism spectrum disorders. Autism and Developmental Disabilities Monitoring Network, 14 sites, United States, 2008. MMWR Surveill Summ 61:1–19
5. Baio J (2014) Prevalence of autism spectrum disorder among children aged 8 years. Autism and Developmental Disabilities Monitoring Network, 11 Sites, United States, 2010. Morb Mortal Wkly Rep 63:1–21
6. Shelton JF, Hertz-Picciotto I, Pessah IN (2012) Tipping the balance of autism risk: potential mechanisms linking pesticides and autism. Environ Health Perspect 120:944–951
7. Geier DA, Kern JK, Garver CR et al (2009) Biomarkers of environmental toxicity and susceptibility in autism. J Neurol Sci 280:101–108
8. Napoli E, Hung C, Wong S, Giulivi C (2013) Toxicity of the flame-retardant BDE-49 on brain mitochondria and neuronal progenitor striatal cells enhanced by a PTEN-deficient background. Toxicol Sci 132:196–210
9. Hallmayer J, Cleveland S, Torres A et al (2011) Genetic heritability and shared environmental factors among twin pairs with autism. Arch Gen Psychiatry 68:1095–1102
10. Lord C, Risi S, Lambrecht L et al (2000) The autism diagnostic observation schedule-generic: a standard measure of social and communication deficits associated with the spectrum of autism. J Autism Dev Disord 30:205–223
11. American Psychiatric Association (2000) American Psychiatric Association diagnostic and statistical manual of mental disorders, 4th edn, text revision. American Psychiatric Association, Washington, DC.
12. WHO (1992) International statistical classification of diseases and related health problems, vol 1, Tenth Revision. World Health Organization, Geneva.
13. Robin ED, Wong R (1988) Mitochondrial DNA molecules and virtual number of mitochondria per cell in mammalian cells. J Cell Physiol 136:507–513
14. Shay JW, Pierce DJ, Werbin H (1990) Mitochondrial DNA copy number is proportional to total cell DNA under a variety of growth conditions. J Biol Chem 265:14802–14807
15. Lee HC, Yin PH, Lu CY et al (2000) Increase of mitochondria and mitochondrial DNA in response to oxidative stress in human cells. Biochem J 348:425–432
16. Liu CS, Tsai CS, Kuo CL et al (2003) Oxidative stress-related alteration of the copy number of mitochondrial DNA in human leukocytes. Free Radic Res 37:1307–1317
17. Chou YF, Huang RF (2009) Mitochondrial DNA deletions of blood lymphocytes as genetic markers of low folate-related mitochondrial genotoxicity in peripheral tissues. Eur J Nutr 48:429–436
18. Yeargin-Allsopp M, Rice C, Karapurkar T et al (2003) Prevalence of autism in a US metropolitan area. JAMA 289:49–55
19. Lee HC, Lu CY, Fahn HJ, Wei YH (1998) Aging- and smoking-associated alteration in the relative content of mitochondrial DNA in human lung. FEBS Lett 441:292–296
20. Moreno-Loshuertos R, Acin-Perez R, Fernandez-Silva P et al (2006) Differences in reactive oxygen species production explain the phenotypes associated with common mouse mitochondrial DNA variants. Nat Genet 38:1261–1268
21. Ylikallio E, Tyynismaa H, Tsutsui H et al (2010) High mitochondrial DNA copy number has detrimental effects in mice. Hum Mol Genet 19:2695–2705
22. Napoli E, Wong S, Giulivi C (2013) Evidence of reactive oxygen species-mediated damage to mitochondrial DNA in children with typical autism. Mol Autism 4:2

23. Frustaci A, Neri M, Cesario A et al (2012) Oxidative stress-related biomarkers in autism: systematic review and meta-analyses. Free Radic Biol Med 52:2128–2141
24. Rose S, Melnyk S, Trusty TA et al (2012) Intracellular and extracellular redox status and free radical generation in primary immune cells from children with autism. Autism Res Treat 2012: :986519
25. Napoli E, Wong S, Hertz-Picciotto I, Giulivi C (2014) Deficits in bioenergetics and impaired immune response in granulocytes from children with autism. Pediatrics 133:1405–1410
26. Sajdel-Sulkowska EM, Xu M, McGinnis W, Koibuchi N (2011) Brain region-specific changes in oxidative stress and neurotrophin levels in autism spectrum disorders (ASD). Cerebellum 10:43–48
27. Chauhan A, Gu F, Essa MM et al (2011) Brain region-specific deficit in mitochondrial electron transport chain complexes in children with autism. J Neurochem 117:209–220
28. Rose S, Melnyk S, Pavliv O et al (2012) Evidence of oxidative damage and inflammation associated with low glutathione redox status in the autism brain. Transl Psychiatry 2:e134
29. Tang G, Gutierrez Rios P, Kuo SH et al (2013) Mitochondrial abnormalities in temporal lobe of autistic brain. Neurobiol Dis 54:349–361
30. James SJ, Rose S, Melnyk S et al (2009) Cellular and mitochondrial glutathione redox imbalance in lymphoblastoid cells derived from children with autism. Faseb J 23:2374–2383
31. James SJ, Cutler P, Melnyk S et al (2004) Metabolic biomarkers of increased oxidative stress and impaired methylation capacity in children with autism. Am J Clin Nutr 80:1611–1617
32. Theoharides TC, Asadi S, Patel AB (2013) Focal brain inflammation and autism. J Neuroinflammation 10:46
33. Hertz-Picciotto I, Croen LA, Hansen R et al (2006) The CHARGE study: an epidemiologic investigation of genetic and environmental factors contributing to autism. Environ Health Perspect 114:1119–1125
34. Wong S, Napoli E, Krakowiak P et al (2016) Role of p53, mitochondrial DNA deletions and paternal age in autism. Pediatrics e20151888
35. Graf WD, Marin-Garcia J, Gao HG et al (2000) Autism associated with the mitochondrial DNA G8363A transfer RNA(Lys) mutation. J Child Neurol 15:357–361
36. Marin-Garcia J, Goldenthal MJ, Filiano JJ (2002) Cardiomyopathy associated with neurologic disorders and mitochondrial phenotype. J Child Neurol 17:759–765
37. Filipek PA, Juranek J, Smith M et al (2003) Mitochondrial dysfunction in autistic patients with 15q inverted duplication. Annals Neurol 53:801–804
38. Pons R, Andreu AL, Checcarelli N et al (2004) Mitochondrial DNA abnormalities and autistic spectrum disorders. J Pediatr 144:81–85
39. Weissman JR, Kelley RI, Bauman ML et al (2008) Mitochondrial disease in autism spectrum disorder patients: a cohort analysis. PLoS One 3:e3815
40. Shadel GS (2008) Expression and maintenance of mitochondrial DNA: new insights into human disease pathology. Am J Pathol 172:1445–1456
41. Elliott HR, Samuels DC, Eden JA et al (2008) Pathogenic mitochondrial DNA mutations are common in the general population. Am J Hum Genet 83:254–260
42. Schapira AHV Mitochondrial diseases. The Lancet 379:1825–1834
43. DiMauro S, Andreu AL (2001) Mutations in mitochondrial DNA as a cause of exercise intolerance. Ann Med 33:472–476
44. Ali IU, Schriml LM, Dean M (1999) Mutational spectra of PTEN/MMAC1 gene: a tumor suppressor with lipid phosphatase activity. J Natl Cancer Inst 91:1922–1932
45. Zhou XP, Waite KA, Pilarski R et al (2003) Germline PTEN promoter mutations and deletions in Cowden/Bannayan-Riley-Ruvalcaba syndrome result in aberrant PTEN protein and dysregulation of the phosphoinositol-3-kinase/Akt pathway. Am J Hum Genet 73:404–411
46. Delatycki MB, Danks A, Churchyard A et al (2003) De novo germline PTEN mutation in a man with Lhermitte-Duclos disease which arose on the paternal chromosome and was transmitted to his child with polydactyly and Wormian bones. J Med Genet 40:e92

47. Rickle A, Behbahani H, Ankarcrona M et al (2006) PTEN, Akt, and GSK3beta signalling in rat primary cortical neuronal cultures following tumor necrosis factor-alpha and trans-4-hydroxy-2-nonenal treatments. J Neurosci Res 84:596–605
48. Merks JH, de Vries LS, Zhou XP et al (2003) PTEN hamartoma tumour syndrome: variability of an entity. J Med Genet 40:e111
49. Kerr F, Rickle A, Nayeem N et al (2006) PTEN, a negative regulator of PI3 kinase signalling, alters tau phosphorylation in cells by mechanisms independent of GSK-3. FEBS Lett 580:3121–3128
50. Griffin RJ, Moloney A, Kelliher M et al (2005) Activation of Akt/PKB, increased phosphorylation of Akt substrates and loss and altered distribution of Akt and PTEN are features of Alzheimer's disease pathology. J Neurochem 93:105–117
51. Eng C (2003) PTEN: one gene, many syndromes. Hum Mutat 22:183–198
52. Butler MG, Dasouki MJ, Zhou XP et al (2005) Subset of individuals with autism spectrum disorders and extreme macrocephaly associated with germline PTEN tumour suppressor gene mutations. J Med Genet 42:318–321
53. Goffin A, Hoefsloot LH, Bosgoed E et al (2001) PTEN mutation in a family with Cowden syndrome and autism. Am J Med Genet 105:521–524
54. Spinelli L, Black FM, Berg JN et al (2015) Functionally distinct groups of inherited PTEN mutations in autism and tumour syndromes. J Med Genet 52:128–134
55. Clipperton-Allen AE, Page DT (2014) Pten haploinsufficient mice show broad brain overgrowth but selective impairments in autism-relevant behavioral tests. Hum Mol Genet 23:3490–3505
56. Zhou J, Parada LF (2012) PTEN signaling in autism spectrum disorders. Curr Opin Neurobiol 22:873–879
57. Kwon CH, Luikart BW, Powell CM et al (2006) Pten regulates neuronal arborization and social interaction in mice. Neuron 50:377–388
58. Zhou J, Blundell J, Ogawa S et al (2009) Pharmacological inhibition of mTORC1 suppresses anatomical, cellular, and behavioral abnormalities in neural-specific Pten knock-out mice. J Neurosci 29:1773–1783
59. Napoli E, Ross-Inta C, Wong S et al (2012) Mitochondrial dysfunction in Pten haplo-insufficient mice with social deficits and repetitive behavior: interplay between Pten and p53. PLoS One 7:e42504
60. Chaste P, Leboyer M (2012) Autism risk factors: genes, environment, and gene-environment interactions. Dialogues Clin Neurosci 14:281–292
61. Gill M (2012) Developmental psychopathology: the role of structural variation in the genome. Dev Psychopathol 24:1319–1334
62. McGrew SG, Peters BR, Crittendon JA, Veenstra-Vanderweele J (2012) Diagnostic yield of chromosomal microarray analysis in an autism primary care practice: which guidelines to implement? J Autism Dev Disord 42:1582–1591
63. Miller DT, Shen Y, Wu BL (2012) Oligonucleotide microarrays for clinical diagnosis of copy number variation and zygosity status. Curr Protoc Hum Genet Chapter 8: Units 8–12
64. Lacaria M, Spencer C, Gu W et al (2012) Enriched rearing improves behavioral responses of an animal model for CNV-based autistic-like traits. Hum Mol Genet 21:3083–3096
65. Moreno-Garcia M, Sanchez del Pozo J, Cruz-Rojo J et al (2012) Array-based characterization of an interstitial de novo deletion of chromosome 4q in a patient with a neuronal migration defect and hypocalcemia plus a literature review. Clin Dysmorphol 21:172–176
66. Philippe A (2012) Genetics of autism spectrum disorders. Medecine Therapeutique Pediatrie 15:194–201
67. Frye RE (2012) Mitochondrial disease in 22q13 duplication syndrome. J Child Neurol 27:942–949
68. Malenfant P, Liu X, Hudson ML et al (2012) Association of GTF2i in the Williams-Beuren syndrome critical region with autism spectrum disorders. J Autism Dev Disord 42:1459–1469

69. He WZ, Liu WQ, Zhong XQ et al (2012) Analysis of de novo copy number variations in a family affected with autism spectrum disorders using high-resolution array-based comparative genomic hybridization. Chinese J Med Genet 29:266–269
70. McDonald TA (2002) A perspective on the potential health risks of PBDEs. Chemosphere 46:745–755
71. Betts KS (2002) Rapidly rising PBDE levels in North America. Environ Sci Technol 36:50A–52A
72. Schecter A, Pavuk M, Papke O et al (2003) Polybrominated diphenyl ethers (PBDEs) in U.S. mothers' milk. Environ Health Perspect 111:1723–1729
73. Petreas M, Nelson D, Brown FR et al (2011) High concentrations of polybrominated diphenylethers (PBDEs) in breast adipose tissue of California women. Environ Int 37:190–197
74. Bernier FP, Boneh A, Dennett X et al (2002) Diagnostic criteria for respiratory chain disorders in adults and children. Neurology 59:1406–1411
75. Giulivi C, Napoli E, Schwartzer J et al (2013) Gestational exposure to a viral mimetic poly(i:C) results in long-lasting changes in mitochondrial function by leucocytes in the adult offspring. Mediators Inflamm 2013:602–609
76. Alexander FE, Patheal SL, Biondi A et al (2001) Transplacental chemical exposure and risk of infant leukemia with MLL gene fusion. Cancer Res 61:2542–2546
77. Ma X, Buffler PA, Gunier RB et al (2002) Critical windows of exposure to household pesticides and risk of childhood leukemia. Environ Health Perspect 110:955–960
78. Woodruff TJ, Axelrad DA, Kyle AD et al (2004) Trends in environmentally related childhood illnesses. Pediatrics 113:1133–1140
79. Kim KH, Bose DD, Ghogha A et al (2011) Para- and ortho-substitutions are key determinants of polybrominated diphenyl ether activity toward ryanodine receptors and neurotoxicity. Environ Health Perspect 119:519–526
80. Dingemans MM, van den Berg M, Westerink RH (2011) Neurotoxicity of brominated flame retardants. (In) direct effects of parent and hydroxylated polybrominated diphenyl ethers on the (developing) nervous system. Environ Health Perspect 119:900–907
81. Belmonte MK, Bourgeron T (2006) Fragile X syndrome and autism at the intersection of genetic and neural networks. Nat Neurosci 9:1221–1225
82. Onore C, Careaga M, Ashwood P (2012) The role of immune dysfunction in the pathophysiology of autism. Brain Behav Immun 26:383–392
83. Brown AS, Derkits EJ (2010) Prenatal infection and schizophrenia: a review of epidemiologic and translational studies. Am J Psychiatry 167:261–280
84. Brown AS, Patterson PH (2011) Maternal infection and schizophrenia: implications for prevention. Schizophr Bull 37:284–290
85. Patterson PH (2011) Maternal infection and immune involvement in autism. Trends Mol Med 17:389–394
86. Patterson PH (2012) Maternal infection and autism. Brain Behav Immun 26:393
87. Brown AS, Begg MD, Gravenstein S et al (2004) Serologic evidence of prenatal influenza in the etiology of schizophrenia. Arch Gen Psychiatry 61:774–780
88. Brown AS, Hooton J, Schaefer CA et al (2004) Elevated maternal interleukin-8 levels and risk of schizophrenia in adult offspring. Am J Psychiatry 161:889–895
89. Atladottir HO, Thorsen P, Ostergaard L et al (2010) Maternal infection requiring hospitalization during pregnancy and autism spectrum disorders. J Autism Dev Disord 40:1423–1430
90. Tinsley JH, Chiasson VL, Mahajan A et al (2009) Toll-like receptor 3 activation during pregnancy elicits preeclampsia-like symptoms in rats. Am J Hypertens 22:1314–1319
91. Chatterjee P, Chiasson VL, Kopriva SE et al (2011) Interleukin 10 deficiency exacerbates toll-like receptor 3-induced preeclampsia-like symptoms in mice. Hypertension 58:489–496
92. Cipolla MJ, Houston EM, Kraig RP, Bonney EA (2011) Differential effects of low-dose endotoxin on the cerebral circulation during pregnancy. Reprod Sci 18:1211–1221
93. Abrahams VM, Bole-Aldo P, Kim YM et al (2004) Divergent trophoblast responses to bacterial products mediated by TLRs. J Immunol 173:4286–4296

94. Pineda A, Verdin-Teran SL, Camacho A, Moreno-Fierros L (2011) Expression of toll-like receptor TLR-2, TLR-3, TLR-4 and TLR-9 is increased in placentas from patients with preeclampsia. Arch Med Res 42:382–391
95. Walker CK, Anderson KW, Milano KM et al (2013) Trophoblast inclusions are significantly increased in the placentas of children in families at risk for autism. Biol Psychiatry 74:204–211
96. West AP, Shadel GS, Ghosh S (2011) Mitochondria in innate immune responses. Nat Rev Immunol 11:389–402
97. Arnoult D, Soares F, Tattoli I, Girardin SE (2011) Mitochondria in innate immunity. EMBO Rep 12:901–910
98. Cloonan SM, Choi AM (2012) Mitochondria: commanders of innate immunity and disease? Curr Opin Immunol 24:32–40
99. Fossati G, Moulding DA, Spiller DG et al (2003) The mitochondrial network of human neutrophils: role in chemotaxis, phagocytosis, respiratory burst activation, and commitment to apoptosis. J Immunol 170:1964–1972
100. Genestier AL, Michallet MC, Prevost G et al (2005) Staphylococcus aureus Panton-Valentine leukocidin directly targets mitochondria and induces Bax-independent apoptosis of human neutrophils. J Clin Invest 115:3117–3127
101. Scheel-Toellner D, Wang K, Craddock R et al (2004) Reactive oxygen species limit neutrophil life span by activating death receptor signaling. Blood 104:2557–2564
102. Maianski NA, Roos D, Kuijpers TW (2003) Tumor necrosis factor alpha induces a caspase-independent death pathway in human neutrophils. Blood 101:1987–1995
103. Burgueno AL, Cabrerizo R, Gonzales-Mansilla N et al (2013) Maternal high-fat intake during pregnancy programs metabolic-syndrome-related phenotypes through liver mitochondrial DNA copy number and transcriptional activity of liver PPAR-GC1A. J Nutr Biochem 24:6–13
104. Garcia AP, Priego T, Palou M et al (2013) Early alterations in plasma ghrelin levels in offspring of calorie-restricted rats during gestation may be linked to lower sympathetic drive to the stomach. Peptides 39:59–63
105. Garay PA, Hsiao EY, Patterson PH, McAllister AK (2013) Maternal immune activation causes age- and region-specific changes in brain cytokines in offspring throughout development. Brain Behav Immun 31:54–68
106. Freeman J, Veggiotti P, Lanzi G et al (2006) The ketogenic diet: from molecular mechanisms to clinical effects. Epilepsy Res 68:145–180
107. Halestrap AP (2013) Monocarboxylic acid transport. Compr Physiol 3:1611–1643
108. Olpin SE (2004) Implications of impaired ketogenesis in fatty acid oxidation disorders. Prostagl Leukot Essent Fatty Acids 70:293–308
109. Hartman AL, Gasior M, Vining EP, Rogawski MA (2007) The neuropharmacology of the ketogenic diet. Pediatr Neurol 36:281–292
110. Kossoff EH, Wang HS (2013) Dietary therapies for epilepsy. Biomed J 36:2–8
111. Marsh EB, Freeman JM, Kossoff EH et al (2006) The outcome of children with intractable seizures: a 3- to 6-year follow-up of 67 children who remained on the ketogenic diet less than one year. Epilepsia 47:425–430
112. Rogovik AL, Goldman RD (2010) Ketogenic diet for treatment of epilepsy. Can Fam Physician 56:540–542
113. Coppola G, Verrotti A, Ammendola E et al (2010) Ketogenic diet for the treatment of catastrophic epileptic encephalopathies in childhood. Eur J Paediatr Neurol 14:229–234
114. El-Gharbawy AH, Boney A, Young SP, Kishnani PS (2011) Follow-up of a child with pyruvate dehydrogenase deficiency on a less restrictive ketogenic diet. Mol Genet Metab 102:214–215
115. Baranano KW, Hartman AL (2008) The ketogenic diet: uses in epilepsy and other neurologic illnesses. Curr Treat Options Neurol 10:410–419

116. Stafstrom CE, Rho JM (2012) The ketogenic diet as a treatment paradigm for diverse neurological disorders. Front Pharmacol 3:59
117. Evangeliou A, Vlachonikolis I, Mihailidou H et al (2003) Application of a ketogenic diet in children with autistic behavior: pilot study. J Child Neurol 18:113–118
118. Herbert MR, Buckley JA (2013) Autism and dietary therapy: case report and review of the literature. J Child Neurol 28:975–982
119. Ahola-Erkkila S, Carroll CJ, Peltola-Mjosund K et al (2010) Ketogenic diet slows down mitochondrial myopathy progression in mice. Hum Mol Genet 19:1974–1984
120. Bough KJ, Wetherington J, Hassel B et al (2006) Mitochondrial biogenesis in the anticonvulsant mechanism of the ketogenic diet. Ann Neurol 60:223–235
121. Hughes SD, Kanabus M, Anderson G et al (2014) The ketogenic diet component decanoic acid increases mitochondrial citrate synthase and complex I activity in neuronal cells. J Neurochem 129:426–433
122. Tachibana M, Sparman M, Sritanaudomchai H et al (2009) Mitochondrial gene replacement in primate offspring and embryonic stem cells. Nature 461:367–372
123. DiMauro S, Hirano M, Schon EA (2006) Approaches to the treatment of mitochondrial diseases. Muscle Nerve 34:265–283
124. Kerr DS (2010) Treatment of mitochondrial electron transport chain disorders: a review of clinical trials over the past decade. Mol Genet Metab 99:246–255
125. Schon EA, DiMauro S, Hirano M, Gilkerson RW (2010) Therapeutic prospects for mitochondrial disease. Trends Mol Med 16:268–276
126. Wenz T, Diaz F, Spiegelman BM, Moraes CT (2008) Activation of the PPAR/PGC-1alpha pathway prevents a bioenergetic deficit and effectively improves a mitochondrial myopathy phenotype. Cell Metab 8:249–256
127. Mootha VK, Handschin C, Arlow D et al (2004) Erralpha and Gabpa/b specify PGC-1alpha-dependent oxidative phosphorylation gene expression that is altered in diabetic muscle. Proc Natl Acad Sci U S A 101:6570–6575
128. Wenz T, Rossi SG, Rotundo RL et al (2009) Increased muscle PGC-1alpha expression protects from sarcopenia and metabolic disease during aging. Proc Natl Acad Sci U S A 106:20405–20410
129. Ikeuchi M, Matsusaka H, Kang D et al (2005) Overexpression of mitochondrial transcription factor a ameliorates mitochondrial deficiencies and cardiac failure after myocardial infarction. Circulation 112:683–690
130. Marriage BJ, Clandinin MT, Macdonald IM, Glerum DM (2004) Cofactor treatment improves ATP synthetic capacity in patients with oxidative phosphorylation disorders. Mol Genet Metab 81:263–272
131. Urnov FD, Rebar EJ (2002) Designed transcription factors as tools for therapeutics and functional genomics. Biochem Pharmacol 64:919–923
132. Porteus MH, Carroll D (2005) Gene targeting using zinc finger nucleases. Nat Biotechnol 23:967–973

Chapter 22
Systemic Oxidative Stress in Patients with Neurodegenerative Diseases

Marisa G. Repetto and Alberto Boveris

Abstract Oxidative stress and oxidative damage have been recognized in the brain of patients with neurodegenerative diseases since the early stages of the diseases. Oxidative stress and damage have been reported in patients of Alzheimer disease, Parkinson disease, amyotrophic lateral sclerosis and vascular dementia. Products of free-radical reactions in the brain, such as lipid peroxidation products and carbonyl groups in small peptides, were determined in blood, plasma, serum and cerebrospinal fluid of patients with the mentioned diseases. There is a clear evidence of a link between brain oxidative stress and damage and circulating indicators of such damage. Body fluids from living patients represent the best source of information about brain metabolism in neurodegenerative diseases. Cerebrospinal fluid provides a unique window of brain status for neuronal cell and brain tissue alterations. The ideal biomarker of brain oxidative stress should be determined in blood or plasma that are easy to collect.

Keywords Oxidative stress • Systemic oxidative stress • Neurodegenerative diseases • Alzheimer disease • Parkinson disease • Amyotrophic lateral sclerosis • Vascular dementia • Plasma biomarkers

1 Introduction

At present, chronic neurodegenerative diseases (ND) have an enormous impact in human health. The World Alzheimer's Report informed that there were 36 million people living with dementia worldwide in 2010, increasing to 66 million by 2030 and to 115 million by 2050. The number of demented patients will rise from 2001 to 2040 by 80–190 % in Europe, North America and the Western Pacific, while in Latin America, India, China, North Africa and the Middle East the increase will be

M.G. Repetto (✉) • A. Boveris
School of Pharmacy and Biochemistry, Department of General Chemistry, General and Inorganic Chemistry Division, Institute of Biochemistry and Molecular Medicine (IBIMOL, UBA-CONICET), University of Buenos Aires, Buenos Aires, Argentina
e-mail: mrepetto@ffyb.uba.ar

© Springer International Publishing Switzerland 2016
R.J. Gelpi et al. (eds.), *Biochemistry of Oxidative Stress*, Advances in Biochemistry in Health and Disease 16, DOI 10.1007/978-3-319-45865-6_22

more than 300 % [1]. Causes of neuronal death in ND are multifactorial and develop as complex networks of pathological events that are interconnected and generate processes leading to increased damage and neuronal death. Several mechanisms were proposed to explain these pathologies: oxidative stress (OS), inflammation, mitochondrial dysfunction, toxic effects of transition metals, depletion of intracellular thiols and redox dysregulation [2, 3]. Most of these mechanisms are closely related. Neuronal death in ND may represent a form of cell demise, that is neither classical necrosis nor apoptosis and that may be triggered during the terminal phase of the chronic progressive disorders [4, 5].

2 The Concept of Cellular Oxidative Stress

Sies [6] coined the concept of OS and considered it as an unbalance between oxidant production and antioxidant defences that may lead to oxidative damage to biomolecules and to cell death. The concept was immediately accepted, adopted and successfully used to design and explain experiments. Moreover, the concept was extended to organs and tissues and even to whole organisms. From a biochemical point of view, cellular OS is described as a situation in which increased steady state concentrations of the chemical species of the free radical chain reaction produce an increase in the rate of the chain reaction and in endogenous antioxidant consumption [7–9]. The "redox hypothesis" updated the classic concept of OS considering that it is due to an oxidative process that alters the redox balance of the thiol groups of low molecular weight components, such as glutathione (GSH) and proteins involved in signal pathways and regulation of physiological functions [10, 11].

3 The Concept of Systemic Oxidative Stress

The concept of systemic oxidative stress strictly applies only to situations in which the whole body has increased levels of reactive oxygen species (ROS) as in exposures to radiation. In general, systemic, circulating or peripheral oxidative stress, all synonyms, is understood as increased levels of ROS and related products in blood, plasma or serum. ROS is a very used term but the meaning is not totally clear. Originally in 1982, Seim [12] referred the term to the products of the partial reduction of oxygen: O_2^-, superoxide radical; H_2O_2, hydrogen peroxide; and HO•, hydroxyl radical, based on the fact that the observed biological effect was similar without discrimination of the responsible chemical species. Later, the meaning of ROS started to include molecules involved in the free radical chain reaction that links O_2^-, H_2O_2 and HO• with the free-radical mediated process of lipid peroxidation. These molecules are R•, alkyl radicals; ROO•, peroxyl radicals; ROOH, organic hydroperoxide; and $^1O_2$, singlet oxygen. The close experimental relationship between the free radical reactions of oxygen intermediates and the process of lipid peroxidation was early and always observed [13, 14]. Some of the products of lipid peroxidation, such as malonaldehyde, usually determined as TBARS (thiobarbituric acid reactive substances), 4-HO-nonenal, aldehydes, etc., are sometimes also considered as ROS. The important fact for the interest of this chapter is that some of the mentioned molecules, as H_2O_2, ROOH (especially short chain ROOH), $^1O_2$, TBARS, 4-HO-nonenal and other small aldehydes are able to diffuse outside from cells. In the brain, these molecules reach blood and cerebrospinal fluid (CSF) and report brain oxidative status. Indicators of brain oxidative condition for ND diagnosis, prevention, and control of disease progression are of major importance in ND medicine. Early diagnosis and new therapeutic strategies may delay or even prevent the irreversible consequences of these pathologies.

4 Neurodegenerative Disorders

Brain OS and oxidative damage have been reported for important ND such as Alzheimer disease (AD), Parkinson disease (PD), amyotrophic lateral sclerosis (ALS), and vascular dementia (VD). Each disease has its own neuropathological characteristics but it has been long postulated that in all cases OS contributes to neuronal death. Amyloid deposits with amyloid beta (Aβ) aggregates are present in AD and PD [15] and α-synuclein is characteristic of PD [16]. The increasingly recognized participation of OS in ND generated a great interest in the research community. At the beginning of 2016, PubMed citations for OS and AD are 5548, for OS and PD are 2260, for OS and ALS are 1071 and for OS and VD are 269, with a total of 9148 references for the four diseases.

4.1 Alzheimer Disease

Currently, 26.6 million people worldwide have AD and this figure could rise to 100 million people by 2050 [17]. The estimated annual incidence and prevalence of AD rises with age. Incidence rates up between approximately 0.4 % in people aged 65–69 up to nearly 10 % in people aged over 90 have been reported, an a prevalence between 2 % in people aged 65–69 up to more than 25 % in people aged over 90 years [1].

4.2 Parkinson Disease

It is a long known neurodegenerative pathology with an incidence of 5–24 people per 100,000/year and is characterized by extrapiramidal movement disturbance and progressive motor dysfunction due to a selective loss of dopaminergic neurons from the substantia nigra that extends to striatum and produces dementia.

4.3 Amyotrophic Lateral Sclerosis

This disease is a motoneuron pathology that progresses over several years and produces patient death. The etiology is complex; with 5–10 % of the cases as the familial form, and with autosomal mutations of which 15–20 % are in the superoxide dismutase 1 gene. Sporadic ALS with a worldwide incidence of 1–2 per 100,000 is poorly understood in environmental terms [18].

4.4 Vascular Dementia

The disease, also called vascular cognitive impairment, is the second most common cause of human dementia and occurs when brain blood flow is reduced. The main symptoms are memory loss, confusion, mood changes and difficulty with daily tasks.

5 Oxidative Stress Mechanisms in Neurodegenerative Processes

OS and oxidative damage have been reported in AD, PD, familial ALS and vascular dementia (VD) with increased processes of lipid peroxidation, protein oxidation, nucleic acid oxidation and depletion of antioxidants [19–22]. The brain intracellular processes associated to oxidative damage are: protein aggregation, mitochondrial dysfunction, metal dyshomeostasis, redox signaling dysregulation, and inflammation. These processes lead to oxidative neuronal damage and incidence and progression of ND [10, 23–30] (Fig. 22.1). The diffusible products of oxidative damage reach blood, directly or through CSF.

Concerning ND etiology, it is not clear if free radical-induced OS is the primary and initiating process causing neurodegeneration. However, it is clear that OS is involved in the propagation of cellular damage that leads to neuropathology. Therefore, there is no need for a cascade of events initiated by OS; rather it seems that OS is part of the events involved in the diseases and neuronal cell death [23, 31].

6 Systemic Oxidative Stress and Neurodegeneration

It has been postulated that increased brain OS is a consequence of an oxidative burden in the whole body of AD patients. Accordingly AD might not be exclusively a primary neurological pathology, but rather a systemic oxidative disorder [32–35]. It

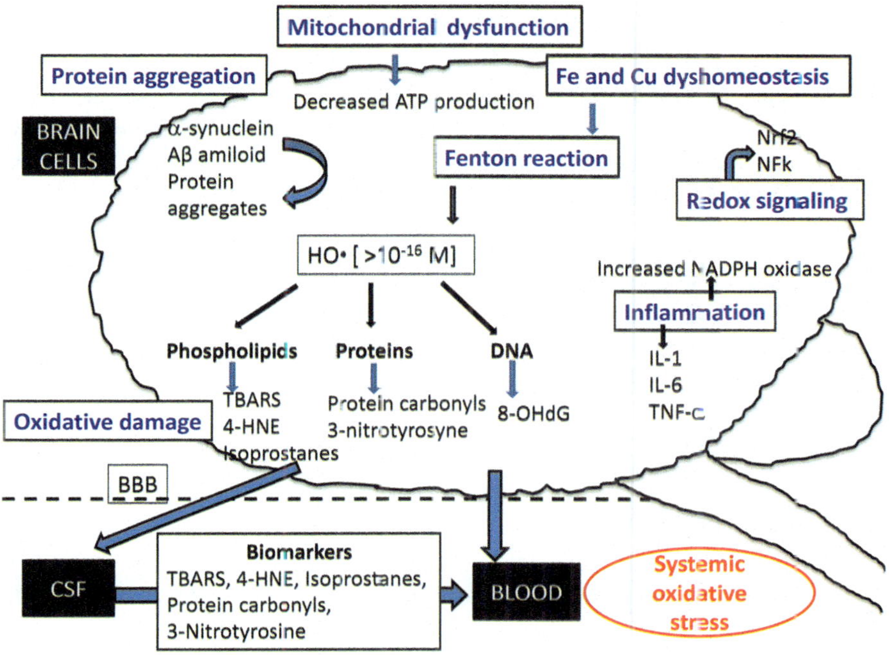

Fig. 22.1 Brain cellular oxidative stress and damage and systemic oxidative stress in neurodegenerative diseases

is difficult to establish whether OS is an early component in AD or a final step in the process. Some studies showed that OS preceded chronologically the Aβ deposits, and the onset of clinical symptoms [36]. In brain, the production of O_2^-, H_2O_2 and HO• leads to the free radical mediated processes of lipid peroxidation, and of protein and nucleic acid oxidation. Protein carbonyls and 3-nitrotyrosine produced in protein oxidation are secreted to the CSF and end up in the blood. The products of free-radical mediated lipid peroxidation (malonaldehyde, TBARS, 4-HO-nonenal, isoprostanes, lipid hydroperoxides and aldehydes) are small, lipophilic compounds that diffuse to blood directly from the tissue or through CSF (Fig 22.1).

There are two hypotheses for the observed close relationship between peripheral OS and ND. One of them is that OS begins in the brain, and that oxidized biomolecules are transported to peripheral blood. The other hypothesis indicates that OS initially develops in peripheral tissues with various causes, and results in a reduction of antioxidants in the central nervous system that leads to oxidative damage and neurodegeneration. There is evidence that point out to a role of iron and copper in the pathogenesis of ND. On the one hand, the increased metal concentrations in plasma and CSF observed in AD [37, 38] and in the other hand, the immediate brain uptake of iron and copper after metal overloads that start the processes of lipid peroxidation and protein oxidation [7, 23].

7 Biomarkers for Neurodegenerative Diseases

The National Institutes of Health defines a biomarker as "a characteristic that is objectively measured and evaluated as an indicator of a normal biologic or biochemical process, a pathogenic process, or a pharmacologic response". The ideal biomarkers should be non-invasive and cost-affordable. Biomarkers need to combine high sensitivity and specificity with high predictive value, and ideally should be validated against some gold standard of confirmed neuropathology before clinical integration, and whenever is possible, correlate with disease stage and/or disease progression [1]. Biomarkers for ND are essential to facilitate disease diagnosis, ideally at early stages, monitor disease progression, and assess response to existing and future treatments. Biomarkers for ND are classified into brain biomarkers and systemic biomarkers.

7.1 Brain Biomarkers

Brain biomarkers are neuroimaging, post mortem histology, and biomarkers in the CSF (Table 22.1).

7.1.1 Neuroimaging

These techniques assess regional structure, function and biochemistry of the brain, as well as brain dysfunction. These methodologies correlate well with the clinical condition in AD patients, with neurological diagnosis based on behavioral determinations, which are available only at specialized dementia clinics [39]. However, these approaches have significant limitations because they are expensive and even invasive for patients.

7.1.2 Histology

The major facts in the post-mortem brain of AD and PD patients are the accumulation of protein aggregations (Aβ and α-synuclein), oxidized biomolecules and enzymes that have lost or decreased activity.

Table 22.1 Brain biomarkers in patients with Alzheimer disease (AD), Parkinson disease (PD) or amyotrophic lateral sclerosis (ALS)

Biomarker	AD	PD	ALS
Neuroimaging (in vivo)	Brain volumetric measurements, by structural magnetic resonance imaging (MRI) of whole brain and specific regions; amyloid tracer imaging positron emission tomography (PET) using the Pittsburg compound B and florbetapir, 18C-fluorodeoxyglucose and functional MRI [3]		
Hallmarks (post-mortem) Protein structures	Senile plaques	Cytoplasmic α-synuclein in cortex, anterior thalamus, hypothalamus, amygdala and basal forebrain.	
	Neurofibrillary tangles formed by Aβ deposits.		
	Hyper-phosphorylated forms of microtubule-associated protein tau [15, 23, 41]	Lewy bodies in surviving neurons	
Enzymes and molecules	Decreased mitochondrial complex II and IV activities, citochrome-oxidase content and ATP production [41]	Decreased GSH	Accumulation of 4-HO-nonenal
	Inhibition of α-ketoglutarate dehydrogenase.	Increased Fe	Protein glycoxidation products (AGE) [46]
	Nrf2 reduced nuclear levels in cytoplasm of hippocampal neurons [42]	Oxidation of dopamine to 6-hidroxydopamine	Impaired mitochondria and respiratory chain complexes [47]
	Increased TBARS, 4-HO-nonenal, protein carbonyls, neurofibrilary tangles and senile plaques, in amygdala, hippocampus, parahippocampal gyrus and spinal cord	Increased TBARS, 4-HO-nonenal, protein carbonyls, and DNA oxidation in substantia nigra [45]	
	Increased DNA oxidation (8-OHdG)		
	Increased Fe, Cu, Hg.		
	Incresed AGE and peroxynitrite, heme oxygenase-1, and SOD-1 in neurofibrillary tangles		
	Activated microglia [23]		

(continued)

Table 22.1 (continued)

Biomarker	AD	PD	ALS
Biomarkers in CSF	High Aβ levels in asymptomatic preclinical stage		
	Tau and p-tau (stages of high synaptic dysfunction and early neuro-degeneration [43]		
	Incresed acrolein, 4-HO-nonenal and 8-OHdG		
	Decreased non-enzymatic antioxidants		
	Amyloid precursor peptide platelet (AβPP)		
	Autoantibodies to Aβ [44]		

7.1.3 Biomarkers in the CSF

The CSF is a body fluid produced in the brain ventricles that occupies the subarachnoid space and the ventricular system around and inside the brain and spinal cord. CSF is in intimate contact with brain, and there is a substance exchange between the neural environment and CSF. Therefore, CSF is almost an ideal source to identify biomarkers for ND. Levels of 4-HO-nonenal and acrolein in CSF appear as clear factors in AD. More studies are required to justify the routine use of CSF [40].

8 Systemic Biomarkers

The relationship between risk factors, biomarkers of ND and clinical symptoms was carefully analyzed in the last years but no clear conclusion was reached. Blood is the most commonly analyzed biological sample for markers in ND. Blood is a very complex system that reflects the physiological activity and the pathology of body organs and systems, including the central nervous system. In humans, about 500 mL of CSF are daily poured into the blood, making CSF a suitable source of biochemical markers from the brain and for ND [48]. It is worth noting that 4-HO-nonenal concentration in CSF from PD patients is fourfold higher than in healthy subjects, whereas in plasma it is 1.5 times higher [49]. The easy of venipuncture compared to lumbar puncture, makes peripheral blood analysis preferable to evaluate diseases progress or the effect of treatments in clinical situations [3]. An increase in the peripheral biochemical circulating markers of OS would allow establishing a preclinical state of disease, many years before the clinical diagnosis, at a time when future patients appear cognitively normal. Systemic biomarkers may be classified as biomarkers of: oxidative stress and damage; protein aggregates [50]; inflammation [51] and lipoprotein metabolism [52].

Table 22.2 Systemic biomarkers in plasma, serum and blood cells from Alzheimer disease (AD), Parkinson disease (PD) and vascular dementia (VD) patients

Systemic biomarker	AD	PD	VD
Oxidative stress and damage	Increased TBARS and 4-HNE in plasma [53, 54]	Increase in plasma protein carbonyls and Cu	TBARS increased in blood cells and plasma [53]
	Increased acrolein [56]	Increased TBARS in blood cells (parallel to clinical stage) [3]	
	Increased protein carbonyls in plasma and erythrocytes [3]		
Protein aggregates	Plasma Aβ		
	Tau protein in serum or plasma (ultrasensitive immunoassay)		
	ApoE polymorphism, inflammatory markers, altered p53 protein [50]		
	AβPP in platelets [56]		
Inflammation	Increased levels of IL-6, CRP and autoantibodies to Aβ in plasma [44]		High serum ACT and CRP [44]
	High plasma homocysteine [57]		High plasma homocysteine [59]

8.1 Biomarkers of Oxidative Stress and Damage

The plasma indication of oxidative damage is either increased oxidation products (of phospholipids, proteins or DNA) or decreased antioxidants [3, 53–56, 58]. The oxidation products indicative of oxidative stress and damage are: TBARS, equivalent to malonaldehyde, 4-hydroxy-nonenal, protein carbonyl groups, 8-isoprostane, cholesterol oxides and 8-HO-deoxyguanosine (8-HO-dG). TBARS, that accounts for 6–9 % of the O_2 consumed during unsaturated fatty acid peroxidation, and 4-HO-nonenal are excellent indicators of lipid peroxidation. Protein carbonyls are sensitive indicators of cellular oxidative damage. Hydroperoxide-induced chemiluminescence (CL-BOOH) is a very sensitive assay indicative of α-tocopherol levels. Decreased plasma TRAP is interpreted as decrease in plasma antioxidants. However, measured plasma TRAP levels are about 300 μM, which is more than double the amount of plasma ascorbic acid and vitamin E, the two main plasma antioxidants. Other plasma components, such as urea, also react with the peroxyl radicals of the TRAP assay. Table 22.2 summarizes results on biomarkers in blood of patients of AD, PD and VD.

There are reports on blood cells of increased production of oxidants or decreased antioxidants in ND. Increased production of oxidants has been reported in neutrophyls [19] in PD and in lymphocytes in AD [60]. Erythocytes show decreased vitamin E (CL-BOOH) in AD [21, 61]. Interestingly, red blood cells show increased SOD activity in AD, suggesting an adaptive response of bone marrow [39, 62].

Table 22.3 Biomarkers of systemic oxidative stress in human plasma from patients with neurodegenerative diseases of comparable ages

State/Disease	TBARS (μM)	4-HO-nonenal (μM)	Protein carbonyls (μM)	CL-BOOH (cpm/mg Hb) × 10^{-2}
Healthy controls	2.9 ± 0.1[a]	0.47 ± 0.12[c,d]	0.9 ± 0.1[b,e]	117 ± 7[a]
Alzheimer	3.9 ± 0.1[a]	2.0 ± 0.5[c]	3.5 ± 0.2[b,e]	178 ± 10[a]
	$p < 0.001$	$p < 0.001$	$p < 0.05$	$p < 0.05$
Parkinson	3.5 ± 0.2[a]	0.68 ± 0.15[d]	2.3 ± 0.2[b]	202 ± 10[a]
	$p < 0.05$	$p < 0.05$	$p < 0.05$	$p < 0.05$
Vascular dementia	3.5 ± 0.1[a,b]	ND	2.8 ± 0.2[b]	175 ± 8[a]
	$p < 0.01$		$p < 0.05$	$p < 0.05$

Data from (a) [61], (b) [3], (c) [63], (d) [49] and (e) [64]

Table 22.4 Antioxidants in human plasma of patients with ND comparable ages

State/Disease	Vitamin E (μM)	FRAP (μM)	TRAP (μM)
Healthy controls	24.5 ± 2.9[a]	989 ± 74[a]	344 ± 18[b]
Alzheimer	15.3 ± 1.5[a]	660 ± 71[a]	227 ± 12[b]
	$p < 0.05$	$p < 0.05$	$p < 0.001$
Parkinson	22.8 ± 1.7[a]	823 ± 49[a]	248 ± 18[b]
	NS	$p < 0.05$	$p < 0.01$
Vascular dementia	16.5 ± 1.1[a]	640 ± 39[a]	210 ± 14[b]
	$p < 0.05$	$p < 0.05$	$p < 0.001$

Data from (a) [3], and (b) [61]

8.2 *Biomarkers of Non-enzymatic Antioxidant Defenses*

Increased levels of TBARS, 4-HO-nonenal [61, 63] and protein carbonyls [64] in plasma as well as of CL-BOOH [61] identify a situation of OS (Table 22.3). In the same way, lower levels of plasma antioxidants, such as vitamin E, TRAP and FRAP [3], indicate a situation of OS (Table 22.4). Plasma non-enzymatic antioxidants are decreased in ND, particularly uric acid, vitamins E and A, and α- and β-carotene [53, 54].

The red blood cells of ND patients afford indicators of OS; some of them reflect erythrocyte production in the bone marrow (GR and SOD) and others indicate metabolic conditions related to intracellular GSH (Table 22.5). There is a clear shift of the glutathione system to a more oxidized state (GSH content and the GSH/GSSG ratio) and increased activities of the GR and SOD enzymes [62].

9 Conclusions

Systemic biomarkers of OS in neurological patients are useful indicators of diagnosis and/or disease progression. The direct involvement of free radicals in the etiology and/or the pathogenesis of NDs is difficult because the short lifetimes of reactive

Table 22.5 Antioxidant systems in erytrocytes from ND patients of comparable ages

State/Disease	GSH (μmol/g Hb)	GSH/GSSG	GR (U/g Hb)	SOD (U_{SOD}/mg protein)
Healthy controls	2.9 ± 0.2[a]	8.7 ± 0.3[a]	15.7 ± 2.4[a]	10.2 ± 0.3
Alzheimer	1.6 ± 0.1[a]	3.8 ± 0.1[a]	24.3 ± 1.5[a]	17.8 ± 0.5[b]
	$p < 0.05$	$p < 0.05$	$p < 0.05$	$p < 0.001$
Parkinson	2.3 ± 0.3[a]	3.6 ± 0.1[a]	15.5 ± 2.3[a]	15.8 ± 0.6[b]
	N.S.	$p < 0.05$	N.S.	$p < 0.001$
Vascular dementia	1.8 ± 0.2[a]	4.0 ± 0.02[a]	25.8 ± 1.9[a]	16.7 ± 0.6[b]
	$p < 0.05$	$p < 0.05$	$p < 0.05$	$p < 0.001$

GR glutathione reductase, *SOD* superoxide dismutase
Data from (a) [3] and (b) [61]

oxygen and nitrogen species and the lack of sensitive technology to detect them *in situ* in biological systems. The ideal biomarker of OS may be found in blood samples, which is relatively easy to collect. OS is present in AD patients since the early stages of the disease, so "oxidative biomarkers" may be studied in plasma, serum and CSF. Oxidative-related changes are already present in early stages of neurodegeneration, reflecting the same chronology that in brain, with increased level of biomarkers of phospholipid, protein and DNA oxidation and decreased level of non-enzymatic antioxidants in plasma and serum.

Three finding are important to highlight: the first is that oxidative-related changes are already present in early stages of neurodegeneration, reflecting the same chronology that brain function, the second is the decreased level of non-enzymatic antioxidants (mainly α-tocopherol) in plasma and serum, and the third is a subtle alteration in the metabolism and levels of transition metals.

Acknowledgements The authors acknowledge the selection of patients and the plasma samples provided by Dr. Raúl Dominguez and the SOD measurements and statistics made by Dr. Jorge A. Serra.

References

1. Hampel H, Prvulovic D, Teipel S et al (2011) The future of Alzheimer's disease: the next 10 years. Prog Neurobiol 95:718–728
2. Jankovic J (2008) Parkinson's disease: clinical features and diagnosis. J Neurol Neurosurg Psychiatry 79:368–376
3. Cristalli D, Arnal N, Marra F et al (2012) Peripheral markers in neurodegenerative patients and their first-degree relatives. J Neurol Sci 314:48–56
4. Jellinger KA (2001) Cell death mechanisms in neurodegeneration. J Cell Mol Med 5:1–17
5. Zhang J (2013) Autophagy and mitophagy in cellular damage control. Redox Biol 1:19–23
6. Sies H (1985) Oxidative stress, introductory remarks. Academic, London
7. Halliwell B, Gutteridge J (1989) Lipid peroxidation-. A radical chain reaction. In: Free Radic Biol Med, 2nd edn. Clarendon, Oxford, pp 188–276
8. Sies H (1991) Oxidative stress from basic research to clinical application. Am J Med 91:31–38

9. Boveris A, Cadenas E (1997) Cellular sources and steady-state levels of reactive oxygen species. In: Biadasz-Clerch L, Massaro DJ (eds) Oxygen, Gene Expression, and Cellular Function. Marcel Dekker, NY, pp 1–25
10. Jones D (2008) Radical-free biology of oxidative stress. Am J Physiol Cell Physiol 295:849–868
11. Sies H, Jones DP (2007) Encyclopedia of Stress, Fink G (ed). 2nd edn, vol. 3. Elsevier, Amsterdam, pp 45–48
12. Seim S (1982) Production of reactive oxygen species and chemiluminescence by human monocytes during differentiation and lymphokine activation in vitro. Acta Pathol Microbiol Immunol Scand C 90:179–185
13. Chance B, Sies H, Boveris A (1979) Hydroperoxide metabolism in mammalian organs. Physiol Rev 59:529–625
14. Boveris A, Cadenas E, Reiter R et al (1980) Organ chemiluminescence: noninvasive assay for oxidative radical reactions. Proc Natl Acad Sci 77:347–351
15. Sloane P, Zimmerman S, Suchindran C et al (2002) The public health impact of Alzheimer's disease 2000–2050: potential implication of treatment advances. Annu Rev Public Health 23:213–231
16. Opazzo C, Barría MI, Ruiz FH, Inestrosa NC (2003) Copper reduction by copper binding proteins and its relation to neurodegenerative diseases. Biometals 16:91–98
17. Wimo A, Winblad B, Aguero-Torres H, von Strauss E (2003) The magnitude of dementia occurrence in the world. Alzheimer Dis Assoc Discord 17:63–67
18. Gordon PH (2011) The range and clinical impact of cognitive impairment in French patients with ALS: a cross-sectional study of neuropsychological test performance. Amyotroph Lateral Scler 25:1–15
19. Gatto EM, Carreras MC, Pargament G et al (1996) Neutrophil function, nitric oxide and blood oxidative stress in Parkinson's disease. Mov Disord 11:261–267
20. Famulari A, Marschoff E, Llesuy S et al (1996) The antioxidant enzymatic blood profile in Alzheimer's and vascular diseases. Their association and a possible assay to differentiate demented subjects and controls. J Neurol Sci 141:69–78
21. Repetto M, Reides C, Evelson P et al (1999) Peripheral markers of oxidative stress in probable Alzheimer patients. Eur J Clin Investing 29:643–649
22. Fiszman M, Ricart K et al (2003) Evidences of oxidative stress in familial amyloidotic polyneuropathy Type 1. Arch Neurol 60:593–597
23. Valko M, Leibfritz D, Moncol J et al (2007) Free radicals and antioxidants in normal physiological functions and human disease. Int J Biochem Cell Biol 39:44–84
24. Kozlowski H, Janck-Klos A, Brasun J et al (2009) Copper, iron, and zinc ions homeostasis and their role in neurodegenerative disorders. Coord Chem Rev 253:2665–2685
25. Hardas S, Sultana R, Clark A et al (2013) Oxidative modification of lipoic acid by HNE in Alzheimer diseased brain. Redox Biol 1:80–85
26. Navarro A, Boveris A, Bández MJ et al (2009) Human brain cortex: mitochondrial oxidative damage and adaptive response in Parkinson disease and in dementia with Lewy bodies. Free Radic Biol Med 46:1574–1580
27. Navarro A, Boveris A (2009) Brain mitochondrial dysfunction and oxidative damage in Parkinson's disease. J Bioenerg Biomembr 41:517–521
28. Navarro A, Boveris A (2010) Brain mitochondrial dysfunction in aging, neurodegeneration, and Parkinson's disease. Front Aging Neurosci 2:1–34
29. Repetto MG, Ferrarotti N, Boveris A (2010) The involvement of transition metal ions on iron-dependent lipid peroxidation. Arch Toxicol 84:255–262
30. Musacco-Sebio R, Ferrarotti N, Saporito-Magriña C et al (2014) Rat brain oxidative damage in iron and copper overloads. Metallomics 6:1410–1416
31. Semprine J, Ferrarotti N, Musacco-Sebio R et al (2014) Brain antioxidant response to iron and copper acute intoxications in rats. Metallomics 6:2083–2089

32. Siciliano R, Barone E, Calabrese V et al (2011) Experimental research on nitric oxide and the therapy of Alzheimer disease: a challenging bridge. CNS Neurol Public Health 23:213–231
33. Calabrese V, Cornelius C, Leso V et al (2012) Oxidative stress, glutathione status, sirtuin and celular stress response in type 2 diabetes. Biochim Biophys Acta 1822:729–736
34. Cornelius C, Trovato-Salinaro A, Scuto M et al (2013) cellular stress response, sirtuins and UCP proteins in Alzheimer disease: role of vitagenes. Immun Ageing 10:41–51
35. Trovato Salinaro A, Cornelius C, Koverech G et al (2014) Cellular stress response, redox status, and vitagenes in glaucoma, a systemic oxidant disorder linked to Alzheimer's disease. Front Pharmacol 5:1–8
36. Nunomura A, Perry G, Aliev G et al (2001) Oxidative damage is the earliest event in Alzheimer disease. J Neuropathol Exp Neurol 60:759–767
37. Gerhardsson I, Lundh T, Minthon L, London E (2008) Metal concentrations in plasma and cerebrospinal fluid in patients with Alzheimer disease. Dement Geriatr Cogn Disord 25:508–515
38. Vural H, Demirin H, Kara Y et al (2010) Alterations in plasma magnesium, copper, zinc, iron and selenium concentrations and some related erythrocyte antioxidant enzyme activities in patients with Alzheimer's disease. J Trace Elem Med Biol 24:169–173
39. Serra JA, Marschoff E, Dominguez R et al (2004) Oxidative stress in Alzheimer's and vascular dementias: masking of the antioxidant profiles by concomitant type II diabetes mellitus condition. J Neurol Sci 218:17–24
40. Henriksen K, O'Bryant SE, Hampel H et al (2014) The future of blood-based biomarkers for Alzheimer's disease. Alzheimers Dement 10:115–131
41. Caldeira GI, Ferreira IL, Rego AC (2013) Impaired transcription in Alzheimer's disease: key role in mitochondrial dysfunction and oxidative stress. J Alzheimers Dis 34:115–131
42. Ramsey CP, Glass CA, Montgomery MB et al (2007) Expression of Nrf2 in neurodegenerative diseases. J Neuropath Exp Neurol 66:75–85
43. Cummings JL, Doody R, Clark C (2007) Disease-modifying therapies for Alzheimer disease: challenges to early intervention. Neurol 69:1622–1634
44. Song F, Poljak A, Smythe G, Sachdev P (2009) plasma biomarkers for mild cognitive impairment and Alzheimer's disease. Brain Res Rev 51:69–80
45. Zarkovic K (2003) 4-Hydroxynonenal and neurodegenerative diseases. Mol Aspect Med 24:293–303
46. Casado A, López-Fernández ME, Casado MC, La-Torre R (2008) Lipid peroxidation and antioxidant enzyme activities in vascular and Alzheimer dementias. Neurochem Res 33:450–458
47. Buendia I, Michalska P, Navarro E et al (2015) Nrf2-ARE pathway: an emerging target against oxidative stress and neuroinfalmmation in neurodegenerative diseases. Pharmacol Ther 157:84–104
48. Hye A, Lymham S, Thambisetty M et al (2006) Proteome-based plasma biomarkers for Alzheimer's disease. Brain 129:3042–3050
49. Selley M (1998) 4-Hydroxy-2-nonenal may be involved in the pathogenesis of Parkinson disease. Free Radic Biol Med 25:169–174
50. Randall J, Mortberg E, Provuncher G et al (2013) Tau proteins in serum predict neurological out-come after hypoxic brain injury from cardiac arrest: results of a pilot study. Resuscitation 84:351–356
51. De Luigi A, Fragiacomo C, Lucca U et al (2001) Inflammatory markers in Alzheimer's disease and multi-infarct dementia. Mechanisms Aging Develop 122:1985–1995
52. Solomon A, Karenhot I, Ngandu T et al (2007) Serum cholesterol changes after midlife and late-life cognition: twenty-one-year follow up study. Neurol 68:751–756
53. Schrag M, Mueller C, Oyoyo U et al (2011) Iron, zinc and copper in the Alzheimer's disease brain: a quantitative meta-analysis. Some insight on the influence of citation bias on scientific opinion. Prog Neurobiol 94:296–306
54. Chang Y, Chang W, Tsai N et al (2014) The roles of biomarkers of oxidative stress and antioxidant in Alzheimer's disease: a systematic review. Bio Med Res Intl 2014:182303

55. Bradley M, Markesbery W, Lovell M (2010) Increased levels of 4-hydroxynonenal and acrolein in the brain in preclinical Alzheimer's disease. Free Radic Biol Med 48:1570–1576
56. Guzmán-Martínez L, Farías G, Maccioni R (2012) Emerging noninvasive biomarkers for early detection of Alzheimer's disease. Arch Med Res 43:663–666
57. Shadri S (2006) Elevated plasma homocysteine levels: risk factor or risk marker for the development of dementia and Alzheimer's disease? J Alzheimer Dis 9:393–398
58. Di Domenico F, Coccia R, Buttterfield A, Perluigi M (2011) Circulating biomarkers of protein oxidation for Alzheimer disease: expectations within limits. Biochem Biophys Acta 1814:1785–1795
59. Ravaglia G, Forti P, Maioli F et al (2007) Blood inflammatory markers and risk of dementia: the Conselice study of brain aging. Neurobiol Aging 28:1810–1820
60. Leutner S, Schindowski K, Frolich L et al (2005) Enhanced ROS-generation in lymphocytes of Alzheimer's disease. Pharmacopsychiatry 38:312–315
61. Repetto M (2008) Clinical use of chemiluminescence assays for the determination of systemic oxidative stress. In: Popov I, Lewin G (eds) Handbook of chemiluminescent methods in oxidative stress assessment. Transword Research Network, Kerala, pp 163–194
62. Lustig E, Serra JA, Kohan S et al (1993) Copper-zinc superoxide dismutase activity in red blood cells and serum in demented patients and in aging. J Neurol Sci 115:18–25
63. McGrath LT, McGreenon BM, Brennan S et al (2001) Increased oxidative stress in Alzheimer's disease as assessed with 4-hydroxynonenal but not malonaldehyde. QJ Med 94:485–490
64. Conrad CC, Marshall JM, Talent TL et al (2000) Oxidized protein in Alzheimer's plasma. Biochem Biophys Res Commun 275:678–681

Part IV
Metabolic Syndrome and Cancer

Chapter 23
Oxidative Stress, Metabolic Syndrome and Alzheimer's Disease

Danira Toral-Rios, Karla Carvajal, Bryan Phillips-Farfán, Luz del Carmen Camacho-Castillo, and Victoria Campos-Peña

Abstract Oxidative stress, neuroinflammation, insulin signaling deficiency and vascular alterations are common features of many neurodegenerative disorders. Oxidative damage induces chronic inflammation, this condition may lead to insulin resistance and damage to the blood brain barrier (BBB), leading to an exacerbation of central inflammation, causing cognitive dysfunction eventually resulting in dementia. The major cause of dementia affecting the elderly is Alzheimer's disease, the most common cause of disability characterized by a progressive loss of memory and cognitive function. Neuropathologically, Alzheimer's disease is characterized by abnormal deposition of the amyloid β (Aβ) peptide and intracellular accumulation of neurofibrillary tangles (NFT) of hyper-phosphorylated tau protein. It has been proposed that Aβ and NFT generate the neuronal damage that leads to cognitive failure, through the generation of reactive oxygen species (ROS). Alternatively, it has been suggested that oxidative damage precedes the accumulation of Aβ, NFTand other alterations such as vascular malfunction and metabolic syndrome. This chapter discusses the main molecular mechanisms associated to oxidative stress, metabolic syndrome and their relationship with Alzheimer's disease. We also include promising therapeutic strategies.

Keywords Alzheimer's Disease • Metabolic syndrome • Oxidative stress • Insulin resistance • Amyloid beta (Aβ) • Antioxidant therapies

D. Toral-Rios
Departamento de Fisiología Biofísica y Neurociencias, Centro de Investigación y de Estudios Avanzados del Instituto Politécnico Nacional, ZIP 07360 Mexico City, Mexico

K. Carvajal • B. Phillips-Farfán • L.d.C. Camacho-Castillo
Laboratorio de Nutrición Experimental, Instituto Nacional de Pediatría, ZIP 045302 Mexico City, Mexico

V. Campos-Peña (✉)
Laboratorio Experimental de Enfermedades Neurodegenerativas, Instituto Nacional de Neurología y Neurocirugía, Insurgentes Sur 3877, 14269 Mexico City, Mexico
e-mail: neurovcp@ymail.com

© Springer International Publishing Switzerland 2016
R.J. Gelpi et al. (eds.), *Biochemistry of Oxidative Stress*, Advances in Biochemistry in Health and Disease 16, DOI 10.1007/978-3-319-45865-6_23

1 Introduction

In recent years, it has been proposed that oxidative stress (OS) plays a fundamental role in the development of neurodegenerative diseases like Alzheimer's disease (AD) [1, 2]. OS can be defined as an imbalance between the formation of reactive oxygen species (ROS) (O_2^-, H_2O_2 and $HO^•$) and their destruction by the antioxidant systems [2] (see Chapter 1 by H. Sies in this volume). ROS are usually produced as a consequence of regular physiological metabolism. In the majority of cells, mitochondria are the primary source of ROS production by the mitochondrial respiratory chain complexes. These multi-subunit complexes include NADH-ubiquinone reductase (NADH dehydrogenase, complex I), succinate dehydrogenase (complex II), ubiquinol-cytochrome c reductase (complex III), cytochrome c oxidase (complex IV) and ATP synthase (complex V) [3]. Under physiological conditions ROS have a crucial role in several signaling pathways that are important for survival, such as the induction of the mitogenic response, neurotransmission, blood pressure regulation, smooth muscle relaxation and immune regulation [4, 5]. However, the excessive production of ROS brings about irreversible damage and cellular disruption. In AD, cytochrome oxidase, pyruvate dehydrogenase and α-ketoglutarate dehydrogenase complexes show reduced activity as a result of oxidative damage [6, 7]. The balance between oxidants and antioxidants is not a static condition and a great number of stimuli can interfere with the redox status [8]. Uncontrolled OS has been implicated in several human inflammatory diseases [4, 5] such as hepatitis, asthma, cancer, vasospasms, stroke, retinal damage, arthritis, traumatic brain injury and atherosclerosis [9, 10]. In the same way, recent studies suggest that ROS levels are increased in hypercholesterolemia, hypertension, diabetes mellitus and abdominal obesity induced by high calorie intakes, which are all important components of the metabolic syndrome (MetS) [1, 11–13].

2 Alzheimer Disease and Oxidative Stress

AD is a disorder of the central nervous system (CNS) that results in generalized brain atrophy [14] and is the most common neurodegenerative disease, accounting for 60–70 % of dementia cases. Its etiology is still unclear, but it is characterized by the accumulation of extracellular neuritic plaques (NP) and intracellular neurofibrillary tangles (NFT) [15]. These deposits are mainly located in the entorhinal cortex, the hippocampus and the frontal and temporal lobes and produce an extensive neuronal death and thus the gradual loss of the cognitive function. NP and NTF are capable of generating neuronal damage that leads to cell death and cognitive failure, via the generation of ROS.

2.1 Tau and Oxidative Stress

NFTs are intracellular deposits of paired helical filaments (PHFs). In these filaments the most important molecular marker is tau, a microtubule-associated protein [16]. Tau promotes microtubule (MTs) assembly and stability, thus it is essential for axonal transport in neurons [17, 18]. It has more than 30 phosphorylation sites and its abnormal hyper-phosphorylation destabilizes microtubules by decreasing their binding affinity and results in its aggregation into NFTs [19]. Besides phosphorylation, tau has other post-translational modifications, such as specific truncation in Glu-391 [16]. Prefibrillar stages of tau could be linked to the toxic effects observed in neurons [20]. Tau also induces mitochondrial dysfunction, leading to severe energy dysfunction and the generation of ROS and reactive nitrogen species (RNS) [21]. In the same way, the expression of truncated-tau at Asp-421 induced mitochondrial fragmentation and elevated OS levels in comparison to cells expressing full-length tau [22]. 4-Hydroxynonenal (HNE) contributes to tau hyper-phosphorylation and induces conformational changes favoring the assembly of Tau into NFTs [23]. On other hand, chronic OS increases the levels of tau phosphorylation at paired helical filaments [23]; possibly by increasing the activity of GSK3β which is up-regulated under OS [24, 25]. In response to OS, GSK3β translocates to mitochondria in a kinase activity-dependent manner and enhances production of cytotoxic ROS [26], contributing to neurodegeneration.

2.2 Aβ and Oxidative Stress

NP are mainly composed of aggregates of β-amyloid (Aβ) and result from the sequential proteolysis of the amyloid precursor protein (AβPP) by β-secretase (BACE1) and γ-secretase. *In vitro* and *in vivo* data now support the notion that Aβ accumulation in the brain can directly or indirectly cause free radical-induced stress, favoring the neurodegenerative process observed in AD [14]. Also, OS may induce the overproduction of Aβ peptides through the activation of BACE1 [27]. BACE1 expression is increased in conditions known to involve OS, including ischemia and hypoxia [28]. The Met-35 of the Aβ peptide is critical for Aβ-associated toxicity and OS. This residue can undergo two-electron oxidation to form methionine sulfoxide (MetSOx) [29]; it modulates OS and the neurotoxic properties of Aβ (42 aminoacids). The transition metals Cu^{2+}, Zn^{2+} and Fe^{3+} are altered in AD and are involved with Aβ aggregation and oxidative damage [30]. Aβ catalyses the reduction of Cu^{2+} and Fe^{3+} generating H_2O_2, which is converted to HO• in the presence of Cu^{1+} and Fe^{2+}; the generation of this reactive species leads to the formation of pro-apoptotic lipid peroxidation (LPO) products, such as 4-hydroxynonenal (HNE) and malonaldehyde [14]. Recent studies suggest that low molecular weight oligomers are more toxic than larger Aβ fibrils [31, 32]. Their presence in the plasma membrane contributes to OS and neurotoxicity. Aβ oligomers activate the N-methyl-D-aspartate receptor (NMDA-R), leading to a rapid influx of calcium, which promotes

ROS generation from the NADPH oxidase. In addition, Aβ accumulation in the parenchyma and in blood vessels causes microglial migration and promotes the production of pro-inflammatory cytokines, prostaglandins, NO and ROS which eventually promote neuronal death [33].

3 Metabolic Syndrome

MetS is defined as a cluster of risk factors including insulin resistance (IR), dyslipidemias, abdominal obesity and arterial hypertension. To maintain the stability of the internal environment the body involves multiple systems, which are controlled in limbic brain structures such as the hippocampus, amygdala and prefrontal cortex. In a similar way, insulin, leptin, cortisol and estrogens act by influencing cognitive function, while other neuropeptides, such as orexin, intervene by regulating gastric secretion and gastrointestinal motility [34]. Several studies have shown that diabetes, IR, obesity and dyslipidemias are risk factors for the development of neurodegenerative diseases, including AD [35, 36]. Another remarkable characteristic of MetS and AD is an enhanced production of ROS with augmented levels of oxidized, nitrosylated and carbonylated proteins, lipid peroxidation, as well as RNA and DNA oxidation.

3.1 Insulin Signaling in the CNS and AD

In the liver, insulin allows the uptake and anabolism of glucose, fatty acids and amino acids by the activation of insulin receptors (IRs). It is currently known that insulin and insulin growth factors (IGFs) can exert these functions through the activation of IRs and insulin growth factor receptors type 1 and 2 (IGFR-1, IGFR-2). Insulin and IGFs promote auto-phosphorylation of IRs and IGFR receptors on tyrosine residues, which creates binding sites for SRC homology 2 domains in phosphoinositide 3-kinase (PI3K) and the growth factor receptor-bound protein 2 [37, 38]. The recruitment of PI3Ks results in PKB/Akt activation, which may influence in different forms downstream signaling molecules like GSK-3, RabGTPase, mTOR and FOXO transcription factors [39]. The recruitment of GRB2 can also activate the Ras/mitogen-activated protein kinase (MAPK) pathway [37].The role of insulin signaling is not only limited to maintain peripheral energy homeostasis. In the CNS, it is involved in novel physiologic functions [40] such as proliferation and differentiation in neurogenic zones [41], modulation of synaptogenesis and synaptic function [42], stimulation of plasticity [43], long-term potentiation [44], neuroprotection against mitochondrial damage [45] and antioxidant effects [46]. Studies in human and rodent brain homogenates revealed that insulin/IGFs receptors are distributed in the hippocampus, hypothalamus (arcuate nuclei, ARC), cerebellum, olfactory bulb, striatum and cerebral cortex [47, 48].

The impairment of insulin signaling in the CNS is associated with the development of aging and neurodegenerative disorders [49]. Reduced expression and activation of IRs, IGF-1R and IRS-1 proteins in the hippocampus and hypothalamus were reported in AD cases and this was correlated with cognitive decline [50]. The development of IGF-1 null mice produced hyper-phosphorylated tau at multiple sites [51], generated by impaired PI3K/Akt. Atk inactivates GSK3β, a kinase that contributes to tau hyper-phosphorylation and aggregation in AD [52]. With respect to Aβ, insulin acting on IGF-1 promotes the release of a soluble APP fragment derived from α-secretase activity (sAPPα) by PI3K activation [53]. Some sAPPα functions are neurite growth, modulation of neuronal excitability, synaptic plasticity and synaptogenesis [54]. Moreover, PI3K activation favored Aβ degradation by insulin-degrading enzyme (IDE) in the CNS, which avoids Aβ plaque formation [55]. Also, IGF-1 promotes the formation of complexes between Aβ and megalin/LRP2 that finally enhance Aβ clearance trough the BBB [56]. On other hand, soluble Aβ oligomers induced the internalization of IRs and a consequent reduction of insulin signaling responsiveness [57]. This pathological state is referred as brain insulin resistance (IR), a condition characterized by an inability to maintain glucose homeostasis despite the presence of high concentrations of insulin in the brain. IR promotes OS, mitochondrial damage and apoptosis [58]. The internalization of IRs mediated by Aβ oligomers could be linked to synaptic dysfunction, because it shares common mechanisms with NMDA receptor internalization [59]. Additionally, it has been suggested that abnormal function of IRs enables nuclear translocation of FOXO1, which may generate ER stress and activate stress kinases [60].

Another mechanism of IR was described, in this sense it is well known that Aβ oligomers induce inflammation. When TNF-α levels are elevated, IRS-1 is inhibited by aberrant activation of c-JunN-terminal kinase (JNK) and this blocks downstream insulin signaling, resulting in peripheral IR [61]. Later work showed that ER stress is crucial for IRS-1 inhibition and memory impairment in AD [62]. IR contributes to AD development since the loss of insulin actions renders neurons more sensitive to the toxic effect of Aβ [63]. Moreover, IR accelerates amyloidosis via the translational up-regulation of BACE1 and proteins implicated in plaque production [64]. It is necessary to consider that more reports are constantly showing that there is a strong correlation between some metabolic disorders that course with peripheral IR and the development of AD [65].

3.2 Insulin Resistance and the Metabolic Syndrome

MetS is a complex disorder characterized by IR, abdominal obesity, arterial hypertension and dyslipidemias [66]. A strong hypothesis on the origin of peripheral IR and MetS is that insulin signaling is altered in the CNS, because of the 80 % inhibition of IRs in medial ARC hypothalamic neurons that promotes decreased hepatic insulin sensitivity and leads to hyperphagia in rats [67]. Moreover, chronic ICV infusion of insulin in C57BL/6 mice promotes lipogenesis and peripheral fat accumulation [68]. On the other hand, obesity is considered a major risk factor for the development of peripheral IR. An excess of adipose tissue leads to ER stress, promotes macrophage activation and subsequent release of adipokines into systemic circulation, including TNF-α, IL-6, IL-18, CRP, MCP-1, leptin and resistin [69]. The activation of inflammatory signaling pathways involves serine/threonine kinases that inhibit IRs molecules and lead to a decreased insulin response. Some of these kinases are JNK, and the inhibitors of NF-kB kinase (IKK) and of PKC [70]. Adiponectin, resistin, leptin and lipocalin have also been linked to obesity-induced IR [71]. Peripheral IR impacts on the MetSby generating two conditions that decline the quality of life of patients, pre-diabetes and diabetes, which have been linked with cognitive impairment and AD development [72]. Other consequences of IR in MetS are increased peripheral inflammation and ER OS, which are assumed to increase dyslipidemia, hypertension and even worse IR [73]. This in turn establishes a vicious cycle that represents a major risk for BBB damage, vascular dementia and AD [1]. Different studies agree that physical activity promotes activation of the AMP-activated protein kinase (AMPK) and sirtuin1 and increases insulin sensitivity, as well as reducing the morbidity and mortality associated with cardiovascular diseases and diabetes [74].

4 Mitochondrial-Derived OS Leads to Modified Protein Structure in MetS and Neurodegeneration

Mitochondria participate in maintaining the energy balance and thus the adequate function in the brain. Neurons are especially vulnerable to any alteration or dysfunction of these organelles. In fact, growing evidence shows that mitochondrial dysfunction is implicated in neurological pathologies such as AD and other neurodegenerative diseases [75]. During oxidative energy production, mitochondria become the main source of ROS in the brain via the electron transfer in the respiratory chain. Thus, any perturbation of these processes results in an unregulated ROS production. In fact, excessive mitochondrial ROS generation has been associated to the pathogenesis of neurodegeneration [76]. OS participates in the development of AD by altering a series of cellular and metabolic mechanisms. Besides the well-understood link between mitochondria and Aβ metabolism, OS interferes with protein function by oxidation and modification of protein structure. Proteins undergo redox-mediated changes, named oxidative post-translational modifications (OPTMs), on specific residues, mainly Cys and Tyr, mediated by the produced ROS. OPTMs include S-glutathionilation, S-nitrosylation, N-sulfonylation, nitration and disulfide bond formation. The oxidized amino acids lead to irreversible or reversible protein modifications, that when exacerbated cause alterations in their activity and thus affect brain function [77]. OPTMs occur normally and modulate protein function in a variety of biological processes. In neuronal tissues they are involved in controlling signal transduction, as well as in synaptic mechanisms including exocytosis and neurotransmitter release [78]. The brain is a major producer of nitric oxide (NO), a molecule that participates in important cellular processes. Nonetheless, in the OS situation NO serves as precursor of even more reactive molecules, such as the toxic peroxynitrite ($ONOO^-$), that is responsible of sustaining S-nitrosylation of proteins and thus plays a relevant role in regulating and impairing neuronal function [79]. Among the affected proteins, it has been showed that S-nitrosylation of Cdk5 increases its kinase activity. Cdk5 participates in APP cleavage, NMDAR function, brain development, differentiation and migration of neurons, axonal maturity and neuronal plasticity [80]. Moreover increased Cdk5 activity mediated by S-nitrosylation has been implicated in the development of several neurological disorders, including AD [81].

In fact, augmented S-nitrosylation is a hallmark of neurodegenerative conditions such as AD [82]. It is thought that S-nitrosylation further contributes to brain mitochondrial damage, mainly by disturbing the mitochondrial fission and fusion processes [83]. It has been proposed that such perturbations are mediated by post-transcriptional nitrosylation of Drp1, which increases its GTPase activity and in turn augments mitochondrial fission [79]. Abnormal mitochondrial dynamics have been proposed to be vital in the neurodegenerative steps underlying AD onset since they may cause the mitochondrial dysfunction that precedes the disease [84]. Other neuronal proteins have also been identified as the targets of OPTMs. This is the case for disulphideisomerase (PDI), which acts as a chaperone during the folding of emerging peptides in the endoplasmic reticulum (ER). During OS it is S-nitrosylated, which inhibits its function. It has been reported that this mechanism is implicated in the ER stress, a marker of neurodegenerative diseases [85]. As already mentioned, MetS and AD converge in key cellular and metabolic features, such as excessive ROS production and mitochondrial dysfunction. Hence molecular and cellular disturbances originated during MetS may be critical for neurodegeneration. In particular, managing elevated ROS generation seems to be crucial to avoid or control further oxidative damage, which could prevent AD development and neurodegeneration.

5 Antioxidant Defenses are Compromised in MetS and AD

OS is a consequence of an imbalance between the production of reactive oxygen species (ROS) and their capture and inactivation by the antioxidant systems. The main ROS are superoxide radical (O_2^-), hydrogen peroxide (H_2O_2) and hydroxyl radical ($HO^•$). Physiological concentrations of ROS are generated mostly by the mitochondrial respiratory chain, during exercise and as a defense against microbial infections. In fact, ROS have an important role as signaling molecules involved in apoptosis, proliferation and gene expression [86]. As a counterpart, antioxidant systems are comprised by: (a) antioxidant enzymes such as: superoxide dismutase (SOD), catalase (CAT), and glutathione peroxidase (GPx); (b) endogenous compounds like reduced glutathione (GSH), uric acid and ubiquinol; and (c) exogenous antioxidant molecules, a group that includes vitamin E, vitamin C, β-carotene and resveratrol [86]. Besides dysregulated ROS production, a marked decrease in antioxidant defenses is associated to MetS and AD development. MetS patients have increased NADPH oxidase and myeloperoxidase activities as well as decreased levels of GSH resulting in enhanced ROS concentrations [73, 87]. These factors together with a low activity of SOD, CAT and GPx lead to an elevated OS, predisposing these subjects to damage in organs and tissues. In fact, it has been shown that OS affects the AD brain since increased nucleic acid oxidation, protein oxidation, lipoperoxidation, nitrosylation and carbonylation are observed [88, 89].

6 Pharmacological and Alternative Therapies to Manage Oxidative Damage in MetS and AD

The first line of therapy to control oxidative damage in MetS and AD, is still focused on life style changes, including modifications in food consumption and exercise. Indeed, high-caloric diets in addition to sedentary habits correlate with elevated oxidation markers in plasma and tissues, which could accelerate the development of these diseases [90]. For AD, many efforts have tried to improve or stop neurodegenerationn by using antioxidant treatments such as vitamins (α-tocopherol and retinol), ubiquinone and flavonoids. However, it seems that once AD has begun antioxidant therapies fail to reduce the extent of harm, since improvements inthe health of patients are scarce [91]. Nonetheless, antioxidant use in animal models of AD has some beneficial effects [92]. This discrepancy may be due to the comorbidities associated to AD, such as MetS, or even due to advanced aging [93]. On the contrary, dietary omega-3 fatty acids, resveratrol, curcumin, selenium, vitamins C and E, as well as physical activity readily evoke improvements in the dysfunctional traits of MetS patients [94, 95]. Recently, growing attention has been paid to the gut microbiota in the control of MetS and AD. Prebiotics and probiotics have beneficial effects on the inflammatory state and on IR. In addition, it has been shown that changes in gut microbiota induced by dietary approaches importantly decrease OS in both animals models and in patients suffering MetS [96, 97]. Similarly, the use of antioxidants like resveratrol, induce changes on gut microbiota, associated to decreased fat accumulation in an obesity model caused by high-fat feeding [98]. Moreover, the tight link between AD and MetS has recently revealed a common factor, that is gut microbiota. Studies have revealed that certain gut microbiota profiles in animals prompt early development of MetS and neurodegeneration mainly by augmenting inflammation and disrupting the cell redox balance [99, 100]. Although many approaches may be tried to control MetS, reducing fat mass and improving IR may reduce the source of OS, such as free fatty acids, as well as mitochondrial dysfunction and thus block further oxidative damage to the proteins involved in regulating brain function.

7 Conclusions

Mitochondrial dysfunction along with IR, as well as chronic inflammation are the underlying factors causing degenerative diseases, such as MetS and AD. Indeed, evidence presented in this review indicates that elevated ROS production may be the

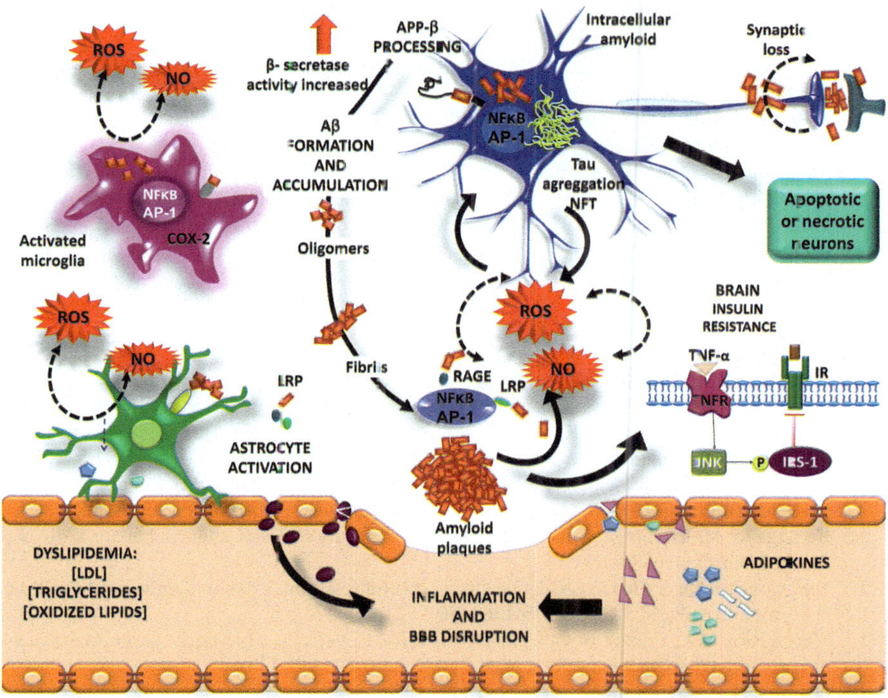

Fig. 23.1 Oxidative Stress (OS), metabolic syndrome (MS) and Alzheimer's disease (AD). Oxidative stress, inflammation and deficiency of insulin signaling, are common features of AD and MetS. The presence of NPs and NFTs generates the activation of astrocytes and microglia, which release proinflammatory cytokines such as tumor necrosis factor (TNF-α) and ROS. TNF- α and ROS increase APP processing by β-secretase increasing amyloid production. In the brain, Aβ oligomers activate the TNF-α receptor (TNFR) inducing the JNK pathway and leading to insulin resistance, which in turn increases the β-secretase activity, resulting in a vicious circle in AD pathology. Therefore, Aβ and IR significantly influence the peripheral and central inflammatory response that eventually converge in a common pathway leading to the neurodegeneration associated with AD and MetS

principal mediator of organ and systemic dysfunction. In conclusion, OS in MetS may degenerate into AD but also, AD-associated OS may aggravate the former. Thus, management of ROS production is mandatory for control and prevention of these pathologies (Fig. 23.1).

References

1. Luque-Contreras D, Carvajal K, Toral-Rios D et al (2014) Oxidative stress and metabolic syndrome: cause or consequence of Alzheimer's disease? Oxid Med Cell Longev 2014:497802
2. Zhao Y, Zhao B (2013) Oxidative stress and the pathogenesis of Alzheimer's disease. Oxid Med Cell Longev 2013:316523

3. Kim GH, Kim JE, Rhie SJ et al (2015) The role of oxidative stress in neurodegenerative diseases. Exp Neurobiol 24:325–340
4. Guerrero CA, Acosta O (2016) Inflammatory and oxidative stress in rotavirus infection. World J Virol 5:38–62
5. Valko M, Leibfritz D, Moncol J et al (2007) Free radicals and antioxidants in normal physiological functions and human disease. Int J Biochem Cell Biol 39:44–84
6. Sultana R, Perluigi M, Butterfield DA (2006) Redox proteomics identification of oxidatively modified proteins in Alzheimer's disease brain and in vivo and in vitro models of AD centered around Abeta(1–42). J Chromatogr B Analyt Technol Biomed Life Sci 833:3–11
7. Bonda DJ, Wang X, Perry G et al (2010) Oxidative stress in Alzheimer disease: a possibility for prevention. Neuropharmacology 59:290–294
8. Luca M, Luca A, Calandra C (2015) The role of oxidative damage in the pathogenesis and progression of Alzheimer's disease and vascular dementia. Oxid Med Cell Longev 2015:504678
9. Oyinloye BE, Adenowo AF, Kappo AP (2015) Reactive oxygen species, apoptosis, antimicrobial peptides and human inflammatory diseases. Pharmaceuticals (Basel) 8:151–175
10. Nordstrom P, Michaelsson K, Gustafson Y et al (2014) Traumatic brain injury and young onset dementia: a nationwide cohort study. Ann Neurol 75:374–381
11. Cifuentes D, Poittevin M, Dere E et al (2015) Hypertension accelerates the progression of Alzheimer-like pathology in a mouse model of the disease. Hypertension 65:218–224
12. Barbagallo M, Dominguez LJ (2014) Type 2 diabetes mellitus and Alzheimer's disease. World J Diabetes 5:889–893
13. Tangvarasittichai S (2015) Oxidative stress, insulin resistance, dyslipidemia and type 2 diabetes mellitus. World J Diabetes 6:456–480
14. Meraz-Rios MA, Franco-Bocanegra D, Toral Rios D et al (2014) Early onset Alzheimer's disease and oxidative stress. Oxid Med Cell Longev 2014:375968
15. Goedert M, Spillantini MG, Jakes R et al (1995) Molecular dissection of the paired helical filament. Neurobiol Aging 16:325–334
16. Wischik CM, Novak M, Edwards PC et al (1988) Structural characterization of the core of the paired helical filament of Alzheimer disease. Proc Natl Acad Sci U S A 85:4884–4888
17. Weingarten MD, Lockwood AH, Hwo SY et al (1975) A protein factor essential for microtubule assembly. Proc Natl Acad Sci U S A 72:1858–1862
18. Kolarova M, Garcia-Sierra F, Bartos A et al (2012) Structure and pathology of tau protein in Alzheimer disease. Int J Alzheimers Dis 2012:731526
19. Goedert M (1999) Filamentous nerve cell inclusions in neurodegenerative diseases: tauopathies and alpha-synucleinopathies. Philos Trans R Soc Lond B Biol Sci 354:1101–1118
20. Patterson KR, Remmers C, Fu Y et al (2011) Characterization of prefibrillar Tau oligomers in vitro and in Alzheimer disease. J Biol Chem 286:23063–23076
21. Rapoport SI (2003) Coupled reductions in brain oxidative phosphorylation and synaptic function can be quantified and staged in the course of Alzheimer disease. Neurotox Res 5:385–398
22. Quintanilla RA, Matthews-Roberson TA, Dolan PJ et al (2009) Caspase-cleaved tau expression induces mitochondrial dysfunction in immortalized cortical neurons: implications for the pathogenesis of Alzheimer disease. J Biol Chem 284:18754–18766
23. Su B, Wang X, Lee HG et al (2010) Chronic oxidative stress causes increased tau phosphorylation in M17 neuroblastoma cells. Neurosci Lett 468:267–271
24. Feng Y, Xia Y, Yu G et al (2013) Cleavage of GSK-3beta by calpain counteracts the inhibitory effect of Ser9 phosphorylation on GSK-3beta activity induced by H(2)O(2). J Neurochem 126:234–242
25. Wakatsuki S, Furuno A, Ohshima M et al (2015) Oxidative stress-dependent phosphorylation activates ZNRF1 to induce neuronal/axonal degeneration. J Cell Biol 211:881–896

26. Kamat PK, Kalani A, Rai S et al (2016) Mechanism of oxidative stress and synapse dysfunction in the pathogenesis of Alzheimer's disease: understanding the therapeutics strategies. Mol Neurobiol 53:648–661
27. Jo DG, Arumugam TV, Woo HN et al (2010) Evidence that gamma-secretase mediates oxidative stress-induced beta-secretase expression in Alzheimer's disease. Neurobiol Aging 31:917–925
28. Zhao J, Fu Y, Yasvoina M et al (2007) Beta-site amyloid precursor protein cleaving enzyme 1 levels become elevated in neurons around amyloid plaques: implications for Alzheimer's disease pathogenesis. J Neurosci 27:3639–3649
29. Butterfield DA, Swomley AM, Sultana R (2013) Amyloid beta-peptide (1–42)-induced oxidative stress in Alzheimer disease: importance in disease pathogenesis and progression. Antioxid Redox Signal 19:823–835
30. Deibel MA, Ehmann WD, Markesbery WR (1996) Copper, iron, and zinc imbalances in severely degenerated brain regions in Alzheimer's disease: possible relation to oxidative stress. J Neurol Sci 143:137–142
31. Roychaudhuri R, Yang M, Hoshi MM et al (2009) Amyloid beta-protein assembly and Alzheimer disease. J Biol Chem 284:4749–4753
32. Yankner BA, Lu T (2009) Amyloid beta-protein toxicity and the pathogenesis of Alzheimer disease. J Biol Chem 284:4755–4759
33. Meraz-Rios MA, Toral-Rios D, Franco-Bocanegra D et al (2013) Inflammatory process in Alzheimer's disease. Front Integr Neurosci 7:1–15
34. Korczynski W, Ceregrzyn M, Kato I et al (2006) The effect of orexins on intestinal motility in vitro in fed and fasted rats. J Physiol Pharmacol 57:S43–S54
35. Solfrizzi V, Panza F, Colacicco AM et al (2004) Vascular risk factors, incidence of MCI, and rates of progression to dementia. Neurology 63:1882–1891
36. Grodstein F, Chen J, Wilson RS et al (2001) Type 2 diabetes and cognitive function in community-dwelling elderly women. Diabetes Care 24:1060–1065
37. Siddle K (2011) Signalling by insulin and IGF receptors: supporting acts and new players. J Mol Endocrinol 47:R1–R10
38. Bevan P (2001) Insulin signalling. J Cell Sci 114:1429–1430
39. Taniguchi CM, Emanuelli B, Kahn CR (2006) Critical nodes in signalling pathways: insights into insulin action. Nat Rev Mol Cell Biol 7:85–96
40. Porte D Jr, Baskin DG, Schwartz MW (2005) Insulin signaling in the central nervous system: a critical role in metabolic homeostasis and disease from C. elegans to humans. Diabetes 54:1264–1276
41. Yuan H, Chen R, Wu L et al (2015) The regulatory mechanism of neurogenesis by IGF-1 in adult mice. Mol Neurobiol 51:512–522
42. Lee CC, Huang CC, Wu MY et al (2005) Insulin stimulates postsynaptic density-95 protein translation via the phosphoinositide 3-kinase-Akt-mammalian target of rapamycin signaling pathway. J Biol Chem 280:18543–18550
43. Chiu SL, Chen CM, Cline HT (2008) Insulin receptor signaling regulates synapse number, dendritic plasticity, and circuit function in vivo. Neuron 58:708–719
44. Dou JT, Chen M, Dufour F et al (2005) Insulin receptor signaling in long-term memory consolidation following spatial learning. Learn Mem 12:646–655
45. Cheng B, Maffi SK, Martinez AA et al (2011) Insulin-like growth factor-I mediates neuroprotection in proteasome inhibition-induced cytotoxicity in SH-SY5Y cells. Mol Cell Neurosci 47:181–190
46. Duarte AI, Santos P, Oliveira CR et al (2008) Insulin neuroprotection against oxidative stress is mediated by Akt and GSK-3beta signaling pathways and changes in protein expression. Biochim Biophys Acta 1783:994–1002
47. Hopkins DF, Williams G (1997) Insulin receptors are widely distributed in human brain and bind human and porcine insulin with equal affinity. Diabet Med 14:1044–1050

48. Pacold ST, Blackard WG (1979) Central nervous system insulin receptors in normal and diabetic rats. Endocrinology 105:1452–1457
49. Blazquez E, Velazquez E, Hurtado-Carneiro V et al (2014) Insulin in the brain: its pathophysiological implications for States related with central insulin resistance, type 2 diabetes and Alzheimer's disease. Front Endocrinol (Lausanne) 5:1–21
50. Talbot K, Wang HY, Kazi H et al (2012) Demonstrated brain insulin resistance in Alzheimer's disease patients is associated with IGF-1 resistance, IRS-1 dysregulation, and cognitive decline. J Clin Invest 122:1316–1338
51. Cheng CM, Tseng V, Wang J et al (2005) Tau is hyperphosphorylated in the insulin-like growth factor-I null brain. Endocrinology 146:5086–5091
52. Hong M, Lee VM (1997) Insulin and insulin-like growth factor-1 regulate tau phosphorylation in cultured human neurons. J Biol Chem 272:19547–19553
53. Adlerz L, Holback S, Multhaup G et al (2007) IGF-1-induced processing of the amyloid precursor protein family is mediated by different signaling pathways. J Biol Chem 282:10203–10209
54. Mattson MP (1997) Cellular actions of beta-amyloid precursor protein and its soluble and fibrillogenic derivatives. Physiol Rev 77:1081–1132
55. Zhao L, Teter B, Morihara T et al (2004) Insulin-degrading enzyme as a downstream target of insulin receptor signaling cascade: implications for Alzheimer's disease intervention. J Neurosci 24:11120–11126
56. Carro E, Spuch C, Trejo JL et al (2005) Choroid plexus megalin is involved in neuroprotection by serum insulin-like growth factor I. J Neurosci 25:10884–10893
57. Zhao WQ, De Felice FG, Fernandez S et al (2008) Amyloid beta oligomers induce impairment of neuronal insulin receptors. FASEB J 22:246–260
58. Kim B, Feldman EL (2012) Insulin resistance in the nervous system. Trends Endocrinol Metab 23:133–141
59. De Felice FG, Vieira MN, Bomfim TR et al (2009) Protection of synapses against Alzheimer's-linked toxins: insulin signaling prevents the pathogenic binding of Abeta oligomers. Proc Natl Acad Sci U S A 106:1971–1976
60. Manolopoulos KN, Klotz LO, Korsten P et al (2010) Linking Alzheimer's disease to insulin resistance: the FoxO response to oxidative stress. Mol Psychiatry 15:1046–1052
61. Bomfim TR, Forny-Germano L, Sathler LB et al (2012) An anti-diabetes agent protects the mouse brain from defective insulin signaling caused by Alzheimer's disease- associated Abeta oligomers. J Clin Invest 122:1339–1353
62. Lourenco MV, Clarke JR, Frozza RL et al (2013) TNF-alpha mediates PKR-dependent memory impairment and brain IRS-1 inhibition induced by Alzheimer's beta-amyloid oligomers in mice and monkeys. Cell Metab 18:831–843
63. Bedse G, Di Domenico F, Serviddio G et al (2015) Aberrant insulin signaling in Alzheimer's disease: current knowledge. Front Neurosci 9:1–13
64. Devi L, Alldred MJ, Ginsberg SD et al (2012) Mechanisms underlying insulin deficiency-induced acceleration of beta-amyloidosis in a mouse model of Alzheimer's disease. PLoS One 7:e32792
65. Watts AS, Loskutova N, Burns JM et al (2013) Metabolic syndrome and cognitive decline in early Alzheimer's disease and healthy older adults. J Alzheimers Dis 35:253–265
66. Guo S (2014) Insulin signaling, resistance, and the metabolic syndrome: insights from mouse models into disease mechanisms. J Endocrinol 220:T1–T23
67. Obici S, Zhang BB, Karkanias G et al (2002) Hypothalamic insulin signaling is required for inhibition of glucose production. Nat Med 8:1376–1382
68. Koch L, Wunderlich FT, Seibler J et al (2008) Central insulin action regulates peripheral glucose and fat metabolism in mice. J Clin Invest 118:2132–2147
69. Ouchi N, Parker JL, Lugus JJ et al (2011) Adipokines in inflammation and metabolic disease. Nat Rev Immunol 11:85–97

70. Wellen KE, Hotamisligil GS (2005) Inflammation, stress, and diabetes. J Clin Invest 115:1111–1119
71. Roberts CK, Hevener AL, Barnard RJ (2013) Metabolic syndrome and insulin resistance: underlying causes and modification by exercise training. Compr Physiol 3:1–58
72. Baker LD, Cross DJ, Minoshima S et al (2011) Insulin resistance and Alzheimer-like reductions in regional cerebral glucose metabolism for cognitively normal adults with prediabetes or early type 2 diabetes. Arch Neurol 68:51–57
73. Bonomini F, Rodella LF, Rezzani R (2015) Metabolic syndrome, aging and involvement of oxidative stress. Aging Dis 6:109–120
74. Kim Y, Park H (2013) Does regular exercise without weight loss reduce insulin resistance in children and adolescents? Int J Endocrinol 2013:402592
75. Onyango IG, Dennis J, Khan SM (2016) Mitochondrial dysfunction in Alzheimer's Disease and the rationale for bioenergetics based therapies. Aging Dis 7:201–214
76. Galasko D, Montine TJ (2010) Biomarkers of oxidative damage and inflammation in Alzheimer's disease. Biomark Med 4:27–36
77. Di Giacomo G, Rizza S, Montagna C et al (2012) Established principles and emerging concepts on the interplay between mitochondrial physiology and S-(De)nitrosylation: Implications in Cancer and Neurodegeneration. Int J Cell Biol 2012:361872
78. Akhtar MW, Sunico CR, Nakamura T et al (2012) Redox regulation of protein function via Cysteine S-Nitrosylation and its relevance to Neurodegenerative Diseases. Int J Cell Biol 2012:463756
79. Nakamura T, Tu S, Akhtar MW et al (2013) Aberrant protein s-nitrosylation in neurodegenerative diseases. Neuron 78:596–614
80. Nakamura T, Cieplak P, Cho DH et al (2010) S-nitrosylation of Drp1 links excessive mitochondrial fission to neuronal injury in neurodegeneration. Mitochondrion 10:573–578
81. Qu J, Nakamura T, Cao G et al (2011) S-Nitrosylation activates Cdk5 and contributes to synaptic spine loss induced by beta-amyloid peptide. Proc Natl Acad Sci U S A 108:14330–14335
82. Zhao QF, Yu JT, Tan L (2015) S-Nitrosylation in Alzheimer's disease. Mol Neurobiol 51:268–280
83. Piantadosi CA (2012) Regulation of mitochondrial processes by protein S-nitrosylation. Biochim Biophys Acta 1820:712–721
84. Wang X, Su B, Zheng L et al (2009) The role of abnormal mitochondrial dynamics in the pathogenesis of Alzheimer's disease. J Neurochem 109:S153–S159
85. Halloran M, Parakh S, Atkin JD (2013) The role of s-nitrosylation and s-glutathionylation of protein disulphide isomerase in protein misfolding and neurodegeneration. Int J Cell Biol 2013:797914
86. Pisoschi AM, Pop A (2015) The role of antioxidants in the chemistry of oxidative stress: A review. Eur J Med Chem 97:55–74
87. Rani V, Deep G, Singh RK et al (2016) Oxidative stress and metabolic disorders: Pathogenesis and therapeutic strategies. Life Sci 148:183–193
88. Butterfield DA (2004) Proteomics: a new approach to investigate oxidative stress in Alzheimer's disease brain. Brain Res 1000:1–7
89. Drake J, Link CD, Butterfield DA (2003) Oxidative stress precedes fibrillar deposition of Alzheimer's disease amyloid beta-peptide (1–42) in a transgenic Caenorhabditis elegans model. Neurobiol Aging 24:415–420
90. Hills AP, Byrne NM, Lindstrom R et al (2013) 'Small changes' to diet and physical activity behaviors for weight management. Obes Facts 6:228–238
91. Teixeira J, Silva T, Andrade PB et al (2013) Alzheimer's disease and antioxidant therapy: how long how far? Curr Med Chem 20:2939–2952
92. Feng Y, Wang X (2012) Antioxidant therapies for Alzheimer's disease. Oxid Med Cell Longev 2012:472932

93. Nunes-Souza V, Cesar-Gomes CJ, Da Fonseca LJ et al (2016) Aging increases susceptibility to high fat diet-induced metabolic syndrome in C57BL/6 Mice: Improvement in Glycemic and Lipid Profile after Antioxidant Therapy. Oxid Med Cell Longev 2016:1987960
94. Strasser B (2013) Physical activity in obesity and metabolic syndrome. Ann NY Acad Sci 1281:141–159
95. Gregorio BM, De Souza DB, De Morais Nascimento FA et al (2016) The potential role of antioxidants in metabolic syndrome. Curr Pharm Des 22:859–869
96. Di Luccia B, Crescenzo R, Mazzoli A et al (2015) Rescue of fructose-induced metabolic syndrome by antibiotics or faecal transplantation in a rat model of obesity. PLoS One 10:e0134893
97. Qiao Y, Sun J, Ding Y et al (2013) Alterations of the gut microbiota in high-fat diet mice is strongly linked to oxidative stress. Appl Microbiol Biotechnol 97:1689–1697
98. Qiao Y, Sun J, Xia S et al (2014) Effects of resveratrol on gut microbiota and fat storage in a mouse model with high-fat-induced obesity. Food Funct 5:1241–1249
99. Naseer MI, Bibi F, Alqahtani MH et al (2014) Role of gut microbiota in obesity, type 2 diabetes and Alzheimer's disease. CNS Neurol Disord Drug Targets 13:305–311
100. Alam MZ, Alam Q, Kamal MA et al (2014) A possible link of gut microbiota alteration in type 2 diabetes and Alzheimer's disease pathogenicity: an update. CNS Neurol Disord Drug Targets 13:383–390

Chapter 24
Oxidative Stress and Antioxidants in Experimental Metabolic Syndrome

Matilde Otero-Losada, Gabriel Cao, Hernán Gómez Llambí, Mariana H. Nobile, Francisco Azzato, and José Milei

> *It is clear that regardless of whether or not people understand what "rich in antioxidants" means, it certainly says 'Buy me, I'm going to help you live forever"*
>
> Phil Lempert

Abstract Metabolic syndrome is a prevalent condition in Western and developing countries (20 to 30 % of adults) that represents a serious public health threat. One of the defects in metabolic syndrome and in its associated diseases such as hypertension, dislipemia, insulin resistance and obesity (mainly visceral), is the increase in general oxidative metabolism with development of oxidative stress. This implies the overproduction (and/or reduced degradation) of reactive oxygen species that usually overrides the physiological antioxidative defense. Thus, there is an increase in the oxidant-to-antioxidant compounds' ratio. In this chapter, we comment and revise some of our evidence on this topic obtained in an experimental model of metabolic syndrome. In particular, we have found that chronic consumption of cola beverages, either sucrose- or artificially-sweetened cola, leads to a condition that reproduces the typical features of human metabolic syndrome. Following the characterization of our model, we have succeeded in replicating this model as shown in our first studies. Here we share some of the most important findings in relation with oxidative metabolism both general and tisular, with particular emphasis on hypertriglyceridemia, pancreatic and renal changes. Oxidative alterations and inflammatory mechanisms are concurrent in otherwise healthy young adult rats following cola beverage drinking for long periods.

Keywords Metabolic syndrome • Hypertriglyceridemia • Insulin resistance • Oxidative stress • Pancreas • Kidneys • Inflammation • Thioredoxins

---

M. Otero-Losada (✉) • G. Cao • H. Gómez Llambí • M.H. Nobile • F. Azzato • J. Milei
Institute of Cardiological Research (ININCA, UBA-CONICET), School of Medicine, University of Buenos Aires and National Research Council, Buenos Aires, Argentina
e-mail: molly1063@gmail.com

© Springer International Publishing Switzerland 2016
R.J. Gelpi et al. (eds.), *Biochemistry of Oxidative Stress*, Advances in Biochemistry in Health and Disease 16, DOI 10.1007/978-3-319-45865-6_24

1 Introduction

Metabolic syndrome (MetS) is a cluster of cardiometabolic risk factors that includes obesity, insulin resistance, hypertension and dyslipidemia, mainly hypertriglyceridemia. The prevalence of metabolic syndrome is dramatically increasing in Western and developing countries [1]. By now, MetS has become a significant public health problem and its prevalence is likely to increase [2]. Approximately 20–30 % of the adult population in most countries is affected by metabolic syndrome [3]. The prevalence is dependent on age, gender, race, and diagnostic criteria [2, 4]. One of the defects in metabolic syndrome and in its associated diseases is a situation of oxidative stress, an excess of reactive oxygen species over antioxidative defense, namely an increase in the oxidant-to-antioxidant ratio. Reactive oxygen species generated by mitochondria, or from other sites within or outside the cell, cause damage to mitochondrial components and initiate degradative processes. Such toxic reactions contribute significantly to the aging process [5]. Aging and age-related diseases in higher organisms are complex processes that are likely controlled by a combination of many different genetic, environmental, nutritional, and pathologic factors. While the specific role of oxidative stress in aging and development of age-related diseases is an area of active investigation, the exact mechanisms that may define this complex relationship are unclear [5]. Metabolic syndrome is often characterized by oxidative stress, a condition in which an imbalance results between the production and inactivation of reactive oxygen species. Reactive oxygen species can best be described as double-edged swords; while they play an essential role in multiple physiological systems, under conditions of oxidative stress, they contribute to cellular dysfunction. Oxidative stress is thought to play a major role in the pathogenesis of a variety of human diseases, including atherosclerosis, diabetes, hypertension, aging, Alzheimer's disease, kidney disease and cancer.

2 Hypertriglyceridemia and Decrease in Endogenous Antioxidants Level in Experimental Metabolic Syndrome

The rising consumption of soft drinks has been linked to metabolic syndrome in humans. However, in spite of experimental data with fructose-enriched diets, little is known on the effects of cola beverages in man or in animal models. Oxidative stress is involved in the pathophysiology and cardiovascular complications of metabolic syndrome [6, 7]. We have reported, and several studies over the years have confirmed, that 6 months of cola drinking is a useful experimental model of MetS. Consistent biochemical and nutritional changes are proof of the reliability, robustness, and reproducibility of our experimental model [7–12]. Characterization of the model included reproduction of most typical features of the MetS condition. Otero-Losada et al. [7] reported the effects of prolonged consumption of cola beverage on oxidative metabolism plasma indicators in adult (2 month-old) male Wistar rats both at the end of a 6 months treatment and after the following 6 months drinking water (wash-out period). The model has a total duration of 12 months from the beginning of the study. The effects of both sucrose-sweetened (regular) cola and non-caloric artificially sweetened (light) cola beverages were evaluated. Plasma concentrations of antioxidants α-tocopherol and ubiquinone-10 (CoQ_{10}) (RP-HPLC) and glucose, triglycerides, total- and HDL- cholesterol (enzymatic assays) were measured in the study [7]. Weight gain (7 %), hypertension (8 %), decreased food ingestion (−31 %) and increased drinking volume (+1.7-fold) were observed after 6-months in rats drinking regular cola (C_6), compared with either water-drinking (W_6) or light-cola drinking rats (L_6) [7]. Six months regular cola drinking (C_6) induced mild hyperglycemia (+15 %), severe hypertriglyceridemia (+3-fold) and a trend to hypercholesterolemia (+32 % total cholesterol) [7]. No changes were found in the HDL-C fraction in C_6. Normoglycemia and an unexpected tendency to hypertriglyceridemia (+2-fold, N.S.) and hypercholesterolemia (+29 % total cholesterol, N.S.) were observed in rats drinking non-caloric cola beverage (L_6) [7].

All the rats were normoglycemic after the wash-out period (C_{12}, L_{12} and W_{12}) (Fig. 24.1). However, hypertriglyceridemia was found in regular cola-drinking and water-drinking rats following wash-out (C_{12} and W_{12}) [7]. The trend to hypertriglyceridemia observed in rats drinking light-cola at the end of treatment, was not found in the same rats following wash-out [7]. Following wash-out, all rats developed hypercholesterolemia compared with the values found at the end of the treatment period while HDL-C levels underwent no major changes [7]. Plasma levels of coenzyme Q10 (CoQ_{10}) were 52 % lower in regular cola-drinking rats compared with water drinking rats at the end of the treatment period (Fig. 24.1a). Based on CoQ_{10} levels, light cola drinking rats were undistinguishable from water drinking or regular cola drinking rats at the end of treatment. No differences in plasma CoQ_{10} level was found among the three groups of rats after wash-out (Fig. 24.1a). Interestingly, rats drinking water during all the experiment showed 46 % lower CoQ_{10} concentration in plasma compared with themselves at the end of the treatment period (i.e. : 6 months earlier when they were younger) (Fig. 24.1a). The plasma levels of α-tocopherol showed variations similar to those of CoQ_{10} (Fig. 24.1b. As observed for CoQ_{10}, α-tocopherol plasma levels decreased by 48 % in rats drinking water during all the experiment, compared with themselves at the end of the previous 6 month period, when they were younger (Fig. 24.1b), life-span being the only concurrent factor.

The "pro-oxidant-to-antioxidants" plasma ratio, a relationship developed for this study and based on the plasma levels of pro-oxidant molecules which undergo metabolic reactions originating reactive oxygen species, and the plasma levels of the circulating antioxidants coenzyme CoQ_{10} and α-tocopherol. The "pro-oxidants-to-antioxidants" plasma ratio dramatically increased in C_6, W_{12}, C_{12} and L_{12} but not in L_6 (Fig. 24.1c). Low plasma coenzyme CoQ_{10} and hypertriglyceridemia were correlated all throughout the study ($p < 0.01$, 2-tailed Pearson correlation).

Coenzyme Q10 deficiency, mitochondrial dysfunction and oxidative damage have been concurrently observed in different conditions [13, 14]. Supplementation with Q_{10} has been reported to reduce hypercholesterolemia [14]. Consumption of

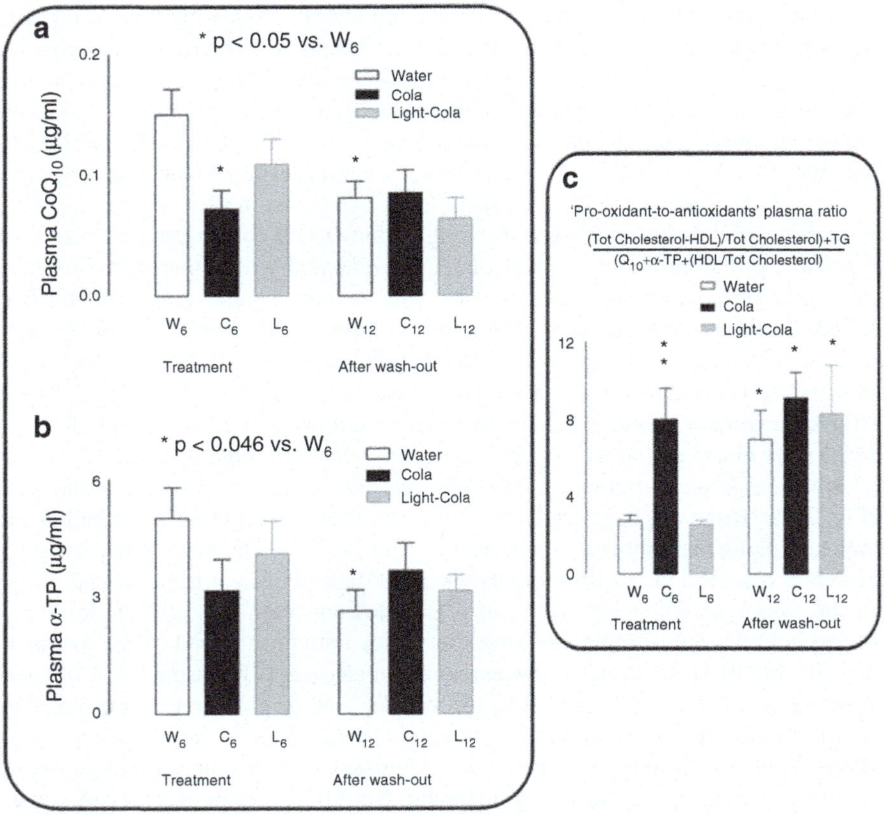

Fig. 24.1 Plasma levels of CoQ_{10} and α-tocopherol, and the "pro-oxidants-to-antioxidants" ratio. * $p < 0.05$, ** $p < 0.01$ vs. W_6. (Modified from Otero-Losada et al. [7])

regular cola for 6 months resulted in weight gain, despite a reduced solid food intake, and in hypertension, hyperglycemia (+15 %), hypertriglyceridemia (3-fold) and low CoQ_{10} plasma levels (−52 %) compared to W_6 and L_6. All parameters were normalized except for hypertriglyceridemia after the 6 months of the wash-out period. Light cola drinking induced neither metabolic nor antioxidants alterations in plasma levels. Twelve months after the beginning of the study, all 3 groups showed low α-tocopherol levels and hypercholesterolemia, results that were unrelated to the treatment and could only be explained by life-span [7]. Twelve months after the beginning of the study all animals had low CoQ_{10} levels. Since α-tocopherol was not substantially affected by treatment, it may be reasonably assumed that α-tocopherol level was preserved at the expense of CoQ_{10} consumption. CoQ_{10} levels have been suggested to be a useful biomarker of oxidative stress. The low plasma levels of CoQ_{10} observed in C_6 rats reflect the decrease in antioxidants after the oxidative stress induced by hypercaloric cola drinking. Long-term ingestion of a hypercalorichyperglycemic diet leads to obesity, to increased lipid peroxidation, and to oxidative stress, the latter condition likely by compromising mitochondrial redox metabolism which plays a crucial role in the pathogenesis of metabolic syndrome [7].

Low CoQ_{10} and α-tocopherol levels have been related with left ventricular function impairment in agreement with our previous echocardiographic observations [7]. The hypothesis that mitochondria are both source and target of free radicals and the knowledge of the important role of CoQ_{10} in mitochondrial function have led others to propose that CoQ_{10} may be involved in the aging mechanisms. Interestingly, ubiquinol-10 supplementation activates mitochondrial functions and decelerate senescence in senescence-accelerated mice [15]. Coenzyme Q10 is a well-known essential electron transporter in complex I (NADH–ubiquinone oxidoreductase), complex II (succinate–ubiquinone oxidoreductase), and complex III (ubiquinol–cytochrome c oxidoreductase) of the mitochondrial respiratory chain [15–17]. The oxidized form of ubiquinone-10 is enzymatically reduced by complexes I and II, to ubiquinol-10 [14, 15, 18], which acts as a fat-soluble antioxidant that effectively protects lipid membranes and lipoproteins from oxidative damage and has beneficial effects in preventing DNA damage [40]. In many tissues, ubiquinol-10 biosynthesis and levels decrease with age, suggesting that ubiquinol-10 may have possible anti-aging effects [15, 19].

Cell aging is thought to be due to cumulative cellular and genomic damage that results in permanent cell-cycle arrest, apoptosis, and senescence [20]. A major source of cellular damage are the reactive oxygen species (ROS), which are mainly generated at complexes I and III of the cellular respiratory chain (2). Indeed the chemical species that is generated at complexes I and III is superoxide radical (O_2^-) which has been clearly identified and its generation was estimated to represent about 1 % of total oxygen consumption [15, 21, 22]. An early study suggested that its generation rate is thought to increase with age, due to the accumulation of damaged mitochondria in a vicious cycle in which ROS-induced mitochondrial impairment results in increased ROS production, which, in turn, leads to further mitochondrial damage in aging [15, 23, 24]. The biochemical profile produced by low-calorie sweet (aspartame) beverage drinking revealed mild changes (triglycerides, total cholesterol) which were not different from those found in the other groups. It seems that is an intermediate condition developed that is not accounted by an extra glucose supply and may result from sweet taste receptors' stimulation. Both sweet taste and energy content are required for a hypothalamic response. Low-calorie sweet beverage is not expected to trigger appropriate anticipatory physiologic responses (cephalic phase insulin release), because aspartame, unlike glucose, does not lead to an early rise in insulin concentrations [7]. There are concerns that consumption of nonnutritive sweeteners promotes energy intake and contributes to obesity by not completely understood mechanisms. In our study, rat chow would provide carbohydrates which in combination with artificial sweeteners might play an active metabolic role within the gastrointestinal tract [7].

The sustained hypertriglyceridemia observed after wash-out could be attributed to the increase in triglycerides associated with life-span [7]. Adaptive changes in rat plasma levels of Co Q_{10} and α-tocopherol were observed after long-term hypercaloric consumption. Concerning with antioxidants levels, long-term hypercaloric consumption mimicked metabolic conditions associated with age [7]. Experimentally-induced metabolic syndrome induces oxidative stress-related biochemical changes resembling those appearing along life-span. Our findings, as shown above, were consistent with and added further support to the idea that the metabolic syndrome is a high-risk condition associated with oxidative metabolism and premature aging-related changes [7].

3 Experimental Metabolic Syndrome Accelerates Atherosclerosis and Increases Arterial Damage in Genetically Atherosclerotic Mice

There is now consensus that atherosclerosis represents a state of heightened oxidative stress characterized by lipid deposits in the vascular wall [25]. Reactive oxygen species (ROS) are key mediators of signaling pathways that underlie vascular inflammation in atherogenesis, starting from the initiation of fatty streak development, through lesion progression, to ultimate plaque rupture. Plaque rupture and thrombosis result in the acute clinical complications of myocardial infarction and stroke. Many data support the notion that ROS released from nicotinamide adenine dinucleotide phosphate (NADPH) oxidase, myeloperoxidase, xanthine oxidase, lipoxygenase, nitric oxide synthase (NOS) and enhanced ROS production from dysfunctional mitochondrial respiratory chain, indeed, have a causative role in atherosclerosis and other vascular diseases. Moreover, oxidative modifications in the arterial wall can contribute to atherosclerosis when the balance between oxidants and antioxidants shifts in favor of the former [25]. We had previously observed and increase in the"pro-oxidants-to-antioxidants" plasma ratio following 6 months of regular cola drinking rats and also in 14 month-old rats associated to the aging process [7]. Hence it was reasonable to study the effects of regular cola drinking in apolipoprotein E deficient (apo$E^{-/-}$) mice that undergo an accelerated aging process. The main components of the metabolic syndrome (obesity, diabetes, hypertension) are known risk factors for atherosclerosis. The increasing incidence of atherosclerosis and metabolic alterations warranted the study of the effects of long-term soft drink consumption on metabolism and atherosclerosis in the genetic deficiency of apolipoprotein E. Apolipoprotein E deficient (apo$E^{-/-}$) mice are a murine model of spontaneous atherosclerosis and metabolic abnormalities [9]. Regular cola drinking for 8 weeks (C, sucrose sweetened) resulted in hyperglycemia in apolipoprotein E deficient (Apo$E^{-/-}$) mice which largely accounted for the observed increase in the non-HDL cholesterol fraction. These changes reversed after a 8-week washout period (mice were switched from cola drinking to water drinking for 8 weeks). Differently, light cola drinking (aspartame–acesulfame K sweetened) induced a mild decrease in glycemia, with hypercreatininemia, hyperuremia and an increase in aspartate aminotransferase (AST), all of which reversed after washout except for hypercreatininemia. Over the time of the study, hypertriglyceridemia and hyperuremia developed in all groups irrespective of drink treatment likely as a result of the aging process which is accelerated in ApoE$^{-/-}$ mice [9].

Acesulfame K and other non-nutritive sweeteners (but not aspartame), have been reported to activate enteroendocrine sweet taste receptors and release incretins which stimulate pancreatic insulin secretion. Reasonably, this mechanism might help to understand the decrease in glycemia observed in the L group. Likewise, phenylalanine that accounts for 40 % of metabolized aspartame, acting synergically with solid food and stimulating insulin release might participate in the decrease in glycemia in L group [9]. Reduced creatinine clearance has been reported in 8 week-old (young) ApoE$^{-/-}$ mice indicating some degree of vulnerability in glomerular filtration. Oxidative stress is associated with either high uremia levels or methanol occurrence in blood. In L mice, uremic toxicity and increased oxidative stress correlate with a pro-inflammatory condition [9].

4 Partial Exhaustion of β-Cell Population in Langerhans Islets as a Model of Metabolic Syndrome

In other study, we reported the effects of long-term cola beverage drinking on glucose homeostasis, endocrine pancreas function and thioredoxins tissue expression and morphology in rats [11]. Wistar rats drank: water (group W), regular cola beverage (group C, sucrose sweetened) or light cola beverage (group L, artificially sweetened). After 6 months, 50 % of the animals in each group were euthanized and the remaining animals consumed water for the next 6 months when euthanasia was performed. Biochemical assays, insulinemia determination, estimation of insulin resistance (HOMA-IR), morphometry and immunohistochemistry evaluations were performed in pancreas. In this study, hyperglycemia (16 %), CoQ_{10} decrease (−52 %), strong hypertriglyceridemia (2.8-fold), hyperinsulinemia (2.4-fold) and HOMA-IR increase (2.7-fold) were observed in regular cola (sucrose sweetened) drinking rats [11].

In this study, group C showed a decrease in the number of α cells (−42 %) and β cells (−58 %) and a moderate increase in α cells size after wash-out (+14 %) as shown in Fig. 24.2. Group L showed reduction in β cells size (−9 %) and only after wash-out (L_{12}) a 19 % increase in size with 35 % decrease in the number of α cells. Groups C and L showed increase in α/β-cell ratio which was irreversible only in C (α/β = +38 % in C6, +30 % in C_{12}, vs. W_6) (Fig. 24.2).

Regular cola induced a robust, striking and irreversible (did not normalize after wash-out) increase in the cytoplasmic expression of Trx-1 (thioredoxin-1) (2.25-fold following 6 months' drinking, 2.7-fold after wash-out) and Prx-2 (peroxiredoxin-2) (3-fold following 6 months' drinking, 2-fold after wash-out). Light cola induced increases in Trx1 (3-fold) and Prx2 (2-fold) after wash-out (L_{12} vs. W_{12}) (Table 24.1) [11].

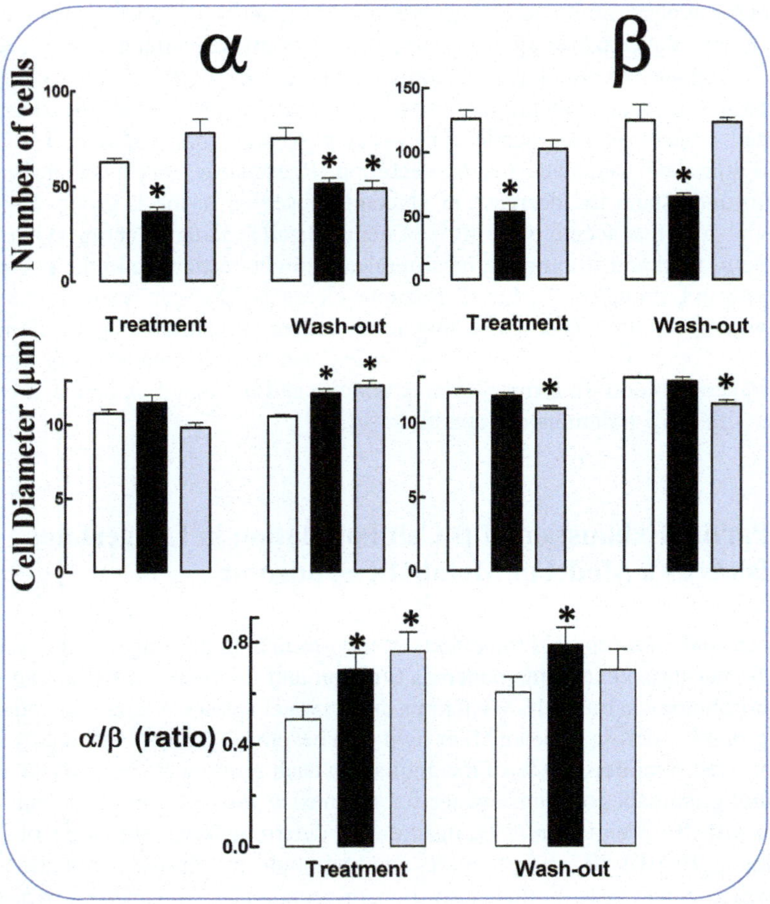

Fig. 24.2 α- and β-cells in pancreatic islets: number, size and α/β-cell ratio. White columns (water, W group), black columns (regular cola, C group), gray columns (light cola, L group). Treatment period, months 0–6 of the study. Wash out period (months 7–12 of the study). Values are mean values ± SEM. *$p < 0.05$ compared with W in the period (Modified from Otero-Losada et al. [11])

Plasma level of CoQ_{10} was lower in C_6 compared with W_6 (−52 %) and an unexpected trend to a decrease in CoQ_{10} level was observed in L group all over the study. Interestingly, age was associated with a decrease in plasma CoQ_{10} levels (−46 % in W_{12} vs. W_6) [11]. Glucotoxicity may contribute to the loss of β cell function with depletion of insulin content. Oxidative stress, suggested by increased expression of thioredoxins and low circulating levels of CoQ_{10}, may follow sustained hyperglycemia. A likely similar panorama may result from the effects of artificially sweetened cola though via other downstream routes [11]. This paper addresses the effects of cola drink consumption on endocrine pancreas function and morphology regarding glucose homeostasis in rats. The contribution of cell proliferation, apoptosis and/or

Table 24.1 Quantitative morphology and immunochemistry data (A = area)

	Treatment			Washout		
Group	Water	Cola	Light-cola	Water	Cola	Light-cola
α/β cell ratio	0.4±0.1	0.8±0.1**	0.6±0.1*	0.4±0.1	0.7±0.1**	0.5±0.2
A_{islet} (10^4 μ^2)	2.2±0.2	1.8±0.2	2.4±0.3	2.5±0.4	2.3±0.2	2.5±0.3
Aα (10^4 μ^2/islet)	0.6±0.1	0.4±0.1*	0.6±0.1	0.7±0.1	0.6±0.1	0.6±0.1
Aβ (10^4 μ^2/islet)	1.5±0.2	0.6 ±0.1	1.0±0.1	1.7±0.2	0.8±0.1**	1.3±0.1
A_{PCNA} (10^4 μ^2/islet)	0.05±0.02	0.02±0.01*	0.05±0.02	0.04±0.01	0.09±0.02	0.08±0.02
A_{TRX} (10^4 μ^2/islet)	0.4±0.1	0.9±0.1**	0.4±0.1	0.3±0.1	0.9±0.1**	1.0±0.1
A_{PRX} (10^4 μ^2/islet)	0.2±0.1	0.9±0.1**	0.2±0.1	0.4±0.1	0.8±0.1**	0.8±0.1**

Values are means ± SEM. Differences were analyzed using the Kruskal-Wallis and the Dunn's multiple comparison test. *$p < 0.01$ **$p < 0.001$ vs. water, within the period (Modified from Otero-Losada et al. [11])

oxidative stress to the observed changes is evaluated in insular α and β cells in pancreas [11]. It is known that sustained hyperglycemia may lead to insulin resistance and type 2 diabetes as well depending on genetic and epigenetic background. Type 2 diabetes has been associated with oxidative stress and a generalized inflammatory condition [11]. Metabolic syndrome and diabetes are known risk factors for cardiovascular disease, which is the leading cause of death in modern Western societies. Soft drink consumption has been related to obesity and to an increased risk of metabolic syndrome. Individuals consuming >500 mL/day of soft drink had a higher prevalence of metabolic syndrome than those consuming <350 mL/day (1 drink serving). It is widely accepted that sugar-sweetened soft drink consumption increases the risk of metabolic disorders. Unlike sugar, artificial sweeteners are usually considered safe and beneficial owing to their low caloric content and artificially sweetened beverages are marketed as low-calorie substitutes to prevent beverage-associated weight gain. While it is undeniable that artificial sweeteners do not add extra calories, they may pose to other risks instead [11].

The HOMA-IR (homeostatic model assessment of insulin resistance) index was used to estimate insulin resistance and was calculated using the validated formula for Wistar rats [11]: HOMA-IR = (I x G) /k, where I = fasting insulinemia, G = fasting glycemia and k is a constant value k = 405 (if G is expressed in mg/dL) or k = 22.5 (if G is expressed in mmol/L) [11]. Regular cola drinking caused an irreversible decrease in the number of both pancreas α cells (−42 % C_5, −32 % C_{12}) and β cells (−58 % C_6, −42 % C12) and a moderate increase in the size of α cells after wash-out (+14 %, p < 0.001 C_{12}, +7 % NS, C_6). Light cola drinking affected β cells leading to irreversible reduction in size (−9 % in L_6; −14 % in L_{12}) and affected α cells as well causing a 19 % increase in size ($p < 0.0001$) and a 35 % reduction in number ($p < 0.01$) only after wash-out (L_{12}). Accordingly, the α/β-cell ratio increased following consumption of one or another type of cola beverage. However, the effect of regular cola was irreversible (α/β-cell ratio = +38 % in C_6, +30 % in C_{12}, $p < 0.001$ vs. W_6 and W_{12} respectively) while the effect of light cola was not (α/β-cell ratio = +52 % in L_6 vs. W_6; +15 % in L_{12} vs. W_{12}). Treatment with cola drinks did not substantially affect the size of the islet over the study time.

Insulin immunolabeling decreased: −59.7 % in C_6, −50.3 % in C_{12} and −33 % in L_6 compared with age-matched W groups. Glucagon immunopositivity was lower in C_6 compared with W_6 (−33 %). Neither regular cola nor light cola drinking modified the nuclear expression of PCNA and caspase-3: isolated positive nuclei represented <1 % of the total population of insular cells. However, lower cytoplasmic immunopositivity for PCNA was found in C (−60 % in C_6, −78 % in C_{12}) and in L_{12} (−80 %) compared with age-matched W group. Regular cola drinking strikingly increased the cytoplasmic expression of Trx1 (2.3 fold in C_6 vs. W_6; 2.7 fold in C_{12} vs. W_{12}) and Prx2 (3-fold in C_6 vs. W_6; 2-fold in C_{12} vs. W_{12}). Light cola drinking induced a remarkable increase in Trx1 (3-fold) and Prx2 (2-fold) after wash-out (L_{12} vs. W_{12}). A correlation was found between HOMA-IR and the cytoplasmic expression of TRX1, Prx2 and PCNA (r, and % mutually explained variation: 0.839, 71 %, $p < 0.037$ for Trx1; 0.878, 77 %, $p < 0.022$ for Prx2; −0.893, 79 %, $p < 0.017$) for all over the study time.

The nuclear immunolabeling for caspase-3 and PCNA were very low, indicating negligible apoptotic and proliferative activity. The cytoplasmic expression of PCNA, specially observed at the islet periphery, suggests other functions than synthesis and reparation of deoxyribonucleic acid (DNA) which is related to the nuclear expression of PCNA. Recent evidence indicates that cytoplasmic PCNA stimulates glycolysis via activation of glyceraldehyde-3-phosphate dehydrogenase. Glutathione synthesis could be indirectly reduced leading to oxidative stress as described in neurons. An inverse pattern of immunolabeling for cytoplasmic PCNA, Trx1 and Prx2 was observed suggesting increased glutathione availability to thioredoxin and peroxiredoxin in this context. The hypothesis of an oxidative microenvironment is reinforced by correlation found for Trx1, Prx2 and cytoplasmic expression of PCNA with HOMA-IR. Glutathione and thioredoxin systems are known to act in concert and insulin resistance and diabetes are associated with decreased antioxidant capacity [11]. The present findings are compatible with a glucotoxic loss of β cell function and depletion of insulin content. Multiple signaling pathways contribute to the adverse effects of glucotoxicity. The increased expression of thioredoxins and the reduction in plasma levels of CoQ_{10} observed after cola drinking is consistent with an oxidative stress condition resulting from hyperglycemia. Hyperglycemia induces synthesis of reactive oxygen species by glucose oxidation, leading to an increased production of advanced glycosylation end products, as well as inflammation and oxidative stress. On the other hand, artificial sweeteners readily originate advanced glycosylation end products with pro-oxidative and inflammatory effects. Failure to maintain a functional β-cell population is a serious problem. When survival of overworking β cells is compromised, sparing the survivor cells may be a solution. Then, β cells may lose their mature identity and dedifferentiate to an insulin-negative neurogenin 3-positive stage (transcription factor neurogenin 3 is predominant during endocrine pancreas embryogenesis). If survival-threatening conditions disappear, dedifferentiated cells can evolve to mature insulin-positive β cells. Recent evidence demonstrates β cell dedifferentiation, rather than apoptosis, is the main mechanism of loss of insulin-positive cells. This mechanism may help explain the gradual decrease in β cell mass in long-standing diabetes and recovery of β cell function in type 2 diabetes following insulin therapy [11]. Remarkable plasticity has been reported in murine β cells which can largely adapt to altered insulin demand all over life-time. The increase in α/β-cell ratio is interpreted as the result of reprogramming or transdifferentiation of β to α cell. Endocrine pancreas might cope with a compromising environment (oxidative stress, hyperglycemia) by triggering β and α mutual transdifferentiation, not only shortening the time delay to satisfy metabolic demand but reducing energy costs as well, an advantageous solution from the biological viewpoint. This is in agreement with recent findings pointing to the so called islet remodeling under critical situations in which insular endocrine cells should give fast response to different metabolic situations. The possible existence of a β-α or α-β switches is indeed more economic and faster than cell proliferation and it may sustain a chronically elevated hormone production [11].

The sweetener aspartame (L-aspartyl-L-phenylalanine methyl ester) and the caramel colorant are rich in advanced glycation end products that potentially increase insulin resistance and inflammation. During regular soft drinks consumption, fat accumulates in the liver by the primary effect of fructose which increases lipogenesis, and in the case of diet soft drinks, by the additional contribution of aspartame sweetener and caramel colorant. Aspartame before glucose ingestion augments glucagon-like peptide-1 (GLP-1) secretion and contributes to obesity, insulin resistance and type 2 diabetes. Aspartame is absorbed from the intestine and broken down to phenylalanine (50 %), aspartic acid (40 %) and methanol (10 %) in the liver. Methanol is converted to formate, formaldehyde, diketopiperazine (a carcinogen) and a number of other highly toxic derivatives. These and other issues have raised concern about the use of aspartame ever since its approval by the U.S. Food and Drug Administration in 1974 [11]. This study shows that artificially sweetened cola may lead to health problems as well. The effects of light-cola drinking, produce: striking increases in Trx1 and Prx2, a robust trend to hypertriglyceridemia, a decrease in the plasma level of CoQ10 and an increase in HOMA-IR, a reduction in β cells' size, an increase in α/β-cell ratio with an increase in α cells' size, and a decrease in α cells number, indicate the development of an oxidative stress condition as well. Our findings are consistent with the report that artificial sweeteners alter gut microbiota, induce glucose intolerance and increased susceptibility to metabolic disease. Propionate, a bacterial end product of aspartame in the gut, is a highly gluconeogenic substrate and may contribute to an increased susceptibility to metabolic disease [11].

In conclusion, we report evidence supporting that chronic cola consumption impairs pancreatic storage of insulin and glucagon, increases α/β-cell ratio and causes a striking increase in triglycerides and oxidative stress. The used experimental model opens new avenues to improve our knowledge of the metabolic syndrome and the associated decline in pancreatic function following long-term ingestion of cola drinks [11]. Although the exact molecular mechanisms underlying the development of diabetes type 2 are unknown, there is growing evidence that excess generation of reactive oxygen species (ROS), largely due to hyperglycemia, causes oxidative stress in a variety of tissues. Oxidative stress may result from either an increase in free radical production, or a decrease in endogenous antioxidant defenses, or both. ROS and reactive nitrogen species (RNS) are products of cellular metabolism and are well recognized for their dual role as both deleterious and beneficial species. In type 2 diabetic patients, oxidative stress is closely associated with chronic inflammation. Multiple signaling pathways contribute to the adverse effects of glucotoxicity on cellular functions. There are many endogenous factors (antioxidants, vitamins, antioxidant enzymes, metal ion chelators) that can serve as modulators of the production and action of ROS. Clinical trials that investigated the effect of antioxidant vitamins on the progression of diabetic complications gave negative or inconclusive results. This lack of efficacy might also result from the fact that they were administered at a time when irreversible alterations in the redox status are already established. Another strategy to modulate oxidative stress is to exploit the pleiotropic properties of drugs directed primarily at other targets and acting as indirect antioxidants [26, 27]. In diabetic patients, oxidative stress is induced by excessive ROS and RNS. It is considered that high levels of ROS result in β-cell dysfunction. On the other hand, Nrf2 activates cytoprotective pathways against oxidative injury [27].

5 Metabolic Syndrome Involves Alterations in the Cardiorenal Axis

Recently we have revised the cardiorenal involvement in the metabolic syndrome induced by cola drinking in rats with particular emphasis on proinflammatory cytokines and impaired antioxidative protection [12]. Considering that the pathophysiology of renal complications associated with metabolic syndrome (MetS) is not clear, we reported experimental evidence confirming renal histopathology in relation with proinflammatory mediators and oxidative metabolism in MetS induced by cola drinking in rats [12]. Hypertriglyceridemia and oxidative stress were found to be key factors in this regard. Hypertriglyceridemic toxicity in the context of defective antioxidant/anti-inflammatory protection due to low CoQ_{10} levels might play an important role in the cardiorenal disorder induced by chronic cola drinking in rats [12]. The dramatic increase in pro-inflammatory cytokines IL-6 and TNFα in renal tubules induced by cola drinking rats as reported in that study (Figs. 24.3 and 24.4) [12], should result in tubular derangement and dysfunction if sustained in time.

In this study, correlations were found for changes in LV dimensions with IL-6 (74 %, $r = 0.86$, $p < 0.001$) and TNFα (52 %, $r = 0.72$, $p < 0.001$) (Fig. 24.3). Controlling for either TG or Q_{10} values individually, the strength of correlations was

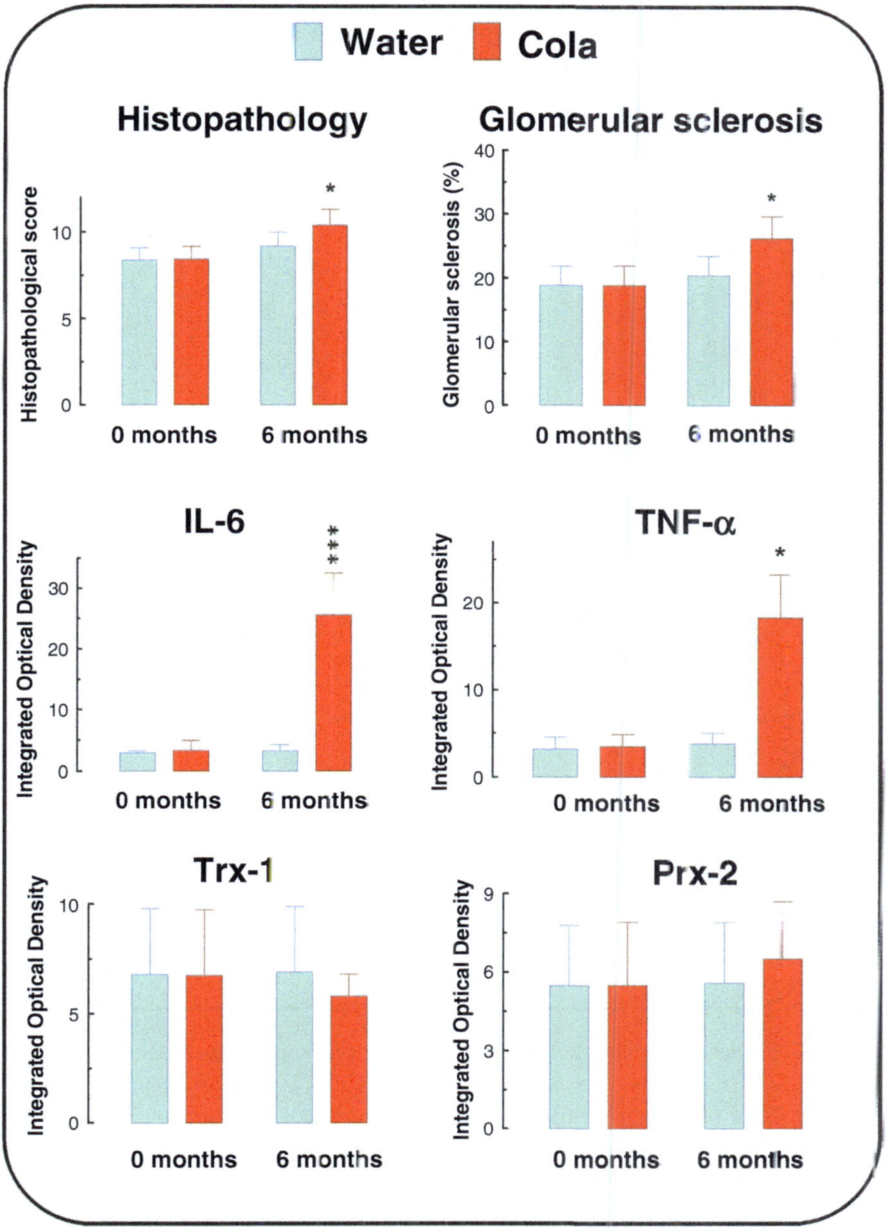

Fig. 24.3 Renal glomerular pathology, inflammatory cytokines and redoxins in renal tubules before and after cola treatment. IL-6: interleukin-6, TNF-α: tumor necrosis factor-alpha, Trx-1: thioredoxin-1, Prx-2: peroxiredoxin-2 (Modified from Otero-Losada et al. [12])

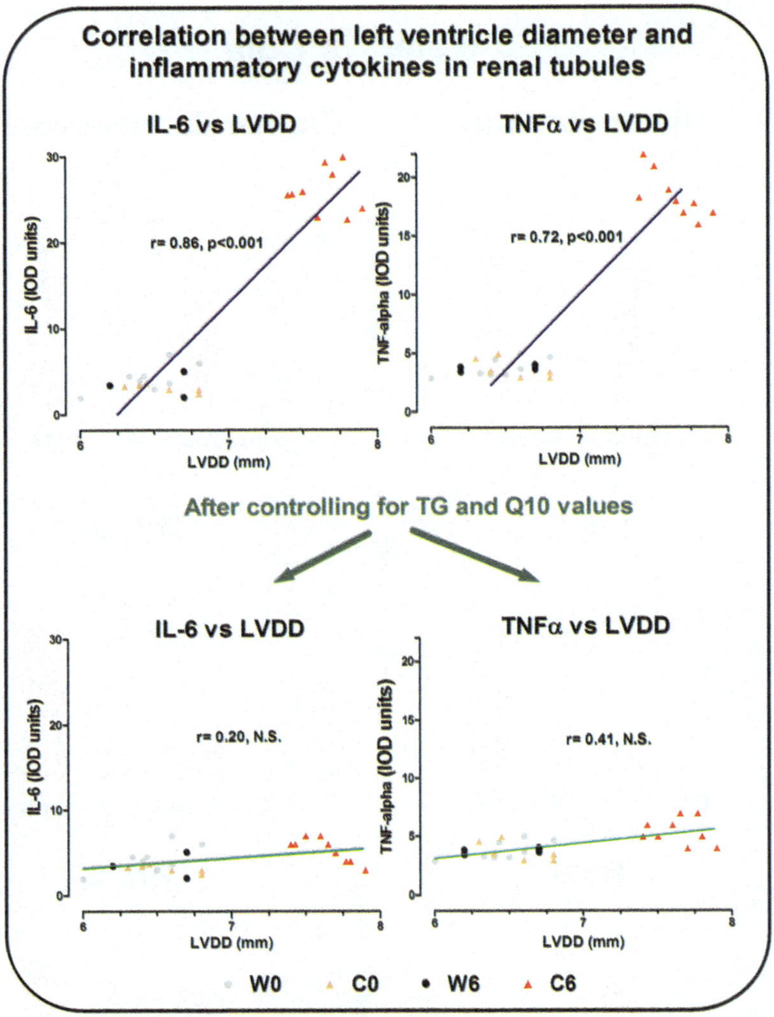

Fig. 24.4 Cardiorenal correlations. Dependence on triglyceride and CoQ_{10} levels

reduced to (% of mutually explained variance) 22 % for IL-6 and 14 % for TNFα). Moreover, the control for both TG and/or Q_{10} levels altogether, abolished any correlation previously observed for LV dimensions with IL-6 ($r = 0.20$, NS) and TNFα ($r = 0.41$, NS) (Fig. 24.4).

Experimental and clinical studies have suggested a correlation between the progression of renal disease and dyslipidemia. Hypertriglyceridemia elicits inflammatory responses in different tissues and is known to affect morphological integrity of kidney [12]. Dyslipidemia and lipotoxicity-induced insulin resistance, inflammation and oxidative stress are the key pathogeneses of renal damage in type 2 diabetes.

Interestingly, we have reported development of insulin resistance in cola drinking rats. In reasonable agreement with present evidence and previous reports, we suggest that MetS induced by cola drinking affects kidney structure and increases pro-inflammatory cytokines in renal tubules in rats, and that hypertriglyceridemia and low levels of Q_{10} may play crucial roles in determining the pathophysiology of the cardiorenal axis. High content of advanced glycation end products (AGEs) in caramel colorant in cola beverages, further perpetuates oxidative stress, contributing to the increase in proinflammatory cytokines in renal tubules and may be involved in the progression to chronic kidney disease as one of the complications of MetS in cola beverages consumers [12]. As a final comment, according to the state of the art in the field, development of new drugs that target key sources of ROS producing enzymes and mimic the physio-pharmacological effects of endogenous antioxidants, appears to be relevant for the treatment and hopefully, for the prevention as well, of the cardiometabolic illnesses of the XXI century.

6 Conclusions

Presently we report evidence supporting that chronic cola consumption impairs pancreatic storage of insulin and glucagon, increases α/β-cell ratio and causes a striking increase in triglycerides and oxidative stress. The used experimental model opens new avenues to improve our knowledge of the metabolic syndrome and the associated decline in pancreatic function following long-term ingestion of cola drinks. Although the exact molecular mechanisms underlying the development of diabetes type 2 are unknown, there is growing evidence that excess generation of reactive oxygen species (ROS), largely due to hyperglycemia, causes oxidative stress in a variety of tissues. Not coincidentally, we also found disturbance in the cardiorenal axis and homeostasis, with increase in the immunohistochemical expression of antioxidant response, namely thioredoxin expression.

References

1. Bonomini F, Rodella LF, Rezzani R (2015) Metabolic syndrome, aging and involvement of oxidative stress. Aging Dis 6:109–120
2. Robberecht H, Hermans N (2016) Biomarkers of metabolic syndrome: biochemical background and clinical significance. Metab Syndr Relat Disord 14:47–93
3. Grundy SM (2008) Metabolic syndrome pandemic. Arterioscler Thromb Vasc Biol 28:629–636
4. Eckel RH, Grundy SM, Zimmet PZ (2005) The metabolic syndrome. Lancet 365:1415–1428
5. Bulteau AL, Szweda LI, Friguet B (2006) Mitochondrial protein oxidation and degradation in response to oxidative stress and aging. Exp Gerontol 41:653–657
6. Roberts CK, Sindhu KK (2009) Oxidative stress and metabolic syndrome. Life Sci 84:705–712

7. Otero-Losada M, Grana DR, Müller A et al (2011) Lipid profile and plasma antioxidant status in sweet carbonated beverage-induced metabolic syndrome in rat. Int J Cardiol 146:106–109
8. Milei J, Otero-Losada M, Gómez Llambí H et al (2011) Chronic cola drinking induces metabolic and cardiac alterations in rats. World J Cardiol 3:111–116
9. Otero-Losada M, Loughlin SM, Rodríguez-Granillo G et al (2013) Metabolic disturbances and worsening of atherosclerotic lesions in ApoE$^{-/-}$ mice after cola beverages drinking. Cardiovasc Diabetol 12:57. doi:10.1186/1475-2840-12-57
10. Otero-Losada M, Cao G, Mc Loughlin S et al (2014) Rate of atherosclerosis progression in ApoE−/− mice long after discontinuation of cola beverage drinking. PLoS One 9:e89838
11. Otero-Losada M, Cao G, González J et al (2015) Morphological changes in endocrine pancreas following cola drink consumption in rats. PLoS One 10:e0118700
12. Otero-Losada M, González J, Ottaviano G et al (2016) Cardiorenal involvement in the metabolic syndrome induced by cola drinking in rats. Proinflammatory cytokines and impaired antioxidative protection. Mediators Inflamm 2016: 0000000 (in press)
13. Henchcliffe C, Beal MF (2008) Mitochondrial biology and oxidative stress in Parkinson disease pathogenesis. Nat Clin Pract Neurol 4:600–609
14. Miyamae T, Seki M, Naga T et al (2013) Increased oxidative stress and coenzyme Q_{10} deficiency in juvenile fibromyalgia: amelioration of hypercholesterolemia and fatigue by ubiquinol-10 supplementation. Redox Rep 18:12–19
15. Tian G, Sawashita J, Kubo H et al (2014) Ubiquinol-10 supplementation activates mitochondria functions to decelerate senescence in senescence-accelerated mice. Antioxid Redox Signal 20:2606–2620
16. Lenaz G (1998) Role of mitochondria in oxidative stress and ageing. Biochim Biophys Acta 1366:53–67
17. Lenaz G, Fato R, Castelluccio C et al (1993) The function of coenzyme Q in mitochondria. Clin Investig 71:S66–S70
18. Tomasetti M, Littarru GP, Stocker R, Alleva R (1999) Coenzyme Q10 enrichment decreases oxidative DNA damage in human lymphocytes. Free Radic Biol Med 27:1027–1032
19. Aberg F, Appelkvist EL, Dallner G, Ernster L (1992) Distribution and redox state of ubiquinones in rat and human tissues. Arch Biochem Biophys 295:230–234
20. Rossi DJ, Jamieson CH, Weissman IL (2008) Stems cells and the pathways to aging and cancer. Cell 132:681–696
21. Fridovich I (2004) Mitochondria: are they the seat of senescence? Aging Cell 3:13–16
22. Murphy MP (2009) How mitochondria produce reactive oxygen species. Biochem J 417:1–13
23. Chabi B, Ljubicic V, Menzies KJ et al (2008) Mitochondrial function and apoptotic susceptibility in aging skeletal muscle. Aging Cell 7:2–12
24. Finkel T, Deng CX, Mostoslavsky R (2009) Recent progress in the biology and physiology of sirtuins. Nature 460:587–591
25. Bonomini F, Tengattini S, Fabiano A, Bianchi R et al (2008) Atherosclerosis and oxidative stress. Histol Histopathol 23:381–390
26. Wild S, Roglic G, Green A et al (2004) Global prevalence of diabetes: estimates for the year 2000 and projections for 2030. Diabetes Care 27:1047–1053
27. Rochette L, Zeller M, Cottin Y, Vergely C (2014) Diabetes, oxidative stress and therapeutic strategies. Biochim Biophys Acta 1840:2709–2729

Chapter 25
Mitochondrial Dynamics Regulates Oxidative Metabolism in Leydig Tumor Cells

Cecilia Poderoso, Cristina Paz, Katia E. Helfenberger, and Ernesto J. Podestá

Abstract Steroidogenesis in Leydig cells is predominantly regulated by trophic hormone LH, whose signaling pathway involves PKA activation. PKA-mediated phosphorylation is involved in the induction and activation of steroidogenic acute regulatory (StAR) protein, which in turn activates the delivery of cholesterol from the outer to the inner mitochondrial membrane, the rate-limiting step in the biosynthesis of all steroid hormones. The access of cholesterol to the inner mitochondrial membrane allows its conversion to pregnenolone by cytochrome P450 side-chain cleavage enzyme. StAR-mediated access of cholesterol to the mitochondria and pregnenolone synthesis are common steps in the hormone-stimulated biosynthesis of steroid hormones. Mitochondria are cellular organelles with crucial roles including ATP synthesis, metabolic integration, ROS synthesis and apoptosis, all processes linked to steroidogenesis rates. Mitochondrial dynamics is then key to normal cell movement and function. In this context, this chapter focuses on reviewing the contribution of mitochondrial dynamics to Leydig cell function and metabolism, including efficient steroid production, and intends to highlight the mechanisms underlying mitochondrial changes and the correct localization of key mitochondrial proteins to achieve maximal steroidogenesis after hormone stimulation.

Keywords Mitochondria • Leydig cells • Mitofusin 2 • Steroidogenesis • Mitochondrial fusion • Protein kinases • MAM • ERK1/2 • Steroidogenic acute regulatory protein • Protein phosphatases

---

C. Poderoso (✉) • C. Paz • K.E. Helfenberger • E.J. Podestá
Departamento de Bioquímica Humana. Facultad de Medicina. Instituto de Investigaciones Biomédicas (INBIOMED, UBA-CONICET), Universidad de Buenos Aires, Ciudad Autónoma de Buenos Aires, Buenos Aires, Argentina
e-mail: cpoderoso@fmed.uba.ar

© Springer International Publishing Switzerland 2016
R.J. Gelpi et al. (eds.), *Biochemistry of Oxidative Stress*, Advances in Biochemistry in Health and Disease 16, DOI 10.1007/978-3-319-45865-6_25

1 Introduction

Steroid hormones are required for normal reproductive function and body homeostasis and are synthesized in steroidogenic cells of the adrenal gland, ovary, testis, placenta and brain under the regulation of trophic hormones. In adrenal cortex, adrenocorticotropin hormone (ACTH) stimulates zone fasciculata (ZF) cells while angiotensin II (Ang II) promotes steroidogenesis in zone glomerulosa (ZG) cells. Steroid synthesis in testicular Leydig and ovarian cells is stimulated by luteinizing hormone (LH) and follicle stimulating hormone (FSH), respectively. Steroidogenic hormones activate G protein-coupled receptors resulting in the activation of phospholipase C (Ang II) or adenylyl cyclase (ACTH, LH and FSH).

The principal role of Leydig cells is to produce androgens, which participate in the maturation of the sex organs and male secondary sex characteristics. In human, the main androgen produced is testosterone, which is rapidly synthesized and released from Leydig cells after LH stimulation. LH promotes an increase in cAMP levels and the consequent activation of cAMP-dependent protein kinase (PKA) [1–3] and protein phosphorylation, a pathway also activated by ACTH [4]. Thus, PKA-dependent protein phosphorylation of key regulatory proteins is necessary for the delivery of cholesterol from the outer (OMM) to the inner mitochondrial membrane (IMM), regarded as the rate-limiting step in steroid production [5–7]. The first series of steps in the biosynthesis of steroids hormones from cholesterol involves side-chain hydrolisis to yield a C_{21} intermediate, pregnenolone. This is the first enzymatic reaction in the synthesis of all steroids and is catalyzed by cytochrome P450scc located in the IMM, although it is the access of cholesterol to the active site of this enzyme that determines the rate of steroids production. In the gonads, pregnenolone may be converted to progesterone (also C_{21}) in a pathway called the Δ^4 pathway.

The transport of cholesterol across the intermembrane space is tightly regulated and provides the substrate for all steroid hormones synthesis [8, 9]. Therefore, an important issue to elucidate in the field of endocrinology is the link between PKA-dependent phosphorylation and the regulation of cholesterol transport. Studies in this issue have been conducted using mouse and rat primary cultured Leydig cells and also several murine cell lines. More than 30 years ago, a report showed the establishment of a mouse Leydig tumor cell line (designated MA-10) which retained many of the properties of normal Leydig cells. These properties include the expression of a functional LH receptor (LHR) which translates in the cells the LH binding into an increase in cAMP and steroid biosynthesis. In spite of the loss of 17α-hydroxylase/C17–20 lyase (resulting in a shift from androgen to progestin production), MA-10 cells have gained wide acceptance as an appropriate model system to study the actions of LH on steroidogenesis and other aspects of the differentiated functions of Leydig cells [10]. Nowadays, MA-10 cells are one of the experimental model of choice when studying signaling transduction cascades stimulated by tropic hormones in Leydig cells. In the other hand, chorionic gonadotropin (CG) is also recognized by the LHR, thus the system MA-10 Leydig cells under the stimulus of either LH or human CG (hCG) is commonly used as experimental model. A permeant analogue of the second messenger, cAMP is also a frequent stimulus used for *in vitro* studies. A measurement of the steroidogenic capacity of this system includes the quantification of progesterone released to the culture media.

As the key point for the regulation of steroid hormone biosynthesis is that it is located in mitochondria; the process occurs within the organelle. Then the respiratory chain, oxidative phosphorylation and reactive oxigen species (ROS) production are intimately linked to steroid production. After many years of research on phosphoproteins that facilitate cholesterol transport from the OMM to the IMM, it is now well established that steroidogenic acute regulatory (StAR) protein is an obligatory cholesterol transporter. StAR is synthesized as a 37 kDa preprotein with a typical mitochondrial leader sequence; upon hormonal stimulation it is imported to the mitochondria to yield a mature 30 kDa mitochondrial protein [11, 12]. In Leydig cells, StAR expression is hormonally regulated by LH/hCG at different levels, like its activity [13, 14] and gene induction [11, 12, 15–19]. StAR mediates the rapid flow of cholesterol from the OMM to the IMM, enabling steroidogenic cells to make a large amount of steroids in a short period of time. A small amount of StAR has been reported to elicit cholesterol transport enough to achieve the maximal rate of steroid synthesis [20]. It is accepted that hormone stimulation of steroid synthesis in adrenal ZF, ZG, and testicular Leydig cells involves the release of arachidonic acid (AA) [21]. Subsequent AA metabolism by lipoxygenase or epoxygenase pathways has been implicated in the regulation of steroid synthesis in adrenal and Leydig cells through the induction of StAR [22, 23]. In MA-10 Leydig cells, our group has described a hormonally regulated pathway for the generation and export of AA in mitochondria which regulates StAR protein induction and steroid synthesis [24, 25]. Intramitochondrial AA is generated by the concerted action of acyl-coenzyme A (CoA) synthetase (Acsl4) and a mitochondrial acyl-CoA thioesterase (Acot2) and AA released from mitochondria participates, through its lipoxygenated metabolites, in the induction of StAR protein [22].

Mitochondrial dynamics allows mitochondrial replication, repair of defective mitochondria, selective elimination of depolarized mitochondria via mitophagy and propagation of intra-mitochondrial calcium waves [26]. Depending on cell type, mitochondria may be highly dynamic, thus undergoing frequent cycles of fusion and fission [27–30]. The two opposing events, fission and fusion, control mitochondrial morphology and are modulated by organelle-associated proteins and by energy substrates [31]. As mitochondrial morphology and function are crucial for central cellular metabolism and for specialized pathways, mitochondrial dynamics may be conceived to have a role in cholesterol transport to the IMM. This chapter then focuses on the role of mitochondrial fusion in steroid synthesis, the underlying involved mechanisms and the impact of mitochondrial rearrangements on the accurate intracellular localization of key steroidogenic proteins, as StAR and Acsl4. It is intended to highlight the impact of mitochondrial dynamics on the metabolism of steroidogenic cells and cholesterol transport, which are essential processes involved in embryogenesis, development, reproduction and salt balance.

2 Proteins Involved in Mitochondrial Dynamics

Recent advances in mitochondrial imaging have revealed that mitochondria are highly dynamic in many cellular types. They can undergo fission/fusion processes modulated by various mitochondria-associated proteins and also by conformational transitions in the IMM. Moreover, precise mitochondrial distribution can be achieved by their movement along the cytoskeleton, recruiting various connector and motor proteins. Such movement is evident in cell types ranging from yeast to mammalian cells and serves to direct mitochondria to specific regions of high ATP demand or to transport mitochondria destined for elimination. The existing data also demonstrate that several aspects of mitochondrial dynamics, morphology, regulation and intracellular organization are cell type-/tissue-specific.

In many types of cells like neurons, pancreatic cells and cardiac muscle cells (HL-1 cell line), the complex dynamics of mitochondria include fission, fusion, and small and large movements of mitochondria, branching in the mitochondrial network and long-distance intracellular translocation of mitochondria. Alternatively, mitochondria can be rather fixed in other cells and tissues, like adult cardiomyocytes or skeletal muscle [32]. These variations and the cell-type specificity of mitochondrial dynamics could be related to specific cellular functions and demands, also indicating a significant role of integrations of mitochondria with other intracellular systems like cytoskeleton, nucleus and endoplasmic reticulum (ER). Mammalian Mitofusin 1 (Mfn1) and Mitofusin 2 (Mfn2) belong to the GTPase family of proteins [33] and are required for OMM fusion [34]. Mfn1 and 2 are implicated in the modulation of mitochondria–mitochondria and ER–mitochondria interactions. Both proteins are located on the OMM, mediating mitochondrial fusion in concert with another GTPase, optic atrophy 1 (OPA1), on the IMM. Mfn1 and 2 are extensively expressed, as demonstrated in brain (mainly Mfn2), liver, adrenal glands and testis and it has been proven that Mfn2 is enriched at contact sites between ER and mitochondria, which are named mitochondria-associated ER membranes or MAM [35]. Indeed, this specific site is considered as a unique subdomain with a vast importance in regulation of calcium signaling, mitochondrial bioenergetics, apoptosis and lipid metabolism [35–38]. Even though the existence and function of MAM have been extensively studied, little is known about the mechanisms triggering its formation. Interestingly, a key enzyme in steroidogenesis such as Acsl4 is localized in the MAM subdomain [39]. Ubiquitous knockout of the Mfn1 or Mfn2 gene results in embryonic lethality during midgestation [40] due to placental dysfunction. Mice ubiquitously lacking Mfn1 are apparently healthy, whereas mice without Mfn2 die in the early postnatal period and show severe defects in movement and balance [40]. These data suggest that Mfn1 and 2 functions are not redundant even though both proteins are mandatory in mitochondrial fusion development.

Mitochondrial fission requires dynamin-related protein 1 (Drp1), a cytosolic protein which is recruited to the OMM by a poorly characterized multiprotein complex. Drp1 is required for mitochondrial division in mammalian cells. Alterations in mitochondrial dynamics are associated with maintaining stemness properties of mammalian epithelial stem-like cells [41]. Moreover, Drp1 is a newly discovered therapeutic target for tumor initiation, migration, proliferation, and chemosensitivity [42–45]. In 2013, Zhao et al. reported that cancer cell migration and invasion were regulated by mitochondrial dynamics [42]. Recently, Han et al. reported that Drp1-dependent mitochondrial fission not only regulates hypoxia-induced migration of breast cancer cells but also facilitates its sensitivity to chemotherapeutic agents [46, 47]. In addition, work by Merrill and colleagues in neurons has shown that PKA recruitment to mitochondria results in mitochondrial elongation as consequence of the phosphorylation-mediated inhibition of Drp1 [48]. On the other hand, Drp1 phosphorylation by PKCδ at Ser 579 increases mitochondrial fragmentation [49]. These results idicate that phosphorylation of key mitochondrial dynamics proteins plays a crucial role in mitochondrial morphology maintenance.

3 Mitochondrial Dynamics in Physiologic and Pathologic Metabolism

Fusion and fission continuously change mitochondrial shape under physiological (e.g. cellular division) [50, 51] and pathophysiological (*e.g.* apoptosis, various stresses) conditions [29, 52]. Defects in fusion frequently cause fragmentation of the mitochondrial network [53, 54], whereas the decrease in mitochondrial fission results in formation of the network of excessively elongated and interconnected organelles [55]. Under physiological conditions, however, mitochondrial fission is counteracted by fusion causing balance and mitochondrial network stability [51]. For example, the increase in Mfn1 and 2 in human skeletal muscle during exercise is described as a mechanism that may collaborate against insulin resistance syndrome [56]. It is also known that alterations in the metabolic conditions of the cell can remarkably modulate fusion and fission machineries and mitochondria may adapt frequently to changing physiological conditions and modify their behavior accordingly [31, 57]. Experiments performed in mouse embryonic fibroblasts (MEFs) cell lines generated from immortalized Mfn1 and Mfn2 knockout mice, have shown that hetero-oligomeric complexes between Mfn1 and Mfn2 are important for control of mitochondrial fusion [58]. Despite its well-established role in mitochondrial fusion, a growing body of evidence suggests that Mfn2 has additional functions, such as ER–mitochondria tethering [35] and control of mitochondrial transport in axons [59], which may explain some of the disease manifestations in humans with Mfn2 mutations [60]. The different functions of Mfn2 are supported by *in vivo* studies in mice, showing that Mfn2 is required for normal glucose homeostasis [61] and axonal projections of dopaminergic neurons [62, 63]. In addition, mutations in Mfn2 cause Charcot-Marie-Tooth type 2A (CMT2A), an autosomal dominant disease, characterized by chronic axonal neuropathy [60, 64, 65]. As mitochondrial mobility and transport are key elements to the functional health of the extended neuronal axons, particularly in peripheral nerves, an alteration in these mitochondrial parameters could be a possible mechanism of action in CMT2A [60].

It is widely recognized that mitochondrial dysfunction contributes to impaired myocardial energetics and increased oxidative stress in cardiomyopathies, cardiac ischemic damage and heart failure (HF). Existing studies have shown that defects in mitochondrial dynamics affecting biogenesis and turnover are linked to cardiac senescence and HF which proves the relevance of the study of mitochondrial dynamics to develop potential strategies for targeting mitochondria and their utility in HF therapy [66]. Mfn2 deficiency reduces fuel oxidation and mitochondrial membrane potential and represses nuclear-encoded subunits of the respiratory chain complexes I, II, III and IV. This Mfn2 loss-of–function could participate in triggering the pathological mechanisms that cause disease in highly oxidative tissues such as nervous system, skeletal muscle or heart [67]. However, the molecular basis for the mitochondrial dysfunction, caused by loss of Mfn2, is not completely understood. Recently, it has been shown that Mfn2 knockout of MEFs cell lines causes depletion of mitochondrial coenzyme Q, which, in turn, leads to respiratory chain dysfunction. Addition of coenzyme Q to these Mfn2 knockout cells or isolated mitochondria partially rescues the respiratory chain dysfunction, which suggests a therapeutic strategy for intervention in affected patients [68]. Mitochondrial dynamics also participates in the regulation of apoptosis. Mfn2 expression has an apoptotic effect in the mitochondrial apoptotic pathway [40] and Mfn2 deficiency profoundly increases the mitochondrial Bax/Bcl-2 ratio, indicating that Mfn2 deficiency potentially induces apoptosis in the mouse embryo through the Bcl-2 and Bax pathway [69].

4 Fusion of Mitochondria in Leydig Cells and Steroid Biosynthesis

Despite the importance of mitochondrial dynamics in several metabolic and hormonal conditions, very little is known about its role in steroidogenic tissues. As a first approach, a paper published 30 years ago showed that in adrenal cells, mitochondria displayed a change in morphology and intracellular localization after hormone stimulation by cAMP [70]. More recently, mitochondria were shown to move across H295R adrenocortical cells after ACTH stimulation in a PKA-dependent manner. The authors investigated the role of ACTH/cAMP-stimulated mitochondrial trafficking in regulating cortisol production and observed that this microtubule-dependent movement could be involved in the transport of substrates between the ER and mitochondria [71]. Besides fusion and fission, mitochondrial biogenesis is a central process in stimulated-steroidogenesis in Leydig cells. Under stress conditions, mitochondrial biogenesis elicits an adaptative response essential to maintain testosterone production and preserve basic steroidogenesis activity [72]. Our group has demonstrated that steroid synthesis depends on changes in mitochondrial fusion which is regulated by hormone levels. In the experiments, two categories were clearly distinguished: punctuated and fused mitochondria. Also, control cells were observed to present mainly a punctuated pattern that changed to the fused type after hormonal stimulation, a finding confirmed through electron microscopy. In basal conditions, mitochondria have an orthodox structure with narrow cristae and a more spherical shape, in the punctuated shape category described above. In hormone-stimulated cells, mitochondria are larger and appear elongated and tubular as compared to the round mitochondria found in control cells. Interestingly, ultrastructural studies show that large portions of filamentous ER appear in close proximity of the enlarged mitochondria in steroid producing cells. Interestingly, our group demonstrated for the first time that Mfn2 is promptly up-regulated after steroidogenic stimuli, suggesting that mitochondrial dynamics might be involved in steroidogenesis. Furthermore, blocking mitochondrial fusion by knocking down Mfn2 expression has a negative impact on steroid synthesis [73]. In agreement with the work of Li and Sewer, mitochondrial fusion might represent a limiting step in the onset of processes that require transport of intermediate products, e.g. liposoluble steroid hormones between organelles, mitochondria and ER probably involving MAM subdomains.

It is known that phospho/dephosphorylation events are obligatory in cholesterol transport to the IMM, and that PKA is one of the proteins required for this process. Furthermore, our group has determined that mitochondrial fusion depends on PKA activity. Extracellular-regulated kinase (ERK1/2) activity is essential for steroidogenesis; however, the inhibition of ERK1/2 phosphorylation had no effect on fused mitochondria under hCG or cAMP stimulation [73] which indicates that mitochondrial fusion is an event upstream ERK1/2 phosphorylation but downstream PKA activity in Leydig cells (Fig. 25.1).

5 ATP Synthesis and Mitochondrial Electrochemical Membrane Potential in Mitochondrial Fusion and Steroidogenesis

It has been extensively documented that the maintenance of adequate mitochondrial values of electrochemical membrane potential ($\Delta\Psi m$) and pH as well as an active ATP synthesis are required for acute steroid biosynthesis. Pharmacological agents known as mitochondrial disrupters, such as carbonyl cyanide m-chlorophenylhydrazone (CCCP) and valinomycin, abolish the mitochondrial H^+ gradient and directly affect mitochondrial ATP production, by uncoupling electron transport from ATP synthesis. Studies performed using such agents have led to the conclusion that mitochondria must be energized, polarized, and actively respiring to support Leydig cell steroidogenesis [74]. Also, it is now well recognized that alterations in mitochondrial functional conditions are reflected in the rate of steroid production. $\Delta\Psi m$ is one of the most important indicators of mitochondrial function in cells [75] and affects steroid synthesis by blocking protein processing into mitochondria. [74, 76]. King et al. analyzed the effects of disruption of the $\Delta\Psi m$ by CCCP and

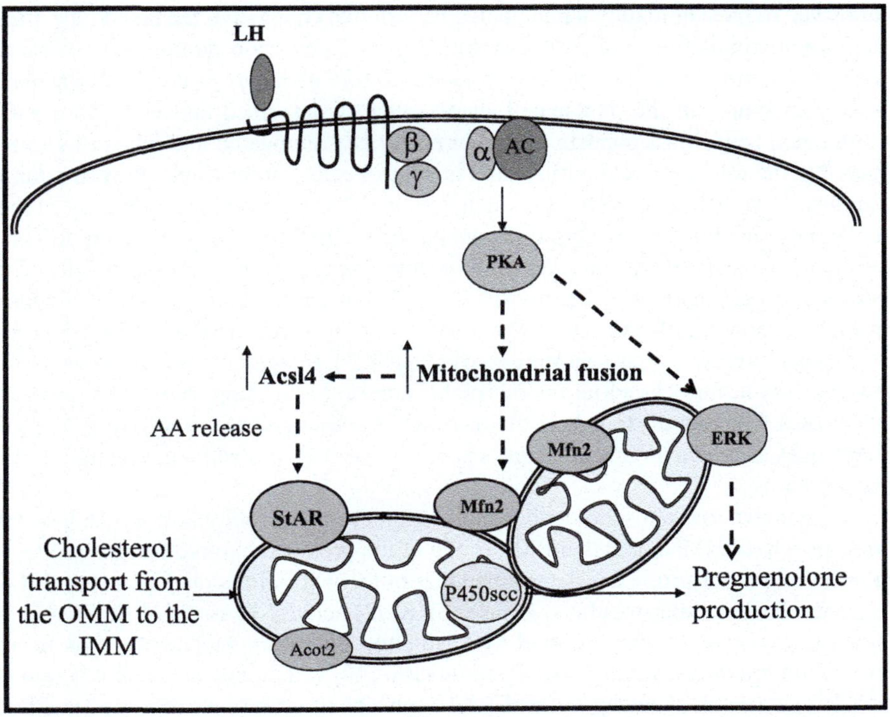

Fig. 25.1 Scheme of the LH cascade showing the recruitment of StAR to the mitochondria, increase in Acsl4 levels; activation of PKA and ERK, and mitochondrial fusion involvement on steroidogenesis. α, β, γ: subunits of the G-protein coupled to the LH receptor. *AC* adenylyl ciclase

valinomycin on progesterone production by MA-10 Leydig cells and showed that both steroidogenesis and mitochondrial import of StAR are inhibited by these agents. The study concludes that mitochondrial membrane potential is obligatory for StAR import into the mitochondria and, in turn, for steroidogenesis [76]. Mitochondrial dynamics is also regulated by the electron transport chain. The pumping of protons across the IMM provokes a typical pH gradient in mitochondria which is a measurement of mitochondrial function and is essential for mitochondrial fusion. The dissipation of the H^+ gradient inhibits fusion, likely by either the selective proteasome-dependent turnover of Mfn1 and Mfn2 or proteolytic processing of OPA1 [77]. In addition, mitochondrial fusion requires an intact $\Delta\Psi m$ in several cell lines [76, 78]. In MA-10 Leydig cells the disruption of $\Delta\Psi m$ by CCCP and valinomycin decreases mitochondrial fusion and steroidogenesis. The CCCP and valinomycin effects are exerted by a cAMP-stimulated pathway in MA-10 Leydig cells but not in control cells [73]. While CCCP-treated cells maintained a punctuated mitochondrial shape similar to that of control cells, even after cAMP treatment, valinomycin treatment led to a different mitochondrial rearrangement, previously described as "swelling shape". This type of mitochondrial rearrangement has been previously associated with severe mitochondrial dysfunction [79]. The inhibition of mitochondrial fusion by CCCP suggests that mitochondrial integrity is strictly necessary for mitochondrial fusion after cAMP treatment. The observed inhibitory effect of CCCP is transient, as the removal of the compound restores cAMP-stimulated mitochondrial fusion and steroid production. This indicates that CCCP does not affect irreversibly any other mitochondrial or cellular parameter. Just as the disruption of $\Delta\Psi m$ has a negative impact on mitochondrial fusion, Mfn2 deficiency results in a low $\Delta\Psi m$ and failure to meet cellular activity demands, which indicates that $\Delta\Psi m$ and mitochondrial fusion are intimately linked to support the cellular metabolic activity and that any alteration in one of the processes will affect the other.

The relevance of ATP synthesis in the mitochondrial fusion onset stimulated by cAMP has been analyzed through the use of oligomycin, a very well known mitochondrial ATP synthesis inhibitor. The presence of oligomycin in the culture media exerts no inhibitory effect on mitochondrial fusion in MA-10 Leydig cells, which suggests that mitochondrial fusion requires an intact $\Delta\Psi m$, as described previously but it is independent of ATP synthesis [73]. In this regard, it has been reported that GTP and not ATP is required for *in vitro* mitochondrial fusion in yeast [80]. Given that ATP is required for steroid synthesis [81], the results from our group suggest that mitochondrial fusion is a previous step in steroidogenesis.

Reactive oxygen species (ROS) are involved in a variety of pathophysiological conditions in the testis, and oxidative stress is known to inhibit ovarian and testicular steroidogenesis. Diemer et al. showed that ROS caused a significant dissipation of $\Delta\Psi m$ and an oxidative stress mediated alteration of mitochondrial function, that results in the inhibition of StAR protein expression, its processing and of cholesterol transfer activity in the corpus luteum and MA-10 Leydig cells. These findings confirm earlier studies that indicate the requirement of an intact $\Delta\Psi m$ for StAR protein function in cholesterol transport. The significant reduction in the 30-kDa mature form of StAR, cessation of cholesterol transport, and loss of $\Delta\Psi m$ are consistent with a marked mitochondrial perturbation by oxidative stress, a condition that is probably involved in testicular pathophysiological events such as infection, reperfusion injury, aging, cryptorchidism, and varicocele [82].

6 Participation of Mfn2 in the Localization of Reticular and Mitochondrial Proteins After Hormone Stimulation

The presence of a multi-protein complex in the OMM has functional repercussions for steroidogenesis in Leydig cells. It has been proposed that cholesterol transfer from OMM to IMM occurs at specialized contact sites bridging the two membranes where several crucial proteins reside. An 800-kDa mitochondrial bioactive complex called transduceosome has been described to contain the OMM translocator protein (18 kDa, TSPO), the voltage-dependent anionic channel (VDAC), the cytochrome P450scc (CYP11A1), ATPase family AAA domain-containing protein 3A

(ATAD3A) and OPA1. These results identify a bioactive, multimeric protein complex spanning the OMM and IMM unit that is responsible for the hormone-induced import, segregation, targeting, and metabolism of cholesterol [83]. These participant proteins are essential for cholesterol transport to the IMM and steroidogenesis, in particular VDAC type1 interacts with bioactive phosphorylated StAR on the OMM, facilitating its activity [84]. It is worth mentioning that the role of TSPO in steroid biosynthesis has been recently ruled out [85].

The reorganization of organelles and contact between membranes can be a primary process in steroid production and secretion through the plasma membrane, as the localization of several enzymes is the clue to ensure appropriate steroidogenesis rates. Out group described the role of members of the mitogen-activated kinases (MAPKs) family, ERK1/2 and its upstream kinase MEK1/2 in the hCG/LH stimulation of StAR protein activity and steroid synthesis. Several groups, including our group, have shown that ERK1/2 and MEK1/2 are targeted to mitochondria in different tissues, particularly to the OMM [86, 87]. Our work has demonstrated a temporal mitochondrial ERK1/2 activation, which is obligatory for PKA-mediated steroidogenesis in MA-10 Leydig cells [88]. StAR gene transcription is increased in a PKA-dependent manner [89, 90], although other kinases such as ERK1/2 also regulate StAR protein levels and steroidogenesis by genomic and non-genomic effects [91–96]. Examining StAR protein structure, our group found a consensus sequence that would allow the docking of StAR protein to ERK1/2 and a consensus site for ERK1/2 phosphorylation. In support of a predicted StAR-ERK interplay, a direct interaction was observed between ERK1/2 and StAR proteins and the phosphorylation of StAR by ERK *in vitro*. Our studies showed that, in a cell free assay, several sequential events in mitochondria such as PKA-dependent MEK phosphorylation, ERK1/2 activity and StAR protein phosphorylation, increase cholesterol transport to mitochondria and progesterone synthesis [88]. Mitochondrial ERK1/2 thus has a functional interaction with StAR protein, MEK1/2 and PKA, suggesting the relevance of the above mentioned multiprotein complex activity. Indeed, StAR protein is phosphorylated by ERK in the serine at the position 232 and the mutation of this specific serine abrogates StAR phosphorylation and function [88]. The increase in StAR mitochondrial levels after hCG and cAMP stimulation in MA-10 Leydig cells is significantly diminished when the Mfn2 protein is knocked down by means of a specific interference RNA, with no effects on StAR basal levels. A similar result has been observed for phosphorylated mitochondrial ERK and, as expected, the knock down of Mfn2 also inhibited progesterone biosynthesis [97]. Mitochondrial fusion reduction correlates with a decrease in both mitochondrial StAR protein and in StAR mRNA levels. A direct role of Mfn2 on StAR gene expression is unknown to date, although a previous report shows that overexpression or knockout of Mfn2 modulates and affects mRNA levels of several mitochondrial proteins [98, 99], then a stabilizing effect of mitochondrial fusion on StAR mRNA cannot be excluded. Interestingly, a report has proposed that levels of the different transcripts of StAR mRNA are regulated by stabilization mediated by cAMP in MA-10 cells [100, 101]. StAR mRNA binds a mitochondrial AKAP1 stabilizing the translational complex at this organelle. Therefore, mitochondrial fusion could mediate the interaction between StAR mRNA and the AKAP1 in mitochondria, thus stabilizing and increasing StAR mRNA levels.

Another key enzyme in steroid synthesis regulation is Acsl4, previously mentioned in this chapter. Interestingly, our work has demonstrated that mitochondrial fusion is required for the localization of Acsl4 at the MAM subdomain after hormone stimulation [97]. Therefore, mitochondrial fusion could participate in StAR induction through Acsl4 activity and localization at the MAM subdomain. In this way, Mfn2 is rendered necessary for the complete ensemble of the mitochondrial multiprotein complex as a crucial event for Leydig cells endocrine function.

7 Phosphatases Involved in Steroidogenesis and Mitochondrial Fusion Regulation

Although it is generally agreed that tyrosine phosphorylation is regulated by the balanced action of protein tyrosine kinases and protein tyrosine phosphatases (PTPs), proportionately much more research has focused on protein tyrosine kinases than on their counteracting phosphatases. Important findings have now led to the emerging recognition that PTPs play specific and active dominant roles in setting the levels of tyrosine phosphorylation in cells and in the regulation of many physiological processes [102]. Whereas PTPs were initially regarded as household enzymes with constitutive activity and capable of all substrate dephosphorylation, evidence in favor of a tight regulation of PTP activity by various mechanisms is now accumulating. Like protein phosphorylation, dephosphorylation by PTPs is required in each cell compartment in a specific manner. Protein-protein interaction domains and compartment-specific targeting domains in PTPs serve to achieve the required PTP localization [102].

Src homology domain (SH2) containing PTP (SHP2) is a ubiquitously expressed, non-receptor, classical PTP that plays an essential role in many organisms from lower eukaryotes to mammals [103]. In addition, SHP2 is one of the PTPs that promote the activation, rather than the down-regulation, of intracellular signaling pathways [104]. Rocchi et al. have demonstrated that SHP2 expression in bovine adrenocortical cells and its activation by ACTH throught PKA-dependent phosphorylation [105]. SHP2 is also expressed in MA-10 Leydig cells, where it plays an obligatory role in steroidogenesis [106]. The specific pharmacological inhibition of SHP2 or its downregulation by interference RNA reduces Acsl4 protein and mRNA levels, while the overexpression of an active form of SHP2 increases Acsl4 protein levels in MA-10 Leydig cells. It is worth pointing out that SHP2 has to be activated through a cAMP-dependent pathway to exert its effect on Acsl4 levels. SHP2 participates in the production and metabolism of AA through the action on Acsl4 and consequently in the steroidogenic capacity of MA-10 Leydig cells. Indeed, the overexpression or knockdown of SHP2 leads to increased or decreased steroid production, respectively [106]. Interestingly, SHP2 is a crucial partner in hormone

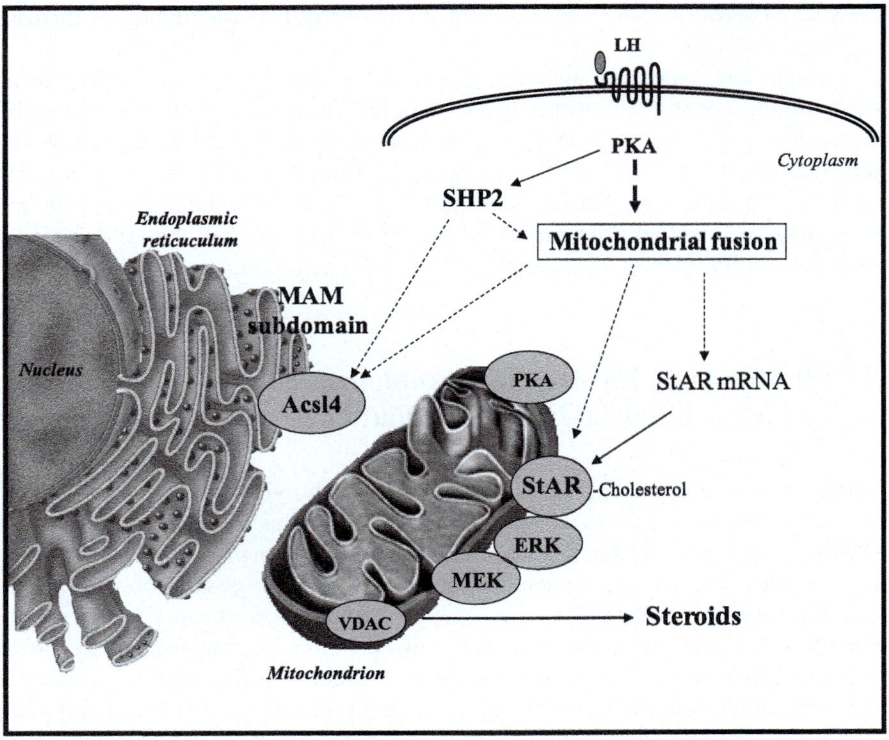

Fig. 25.2 Model of the proposed interactions between organelles and key steroidogenic proteins mediated by mitochondrial fusion activity. Kinases and phosphatases involved in stimulated steroid synthesis are shown.

induced mitochondrial fusion in MA-10 Leydig cells. SHP2 is at least one of the PTP that participates in mitochondrial fusion, as the knockdown of the phosphatase abrogates mitochondrial fusion stimulated by hCG/cAMP [73] (Fig. 25.2).

Steroidogenic hormones promote a notable change in cell shape involving the reorganization of the actin cytoskeleton, which precedes steroid biosynthesis and secretion [107]. Paxillin is a focal adhesion protein that is rapidly tyrosine dephosphorylated after the ACTH/cAMP cascade activation in adrenocortical cells [108, 109]. Moreover, ACTH promotes SHP2 activation in a PKA-dependent manner [105] and paxillin is dephosphorylated by SHP2 in MCF-7 cells to regulate cell motility [110]. Thus, it is conceivable that SHP2, through paxillin dephosphorylation, modulates the actin cytoskeleton with a direct relationship with mitochondrial subcellular distribution and fusion. In addition, SHP2 knockdown reduces the recruitment of ERK1/2 to mitochondria and its phosphorylation [73], suggesting that SHP2 activity affects mitochondrial fusion and, consequently ERK recruitment and activation in mitochondria. Mitochondrial ERK is transiently activated after hormone stimulation in MA-10 Leydig cells [88]. Our group has reported that hormone stimulation of these cells promotes the induction of several members of MAP kinase phosphatase (MKP) family [111, 112]. As one of the MAP kinases, MKP-1 is capable of dephosphorylating ERK1/2 and other MAP kinases. MKP-1 protein levels are rapidly increased after stimulation not only in the cytosol but also in the mitochondria of MA-10 Leydig cells [111]. The kinetic profile of MKP-1 accumulation into mitochondria correlates with ERK1/2 dephosphoryation in this organelle, suggesting that MKP-1 induction could impact steroid synthesis through dephosphorylation and inactivation of ERK1/2 and the cessation of ERK-mediated StAR phosphorylation. On the other hand, MKP-1 accumulation outside mitochondria contributes to the turn-off of the hormonal signal. Specifically, the rate of several ERK-dependent processes necessary for Leydig cell functions, as StAR induction and consequent steroid production, are decreased in the same temporal frame of MKP-1 accumulation [111].

8 Conclusions

In this chapter, a large body of evidence is presented on how mitochondrial fusion operates in steroidogenic Leydig cells and the underlying mechanisms. Many of the kinases mentioned here are endogenously localized in the mitochondria without a specific mitochondrial leader peptide. Although there are anchorage proteins for these kinases, how they translocate to different subcellular compartements remains elusive. Then, mitochondrial dynamics, and particularly fusion, might be offering the chance to tether mitochondria to these crucial kinases to conform the mitochondrial complex and regulate steroid biosynthesis onset. The changes observed in mitochondrial fusion might also be central for the formation of the mitochondrial multiprotein complex which delivers cholesterol to the P450scc system. This, in turn, suggests that the dynamics of organelles might represent a limiting step in the onset of processes that require transport of intermediate products between them.

References

1. Ascoli M (1996) Contemporary endocrinology: the leydig cell in health and disease. In: Contemporary endocrinology: the leydig cell in health and disease. Humana Press Inc. Totowa, pp 373–381
2. Cooke BA, Choi MC, Dirami G et al (1992) Control of steroidogenesis in Leydig cells. J Steroid Biochem Mol Biol 43:445–449
3. Saez JM (1994) Leydig cells: endocrine, paracrine, and autocrine regulation. Endocr Rev 15:574–626
4. Garren LD, Gill GN, Masui H et al (1971) On the mechanism of action of ACTH. Recent Prog Horm Res 27:433–478
5. Crivello JF, Jefcoate CR (1980) Intracellular movement of cholesterol in rat adrenal cells. Kinetics and effects of inhibitors. J Biol Chem 255:8144–8151
6. Crivello JF, Gill GN (1983) Induction of cultured bovine adrenocortical zona glomerulosa cell 17-hydroxylase activity by ACTH. Mol Cell Endocrinol 30:97–107

7. Jefcoate CR, McNamara BC, Artemenko I et al (1992) Regulation of cholesterol movement to mitochondrial cytochrome P450scc in steroid hormone synthesis. J Steroid Biochem Mol Biol 43:751–767
8. Privalle CT, Crivello JF, Jefcoate CR (1983) Regulation of intramitochondrial cholesterol transfer to side-chain cleavage cytochrome P-450 in rat adrenal gland. Proc Natl Acad Sci U S A 80:702–706
9. Jefcoate C (2002) High-flux mitochondrial cholesterol trafficking, a specialized function of the adrenal cortex. J Clin Invest 110:881–890
10. Ascoli M (1981) Characterization of several clonal lines of cultured Leydig tumor cells: gonadotropin receptors and steroidogenic responses. Endocrinology 108:88–95
11. Stocco DM, Clark BJ (1996) Regulation of the acute production of steroids in steroidogenic cells. Endocr Rev 17:221–244
12. Stocco DM, Sodeman TC (1991) The 30-kDa mitochondrial proteins induced by hormone stimulation in MA-10 mouse Leydig tumor cells are processed from larger precursors. J Biol Chem 266:19731–19738
13. Arakane F, King SR, Du Y et al (1997) Phosphorylation of steroidogenic acute regulatory protein (StAR) modulates its steroidogenic activity. J Biol Chem 272:32656–32662
14. LeHoux JG, Fleury A, Ducharme L et al (2004) Phosphorylation of the hamster adrenal steroidogenic acute regulatory protein as analyzed by two-dimensional polyacrylamide gel electrophoresis. Mol Cell Endocrinol 215:127–134
15. Clark BJ, Wells J, King SR et al (1994) The purification, cloning, and expression of a novel luteinizing hormone-induced mitochondrial protein in MA-10 mouse Leydig tumor cells. Characterization of the steroidogenic acute regulatory protein (StAR). J Biol Chem 269:28314–28322
16. Clark BJ, Soo SC, Caron KM et al (1995) Hormonal and developmental regulation of the steroidogenic acute regulatory protein. Mol Endocrinol 9:1346–1355
17. Manna PR, Dyson MT, Jo Y et al (2009) Role of dosage-sensitive sex reversal, adrenal hypoplasia congenita, critical region on the X chromosome, gene 1 in protein kinase A- and protein kinase C-mediated regulation of the steroidogenic acute regulatory protein expression in mouse Leydig tumor cells: mechanism of action. Endocrinology 150:187–199
18. Stocco DM, Wang X, Jo Y et al (2005) Multiple signaling pathways regulating steroidogenesis and steroidogenic acute regulatory protein expression: more complicated than we thought. Mol Endocrinol 19:2647–2659
19. Castillo AF, Fan J, Papadopoulos V et al (2011) Hormone-dependent expression of a steroidogenic acute regulatory protein natural antisense transcript in MA-10 mouse tumor Leydig cells. PLoS One 6:e22822
20. Artemenko IP, Zhao D, Hales DB et al (2001) Mitochondrial processing of newly synthesized steroidogenic acute regulatory protein (StAR), but not total StAR, mediates cholesterol transfer to cytochrome P450 side chain cleavage enzyme in adrenal cells. J Biol Chem 276:46583–46596
21. Castilla R, Gadaleta M, Castillo AF et al (2008) New enzymes involved in the mechanism of action of epidermal growth factor in a clonal strain of Leydig tumor cells. Endocrinology 149:3743–3752
22. Wang XJ, Shen CL, Dyson MT et al (2006) The involvement of epoxygenase metabolites of arachidonic acid in cAMP-stimulated steroidogenesis and steroidogenic acute regulatory protein gene expression. J Endocrinol 190:871–878
23. Wang XJ, Dyson MT, Jo Y et al (2003) Involvement of 5-lipoxygenase metabolites of arachidonic acid in cyclic AMP-stimulated steroidogenesis and steroidogenic acute regulatory protein gene expression. J Steroid Biochem Mol Biol 85:159–166
24. Maloberti P, Castilla R, Castillo AF et al (2005) Silencing the expression of mitochondrial acyl-CoA thioesterase I and acyl-CoA synthetase 4 inhibits hormone-induced steroidogenesis. FEBS J 272:1804–1814
25. Castillo AF, Cornejo Maciel F, Castilla R et al (2006) cAMP increases mitochondrial cholesterol transport through the induction of arachidonic acid release inside this organelle in Leydig cells. FEBS J 273:5011–5021

26. Martin SJ (2011) Mitochondrial fusion: bax to the fussure. Dev Cell 20:142–143
27. Chan DC (2006) Mitochondrial fusion and fission in mammals. Annu Rev Cell Dev Biol 22:79–99
28. Herzig S, Martinou JC (2008) Mitochondrial dynamics: to be in good shape to survive. Curr Mol Med 8:131–137
29. Karbowski M, Youle RJ (2003) Dynamics of mitochondrial morphology in healthy cells and during apoptosis. Cell Death Differ 10:870–880
30. Yaffe MP (1999) Dynamic mitochondria. Nat Cell Biol 1:149–150
31. Rossignol R, Gilkerson R, Aggeler R et al (2004) Energy substrate modulates mitochondrial structure and oxidative capacity in cancer cells. Cancer Res 64:985–993
32. Mishra P, Varuzhanyan G, Pham AH et al (2015) Mitochondrial dynamics is a distinguishing feature of skeletal muscle fiber types and regulates organellar compartmentalization. Cell Metab 22:1033–1044
33. Bourne HR, Sanders DA, McCormick F (1990) The GTPase superfamily: a conserved switch for diverse cell functions. Nature 348:125–132
34. Rojo M, Legros F, Chateau D et al (2002) Membrane topology and mitochondrial targeting of mitofusins, ubiquitous mammalian homologs of the transmembrane GTPase Fzo. J Cell Sci 115:1663–1674
35. de Brito OM, Scorrano L (2009) Mitofusin-2 regulates mitochondrial and endoplasmic reticulum morphology and tethering: the role of Ras. Mitochondrion 9:222–226
36. Vance JE, Vance DE (2004) Phospholipid biosynthesis in mammalian cells. Biochem Cell Biol 82:113–128
37. Hayashi T, Rizzuto R, Hajnoczky G et al (2009) MAM: more than just a housekeeper. Trends Cell Biol 19:81–88
38. Rizzuto R, Pinton P, Brini M et al (1999) Mitochondria as biosensors of calcium microdomains. Cell Calcium 26:193–199
39. Lewin TM, Van Horn CG, Krisans SK et al (2002) Rat liver acyl-CoA synthetase 4 is a peripheral-membrane protein located in two distinct subcellular organelles, peroxisomes, and mitochondrial-associated membrane. Arch Biochem Biophys 404:263–270
40. Chen H, Detmer SA, Ewald AJ et al (2003) Mitofusins Mfn1 and Mfn2 coordinately regulate mitochondrial fusion and are essential for embryonic development. J Cell Biol 160:189–200
41. Katajisto P, Dohla J, Chaffer CL et al (2015) Stem cells. Asymmetric apportioning of aged mitochondria between daughter cells is required for stemness. Science 348:340–343
42. Zhao J, Zhang J, Yu M et al (2013) Mitochondrial dynamics regulates migration and invasion of breast cancer cells. Oncogene 40:4814–4824
43. Ferreira-da-Silva A, Valacca C, Rios E et al (2015) Mitochondrial dynamics protein Drp1 is overexpressed in oncocytic thyroid tumors and regulates cancer cell migration. PLoS One 10:e0122308
44. Qian W, Choi S, Gibson GA et al (2012) Mitochondrial hyperfusion induced by loss of the fission protein Drp1 causes ATM-dependent G2/M arrest and aneuploidy through DNA replication stress. J Cell Sci 125:5745–5757
45. Wang JX, Li Q, Li PF (2009) Apoptosis repressor with caspase recruitment domain contributes to chemotherapy resistance by abolishing mitochondrial fission mediated by dynamin-related protein-1. Cancer Res 69:492–500
46. Han XJ, Yang ZJ, Jiang LP et al (2015) Mitochondrial dynamics regulates hypoxia-induced migration and antineoplastic activity of cisplatin in breast cancer cells. Int J Oncol 46:691–700
47. Jia Y, Zhou L, Tian C et al (2015) Dynamin-related protein 1 is involved in micheliolide-induced breast cancer cell death. Onco Targets Ther 8:3371–3381
48. Merrill RA, Dagda RK, Dickey AS et al (2011) Mechanism of neuroprotective mitochondrial remodeling by PKA/AKAP1. PLoS Biol 9:e1000612
49. Qi X, Disatnik MH, Shen N et al (2011) Aberrant mitochondrial fission in neurons induced by protein kinase C{delta} under oxidative stress conditions in vivo. Mol Biol Cell 22:256–265

50. Bereiter-Hahn J, Voth M (1994) Dynamics of mitochondria in living cells: shape changes, dislocations, fusion, and fission of mitochondria. Microsc Res Tech 27:198–219
51. Mozdy AD, Shaw JM (2003) A fuzzy mitochondrial fusion apparatus comes into focus. Nat Rev Mol Cell Biol 4:468–478
52. Perfettini JL, Roumier T, Kroemer G (2005) Mitochondrial fusion and fission in the control of apoptosis. Trends Cell Biol 15:179–183
53. Griparic L, van der Wel NN, Orozco IJ et al (2004) Loss of the intermembrane space protein Mgm1/OPA1 induces swelling and localized constrictions along the lengths of mitochondria. J Biol Chem 279:18792–18798
54. Chen H, Chan DC (2005) Emerging functions of mammalian mitochondrial fusion and fission. Hum Mol Genet 14:R283–R289
55. Smirnova E, Griparic L, Shurland DL et al (2001) Dynamin-related protein Drp1 is required for mitochondrial division in mammalian cells. Mol Biol Cell 12:2245–2256
56. Cartoni R, Leger B, Hock MB et al (2005) Mitofusins 1/2 and ERRalpha expression are increased in human skeletal muscle after physical exercise. J Physiol 567:349–358
57. Benard G, Bellance N, James D et al (2007) Mitochondrial bioenergetics and structural network organization. J Cell Sci 120:838–848
58. Detmer SA, Chan DC (2007) Functions and dysfunctions of mitochondrial dynamics. Nat Rev Mol Cell Biol 8:870–879
59. Misko A, Jiang S, Wegorzewska I et al (2010) Mitofusin 2 is necessary for transport of axonal mitochondria and interacts with the Miro/Milton complex. J Neurosci 30:4232–4240
60. Zuchner S, Mersiyanova IV, Muglia M et al (2004) Mutations in the mitochondrial GTPase mitofusin 2 cause Charcot-Marie-Tooth neuropathy type 2A. Nat Genet 36:449–451
61. Sebastian D, Hernandez-Alvarez MI, Segales J et al (2012) Mitofusin 2 (Mfn2) links mitochondrial and endoplasmic reticulum function with insulin signaling and is essential for normal glucose homeostasis. Proc Natl Acad Sci U S A 109:5523–5528
62. Lee S, Sterky FH, Mourier A et al (2012) Mitofusin 2 is necessary for striatal axonal projections of midbrain dopamine neurons. Hum Mol Genet 21:4827–4835
63. Pham AH, Meng S, Chu QN et al (2012) Loss of Mfn2 results in progressive, retrograde degeneration of dopaminergic neurons in the nigrostriatal circuit. Hum Mol Genet 21:4817–4826
64. Pareyson D (1999) Charcot-Marie-Tooth disease and related neuropathies: molecular basis for distinction and diagnosis. Muscle Nerve 22:1498–1509
65. Boerkoel CF, Takashima H, Garcia CA et al (2002) Charcot-Marie-Tooth disease and related neuropathies: mutation distribution and genotype-phenotype correlation. Ann Neurol 51:190–201
66. Goldenthal MJ (2016) Mitochondrial involvement in myocyte death and heart failure. Heart Fail Rev 21:137–155
67. Pich S, Bach D, Briones P et al (2005) The Charcot-Marie-Tooth type 2A gene product, Mfn2, up-regulates fuel oxidation through expression of OXPHOS system. Hum Mol Genet 14:1405–1415
68. Mourier A, Motori E, Brandt T et al (2015) Mitofusin 2 is required to maintain mitochondrial coenzyme Q levels. J Cell Biol 208:429–442
69. Zhao N, Zhang Y, Liu Q et al (2015) Mfn2 affects embryo development via mitochondrial dysfunction and apoptosis. PLoS One 10:e0125680
70. Soto EA, Kliman HJ, Strauss JF et al (1986) Gonadotropins and cyclic adenosine 3′,5′-monophosphate (cAMP) alter the morphology of cultured human granulosa cells. Biol Reprod 34:559–569
71. Li D, Sewer MB (2010) RhoA and DIAPH1 mediate adrenocorticotropin-stimulated cortisol biosynthesis by regulating mitochondrial trafficking. Endocrinology 151:4313–4323
72. Gak IA, Radovic SM, Dukic AR et al (2015) Stress triggers mitochondrial biogenesis to preserve steroidogenesis in Leydig cells. Biochim Biophys Acta 1853:2217–2227
73. Duarte A, Poderoso C, Cooke M et al (2012) Mitochondrial fusion is essential for steroid biosynthesis. PLoS One 7:e45829

74. Allen JA, Shankara T, Janus P et al (2006) Energized, polarized, and actively respiring mitochondria are required for acute Leydig cell steroidogenesis. Endocrinology 147:3924–3935
75. Acton BM, Jurisicova A, Jurisica I et al (2004) Alterations in mitochondrial membrane potential during preimplantation stages of mouse and human embryo development. Mol Hum Reprod 10:23–32
76. King SR, Liu Z, Soh J et al (1999) Effects of disruption of the mitochondrial electrochemical gradient on steroidogenesis and the Steroidogenic Acute Regulatory (StAR) protein. J Steroid Biochem Mol Biol 69:143–154
77. Song Z, Chen H, Fiket M et al (2007) OPA1 processing controls mitochondrial fusion and is regulated by mRNA splicing, membrane potential, and Yme1L. J Cell Biol 178:749–755
78. Mattenberger Y, James DI, Martinou JC (2003) Fusion of mitochondria in mammalian cells is dependent on the mitochondrial inner membrane potential and independent of microtubules or actin. FEBS Lett 538:53–59
79. Mikkola R, Saris NE, Grigoriev PA et al (1999) Ionophoretic properties and mitochondrial effects of cereulide: the emetic toxin of B. cereus. Eur J Biochem 263:112–117
80. Meeusen S, McCaffery JM, Nunnari J (2004) Mitochondrial fusion intermediates revealed in vitro. Science 305:1747–1752
81. Duarte A, Castillo AF, Castilla R et al (2007) An arachidonic acid generation/export system involved in the regulation of cholesterol transport in mitochondria of steroidogenic cells. FEBS Lett 581:4023–4028
82. Diemer T, Allen JA, Hales KH et al (2003) Reactive oxygen disrupts mitochondria in MA-10 tumor Leydig cells and inhibits steroidogenic acute regulatory (StAR) protein and steroidogenesis. Endocrinology 144:2882–2891
83. Rone MB, Midzak AS, Issop L et al (2012) Identification of a dynamic mitochondrial protein complex driving cholesterol import, trafficking, and metabolism to steroid hormones. Mol Endocrinol 26:1868–1882
84. Bose M, Whittal RM, Miller WL et al (2008) Steroidogenic activity of StAR requires contact with mitochondrial VDAC1 and phosphate carrier protein. J Biol Chem 283:8837–8845
85. Tu LN, Morohaku K, Manna PR et al (2014) Peripheral benzodiazepine receptor/translocator protein global knock-out mice are viable with no effects on steroid hormone biosynthesis. J Biol Chem 289:27444–27454
86. Baines CP, Zhang J, Wang GW et al (2002) Mitochondrial PKC-epsilon and MAPK form signaling modules in the murine heart: enhanced mitochondrial PKC-epsilon-MAPK interactions and differential MAPK activation in PKC-epsilon-induced cardioprotection. Circ Res 90:390–397
87. Alonso M, Melani M, Converso DP et al (2004) Mitochondrial extracellular signal-regulated kinases 1/2 (ERK1/2) are modulated during brain development. J Neurochem 89 248–256
88. Poderoso C, Converso DP, Maloberti P et al (2008) A mitochondrial kinase complex is essential to mediate an ERK1/2-dependent phosphorylation of a key regulatory protein in steroid biosynthesis. PLoS One 3:e1443
89. Jefcoate CR, Lee J, Cherradi N et al (2011) cAMP stimulation of StAR expression and cholesterol metabolism is modulated by co-expression of labile suppressors of transcription and mRNA turnover. Mol Cell Endocrinol 336:53–62
90. Clem BF, Hudson EA, Clark BJ (2005) Cyclic adenosine 3′,5′-monophosphate (cAMP) enhances cAMP-responsive element binding (CREB) protein phosphorylation and phospho-CREB interaction with the mouse steroidogenic acute regulatory protein gene promoter. Endocrinology 146:1348–1356
91. Gyles SL, Burns CJ, Whitehouse BJ et al (2001) ERKs regulate cyclic AMP-induced steroid synthesis through transcription of the steroidogenic acute regulatory (StAR) gene. J Biol Chem 276:34888–34895
92. Hirakawa T, Ascoli M (2003) The lutropin/choriogonadotropin receptor-induced phosphorylation of the extracellular signal-regulated kinases in leydig cells is mediated by a protein kinase a-dependent activation of ras. Mol Endocrinol 17:2189–2200

93. Manna PR, Jo Y, Stocco DM (2007) Regulation of Leydig cell steroidogenesis by extracellular signal-regulated kinase 1/2: role of protein kinase A and protein kinase C signaling. J Endocrinol 193:53–63
94. Martinelle N, Holst M, Soder O et al (2004) Extracellular signal-regulated kinases are involved in the acute activation of steroidogenesis in immature rat Leydig cells by human chorionic gonadotropin. Endocrinology 145:4629–4634
95. Seger R, Hanoch T, Rosenberg R et al (2001) The ERK signaling cascade inhibits gonadotropin-stimulated steroidogenesis. J Biol Chem 276:13957–13964
96. Otis M, Gallo-Payet N (2007) Role of MAPKs in angiotensin II-induced steroidogenesis in rat glomerulosa cells. Mol Cell Endocrinol 265–266:126–130
97. Duarte A, Castillo AF, Podesta EJ et al (2014) Mitochondrial fusion and ERK activity regulate steroidogenic acute regulatory protein localization in mitochondria. PLoS One 9:e100387
98. Papanicolaou KN, Ngoh GA, Dabkowski ER et al (2012) Cardiomyocyte deletion of mitofusin-1 leads to mitochondrial fragmentation and improves tolerance to ROS-induced mitochondrial dysfunction and cell death. Am J Physiol Heart Circ Physiol 302:H167–H179
99. Xu FY, McBride H, Acehan D et al (2010) The dynamics of cardiolipin synthesis post-mitochondrial fusion. Biochim Biophys Acta 1798:1577–1585
100. Zhao D, Duan H, Kim YC et al (2005) Rodent StAR mRNA is substantially regulated by control of mRNA stability through sites in the 3′-untranslated region and through coupling to ongoing transcription. J Steroid Biochem Mol Biol 96:155–173
101. Grozdanov PN, Stocco DM (2012) Short RNA molecules with high binding affinity to the KH motif of A-kinase anchoring protein 1 (AKAP1): implications for the regulation of steroidogenesis. Mol Endocrinol 26:2104–2117
102. den Hertog J, Ostman A, Bohmer FD (2008) Protein tyrosine phosphatases: regulatory mechanisms. FEBS J 275:831–847
103. Adachi M, Iwaki H, Shindoh M et al (1997) Predominant expression of the src homology 2-containing tyrosine phosphatase protein SHP2 in vascular smooth muscle cells. Virchows Arch 430:321–325
104. Matozaki T, Murata Y, Saito Y et al (2009) Protein tyrosine phosphatase SHP-2: a proto-oncogene product that promotes Ras activation. Cancer Sci 100:1786–1793
105. Rocchi S, Gaillard I, van Obberghen E et al (2000) Adrenocorticotrophic hormone stimulates phosphotyrosine phosphatase SHP2 in bovine adrenocortical cells: phosphorylation and activation by cAMP-dependent protein kinase. Biochem J 352:483–490
106. Cooke M, Orlando U, Maloberti P et al (2011) Tyrosine phosphatase SHP2 regulates the expression of acyl-CoA synthetase ACSL4. J Lipid Res 52:1936–1948
107. Schimmer BP (1979) Adrenocortical Y1 cells. Meth Enzymol 58:570–574
108. Han JD, Rubin CS (1996) Regulation of cytoskeleton organization and paxillin dephosphorylation by cAMP. Studies on murine Y1 adrenal cells. J Biol Chem 271:29211–29215
109. Vilgrain I, Chinn A, Gaillard I et al (1998) Hormonal regulation of focal adhesions in bovine adrenocortical cells: induction of paxillin dephosphorylation by adrenocorticotropic hormone. Biochem J 332:533–540
110. Manes S, Mira E, Gomez-Mouton C et al (1999) Concerted activity of tyrosine phosphatase SHP-2 and focal adhesion kinase in regulation of cell motility. Mol Cell Biol 19:3125–3135
111. Brion L, Maloberti PM, Gomez NV et al (2011) MAPK phosphatase-1 (MKP-1) expression is up-regulated by hCG/cAMP and modulates steroidogenesis in MA-10 Leydig cells. Endocrinology 152:2665–2677
112. Mori-Sequeiros GM, Gomez NV, Gorostizaga A et al (2013) MAP kinase phosphatase-3 (MKP-3) is transcriptionally and post-translationally up-regulated by hCG and modulates cAMP-induced p21 expression in MA-10 Leydig cells. Mol Cell Endocrinol 371:174–181

Chapter 26
Cancer and Mitochondria

Juan José Poderoso

Abstract Cancer is a devastating pathology that involves symptoms, discomfort, family and social disintegration, and finally leads to a debilitating condition and death. It was Otto Warburg, the extraordinary German scientist, who opened our eyes towards the notion of the existence of a metabolic disruption in cancer cells. Since Warburg pioneering studies physicians and scientists had considered a role of mitochondria in the genesis of cancer. Mitochondria are ancient rickettsia incorporated as organelles into a primitive eukaryotic recipient cell. Mitochondria changed the primitive anaerobic metabolism to an oxygen dependent respiration that multiplied the energy production in cells from 2 to 36 moles of ATP by mol of utilized glucose. Therefore, this chapter is mainly directed to respond what happens to mitochondria in cancer. We will describe here the mitochondrial alterations reported in this pathology. The reader will realize that we favor the notion that mitochondrial dysfunction and the incorrect use of oxygen are factors associated to a particular cell development and function that at the same time is the mechanism of pathogenesis of a debilitating and mortal illness.

Keywords Oxidative stress • Warburg effect • DNA mutation • 8-HO-deoxyguanine • Cancer and oxidative stress • mtDNA • Homoplasmy • Heteroplasmy • Colon cancer

1 Introduction: The Induction of Cancer

The induction of cancer involves a multistage and multistep process; the early steps are initiation and promotion. Initiation results from a DNA mutation that followed by a round of DNA synthesis results in fixation of the mutation and production of an initiated cell. Initiated mutated cell can occur through the cell interaction with physical carcinogens like UV light or radiation or with chemical carcinogens. Both

J.J. Poderoso (✉)
Instituto de Inmunología, Genética y Metabolismo (INIGEM, UBA–CONICET),
Laboratorio de Metabolismo del Oxígeno, Hospital Universitario,
Universidad de Buenos Aires, Buenos Aires, Argentina
e-mail: jpoderos@fmed.uba.ar

© Springer International Publishing Switzerland 2016
R.J. Gelpi et al. (eds.), *Biochemistry of Oxidative Stress*, Advances
in Biochemistry in Health and Disease 16, DOI 10.1007/978-3-319-45865-6_26

types of agents damage DNA and produce mutations. Mutations can be also acquired by misrepair of damaged DNA. Following the formation of initiated cells, chemicals and metabolites influence clonal growth and tumor promotion that are initially associated to modulation of gene expression and results in an increased cell number by cell division and by a decrease in apoptosis.

The mechanisms by which carcinogens induces their effects have been studied by many years. In rodents, carcinogens produce cellular effects by genotoxic (DNA reactive) or non genotoxic mechanisms. Genotoxic mechanisms imply the interaction of the agents or metabolites with genomic DNA; usually this effect acts at the initiation or progression stage of cancer. Non genotoxic agents may be cytotoxic or mitogenic or may act through receptor–mediated pathways.

Reactive oxygen species (ROS) play an important role in tumor development and are produced either by endogenous factors, including inflammatory cells, or by exogenous sources. Oxidative stress causes DNA damage, chromosome instability, gene mutation and modulation of cell growth, all these pathways resulting in cancer. In the mode of action of hepatic carcinogens, a well known group of substances, a distinction is made between genotoxic (nitrosamines, aflatoxin B, polyaromatic hydrocarbons) and cytotoxic (choloform, melamine, phenobarbital, trichloroethylene, fibrates, and polychlorinated biphenyls) carcinogens [1].

2 Mitochondrial Dysfunction in Cancer

It is still uncertain how changes in mitochondrial mass and function affect the development and biology of cancer or determine the clinical outcome for cancer patients. Abnormal mitochondria were suggested many years ago by Otto Warburg to explain his observation that tumors undergo aerobic glycolysis (the Warburg effect) [2].

Mutations in Krebs cycle enzymes strongly support the notion that mitochondrial metabolism is inherently defective in cancer. However, evidence about dysfunctional mitochondria as the major or unique cause for the Warburg effect is limited. Instead, the altered expression and activity of glycolytic enzymes in cancer has been recognized. For example, altered expression of phosphoglycerate dehydrogenase, triose phosphate isomerase and piruvate kinase reduce the rate of glycolytic flux to piruvate and increase related biosynthetic pathways such as serine synthesis and the pentose pathway. Moreover, the Warburg effect and the increased glucose metabolism produce a redox regulation of cytochrome c and inhibition of apoptosis. In mitochondria most of consumed oxygen is reduced to water. However, 1–2 % is reduced to superoxide (O_2^-) and converted to reactive oxygen species (ROS). The concept of ROS is biologically relevant considering that the involved chemical species produce similar biological effects. From a biochemical point of view, ROS is a group of chemical species that are able to sustain the free-radical mediated process characteristic of aerobic life. These chemical species are: O_2^- (superoxide radical),

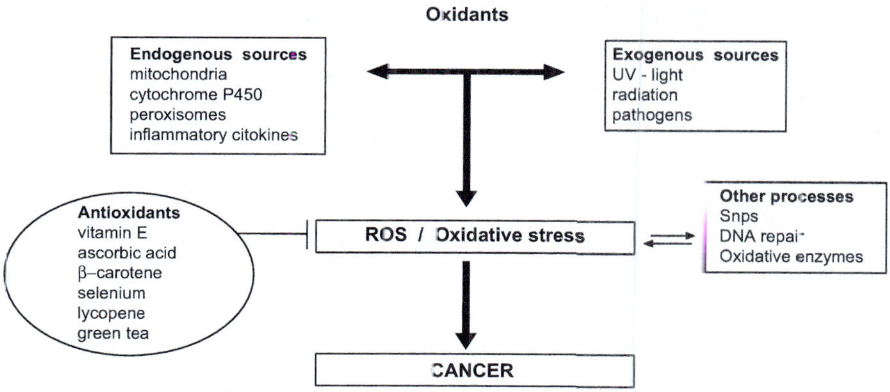

Fig. 26.1 Oxidative stress and human cancer. *Snps* single nucleotide polymorphism

H_2O_2 (hydrogen peroxide), HO• (hydroxyl radical), ROO• (peroxyl radical), ROOH (organic hydroperoxide), and $^1O_2$ (singlet oxygen).

The majority of O_2^- is produced by complex I and complex III in the mitochondrial electron transfer chain. Complex III ROS generation has been associated to the activation of hypoxia inducible factor (HIF) which modulates cell proliferation and angiogenesis.

The role of mitochondrial ROS in cancer remains unsolved but ROS generation is higher in aged and cancer cells than in normal cells. Cellular ROS can be also contributed by inflammatory cells that normally play a role in killing bacteria. Mutational studies have suggested that chronic oxidative stress is associated with carcinogenesis. Examples are ulcerative colitis, colorectal cancer, and the infection by *H. pilori* with the subsequent association to chronic gastritis and gastric carcinoma. Additionally, RNA-induced damage by these oxidative mechanisms may cause errors in protein synthesis or dysregulation of gene expression. In hepatocarcinogenesis, oxidative damage to DNA has also been linked to *p53* and *ras* gene mutations [3].

3 The Hypotheses of Cellular Oxidative Damage as Inducer of Cancer

The concept that cellular oxidative damage is a step in the pathway to cancer is illustrated in Fig. 26.1. The condition of an increase in oxidants, either endogenous or exogenous, or of a decrease in antioxidants, leads equally to oxidative stress and damage. This cellular situation implies an increased oxidative damage to DNA that promotes mutations in the genetic material.

3.1 Molecular Mechanisms of DNA Damage and Mutation

Oxidative DNA damage is a physiological process in aerobic organs. Fraga et al. (1990) have estimated the frequency of DNA damage in human cells as 10^4 lesions per cell per day [4]. Lu et al. considered that hydroxyl radical (HO•) is the predominant ROS that targets DNA while H_2O_2 is less reactive and more diffusible [5]. The free-radical mediated process, a consequence of aerobic life, is a cellular phenomenon that includes all chemical types of cell components: phospholipids, proteins and nucleic acids. Both, DNA and RNA, this latter, coding and non-coding, are involved and likely cause errors in protein synthesis or dysregulation of gene expression. In this way, the free-radical mediated process affords a complete pathogenic mechanism and it has been proposed as an underlying mechanism of several human diseases, as the chronic degeneration of neurons [6]. Lipid peroxidation results in the formation of reactive aldehydes, as malonaldehyde (MDA) and 4-hydroxy-2-nonenal (4-HNE), which have high reactivity with proteins and DNA [7]. While MDA and MDA-MDA dimers are mutagenic in bacterial assays [1], evidence for their contribution to mutation in humans is lacking. A clear link between protein oxidation and carcinogenesis has not been established. In contrast, oxidative damage to DNA has been linked to aflatoxin B-induced *p53* and *ras* gene mutations in hepatic carcinogenesis and in UV-induced mouse and human skin cancers [3, 8].

3.2 DNA Mutagenic Lesions: 8-Hydroxy-Guanine

The production of 8-hydroxy-guanine (8OHdG) with its tautomeric form oxoguanine, is the most studied and common oxidative DNA lesion originated from the reaction of ROS with the DNA bases (Fig. 26.2). Guanine is far more oxidized, due to the much higher oxidation potential of this base relative to cytosine, thymine and adenine. Oxoguanine is mutagenic in bacteria and mammalian cells and is elevated in human cancer cells. 8-HO-2′-deoxyguanine and 8-HO-guanine are biomarkers of the oxidative damage to DNA and RNA nucleic acids, respectively. Based on these evidences, 8OHdG has been widely utilized as a biomarker of DNA oxidative damage and is considered a valid measurement of the cellular load of oxidative stress. Similarly, as 8HOdG is applied to oxidative stress, 8-nitroguanine is applied to the determination of RNS (reactive nitrogen species) that show mutagenic properties by leading G-T transversions. High ROS levels induce apoptosis, necrosis and carcinogenesis with an altered expression of growth factors and proto-oncogenes [9]. Care must be taken in the assay of 8HOdG due to the ease with which it can be oxidized during the extraction and assay procedures [10–12]. Oxidative DNA damage is the major source of mutation in living organisms. More than one hundred oxidative DNA adducts with purine, pyrimidine, and deoxyribose backbones have been identified [13]. In addition, p15INK4B and p16INK4A tumor suppressor genes appear as targets of ROS-induced renal cell carcinoma in rats.

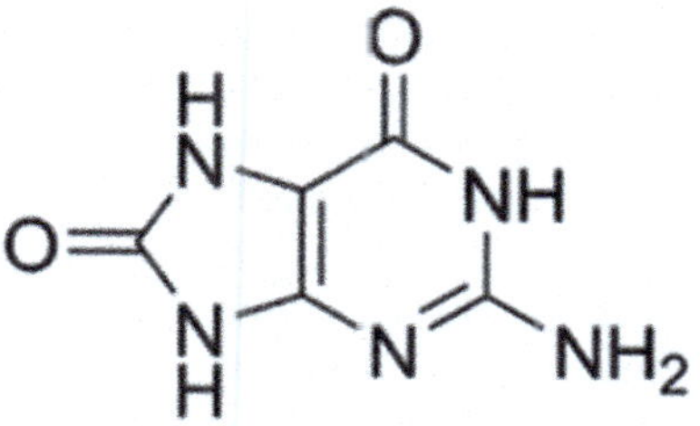

Fig. 26.2 8-HO-guanine. The C at the 8-position, by the vicinity of the N atoms at positions 7 and 9, has a increased reactivity towards oxidants and easily forms the oxo-derivative oxoguanine (shown)

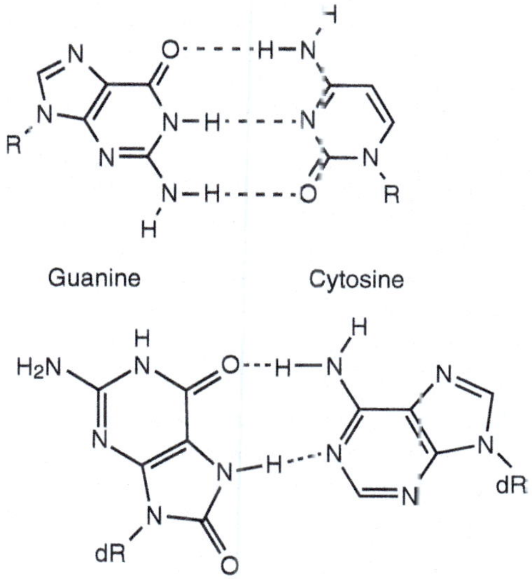

Fig. 26.3 8-oxoG (*syn*) in a Hoogsteen base pair with dA (*anti*). For comparison below is a standard (non-mutagenic) GC base pair with both bases in the *anti* configuration of the bond between base and sugar

ROS-induced DNA damage can result in base modification, deoxyribose modification, DNA cross-linking and single- or double-strand breakage. Genomic instability, replication errors, DNA mutation and cell death occur if the oxidative DNA damage is not repaired prior to DNA replication. Numerous studies have shown that 8OHdG levels are elevated in animal tumor models and in various human cancers [9, 10, 14]. 8-HOdG in its stable syn conformation pairs with both cytosine and adenine. If the A:G mismatch is not repaired, a G:C to T:A transversion will occur, which is commonly found in mutated oncogenes and tumor suppressor genes (Fig. 26.3) [15, 16]. During DNA replication, ROS can also react with dGTP of the nucleotide pool to form 8-HOdG. Therefore, in addition to the G:C to T:A caused by 8-HOdG in the DNA template, it is possible that during DNA replication, the deoxyguanilate of the nucleotide pool would be incorporated into DNA opposite to dC or dA of the template strand, resulting in a A:T to C:G transversion [14]. The measurement of 8-OHdG levels is used to evaluate the load of oxidative stress in cells [11, 17].

In addition, the RNS produced in chronic inflammation, can cause nitrative DNA damage to form 8-nitroguanine, that has been observed in human tissues.

Experimental evidence suggests that 8-nitroguanine is also a mutagenic DNA lesion, which may lead to G:T transversions [18]. The assessment of oxidative DNA damage products (8-HOdG or 8-nitroguanine) in various biological matrixes, such as serum and/or urine, is important to understand the role of oxidative stress in cancer development and disease intervention. Accurate and reliable measurement of oxidative damage to phospholipids, proteins, and DNA are important in the evaluation of the extent and distribution of ROS-induced damage. Although much of the effort has been on the understanding of the effects of ROS on DNA damage and induction of mutations, the epigenetic effects of ROS have also been studied [19].

3.3 Cellular Responses to Oxidative Stress and ROS

A common adaptive response of mammalian cells after oxidative stress and damage is the up regulation of stress-response genes, that encode antioxidant defense enzymes [1, 20]. While high levels of ROS production may lead to the induction of apoptosis or necrosis, increasing evidence demonstrates that low or transient ROS exposure increases cell proliferation, likely through altered expression of growth factors and proto-oncogenes [21]. As described, it is clear that many xenobiotics increase the cellular levels of ROS through various mechanisms. ROS-induced alteration of gene expression involves signaling pathways including cAMP-mediated cascades, calcium-calmodulin pathways, and intracellular levels of nitric oxide. Several studies have also shown that intracellular ROS release calcium from cellular stores, resulting in the activation of kinases, such as protein kinase C (PKC), which regulates a variety of cell functions including cell proliferation, cell cycle, differentiation, cytoskeletal organization, cell migration, and apoptosis. While PKC can be activated by ROS [22], the activation of PKC is required to generate ROS to inhibit gap junction intercellular communications [23]. The activation of PKC seems to be differentially regulated by cellular oxidants: oxidation at the NH_2-terminal regulatory domain activates PKC, whereas oxidation at the COOH terminal domain inactivates PKC [24].

Similarly, H_2O_2 activates protein kinases such as extracellular signal regulated kinase (ERK)1/2, phosphoinositide 3-kinase/serine-threonine kinase (P13K/Akt), protein kinase B (PKB) and protein tyrosine phosphatases (PTPs). Because these pathways regulate cellular migration, proliferation, survival, and death responses, their activation has been suggested as a potential mechanism of ROS-induced carcinogenesis. Cellular oxidants have been related to the activation of transcription factors with significant effects on signaling pathways involving the nuclear factor erythroid 2-related factor 2 (NF-E2/rf2 or Nrf2) [25], mitogen-activated protein (MAP) kinase/AP-1 [26], NF-kB pathways and hypoxia-inducible transcription factor 1a (HIF-1a) [27]. Therefore, the cellular concentration of ROS appears to influence the activation of transcription factors and explains that either cell death or cell proliferation may result from exposure to oxidative stress. Hydrogen peroxide also activates AP-1 mediated by JNK and p38MAP kinase that translocate to the nucleus, phosphorylate c-jun and ATF2 and enhance transcriptional activity. In addition, the effect of H_2O_2 on thioredoxins results in activation of ASK1 and JNK.

An additional effect of cellular oxidants is the activation of transcription factors like Nrf2. The transcription factor Nrf2 is bound to the protein Keap 1 which makes it inactive and favors its degradation. Oxidative stressors and ROS disrupt the binding with Nrf2-Keaps dissociation that favors its degradation and the traffic to the nucleus. A number of stressors like ROS and RNS are implicated in the control of Nrf2.

HIF is a heterodimeric transcription factor that has a role in signaling the cellular oxygen levels. It is composed by two units HIF 1α and HIF 1β. The level of HIF 1α is controlled by cellular oxygen levels. HIF 1α is produced continuously and accumulates in hypoxic cells whereas it is degraded in normoxic cells. HIF-1α has been implicated in ROS-induced carcinogenesis in tumors of bladder, breast and colon, all of them associated to a poor outcome. ROS originated from mitochondrial complex III are required for the hypoxic activation of HIF-1α. The potent carcinogen aflatoxin B is implicated in the ROS effects on p53 and ras mutations that increase the risk of cancer growth.

Alteration of the mitochondrial function has long been suspected to participate in the development and progression of cancer. Almost half a century ago, Warburg focused his research on the existence of mitochondrial alterations in cancer and proposed a mechanism to explain the differences in energy metabolism between normal and cancer cells. He suggested that mitochondrial alterations provide unique therapeutic targets in several cancer types. Since Warburg's proposal, several cancer-related mitochondrial alterations have been identified [28].

To develop its functions, the mitochondrion has its own genome, which is composed by 13 polypeptides that are components of the electron transport chain (ETC), 22 tRNA and 2 rRNA genes for its own protein synthesis. The remaining protein subunits involved in the ETC complexes, along with those required for maintenance of mitochondrial DNA (mtDNA), are nuclear encoded, synthesized in the cytosol, and specifically targeted to mitochondria. Mammalian cells contain 10^3–10^4 copies of mtDNA that replicates independently of nuclear DNA Human mtDNA is a 16.6 kb circular double-stranded DNA molecule that surprisingly is devoid of protective histones, although mitochondrial transcription factor A and single-stranded DNA-binding protein are cooperatively involved in the maintenance of mtDNA. Unlike nuclear DNA, which is inherited from both parents and in which genes are rearranged by recombination, mtDNA is transmitted from mother to offspring. Then, mitochondrial genes have an exclusively maternal inheritance in mammals with clonal mtDNA lineages.

Because of the susceptibility of mtDNA to damage, mitochondria also contain their own DNA repair systems. One important mtDNA repair mechanism is base excision repair (BER), and the detection of various BER enzymes in mitochondria highlights the importance of maintaining mtDNA integrity for normal cellular function. Abnormalities in the mitochondrial genome may also arise from nuclear genetic disorders. It has been well established that mitochondria play a role in the regulation of apoptosis by release of pro-apoptotic agents and/or disruption of cellular energy metabolism.

The mitochondrial ETC is the major source of reactive oxygen species. Increased levels of ROS are clearly associated with different diseases including cancer. ROS start both a free-radical mediated chain reaction and a cascade of redox signaling reactions that conduct to DNA damage. mtDNA is permanently exposed to the ROS produced within mitochondria and is also prone to chemical damage by environmental factors such as UV, cigarette smoke, and ionizing radiation. In this context, mtDNA shows a tenfold higher accumulation of mutations than nuclear DNA. Cancer cells carry mtDNA mutations and altered copy numbers of mtDNA, which affect the expression and activity of the ETC. A number of antioxidant defense mechanisms exist in mitochondria to remove ROS. Cells are normally able to defend themselves against ROS damage through ROS scavengers and antioxidant enzymes such as superoxide dismutases (SOD), catalases, and glutathione peroxidases. These enzymes are often deficient in tumor cells. The mentioned antioxidant enzymes afford the fundamental biological control mechanism. Another of the major antioxidant systems is the thioredoxin system comprised of thioredoxin (TRX) and thioredoxin reductase (TR) [29]. Together they form a powerful antioxidant system, especially active against protein oxidized -SS- groups. The balance between -SH and -SS- is involved in many central intracellular and extracellular processes, such as protection against oxidative stress, anti-apoptotic functions, cell proliferation, redox regulation of gene expression and signal transduction, growth factors and co-cytokine effects, and regulation of the redox state of the extracellular environment. Over recent years this system has increasingly been linked to the development and expression of cancer phenotypes. Immune cytochemical approaches have been used to determine the expression and localization of TRX and TR in primary human cancers, including breast, prostate and colorectal carcinomas and malignant melanoma. In aggressive invasive mammary carcinomas and malignant melanomas, TRX is found over-expressed in the nucleus and the cytosol. In aggressive neoplasic tumor cells, TRX was highly over-expressed, compared to tumors of lesser aggressive nature. Increased levels of TRX positively correlate with TR expression and localization. These immune-cytochemical studies of the expression and localization of TRX and TR in melanomas, thyroid, prostate, and breast carcinomas, extended the range of human cancers with available data on the thioredoxin ystem. The results support the conclusion that aggressive tumors over-express TRX and TR and posses a high proliferation capacity, a low apoptosis rate and an elevated metastatic potential. Therefore, the TRX system seems involved in the processes of oncogenesis and tumor growth and confirms its potential as target for anticancer therapy in a wide range of human tumors. The thioredoxin reductase isoenzymes are homologous to glutathione reductase and contain a conserved C-terminal elongation with a cysteine-selenocysteine sequence forming the redox-active zone. Therefore, it seems clear that persistent cell oxidative stress leads to promotion of cancer growth and metastasis through induction of DNA damage and mutations.

4 Aging and Cancer

It is widely recognized that the incidence of cancer increases with aging and many hypotheses have been put forward to explain this association. Like cancer, aging is a complex process of declination in the organism physiological functions. As it was previously referred, the most accepted theory to explain this phenomenon, applicable to both cancer and aging, is oxidative stress and damage. The biomarkers of oxidative stress are substances that are formed during oxidative damage to phospholipids, proteins, and nucleic acids, and that are normally present in the body fluids of diseased and healthy people, in the latter case at physiological concentrations. Malonaldehyde (TBARS), 4-HO-nonenal and 8-*iso* prostaglandin $F_{2\alpha}$ are the biomarkers of phospholipid oxidative damage. Similarly, protein carbonyl groups, *o*-tyrosine and 3-nitrotyrosine are biomarkers of protein oxidative damage. Concerning DNA damage, as discussed before, 8-HO-deoxyguanine is the most recognized biomarker. It is thought that the concentration of biomarkers increases as the age of people increases. However, the concentration of biomarkers in body fluids is very low and, therefore, it is necessary to use sensitive analytical methods. A combination of HPLC and MS is the best choice to determine biomarker concentrations in people of different ages and to define differences among them [9].

5 mtDNA Homoplasmy and Heteroplasmy

Cells contain multiple copies of mitochondrial genes, a situation which is clearly different from the only 2 copies of each nuclear gene. Homoplasmy is the state in which all cell mitochondria have the same mtDNA genome, either wild type or alternatively a mutated one. As cells contain many mitochondria and each mitochondrion has 2–10 copies of mtDNA, it is likely that wild type and mutant mtDNA would coexist in a state that is defined as heteroplasmy. Hence, the biological impact of a given mtDNA mutation would be different depending on the proportion and copies of the mutant mtDNA in each cell. In the earliest studies no evidence of heteroplasmy was detected, probably due to the low sensitivity of the primitive techniques. However, using massive parallel sequencing-by-synthesis, a marked heteroplasmy in the mtDNA of normal human cells was reported. It is now accepted that mtDNA constantly undergoes mutations, with expansion or loss of mutated mtDNA copies and with permanent changes in the proportion of homoplasmy and heteroplasmy (Fig. 26.4).

Whether mtDNA point mutations occur by simple clonal expansion is debated. Although a large number of mtDNA deletions are capable of clonal expansion in individual cells, point mutations have not been shown to be able to clonally expand or reach homoplasmy, though there are references about this fact in cell cultures and germ cell lines [30, 31]. Indirect evidence supporting the possibility of clonal

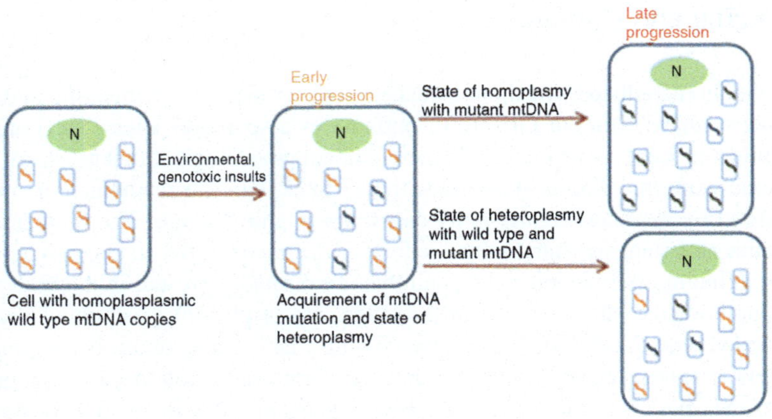

Fig. 26.4 Homoplasmic and heteroplasmic mtDNA mutations due to environmental and genotoxic damages, along with nuclear (N) genetic changes. Cells may incorporate mtDNA mutations and acquire a state of heteroplasmy with both wild-type and mutated mtDNA copies. Further progression of tumor cells may be aided by a homoplasmic bias with mutant mtDNA copies or a mixture of both wild-type and mutant mtDNA, in a state of heteroplasmy

expansion of point mutations comes from genetically engineered heteroplasmic studies in mice.

Considering that a mutation confers cell growth or survival advantage or facilitates mtDNA replication, such mutation is likely to survive through selection and may eventually become dominant and evolve to become a homoplasmic mutant by clonal expansion. It has been indicated that single cells with a mutant mitochondrial genome may acquire a selective growth advantage during tumor evolution, allowing it to become the predominant cell type in the tumor cell population [32]. It is possible that some germ line mtDNA mutations are actually mutations that occurred early during prenatal development of the individual and drifted to homoplasmy. As such, heteroplasmic mutations may reflect an intermediate stage in the process. For a mtDNA mutation to have a significant effect on cell physiology it must reach a threshold level of 60 % or more, depending on the type of mutation. Also, it is insufficient for the cell to accumulate mutations, as different mutations are likely to compensate the deficiencies [33]; however, there are limitations of the trans-complementations [34]. The hypothesis is that mutants would accumulate a single initial mutation via clonal expansion. An alternative explanation for homoplasmic mtDNA has been presented, using a mathematical model that shows that a homoplasmic mitochondrial mutation in a tumor results from the segregation of a mutant genome in the cell generations that occur during tumor development [35]. Whether this model is valid or not is still a topic of debate. A high background of random mutations provides no selective advantage. But there are mutations that alter mitochondrial function and cell physiology in a manner that has a significant effect favoring tumor development.

6 mtDNA Variations in the Progression of Cancer

Mitochondrial cellular content and mutations are emerging as new molecular markers for cancer detection and monitoring. It has been shown that damaged mtDNA in cell lines led to the rapid evolution of homoplasmic mutations. Recent refinements in techniques including next-generation sequencing for the detection of mtDNA content and copy number combined with rapid high-throughput methods of mutation detection, have incited interest in clinical studies in various tissues and body fluids. Initial examination of human bladder, head and neck, breast, and primary lung tumors showed a high frequency of mtDNA mutations. One study revealed a correlation between increased mtDNA copies and the risk of lung cancer in heavy smokers. The mutated mtDNA is easily detected in cancer-paired body fluids (urine, saliva, and sputum) from each cancer and it was 19–220 times more abundant than mutated nuclear p53 DNA. Recent analysis from the laboratories of Chaterjee and collegues detected the presence of clonal mtDNA mutations in histologically normal respiratory mucosa and surgical margins of smoker lung cancers and recurrent head and neck cancer patients [36]. This fact was reported by several investigators, supporting the presence of extensive altered cellular components with aberrant mtDNA, much smaller than in clinical neoplasms. It is critical to know the timing of mtDNA mutation in the course of disease development and progression if such alterations are to be used as markers. mtDNA mutations have been detected in clinical samples obtained from early stage patients, showing that they could be useful in early cancer detection. In a comprehensive analysis, 93 premalignant lesions from the upper aerodigestive tract were examined for mtDNA D-loop mutations; 22 % of hyperplasia, 33 % of mild dysplasia, 35 % of moderate dysplasia, 50 % of severe dysplasia, and 62 % of carcinoma *in situ* lesions carried mutation in the D-loop region [32]. In two other studies, the whole mitochondrial genome and D-loop regions were sequenced in gastric intraepithelial neoplasia and sequence variants were detected in 100 % and 62 % of the cases, respectively [37]. By using D310 mutations as a clonal marker, changes in fine-needle aspirates have been detected in breast and digestive cancer and in urine sediments from patients with bladder and prostate cancer [38]. A study on oral cells from 42 healthy smokers and 30 non-smokers reported higher frequency of mtDNA mutations in the smoker samples. However, no clear connection has been established that smokers with mtDNA mutations are more likely to develop cancer.

Due to their clonal nature and high copy number, mitochondrial mutations could provide a molecular marker for the non-invasive early detection of cancer. A specific mtDNA mutation was clonally detected in bone marrow samples collected at different time points from a leukemia patient [39], with a possible role of this alteration to monitor disease progression. In light of this, it remains important to identify disease-associated clonal mtDNA mutations in different cancers at all stages utilizing appropriate detection platforms. With the advent of high-throughput approaches such as Mito chips, more prospective studies may be carried out to determine the value of these mutations in the early cancer detection [40]. As most cancer cells harbor homoplasmic mitochondrial mutations, their ease of detection in body fluids and minute cellular samples such as biopsies offer opportunities for clinical use. As noted earlier, definitive diagnosis of clonal expansion through mtDNA testing would be helpful to classify difficult biopsies as malignant. Testing mtDNA mutations in blood, urine, saliva, or sputum could be used with imaging studies to detect early cancers. Clonal mtDNA mutations could be followed to indicate recurrence and to monitor responses to therapy. Finally, abrogation of mtDNA mutations in patients undergoing chemoprevention could be used to monitor reversal of early benign lesions. Validation of tumor-associated mtDNA mutations by comparing matched normal and tumor samples, followed by their detection in the available samples, including paired body fluids, will facilitate their utilization in prevention, early detection, and monitoring strategies.

7 mtDNA Mutations in Human Cancer: D-Loop Alterations

At present, whole genome or specific regions of human mtDNA have been sequenced in different tumor types for mutation detection. The mutations detected specifically in the regulatory non-coding D-loop or other coding regions of the mtDNA were examined. The D-loop or displacement loop region is the main non-coding area of the mtDNA molecule and where replication of mtDNA occurs. The region contains promoters for transcription of the 2 strands of mtDNA. D-loop alterations might interfere with initiation of mtDNA replication, although there are no reports of consequences of D-loop mutations [41]. An earlier study from Chaterjee et al. identified the C-tract, a poly-C-mononucleotide repeat located between the nt303 and nt315 (regarded as the D310 segment) within the D-loop as a mutational hot spot in primary tumors [37]. The stretch of cytosines is polymorphic ranging between 7-C to 9-C, with the most common sequence represented by 7-C. Deletion or insertion mutations in this region have been observed in approximately 22 % of the tumors analyzed [42]. To date, frequent D-loop alterations (insertion/deletion) have been identified in all major cancer types. The number of cytosines in the D310 area seems to influence the incidence of mutations. The majority of D310 mutations were observed when the number of cytosines was more than 7 (i.e., 8 or 9). Therefore, this variation in cytosine number explains discrepancies among published results [32].

8 Mitochondrial Dysfunction and Cancer Progression

The classic association between cancer and mitochondrial dysfunction is actually considered an alteration of the role of mitochondria in cellular signaling. For example, today it is understood that mitochondrial oxidative damage and NO and H_2O_2 diffusion are involved in the progression of human colorectal cancer [43, 44]. Mitochondria from human colorectal cancer show a markedly increased oxidative damage with increased content of TBARS and protein carbonyls. Mitochondrial activities, as NADH-cytochrome c reductase and cytochrome c oxidase, were decreased in tumors. Cu-Zn SOD activity was decreased by 42 %, while MnSOD activity did not change and mtNOS activity increased by 46 % in tumors. A direct linear relationship between mtNOS and oxidative damage in tumors and non tumoral adjacent tissues indicates that NO and H_2O_2 diffuse from the tumor to adjacent tissues signaling for cell death.

9 Metabolic Pathways in Cancer Cells

Fast growing and poorly differentiated cancer cells show high aerobic glicolysis. Most of the glycolytic piruvate is deviated to lactate production even in the presence of oxygen (Warburg effect), little piruvate enters the Krebs cycle and is extruded to the cytosol. This effect makes tumor cells adapted to low O_2 utilization. An alternative substrate is glutamine that is deaminated to glutamic acid and transaminated to α-ketoglutarate, this latter able to enter the Krebs cycle. Glutamine is completely oxidized in the Krebs cycle only if acetyl-Co A is available. In cancer cells, malate enters mitochondria and is oxidized to pyruvate by the NADP-dependent malic enzyme. Continuously acting glycolysis and fatty acid oxidation are futile cycles that maintain a negative energy balance leading to the cancer associated caquexia.

10 Intracolonic Free Radicals

Intracolonic free radicals may be an effective cause of carcinogenesis. The relative high level of iron in feces, with the ability of bile pigments to act as iron chelators, support an efficient HO• generation by Fenton chemistry in colon cells. This leads to the formation of active carcinogens, that are correlated with the amount of dietary red meat. ROS, RNS and NO are capable of producing damages in mitochondrial membranes and in DNA in carcinogenesis.

11 Conclusion

Cancer can be viewed as an evolutionary conserved life form rather than as an expression of a definite illness. Therefore, aggressive cancers are considered to fall into three categories: taxonomy (phylogenation), atavism (reprimivitization), and androbustness (adaptive resilience). The explanation is not convergent evolution but rather the release of a highly conserved program developed by the exigencies of the Precambrian period to one version in which cancer cells seem better adapted. Central to this program is the Warburg effect whose malign influence permeates well through aerobic glycolysis to include metabolism and biomass conversions.

References

1. Klaunig JE, Kamendulis LM, Hocevar BA (2010) Oxidative stress and oxidative damage in carcinogenesis. Toxicol Pathol 38:96–109
2. Koppenol WH, Bounds PL, Dang CV (2011) Otto Warburg's contributions to current concepts of cancer metabolism. Nature Review Cancer 11:325–337
3. Shen HM, Ong CN (1996) Mutation of the p53 tumor suppressor gene and ras oncogenes in aflatoxin hepatocarcinogenesis. Mutat Res 366:23–44
4. Fraga CG, Shigenaga MK, Park J-W et al (1990) Oxidative damage to DNA during aging: 8-hydroxy-deoxyguanosine in rat organ DNA and urine. Proc Natl Acad Sci U S A 87:4533–4537
5. Lu J, Tan M, Cai Q (2015) The Warburg effect in tumor progression: mitochondrial oxidative metabolism as an anti-metastasis mechanism. Cancer Lett 356:156–164
6. Nunomura W, Takakuwa Y, Cherr GN et al (2007) Characterization of protein 4.1R in the erythrocytes of zebrafish (Danio rerio): unique binding properties with transmembrane proteins and calmodulin. Comp Biochem Physiol B Biochem Mol Biol 148:124–138
7. Uchida HK, Szweda I, Chae Z et al (1993) Immunochemical detection of 4-hydroxynonenal protein adducts in oxidized hepatocytes. Proc Natl Acad Sci U S A 90:8742–8746
8. Nishigori C, Hattori Y, Toyokuni S (2004) Role of reactive oxygen species in skin carcinogenesis. Antioxid Redox Signal 6:561–570
9. Weiss KM (2005) The phenogenetic logic of life. Nat Rev Genet 6:36–45
10. Miyake T, Reese J, Loch CM et al (2004) Genome-wide analysis of ARS (autonomously replicating sequence) binding factor 1 (Abf1p)-mediated transcriptional regulation in Saccharomyces cerevisiae. J Biol Chem 279:34865–34872
11. Syslová K, Böhmová A, Mikoška M et al (2014) Multimarker screening of oxidative stress in aging. Oxid Med Cell Longev 2014:562860. doi:10.1155/2014/562860
12. Hwang ES, Bowen PE (2007) DNA damage, a biomarker of carcinogenesis: its measurement and modulation by diet and environment. Crit Rev Food Sci Nutr 47:27–50
13. Frenkel K (1992) Carcinogen-mediated oxidant formation and oxidative DNA damage. Pharmac Ther 53:127–166
14. Demple B, Harrison L (1994) Repair of oxidative damage to DNA: enzymology and biology. Annu Rev Biochem 63:915–948
15. Shibutani S, Takeshita M, Grollman AP (1991) Insertion of specific bases during DNA synthesis past the oxidation damaged base 8-oxodG. Nature 349:431–444
16. Gredilla R, Bohr VA, Stevnsner T (2010) Mitochondrial DNA repair and association with aging: an update. Exp Gerontol 45:478–488
17. Gao Y, Dickerson JB, Guo F et al (2004) Rational design and characterization of a Rac GTPase-specific small molecule inhibitor. Proc Natl Acad Sci U S A 101:7618–7623
18. Kawanishi S, Hiraku Y (2006) Oxidative and nitrative DNA damage as biomarker for carcinogenesis with special reference to inflammation. Antioxid Redox Signal 8:1047–1058
19. Evans MD, Dizdaroglu M, Cooke M (2004) Oxidative DNA damage and disease: induction, repair and significance. Mutat Res 567:1–61
20. Liu RH (2004) Potential synergy of phytochemicals in cancer prevention: mechanism of action. J Nutr 134:3479S–3485S
21. Ishikawa K, Takenaga K, Akimoto M et al (2008) ROS generating mitochondrial DNA mutations regulate tumor cell metastasis. Science 320:661–664
22. Frank GD, Mifune M, Inagami T et al (2003) Distinct mechanisms of receptor and nonreceptor tyrosine kinase activation by reactive oxygen species in vascular smooth muscle cells: role of metalloprotease and protein kinase C-delta. Mol Cell Biol 235:1581–1589
23. Lin D, Takemoto DJ (2005) Oxidative activation of protein kinase C gamma through the C1 domain. Effects on gap junctions. J Biol Chem 280:13682–13693

24. Gopalakrishna R, Anderson WB (1989) Ca2+ and phospholipid independent activation of protein kinase C by selective oxidative modification of the regulatory domain. Proc Natl Acad Sci 86:6758–6762
25. Kensler TW, Wakabayashi N, Biswal S (2007) Cell survival responses to environmental stresses via the Keap1-Nrf2-ARE pathway. Annu Rev Pharmacol Toxicol 47:89–116
26. Benhar M, Engelberg D, Levitzki A (2002) ROS, stress-activated kinases and stress signaling in cancer. EMBO Rep 3:420–425
27. Rankin EB, Giaccia AJ (2008) The role of hypoxia-inducible factors in tumorigenesis. Cell Death Differ 15:678–685
28. Warburg O (1956) On the origin of cancer cells. Science 123:309–314
29. Arner ES, Holmgren A (2000) Physiological functions of thioredoxin and thioredoxin reductase. Eur J Biochem 267:6102–6109
30. Dunbar DR, Moonie PA, Jacobs HT et al (1995) Different cellular back-grounds confer a marked advantage to either mutant or wild-type mitochondrial genomes. Proc Natl Acad Sci U S A 92:6562–6566
31. Nishikawa M, Nishiguchi S, Shiomi S et al (2001) Somatic mutation of mitochondrial DNA in cancerous and noncancerous liver tissue in individuals with hepatocellular carcinoma. Cancer Res 61:1843–1845
32. Lievre A, Blons H, Houllier AM et al (2006) Clinicopathological significance of mitochondrial D-Loop mutations in head and neck carcinoma. Br J Cancer 94:692–697
33. Prior SL, Griffiths AP, Baxter JM et al (2006) Mitochondrial DNA mutations in oral squamous cell carcinoma. Carcinogenesis 27:945–950
34. He Y, Wu J, Dressman DC et al (2010) Heteroplasmic mitochondrial DNA mutations in normal and tumour cells. Nature 464:610–614
35. Coller HA, Khrapko K, Bodyak ND et al (2001) High frequency of homoplasmic mitochondrial DNA mutations in human tumors can be explained without selection. Nat Genet 28:147–150
36. Chatterjee A, Dasgupta S, Sidransky D (2011) Mitochondrial subversion in cancer. Cancer Prev Res (Phila) 4:638–654
37. Wheelhouse NM, Lai PB, Wigmore SJ et al (2005) Mitochondrial D-loop mutations and deletion profiles of cancerous and noncancerous liver tissue in hepatitis B virus-infected liver. Br J Cancer 92:1268–1272
38. Gochhait S, Bhatt A, Sharma S et al (2008) Concomitant presence of mutations in mitochondrial genome and p53 in cancer development: a study in north India sporadic breast and esophageal cancer patients. Int J Cancer 123:2580–2586
39. Carew JS, Zhou Y, Albitar M et al (2003) Mitochondrial DNA mutations in primary leukemia cells after chemotherapy: clinical significance and therapeutic implications. Leukemia 17:1437–1447
40. Maitra A, Cohen Y, Gillespie SE et al (2004) The Human MitoChip: a high-throughput sequencing nicroarray for mitochondrial mutation detection. Genome Res 14:812–819
41. Hung WY, Lin JC, Lee LM et al (2008) Tandem duplication- triplication correlated with polycytosine stretch variation in human mitochondrial DNA D-loop region. Mutagenesis 23:137–142
42. Sanchez-Pino MJ, Moreno P, Navarro A (2007) Mitochondrial dysfunction in human colorectal cancer progression. Front Biosci 12:1190–1199
43. Sanchez-Cespedes M, Parrella P, Nomoto S et al (2001) Identification of a nucleotide repeat as a major target for mitochondrial DNA alterations in human tumors. Cancer Res 61:7015–7019
44. Galli S, Labato MI, Bal de Kier JE et al (2003) Decreased mitochondrial nitric oxide activity and hydrogen peroxide relate persistent tumoral proliferation to embryonic behavior. Cancer Res 63:6370–6377

Chapter 27
Tumor Immunology and Immunotherapy in Cancer Patients

Reinhard Marks and Roland Mertelsmann

Abstract Immunotherapy as a strategy for the treatment of cancer has experienced a major breakthrough in the recent years. Starting in the nineteenth century with the first observations by Rudolf Virchow suggesting a link between cancer and inflammation, thoroughly exploration of the immune systems cellular and humoral components together with their molecular mechanisms of activation and responses have paved the way for a deeper understanding of the immunological anti-tumor reactivity. In particular, insights into the mutual interaction between tumor growth and modulation of immune cell activation became of major impact for cancer treatment. Translation of this knowledge into the clinic resulted in a plethora of immunotherapy approaches. In this regard, tumor-specific vaccination, the use of cytokines or adoptive transfer of T cells aimed for direct augmentation of anti-tumor responses. Another approach uses the reversal of tumor induced T cell unresponsiveness by interference with T cell checkpoint inhibitors and their ligands. Especially the latter strategy resulted in until then unobserved tumor control rates in metastatic melanoma, lung cancer and other solid organ tumors. Additional advances could be achieved in redirection of T cells to tumor targets using genetic modification of T cell receptors or a new generation of antibody constructs, with impressive longterm results in hematological malignancies. Therefore, immunotherapy for cancer has been developed from an experimental approach to a new pillar of treatment for malignant diseases.

Keywords Cancer and inflammation • Immunological anti-tumor response • T cell response • Melanoma • Lung cancer • Solid tumors • Antibody constructs • Cancer immunothrapy

R. Marks • R. Mertelsmann (✉)
Department of Hematology/Oncology, University Medical Center,
Hugstetter Str. 55, 79106 Freiburg, Germany
e-mail: roland.mertelsmann@uniklinik-freiburg.de

© Springer International Publishing Switzerland 2016

R.J. Gelpi et al. (eds.), *Biochemistry of Oxidative Stress*, Advances in Biochemistry in Health and Disease 16, DOI 10.1007/978-3-319-45865-6_27

1 Introduction

The recent success of immunotherapy as cancer treatment can be viewed as consequent application of results of decades of research in tumor immunology. Therefore, in 2013 cancer immunotherapy has been attributed as breakthrough of the year by Science magazine [1]. In the beginning of tumor immunology, Rudolf Virchow observed infiltration of immune cells in human tumors [2], suggestive for an immune system response to cancer. The first attempts of immunotherapy as intentional induction of immune responses to cancer can be traced back to the surgeon William Coley. In his work at the New York Cancer Hospital about 1890 he injected a bacterial broth, the so called Coley's toxin, into the sites of unresectable soft tissue tumors [3], thereby observing tumor shrinkage, at least in some patients, due to an inflammatory process. Nevertheless, it took several decades to obtain sufficient knowledge about the cellular composition of the immune system together with the mechanisms of action of the specific immune cells about on one side and the interplay of cancer and the immune system on the other side to be able to induce powerful anti-tumor immes responses resulting in modern immunotherapy as new pillar of cancer treatment.

2 Tumor Immunology

Since the mid nineteenth century, when Rudolf Virchow proposed a link between the immune system and cancer, due to the observation of tumor grow in sites of chronic inflammation. This link has been supported by many epidemiological and experimental data. Central for the former were analyses reporting increased incidences of cancer in patients in clinical situations of immunosuppression [4]. On a pathological level, the multiple interactions between malignant tumor cells and the immune system implement several layers. On one side, certain molecular characteristics of the tumor will induce cellular immune responses, especially T cell answers, which in its entity constitute the inflammatory microenvironment [5]. On the other side, immunity directed against the tumor will result in specific modifications of tumor growth, thereby contributing to a mutual editing of the tumor and the immune system, resulting in control or progression of malignant growth [6].

3 Components of the Anti-tumor Response

More recent data could identify $CD8^+$ and $CD4^+$ T cells as the most important cellular mediators of the anti-tumor response, thereby focusing on adaptive immunity. $CD8^+$ T cells can induce apoptosis/lysis of malignant cells due to cytotoxic granules (e.g., granzymes and perforin) [7]. After activation, differentiated $CD4^+$ T cells will augment the anti-tumor response by their effector functions, in particular secretion of cytokines like interferon- γ (IFN-γ) [8]. In addition, activation of T cells might result in migration and infiltration of T cells into the tumor side. The occurrence of the latter are called tumor infiltrating lymphocytes (TIL) and are associated with a favorable clinical outcome in some tumors [9, 10]. Since the occurrence and the composition of the chronic inflammatory microenvironment surrounding the tumor side most likely reflects innate and adaptive immune responses to the malignant growth, a prognostic value of certain aspects of the microenvironment to the patients' outcome could be assumed. In this microenvironment the distribution of inflammatory cells shows local differences. The invasive margin and the tumor itself are infiltrated by granulocytes, macrophages and mast cells. While a similar pattern can be observed for T cells, NK cells are found not to be directly in contact with the tumor cells [9]. In addition to the cellular composition, the density of infiltrating cells also seems to correlate with patient outcome in certain tumors. In this regard, infiltrates showing increased $CD8^+$ T cell numbers or a so called immune signature (high expression of TH1 cells and cytotoxicity-associated genes) are associated with favourable clinical outcome in many malignant disorders like lung, liver, stomach, colorectal, breast, esophageal, bladder, melanoma, ovarian, and prostate cancers [9]. Interestingely, the same signature appears to be associated with less favorite outcome in aggressive B cell and Hodgkin lymphoma [11,12]. In the latter, the additional prognostic value of the frequency of infiltrating macrophages and the putative association with a less favorable outcome is a matter of debate since years [13].

In addition to the general features of infiltrative changes surrounding tumor cells, $CD8^+$ T cells in particular, but also dendritic cells and B cells, can also be found in many tumors forming specific cell aggregates in the invasive margin, called tertiary lymphoid structures (TLS). This TLS appear as sites resembling ongoing immune responses with induction of germinal centers very similar to lymphoid structures found in other tissues due to infections or chronic inflammatory disorders. It has been hypothesized that TLS represent an active local anti-tumor immune response, where in situ antigen presentation and lymphocyte activation can occur under a protected environment [14] . In favor for TLS as marker of reactive anti-tumor immunity are also data showing a correlation between TLS density and improved clinical outcome in lung cancer [15, 16], renal cell cancer (RCC) [17], melanoma [18], and breast cancer [19, 20]. Naturally occuring regulatory $FoxP3^+$ T cells (Treg), which are key in controlling autoimmunity and T cell homoeostasis by suppresson of activated T cells, can also be found in the tumor environment where they might exert their immunosuppressive function, thereby limiting the activity of tumor-infiltrating cytotoxic T cells [21]. In this regard, analyses on their prognostic value showed mixed results. While $FoxP3^+$ T cells were associated with poor prognosis in hepatocellular cancer and good outcome in colorectal cancer, the impact in other cancers revealed inconsistent results [22, 23].

In a model of tumor induced T cell responses, activation of the T cell will be initiated by antigens expressed by the tumor cell, of which the ones encoded by mutated genes might be of particular importance, since they should increase the immunogenicity of the tumor [24]. Indeed, tumor-specific lymphocytes can often be detected in patients with cancer and TIL can induce tumor cell lysis in vitro and in vivo [25]. A correlation between the mutational load and anti-tumor response could be observed in patients with colorectal cancers, where microsatellite instability (MSI) fosters the expression of thousands of new antigens on tumor cells. Characteristically, MSI+ tumors have a prominent $CD8^+$ T cell infiltration and are associated with favorable clinical outcome [26]. As important mediators of the orchestration of the cellular responses, cytokines and chemokines are centrally involved in induction and maintenance of anti-tumor responses, e.g. IFN-γ, IL-12, CXCL9 and CXCL1 are acting on $CD8^+$ T cells by recruitment to the tumor side and activation [27, 28]. In order to overcome the selection pressure mediated by the immune system anti-tumor response, tumor cells developed certain escape mechanisms. A common motif of these mechanisms is the modulation of immune responses either by masking antigens, which could elicit T cell responses, or by downregulation of immune effector functions, thereby directly acting as immunosuppressants. In more detail, three mechanisms of cancer escape mechanisms could be identified and might be exploited for therapeutic interventions: first, the loss of expression of major histocompatibility complex (MHC) class I molecules on tumor cells reduces T cell directed cytotoxic cell lysis [29]. Second, expression of immunosuppressive molecules by the cancer cell or the tumor microenvironment (*e.g.*, IL-10 and TGF-β) will hamper the cytotoxic and proliferative capacity of T cells [30]. Third, increased surface expression of ligands like PD-L1 and PD-L2 on the cancer cell will induce inhibitory signals on activated T cells mediated by PD-1, resulting in reduced activation and proliferation of $CD8^+$ and $CD4^+$ T cells, and in particular TILs. In addition to the PD-1/PD-L1 axis, similar effects are being proposed for the cancer cell expression of the ligands for TIM-3 and LAG-3, two additional inhibitory receptors expressed on T cells [31]. Interestingely, Treg cells as putative cellular regulator of anti-tumor responses in the microenvironment express high levels of immune checkpoint receptors, including CTLA-4 and PD-1, wherein these inhibitors of effector T cells paradoxically enhance activity of Treg cells [32].

4 Immunology of the Anti-tumor Response

Since decades there is a debate about the proposed function of the immune system in active surveillance of malignant transformation and clonal expansion of cancer cells. Although some parts of the scientific discussion might be determined by the idea, that "good" forces of the immune system should combat the "evil" cancer, a growing body of evidence could be generated, answering the question in a more differentiated way [6]. One approach focussing on the impact of immunocompetence, as defined by proper T cell function in a quantitative and qualitative perspective, on incidences of malignant tumors used experimental data in animals and epidemiological analyses in humans. In murine models of T cell deficiency, extensive studies performed using athymic nude mice failed to provide any evidence of an increased susceptibility to the development of spontaneous tumors [33]. On the other hand, RAG-deficient mice (lacking T, B, and NK-T cells) do spontaneously develop gastrointestinal epithelial malignancies [34]. Mice with severe combined immunodeficiency (SCID), kept under pathogen-free conditions, have a high incidence of spontaneous thymic lymphomas, and about 2 % of breeders developed a variety of non-thymic tumors [35]. Apart from this very artificial model systems, data in immunocompromised humans, *e.g.* patients with immunosuppressive medication after organ or stem cell transplantation, show an increased incidence of tumors such as skin carcinomas and lymphomas [36]. But a matter of debate in this regard is, if this observation is mainly due to an increased susceptibility to reactivation by potentially oncogenic viruses (human papillomavirus, hepatitis C virus, and Epstein-Barr virus). Nevertheless, although the incidences of tumors of epithelial surfaces seem to be increased in patients after hematologic stem cell transplantation, the impact of longterm effects of preparative regimens and procedure induced side effects might outweigh by far post-transplantation immunodeficiency [37].

Another critical issue in anti-tumor responses might be the frequency and activation status of tumor antigen specific T cells. Focussing on spontaneously occurring tumors, expression of tumor specific antigens will be by far exceled by presentation of self-antigens, and hence thymus-derived T cells responding to these antigens would have been deleted in the thymus. In addition, even low-affinity T cells specific for these self-antigens, who have escaped deletion and migrated to the periphery, would be held in check by the growing tumors due to the mechanisms described above. Considering tumor-specific mutations, there are data, that tumors harboring a high frequency of mutations that results in generation of de novo antigens (*e.g.*, BCR-Abl, mutated forms of p53) do appear foreign to the immune system [38, 39]. In this cases, therefore, both T cells and antibodies directed to tumor antigens should theoretically be able to eliminate the tumor. However, in other cases of tumor specific mutations it is not clear, how likely it is, that a point mutation in a single oncogene would give rise to an antigenic peptide, capable to induce a T or B cell response. Apart from tumor specific T cells, in certain cases anti-tumor responses might be elicited by pathogen specific T cells that cross-react with neoantigens acquired through somatic mutation [40]. Nevertheless, the bulk of the observations on immune surveillance have in fact shown that in principle T cells and antibodies are capable to recognize and eliminate tumors [41].

For clinical purposes, considering the outcome of the patient, the interplay of the malignancy and the immune system over time is of extraordinary importance. In order to better understand this interplay and the actual orgin of tumor immunity Schreiber and colleagues [42] suggested a three step model, known as cancer immunoediting hypothesis: elimination, equilibrium, and escape. In the first phase, tumors developing before being clinically apparent are destroyed and eliminated by both innate and adaptive immune mechanisms. If clonal variants of the tumor arise and fail to be eliminated, cells of the adaptive immune system may restrain their growth during the lifetime of their host and a state of equilibrium is reached. Constant immune selective pressure of genetically unstable tumors may, however, lead to clonal variants that can no longer be recognized by effector T cells because they have downregulated antigen or MHC, secreted immunosuppressive cytokines such as TGF-β or IL-10, or recruited Treg cells in their environment. They then enter the escape phase and become clinically apparent.

5 Immunotherapy

In parallel to this growing body of insights into keyplayer and mechanisms of anti-tumor immunity, for many years translational researchers already tried to exploit this accumulating knowledge for clinically relevant cancer therapy in humans (for a review [43]). One major focus in the beginning was the modulation of tumor infiltrating cells in melanoma patients either by adoptive cell therapy, cytokine-treatment with interleukin-2 (IL-2) or interferon-α (IFN-α) or vaccines aimed at stimulating T cell immune responses. However, the recent therapeutic breakthrough of immunotherapy in cancer rests mainly on developments on the manipulation of negative regulators of T cell function by monoclonal antibodies and to lesser extent on dendritic cell therapy. In a first wave, the initiating agents of the immuno-oncology era, ipilimumab (an anti-CTLA-4 antibody) and sipuleucel-T (an autologous dendritic cell therapy), were approved for clinical purposes based on survival improvements in randomized phase III trials in 2010 (sipuleucel-T) and 2011 (ipilimumab). This new developments were soon followed by a second wave of immuno-oncology agents with PD1- and PD-L1-blocking antibodies at the centre. The first PD1-targeted agents, pembrolizumab and nivolumab, were approved by the FDA and the EMA in 2014 (pembrolizumab) and 2015 (nivolumab), and the anti-PD-L1 agents atezolizumab and durvalumab are in pivotal clinical trials. In a second approach to overcome barriers erected by tumor cells and their microenvironment, the development of modern gene transfer technologies facilitated manipulations of autologous T cells to target antigens more efficiently. Expression of CD19-specific chimeric antigen receptors (CARs) enabled the use of this cell product as treatment of pediatric and adult acute lymphoblastic leukemia in 2014.

6 Cytokines and Vaccines

Application of cytokines, recombinant interleukin-2 (IL-2) and interferon (IFN), were in the beginning of clinical immunotherapy. IFN-α has been shown to significantly increase progression free survival (PFS) with less benefit in overall survival (OS) in an adjuvant setting in melanoma patients [45]. In an attempt to induce T cell activation and proliferation, high dose IL-2 has been used in patients with metastatic melanoma and renal cell carcinoma with limited success, particularly in metastatic disease. The treatment was accompanied by severe side effects. Nevertheless, complete response rates up to 10 % [46, 47] have been reported, with even some long lasting remissions [48]. Therefore, immunotherapy by boosting T cell responses had been apparently shown in certain circumstances the potential to induce durable responses even in advanced invasive cancers.

Another way to stimulate the immune system hase been accomplished by several vaccination strategies. Cancer vaccination approaches included peptide and protein vaccines, recombinant vector-based vaccines, dendritic cell vaccines and viruses. In general, vaccination by dermal injection of the antigen together with adjuvant activates resting dendritic cells, followed by migration to the local lymph node and presentation of the antigenic peptide in the context of the MHC. The antigen will be recognized by specific T cell receptors (TCR) on $CD4^+$ or $CD8^+$ T cells resulting in acivation, expansion and acquisition of effector functions of antigen specific T cell clones. MAGE-A3 is a cancer testis antigen [49], that is expressed at significant levels only in the testes, where it cannot be detected by T cells. Therefore, since MAGE-A3 expression is increased in approximately 35 % of lung cancer tumors [50], vaccination with this antigen has been regarded as suitable approach for lung cancer immunotherapy. In a very large phase III study with MAGE-A3-positive early stage non small cell lung cancer (NSCLC) 2312 patients were randomly assigned to receive MAGE-A3 vaccine ($n = 1515$) or placebo ($n = 757$) as adjuvant therapy [51] after surgical resection. Disease-free survival was not significantly different between the two groups.

In addition to lung cancer, MAGE-A3 is also expressed in approximately 65 % of melanomas [50]. In early studies, a fusion protein, composed of MAGE-A3 plus influenza protein D, was administered as injection without adjuvant in a phase I/II trial to 32 patients with metastatic melanoma whose tumors expressed MAGE-A3 [52]. Among the 26 patients who received ≥4 vaccinations, only a single partial response and four mixed responses were observed. In extension of this approach, using a MAGE-A3 vaccine together with AS02B or AS15 adjuvants in another randomized phase II trial resulted in an improved MAGE-A3-specific T cell and antibody responses, as well as objective tumor responses using the AS15 adjuvant [44]. On the basis of these findings, additional phase II and III trials are being conducted for patients with advanced melanoma.

The glycoprotein mucin-1 (MUC-1) is a glycosylated transmembrane protein expressed in normal tissue, especially epithelial cells [53]. However, MUC-1 can be overexpressed or aberrantly glycosylated in cancer, including NSCLC, in which approximately 60–70 % of cases MUC-1 antigen can be detected. To target MUC-1 in NSCLC, a lyophilized liposomal product has been generated by combination of a 25-amino-acid lipopeptide (BLP-25), monophosphoryl lipid A (MPL) and three lipids. In a randomized trial in 171 NSCLC patients, L-BLP-25 has been tested as vaccine compared to best supportive care in those patients whose disease had stabilized after initial treatment with chemotherapy (with or without radiotherapy) [54]. Since cyclophosphamide has been shown to improve immune responses to vaccinations [55], patients in the vaccine arm received also low dose cyclophosphamide to increase vaccination efficacy. The vaccine was well tolerated with few treatment-related adverse events However, median overall survival (primary end point), was not significantly different between arms: 17.2 months in the L-BLP-25 group versus 13 months in the best supportive care arm ($P = 0.066$). Interestingly, patients with advanced locoregional disease (stage IIIB) seemed to show some clinical benefit, with a median survival time of 30.6 months versus 13.3 months. On the basis of these results, a phase III trial was designed to include only patients with potentially curable stage IIIB disease. In this trial, a total of 1,464 patients with unresectable stage III disease and stable or responding disease after definitive concurrent chemotherapy and radiation were randomized to either L-BLP-25 vaccine or observation. Unfortunately, median overall survival was not significantly improved in the whole population (25.6 months vs. 22.3 months) [56].

In patients with renal cell carcinoma a set of nine tumor associated peptides, identified by eluation of surface peptides bound to class MHC I HLA-A*02 molecules, has been used together with GM-CSF as adjuvant. Combined phase I and phase II data on this agent were recently reported [57]. In the phase I part 28 HLA-A*02 patients received up to eight vaccinations, only one patient showed a partial response, 16 patients had disease progression and 11 patients had stable disease. Immune responses to the targeted peptides were detected in several of the treated patients. In the phase II part patients with RCC were randomly assigned to receive multi-peptide vaccination, with or without a single dose of intravenous cyclophosphamide. Again, objective tumour regressions were relatively rare, with only a single partial response among 64 patients. Nevertheless, there was a trend towards improved overall survival in the arm receiving vaccine plus low-dose cyclophosphamide, but this was not statistically significant. The only vaccine up to date to show documented improved overall survival in clinical practice in at least on phase III trial was the dendritic cell vaccine sipuleucel-T in prostate cancer resulting in subsequent FDA approval [58].

In conclusion, cancer vaccines designed to induce T cell responses against tumour cells have been at the centre of cancer immunotherapy for decades but have not yet delivered the desired clinical results [59]. One prominent reason for this may be that these vaccines did not address the role of immunosuppression in cancer; respectively current vaccines may be unable to circumvent effectively the multiple immunosuppressive mechanisms operative in the tumor microenvironment [29], which can now be modulated by checkpoint blockers.

7 Checkpoint Blockade

Full T cell activation activation initiated by recognition of antigen-derived peptides displayed in the context of MHC molecules on the surface of antigen presenting cells (APCs) is depending on two signals. TCR-MHC interaction induces a central stimulatory pathway, but full T cell activation is only achieved in context with costimulatory signals derived from the surface molecule CD28 and its ligands CD80 (B7-1) and CD86 (B7-2) on the APC [60]. Counteracting to the activating T cell responses induced by costimulatory receptors, ligation of negative regulatory surface receptors, the so-called inhibitory checkpoints, leads to downregulation of TCR mediated T cell activation. Therefore, immune checkpoints, in particular CTLA-4 (CD152) and PD-1 (CD279), have been proven to be important regulators of immune homoeostasis and autoimmunity [62].

CTLA-4 is expressed on activated T cells, and its main physiological function is to down-modulate $CD4^+$ T helper cell activity by competing with CD28 for the ligands CD80 and CD86 with a 10–100 fold higher affinity. In addition, it increases $CD4^+CD25^+FoxP3^+$ Treg-cell-mediated immunosuppression [62]. CTLA-4 acts during the early stages of activation of naive and memory T cells [63]. In crucial experiments Allison and colleagues [64] could prove the therapeutic potential of immunomodulation by blockade of CTLA-4 action by monoclonal antibodies. In their preclinical studies, using transplantable murine colon carcinoma and fibrosarcoma lines, blocking CTLA-4 induced anti-tumor T cells to acquire effector function.

PD-1 is expressed not only following T cell activation but also on activated B cells and NK cells. PD-1 expression can be regulated by interferon-γ during chronic antigen-induced T cell stimulation and on exhausted T cells, thereby limiting the activity of T cells that have already been activated due to different immune reactions [65, 66]. Furthermore, signals from inflammatory tissues induce the expression of the PD-1 ligands, PD-L1 (B7-H1) or PD-L2 (B7-DC), that downregulate T cell activity, thereby limiting collateral damage and even autoimmunity. As with CTLA-4, mutations in PD-1 have also been associated with autoimmunity in humans [67]. The exploitation of this regulatory receptors as targets for immunotherapy were driven by the hypothesis that antagonists directed to these immune inhibitory checkpoints should induce tumorspecific T cell proliferation, resulting in improved tumor control. The hypothesis of PD-1 modulated tumorspecific T cell anergy was supported by demonstrating the expression of the ligand PD-L1 on the surface of many different cancer cells [68]. Thereby, interfering with the PD-L1/PD-1 axis might overcome this anergy and improve tumor-reactive T cell responses. Furthermore, PD-L1 on tumor cells and the surrounding stromal cells is upregulated by interferon-γ secreted by the activated T cells that have infiltrated the tumor environment [69]. In consequence, blocking monoclonal antibodies directed against CTLA-4 and PD-1 or PD-L1 have entered clinical trials. Two anti-CTLA-4 antibodies have shown impressive responses in clinical trials: ipilimumab and tremelimumab. As a proof of principle for checkpoint modulation, ipilimumab produced significant survival benefits in metastatic melanoma in two landmark phase III studies in pre-treated and untreated patients, reducing the risk of death by 28–38 % [70, 71]. In addition, the survival curves in both trials showed a plateau, suggesting that at least some patients experience long-term survival in metastatic melanoma, which have not been seen in this disease besides some observations with IL-2 treatment.

Nevertheless, when ipilimumab and tremelimumab were initially investigated under the chemotherapy drug-development paradigm, responses according to criteria such as the Response Evaluation Criteria in Solid Tumors (RECIST) were low and side effects were inflammatory in nature, but different to those observed with other cancer drugs, and, in some cases, these side effects were severe [72, 73].

Capturing the observed durable responses was of particular interest and the altered survival end points (*e.g.* 1-year or 2-year survival) added meaningful information to the standard measure of median survival. Eventually, new criteria for clinical response in immunotherapy with checkpoint inhibitors termed immune-related response criteria (irRC), and associated clinical end points, such as immune-related overall response rate (irORR), immune-related disease control rate (irDCR), immune-related progression-free survival (irPFS) have been developed [74, 75]. In addition, the "new" specific toxicities observed during immunotherapy, like diarrhea, pneumonitis, hepatitis, autoimmune endocrinopathy and other autoimmune-inflammatory side effects, have been systemically characterized and termed immune-related adverse events (irAEs) [76]. As a major advantage in the field, activity of checkpoint inhibitors could be found outsite malignant melanoma. In patients with advanced non-small cell lung cancer (NSCLC) ipilimumab improved irPFS significantly (median 5.7 vs. 4.6 months) compared to chemotherapy [77] with some improvement in overall survival.

Several anti-PD-1/PD-L1 antibodies are being tested in clinical trials: nivolumab, pembrolizumab, pidilizumab, and AMP-224, targeting PD-1; and BMS-935559, MEDI4736, MPDL3280A, and MSB0010718C targeting PD-L1. While the occurrence of immune-related side effects appeared to be less frequent in patients treated with PD-1 inhibitors, recent phase III studies of nivolumab in patients with untreated metastatic melanoma showed again superiority over standard dacarbazine chemotherapy (OS advantage after 1 year of 72.9 % vs. 42.1 %) with even higher rates of survival observed in patients treated with nivolumab compared to those previously reported with ipilimumab [78, 79]. Similar results could be shown with pembrolizumab in a recent phase III study [80]. Since (re)-activation of tumor responsive T cells might be a general therapeutic approach to cancer independent of the actual malignant entity, several additional trials have investigated PD-1 inhibition in other solid tumors and hematologic diseases.

Very impressive data in this regard could be generated for NSCLC. In a landmark phase III trial 272 patients with advanced NSCLC were randomly assigned to receive nivolumab or docetaxel. Median OS improved significantly for nivolumab treated patients (9.2 vs. 6.0 months, $p < 0.001$), challenging chemotherapy as second line treatment in advanced NSCLC [81]. A second phase III trial could confirm the OS advantage of nivolumab [82]. Interestingly, in this trial PFS for nivolumab treated patients was even inferior to docetaxel, arguing that the observed improvement of overall OS reflects the oustanding durable responses in a subgroup of patients treated with PD-1 inhibition. Similar results could be obtained with pembrolizumab [83].

In addition to melanoma and lung cancer, PD-1 inhibition showed activity in several other cancers [43] with most mature data for metastatic renal cell carcinoma [84], colon cancer [61], prostate cancer [86] and relapsed Hodgkins lymphoma [87], in each clinical setting superior to established treatment regimens. Therefore, checkpoint blockade will most likely change the standard of care for treatment in this advanced disease situations. With these impressive results, combinations of checkpoint inhibitory antibodies with the current standard agents or other checkpoint inhibitors have been investigated. Again, in untreated metastatic melanoma patients combining ipilimumab and nivolumab resulted in increased objective response rate up to 61 % (11 % with ipilimumab monotherapy. Nevertheless, in this combination grade 3-4 adverse events have been observed in 54 % patients with 38 % treatment discontinuation because of side effects. This data emphasize the potential of checkpoint inhibition for, although not chemotherapeutic in nature and still serious, "new" side effects. Several additional trials for combinations in melanoma but also other cancer are ongoing.

In another approach to target the PD-1/PD-L1 axis, a monoclonal antibody against PD-L1, atezolizumab, was tested as second-/third-line therapy in NSCLC in a phase II trial. Comparing with docetaxel PD-L1 inhibition resulted in improved median OS (12.6 months vs. 9.7 months, p = 0.04), with no difference in PFS and objective response rates [88]. Interestingly, in a small series PD-L1 inhibition showed activity in metastatic bladder cancer [89]. Again, a plethora of trials in several cancers are ongoing. In addition, inhibitors against other immune checkpoint regulators expressed on the cell surface, such as TIM3 and LAG3, and agonist antibodies acting on T cell costimulatory receptors, such as OX40 and CD137, are being developed for cancer immunotherapy. The impressive activity of inhibitory checkpoint blockade in controlling malignant growth among many different cancer types underscores the universal nature of the checkpoint-modulatory antibodies. Their mechanisms of action are entirely centred on the immune system and are independent of cancer histology or specific mutations. Therefore, attempts to identify biomarkers as predictors of response focus also on immune system parameters and the tumor microenvironment [87, 90, 91] or immune cells [92, 93]. Unfortunately, up to date no universal markers could be identified [43].

8 Adoptive T Cell Therapy and T Cell Activation by CARS and BiTE

The idea of eliciting a long lasting tumor directed T cell response by adoptive transfer could act as cancer vaccines inspired several approaches of T cell cancer immunotherapy. Two of them, using the transfer of the patients own reactive or genetically modified T cells will be discussed. For many years Rosenberg and colleagues focused directly on the patient's autologous tumor-infiltrating lymphocytes (TILs) when those exist and were accessible. A major advantage of using this lymphocytes would be the wide range of defined, like MELAN-A, gp100, MAGE-A3, NY-ESO-1, and, quantitatively more prominent, as yet undefined antigens being recognized by this T cells. In TIL therapy, T cells retrieved from surgical melanoma specimens are expanded in vitro under activating conditions and reinfused with IL-2 [94] and/or after host lymphoablative preconditioning [95] with cyclophosphamide and fludarabine chemotherapy. Multiple clinical trials in metastatic malignant melanoma could demonstrate meaningful and durable clinical response rates, with overall response rates up to 50 % and complete remisson rates of 20 % in these heavily pretreated patients [7, 96–98]. Nevertheless, this promising data need to be confirmed in larger trials with more consistency among the protocols used for TIL expansion. Further development in human gene transfer technologies enabled the transduction of human T lymphocytes using retroviral vectors [99, 100]. Therefore, transduction of antigen receptors became possible. Next to approaches using specific but physiological TCR [101], which have been used in melanoma [102], targeting NY-ESO-1, and other clinical scenarios, another technique of transducing synthetic receptors now known as CARs set out to revolutionize the field of adoptive cell transfer. CARs are artificial, composite chimeric receptors for antigen that integrate principles of B cell and T cell antigen recognition. Therefore CAR enable antigen recognition without MHC restriction, thereby overcoming the long known obstacles of adoptive T cell transfer experiments. It also makes those cells capable of antigen recognition, even if no MHC expression is detectable, as described as putative tumor escape mechanism. Once antigen has been recognized, T cell function will be executed, clonal proliferation occurs and longterm persistence of CAR cells can be observed [103]. With this technique, using CD19 as antigen, very impressive results in CLL and acute lymphoblastic leukemia with durable response rates as high as 90 % in some studies were reported [85, 104–108].

Another very important line of development in the immunotherapy of cancer with regard to stimulation of tumor directed T cell responses was the generation of new antibody constructs, the so called bi-specific T cell engager (BiTE). Not a complete antibody, BiTE structure is composed of two antigen recognizing scFv-fragments linked by a peptide bridge, thereby capable of binding to two different antigens. Since one antigen recognition side specifically binds to CD3, these antibodies can transiently induce a cytolytic synapse between a cytotoxic T cell and the cancer target cell. Consequently, granzyme containing granules and the pore-forming protein perforin fuse with the T-cell membrane and discharge their toxic content. By this mechanism, BiTE molecules can engage all cytotoxic T cells of a patient for redirected lysis of tumor cells. The first of this bispecific BiTE antibody construct tested, blinatumomab, shows a specifity for CD19 and CD3. Therefore, it was used for targeting $CD19^+$ B cell malignancies, showing impressive results in relapsed/refractory ALL [109, 110] and diffuse large B cell lymphoma [111]. The activity as salvage therapy in ALL led to a FDA approval in 2015. The last years have seen a plethora of new therapeutic strategies in immunotherapy to cancer. Both, adoptive cell transfer and the use of antibodies focus on T cell directed anti-tumor responses. After many years of setbacks, consequent exploration of the cellular and molecular determinants of the anti-cancer immune response could be finally translated into effective treatments with impressive results in hard to treat solid tumors. The excitement of cancer physicians being provoked by these observations today is a good reminiscence of what William Cooley might have felt, when he saw shrinkage of unresectable tumors by his "toxin".

9 Conclusions

In the beginning of clinical immunotherapy strategies Wiliam Coley initiated anti-cancer treatment by deliberately inducing inflammatory responses. As a result of deeper insights into cellular and molecular mechanisms of the immunological anti-tumor response, cytokines and vaccination were used to augment the tumor directed immune reactivity. Nevertheless, basis for the recently observed major breakthrough was the exploration of T cell anti-tumor responses on a cellular and molecular level. Exploitation of the tumor microarchitecture led to adoptive T cell transfer approaches and redirection of T cells to tumor targets using genetic modification of T cell receptors or a new generation of antibody constructs. Eventually, modulation of molecular mechanism of tumor induced T cell unresponsiveness by antibodies, in particular interference with CTLA-4 and PD-1/PD-L1 pathways, has been shown to be effective in a wide variety of tumors including malignant melanoma and lung cancer, underscoring this approach as putative universal "achilles heel" in human cancer. In parallel to this stepwise development of immunotherapy, clinical results showed continuous improvements up to until then not considered possible longterm results in tumor control. Therefore, the enthusiasm and promises of modern immunotherapy exerts on todays oncologists likely reflect what Coley might have experienced 100 years ago.

References

1. Couzin-Frankel J (2013) Breakthrough of the year 2013. Cancer immunotherapy. Science 342:1432–1433
2. Balkwill F, Mantovani A (2001) Inflammation and cancer: back to Virchow? Lancet 357:539–545
3. Coley WB (1891) II. Contribution to the knowledge of Sarcoma. Ann Surg 14:199–220
4. Bower M, Palmieri C, Dhillon T (2006) AIDS-related malignancies: changing epidemiology and the impact of highly active antiretroviral therapy. Curr Opin Infect Dis 19:14–19
5. Giraldo NA, Becht E, Vano Y et al (2015) The immune response in cancer: from immunology to pathology to immunotherapy. Virchows Arch 467:127–135
6. Miller JF, Sadelain M (2015) The journey from discoveries in fundamental immunology to cancer immunotherapy. Cancer Cell 27:439–449
7. Restifo NP, Dudley ME, Rosenberg SA (2012) Adoptive immunotherapy for cancer: harnessing the T cell response. Nat Rev Immunol 12:269–281
8. Hung K, Hayashi R, Lafond-Walker A et al (1998) The central role of CD4(+) T cells in the antitumor immune response. J Exp Med 188:2357–2368
9. Fridman WH, Pages F, Sautes-Fridman C et al (2012) The immune contexture in human tumours: impact on clinical outcome. Nat Rev Cancer 12:298–306
10. Galon J, Costes A, Sanchez-Cabo F et al (2006) Type, density, and location of immune cells within human colorectal tumors predict clinical outcome. Science 313:1960–1964
11. Muris JJ, Meijer CJ, Cillessen SA et al (2004) Prognostic significance of activated cytotoxic T-lymphocytes in primary nodal diffuse large B-cell lymphomas. Leukemia 18:589–596

12. Scott DW, Chan FC, Hong F et al (2013) Gene expression-based model using formalin-fixed paraffin-embedded biopsies predicts overall survival in advanced-stage classical Hodgkin lymphoma. J Clin Oncol 31:692–700
13. Kamper P, Bendix K, Hamilton-Dutoit S et al (2011) Tumor-infiltrating macrophages correlate with adverse prognosis and Epstein-Barr virus status in classical Hodgkin's lymphoma. Haematologica 96:269–276
14. Dieu-Nosjean MC, Goc J, Giraldo NA et al (2014) Tertiary lymphoid structures in cancer and beyond. Trends Immunol 35:571–580
15. Dieu-Nosjean MC, Antoine M, Danel C et al (2008) Long-term survival for patients with non-small-cell lung cancer with intratumoral lymphoid structures. J Clin Oncol 26:4410–4417
16. Goc J, Germain C, Vo-Bourgais TK et al (2014) Dendritic cells in tumor-associated tertiary lymphoid structures signal a Th1 cytotoxic immune contexture and license the positive prognostic value of infiltrating CD8+ T cells. Cancer Res 74:705–715
17. Giraldo NA, Becht E, Pages F et al (2015) Orchestration and prognostic significance of immune checkpoints in the microenvironment of primary and metastatic renal cell cancer. Clin Cancer Res 21:3031–3040
18. Cipponi A, Mercier M, Seremet T et al (2012) Neogenesis of lymphoid structures and antibody responses occur in human melanoma metastases. Cancer Res 72:3997–4007
19. Gu-Trantien C, Loi S, Garaud S et al (2013) CD4(+) follicular helper T cell infiltration predicts breast cancer survival. J Clin Invest 123:2873–2892
20. Martinet L, Garrido I, Filleron T et al (2011) Human solid tumors contain high endothelial venules: association with T- and B-lymphocyte infiltration and favorable prognosis in breast cancer. Cancer Res 71:5678–5687
21. Curiel TJ (2007) Tregs and rethinking cancer immunotherapy. J Clin Invest 117:1167–1174
22. de Leeuw RJ, Kost SE, Kakal JA et al (2012) The prognostic value of FoxP3+ tumor-infiltrating lymphocytes in cancer: a critical review of the literature. Clin Cancer Res 18:3022–3029
23. Zhang X, Kelaria S, Kerstetter J et al (2015) The functional and prognostic implications of regulatory T cells in colorectal carcinoma. J Gastrointest Oncol 6:307–313
24. Matsushita H, Vesely MD, Koboldt DC et al (2012) Cancer exome analysis reveals a T-cell-dependent mechanism of cancer immunoediting. Nature 482:400–404
25. Lennerz V, Fatho M, Gentilini C et al (2005) The response of autologous T cells to a human melanoma is dominated by mutated neoantigens. Proc Natl Acad Sci USA 102:16013–16018
26. Cunningham D, Atkin W, Lenz HJ et al (2010) Colorectal cancer. Lancet 375:1030–1047
27. de Chaisemartin L, Goc J, Damotte D et al (2011) Characterization of chemokines and adhesion molecules associated with T cell presence in tertiary lymphoid structures in human lung cancer. Cancer Res 71:6391–6399
28. Mlecnik B, Tosolini M, Charoentong P et al (2010) Biomolecular network reconstruction identifies T-cell homing factors associated with survival in colorectal cancer. Gastroenterology 138:1429–1440
29. Drake CG, Jaffee E, Pardoll DM (2006) Mechanisms of immune evasion by tumors. Adv Immunol 90:51–81
30. Zou W (2005) Immunosuppressive networks in the tumour environment and their therapeutic relevance. Nat Rev Cancer 5:263–274
31. Topalian SL, Drake CG, Pardoll DM (2015) Immune checkpoint blockade: a common denominator approach to cancer therapy. Cancer Cell 27:450–461
32. Simpson TR, Li F, Montalvo-Ortiz W et al (2013) Fc-dependent depletion of tumor-infiltrating regulatory T cells co-defines the efficacy of anti-CTLA-4 therapy against melanoma. J Exp Med 210:1695–1710
33. Rygaard J, Povlsen CO (1976) The nude mouse vs. the hypothesis of immunological surveillance. Transplant Rev 28:43–61

34. Shankaran V, Ikeda H, Bruce AT et al (2001) IFNgamma and lymphocytes prevent primary tumour development and shape tumour immunogenicity. Nature 410:1107–1111
35. Huang P, Westmoreland SV, Jain RK et al (2011) Spontaneous nonthymic tumors in SCID mice. Comp Med 61:227–234
36. Penn I, Starzl TE (1973) Immunosuppression and cancer. Transplant Proc 5:943–947
37. Roziakova L, Bojtarova E, Mistrik M et al (2011) Secondary malignancies after hematopoietic stem cell transplantation. Neoplasma 58:1–8
38. Coulie PG, Van den Eynde BJ, van der Bruggen P et al (2014) Tumour antigens recognized by T lymphocytes: at the core of cancer immunotherapy. Nat Rev Cancer 14:135–146
39. Gubin MM, Zhang X, Schuster H et al (2014) Checkpoint blockade cancer immunotherapy targets tumour-specific mutant antigens. Nature 515:577–581
40. Snyder A, Makarov V, Merghoub T et al (2014) Genetic basis for clinical response to CTLA-4 blockade in melanoma. N Engl J Med 371 2189–2199
41. Boon T, Van den Eynde BJ, Hirsch H et al (1994) Genes coding for tumor-specific rejection antigens. Cold Spring Harb Symp Quant Biol 59:617–622
42. Schreiber RD, Old LJ, Smyth MJ (2011) Cancer immunoediting: integrating immunity's roles in cancer suppression and promotion. Science 331:1565–1570
43. Hoos A (2016) Development of immuno-oncology drugs - from CTLA4 to PD1 to the next generations. Nat Rev Drug Discov 15:235–247
44. Kruit WH, Suciu S, Dreno B et al (2013) Selection of immunostimulant AS15 for active immunization with MAGE-A3 protein: results of a randomized phase II study of the European Organisation for Research and Treatment of Cancer Melanoma Group in Metastatic Melanoma. J Clin Oncol 31:2413–2420
45. Wheatley K, Ives N, Hancock B et al (2003) Does adjuvant interferon-alpha for high-risk melanoma provide a worthwhile benefit? A meta-analysis of the randomised trials. Cancer Treat Rev 29:241–252
46. Atkins MB, Lotze MT, Dutcher JP et al (1999) High-dose recombinant interleukin 2 therapy for patients with metastatic melanoma: analysis of 270 patients treated between 1985 and 1993. J Clin Oncol 17:2105–2116
47. Rosenberg SA (2014) IL-2: the first effective immunotherapy for human cancer. J Immunol 192:5451–5458
48. Hughes T, Klairmont M, Broucek J et al (2015) The prognostic significance of stable disease following high-dose interleukin-2 (IL-2) treatment in patients with metastatic melanoma and renal cell carcinoma. Cancer Immunol Immunother 64:459–465
49. Simpson AJ, Caballero OL, Jungbluth A et al (2005) Cancer/testis antigens, gametogenesis and cancer. Nat Rev Cancer 5:615–625
50. Brichard VG, Lejeune D (2007) GSK's antigen-specific cancer immunotherapy programme: pilot results leading to Phase III clinical development. Vaccine 25(S2) B61–B71
51. Vansteenkiste JF, Cho BC, Vanakesa T et al (2016) Efficacy of the MAGE-A3 cancer immunotherapeutic as adjuvant therapy in patients with resected MAGE-A3-positive non-small-cell lung cancer (MAGRIT): a randomised, double-blind, placebo-controlled, phase 3 trial. Lancet Oncol 17:822–835
52. Kruit WH, van Ojik HH, Brichard VG et al (2005) Phase 1/2 study of subcutaneous and intradermal immunization with a recombinant MAGE-3 protein in patients with detectable metastatic melanoma. Int J Cancer 117:596–604
53. Sangha R, Butts C (2007) L-BLP25: a peptide vaccine strategy in non small cell lung cancer. Clin Cancer Res 13:4652–4654
54. Butts C, Murray N, Maksymiuk A et al (2005) Randomized phase IIB trial of BLP25 liposome vaccine in stage IIIB and IV non-small-cell lung cancer. J Clin Oncol 23:6674–6681
55. North RJ (1982) Cyclophosphamide-facilitated adoptive immunotherapy of an established tumor depends on elimination of tumor-induced suppressor T cells. J Exp Med 155:1063–1074

56. Butts C, Socinski MA, Mitchell PL et al (2014) Tecemotide (L-BLP25) versus placebo after chemoradiotherapy for stage III non-small-cell lung cancer (START): a randomised, double-blind, phase 3 trial. Lancet Oncol 15:59–68
57. Walter S, Weinschenk T, Stenzl A et al (2012) Multipeptide immune response to cancer vaccine IMA901 after single-dose cyclophosphamide associates with longer patient survival. Nat Med 18:1254–1261
58. Kantoff PW, Higano CS, Shore ND et al (2010) Sipuleucel-T immunotherapy for castration-resistant prostate cancer. N Engl J Med 363:411–422
59. Goldman B, De Francesco L (2009) The cancer vaccine roller coaster. Nat Biotechnol 27:129–139
60. Harding FA, McArthur JG, Gross JA et al (1992) CD28-mediated signalling co-stimulates murine T cells and prevents induction of anergy in T-cell clones. Nature 356:607–609
61. Le DT, Uram JN, Wang H et al (2015) PD-1 blockade in tumors with mismatch-repair deficiency. N Engl J Med 372:2509–2520
62. Pardoll DM (2012) The blockade of immune checkpoints in cancer immunotherapy. Nat Rev Cancer 12:252–264
63. Krummel MF, Allison JP (2011) Pillars article: CD28 and CTLA-4 have opposing effects on the response of T cells to stimulation. J Immunol 187:3459–3465
64. Leach DR, Krummel MF, Allison JP (1996) Enhancement of antitumor immunity by CTLA-4 blockade. Science 271:1734–1736
65. Keir ME, Butte MJ, Freeman GJ et al (2008) PD-1 and its ligands in tolerance and immunity. Annu Rev Immunol 26:677–704
66. Perez-Gracia JL, Labiano S, Rodriguez-Ruiz ME et al (2014) Orchestrating immune checkpoint blockade for cancer immunotherapy in combinations. Curr Opin Immunol 27:89–97
67. Watanabe N, Nakajima H (2012) Coinhibitory molecules in autoimmune diseases. Clin Dev Immunol 2012:269756
68. Festino L, Botti G, Lorigan P et al (2016) Cancer treatment with anti-PD-1/PD-L1 agents: is PD-L1 expression a biomarker for patient selection? Drugs 76:925–945
69. Iwai Y, Ishida M, Tanaka Y et al (2002) Involvement of PD-L1 on tumor cells in the escape from host immune system and tumor immunotherapy by PD-L1 blockade. Proc Natl Acad Sci USA 99:12293–12297
70. Hodi FS, O'Day SJ, McDermott DF et al (2010) Improved survival with ipilimumab in patients with metastatic melanoma. N Engl J Med 363:711–723
71. Robert C, Thomas L, Bondarenko I et al (2011) Ipilimumab plus dacarbazine for previously untreated metastatic melanoma. N Engl J Med 364:2517–2526
72. O'Day SJ, Maio M, Chiarion-Sileni V et al (2010) Efficacy and safety of ipilimumab monotherapy in patients with pretreated advanced melanoma: a multicenter single-arm phase II study. Ann Oncol 21:1712–1717
73. Weber JS, O'Day S, Urba W et al (2008) Phase I/II study of ipilimumab for patients with metastatic melanoma. J Clin Oncol 26:5950–5956
74. Hoos A, Eggermont AM, Janetzki S et al (2010) Improved endpoints for cancer immunotherapy trials. J Natl Cancer Inst 102:1388–1397
75. Wolchok JD, Hoos A, O'Day S et al (2009) Guidelines for the evaluation of immune therapy activity in solid tumors: immune-related response criteria. Clin Cancer Res 15:7412–7420
76. Della Vittoria-Scarpati G, Fusciello C, Perry F et al (2014) Ipilimumab in the treatment of metastatic melanoma: management of adverse events. Onco Targets Ther 7:203–209
77. Lynch TJ, Bondarenko I, Luft A et al (2012) Ipilimumab in combination with paclitaxel and carboplatin as first-line treatment in stage IIIB/IV non-small-cell lung cancer: results from a randomized, double-blind, multicenter phase II study. J Clin Oncol 30:2046–2054
78. Robert C, Long GV, Brady B et al (2015) Nivolumab in previously untreated melanoma without BRAF mutation. N Engl J Med 372:320–330

79. Weber JS, D'Angelo SP, Minor D et al (2015) Nivolumab versus chemotherapy in patients with advanced melanoma who progressed after anti-CTLA-4 treatment (CheckMate 037): a randomised, controlled, open-label, phase 3 trial. Lancet Oncol 16:375–384
80. Robert C, Schachter J, Long GV et al (2015) Pembrolizumab versus ipilimumab in advanced melanoma. N Engl J Med 372:2521–2532
81. Brahmer J, Reckamp KL, Baas P et al (2015) Nivolumab versus docetaxel in advanced squamous-cell non-small-cell lung cancer. N Engl J Med 373:123–135
82. Borghaei H, Paz-Ares L, Horn L et al (2015) Nivolumab versus docetaxel in advanced non-squamous non-small-cell lung cancer. N Engl J Med 373:1627–1639
83. Herbst RS, Baas P, Kim DW et al (2016) Pembrolizumab versus docetaxel for previously treated, PD-L1-positive, advanced non-small-cell lung cancer: a randomised controlled trial. Lancet 387:1540–1550
84. Motzer RJ, Escudier B, McDermott DF et al (2015) Nivolumab versus everolimus in advanced renal-cell carcinoma. N Engl J Med 373:1803–1813
85. Lee DW, Kochenderfer JN, Stetler-Stevenson M et al (2015) T cells expressing CD19 chimeric antigen receptors for acute lymphoblastic leukaemia in children and young adults: a phase 1 dose-escalation trial. Lancet 385:517–528
86. Kwon ED, Drake CG, Scher HI et al (2014) Ipilimumab versus placebo after radiotherapy in patients with metastatic castration-resistant prostate cancer that had progressed after docetaxel chemotherapy: a multicentre, randomised, double-blind, phase 3 trial. Lancet Oncol 15:700–712
87. Ansell SM, Lesokhin AM, Borrello I et al (2015) PD-1 blockade with nivolumab in relapsed or refractory Hodgkin's lymphoma. N Engl J Med 372:311–319
88. Fehrenbacher L, Spira A, Ballinger M et al (2016) Atezolizumab versus docetaxel for patients with previously treated non-small-cell lung cancer: a multicentre, open-label, phase 2 randomised controlled trial. Lancet 387:1837–1846
89. Powles T, Eder JP, Fine GD et al (2014) MPDL3280A (anti-PD-L1) treatment leads to clinical activity in metastatic bladder cancer. Nature 515:558–562
90. Taube JM, Klein A, Brahmer JR et al (2014) Association of PD-1, PD-1 ligands, and other features of the tumor immune microenvironment with response to anti-PD-1 therapy. Clin Cancer Res 20:5064–5074
91. Topalian SL, Hodi FS, Brahmer JR et al (2012) Safety, activity, and immune correlates of anti-PD-1 antibody in cancer. N Engl J Med 366:2443–2454
92. Shin DS, Ribas A (2015) The evolution of checkpoint blockade as a cancer therapy: what's here, what's next? Curr Opin Immunol 33:23–35
93. Tumeh PC, Harview CL, Yearley JH et al (2014) PD-1 blockade induces responses by inhibiting adaptive immune resistance. Nature 515:568–571
94. Rosenberg SA, Spiess P, Lafreniere R (1986) A new approach to the adoptive immunotherapy of cancer with tumor-infiltrating lymphocytes. Science 233:1318–1321
95. Gattinoni L, Powell DJ Jr, Rosenberg SA et al (2006) Adoptive immunotherapy for cancer: building on success. Nat Rev Immunol 6:383–393
96. Besser MJ, Shapira-Frommer R, Treves AJ et al (2010) Clinical responses in a phase II study using adoptive transfer of short-term cultured tumor infiltration lymphocytes in metastatic melanoma patients. Clin Cancer Res 16:2646–2655
97. Dudley ME, Yang JC, Sherry R et al (2008) Adoptive cell therapy for patients with metastatic melanoma: evaluation of intensive myeloablative chemoradiation preparative regimens. J Clin Oncol 26:5233–5239
98. Ellebaek E, Iversen TZ, Junker N et al (2012) Adoptive cell therapy with autologous tumor infiltrating lymphocytes and low-dose Interleukin-2 in metastatic melanoma patients. J Transl Med 10:169–175
99. Bunnell BA, Muul LM, Donahue RE et al (1995) High-efficiency retroviral-mediated gene transfer into human and nonhuman primate peripheral blood lymphocytes. Proc Natl Acad Sci USA 92:7739–7743

100. Mavilio F, Ferrari G, Rossini S et al (1994) Peripheral blood lymphocytes as target cells of retroviral vector-mediated gene transfer. Blood 83:1988–1997
101. Stone JD, Kranz DM (2013) Role of T cell receptor affinity in the efficacy and specificity of adoptive T cell therapies. Front Immunol 4:244–252
102. Robbins PF, Morgan RA, Feldman SA et al (2011) Tumor regression in patients with metastatic synovial cell sarcoma and melanoma using genetically engineered lymphocytes reactive with NY-ESO-1. J Clin Oncol 29:917–924
103. Porter DL, Hwang WT, Frey NV et al (2015) Chimeric antigen receptor T cells persist and induce sustained remissions in relapsed refractory chronic lymphocytic leukemia. Sci Transl Med 7:303ra139
104. Brentjens RJ, Davila ML, Riviere I, et al (2013) CD19-targeted T cells rapidly induce molecular remissions in adults with chemotherapy-refractory acute lymphoblastic leukemia. Sci Transl Med 5:177ra138
105. Davila ML, Riviere I, Wang X et al (2014) Efficacy and toxicity management of 19-28z CAR T cell therapy in B cell acute lymphoblastic leukemia. Sci Transl Med 6:224ra225
106. Grupp SA, Kalos M, Barrett D et al (2013) Chimeric antigen receptor-modified T cells for acute lymphoid leukemia. N Engl J Med 368:1509–1518
107. Maude SL, Frey N, Shaw PA et al (2014) Chimeric antigen receptor T cells for sustained remissions in leukemia. N Engl J Med 371:1507–1517
108. Porter DL, Levine BL, Kalos M et al (2011) Chimeric antigen receptor-modified T cells in chronic lymphoid leukemia. N Engl J Med 365:725–733
109. Topp MS, Gokbuget N, Zugmaier G et al (2014) Phase II trial of the anti-CD19 bispecific T cell-engager blinatumomab shows hematologic and molecular remissions in patients with relapsed or refractory B-precursor acute lymphoblastic leukemia. J Clin Oncol 32:4134–4140
110. Topp MS, Kufer P, Gokbuget N et al (2011) Targeted therapy with the T-cell-engaging antibody blinatumomab of chemotherapy-refractory minimal residual disease in B-lineage acute lymphoblastic leukemia patients results in high response rate and prolonged leukemia-free survival. J Clin Oncol 29:2493–2498
111. Viardot A, Goebeler ME, Hess G et al (2016) Phase 2 study of the bispecific T-cell engager (BiTE) antibody blinatumomab in relapsed/refractory diffuse large B-cell lymphoma. Blood 127:1410–1416

Index

© Springer International Publishing Switzerland 2016

R.J. Gelpi et al. (eds.), *Biochemistry of Oxidative Stress*, Advances in Biochemistry in Health and Disease 16, DOI 10.1007/978-3-319-45865-6

N

R

S

The manufacturer's authorised representative in the EU is Springer Nature Customer Service Centre GmbH, Europaplatz 3, 69115 Heidelberg, Germany. If you have any concerns regarding our products, please contact ProductSafety@springernature.com

Printed and bound by CPI Group (UK) Ltd, Croydon, CR0 4YY
30/09/2026
02225501-0012